Oilseed Crops

Oilseed Crops

Second Edition

E.A. Weiss
Consultant in Tropical Agriculture,
Victoria, Australia

Foreword by Sir Charles Pereira, FRS

Blackwell
Science

Editorial Offices:
Osney Mead, Oxford OX2 0EL
25 John Street, London WC1N 2BL
23 Ainslie Place, Edinburgh EH3 6AJ
350 Main Street, Malden
MA 02148 5018, USA
54 University Street, Carlton
Victoria 3053, Australia
10, rue Casimir Delavigne
75006 Paris, France

Other Editorial Offices:

Blackwell Wissenschafts-Verlag GmbH
Kurfürstendamm 57
10707 Berlin, Germany

Blackweell Science KK
MG Kodenmacho Building
7-10 Kodenmacho Nihombashi
Chuo-ku, Tokyo 104, Japan

First published 1983 by Longman Group Ltd; second edition published by Blackwell Science 2000

Set in Times
by Best-set Typesetter Ltd., Hong Kong
Printed and bound in Great Britain by MPG Books Ltd, Bodmin, Cornwall

DISTRIBUTORS

Marston Book Services Ltd
PO Box 269
Abingdon
Oxon OX14 4YN
(*Orders:* Tel: 01235 465500
Fax: 01235 465555)

USA
Blackwell Science, Inc.
Commerce Place
350 Main Street
Malden, MA 02148 5018
(*Orders:* Tel: 800 759 6102
781 388 8250
Fax: 781 388 8255)

Canada
Login Brothers Book Company
324 Saulteaux Crescent
Winnipeg, Manitoba R3J 3T2
(*Orders:* Tel: 204 837-2987
Fax: 204 837-3116)

Australia
Blackwell Science Pty Ltd
54 University Street
Carlton, Victoria 3053
(*Orders:* Tel: 03 9347 0300
Fax: 03 9347 5001)

A catalogue record for this title is available from the British Library

ISBN 0-632-05259-7

Library of Congress Cataloging-in-Publication Data is available

For further information on Blackwell Science, visit our website:
www.blackwell-science.com

Contents

Foreword to the second edition

This new edition of a major treatise on a critically important group of tropical crops makes an important contribution to both science and practice. The information is meticulously researched and lucidly presented. The author's long and successful managerial experience in the growing of oilseed crops has enabled him to relate the results of research to the field problems of crop production.

The first edition, in 1983, was both authoritative and outstandingly practical as a source of information and of agricultural guidance. In the sixteen years between the two editions the importance of the information has increased as the world production of oilseed crops has tripled, as has the production of oils and of the protein-rich meals that remain after crushing. World trade in these products for food and feed has correspondingly increased.

The oilseeds are grown over the whole range of scales from enterprises harvesting thousands of hectares in the USA and in Brazil and Argentina, to peasant farmers in West Africa. For the developing countries oilseeds are important cash crops in the transition from subsistence agriculture to market economies: population growth is now rendering this transition essential for survival.

The new edition is an important reference book for the professional adviser in tropical agriculture, a useful source of information for the research worker and will be profitable reading for those involved in the commercial management of the massive international flows of oilseeds, oils and meal. All readers will appreciate the formidable amount of work that the author has put into this volume and the readable way in which he has arranged a great harvest of current information.

Charles Pereira
Teston, Kent
1999

Acknowledgements

The author wishes to acknowledge with thanks, permission to use the following figures or photographs. The remainder are his own or have been in his possession for many years and the original donors cannot be traced.

American Soyabean Association, 9.10; Anderson-IBEC Inc., Ohio, USA, 9.4, 9.5, 9.6, 9.9; AVRDC, Shanhua, Taiwan, 6.10, 6.12; Professor R.D. Brigham, Texas A & M University, USA, 2.8; Dr A.H. Chaudrey, Tandojam, India, 7.2, 7.12, 7.13; Mr N.E. Delaney, Kingaroy, Queensland, Australia, 7.8; Dr W.E. Domingo, La Mesa, California, USA, 2.10, 2.6; Elsevier Pub. Co., Amsterdam, Holland, 7.3; FMC Corp, Louisiana, USA, 6.11; Mr P. Gillier IRHO, Paris, France, 5.6; Department of Primary Industry, Queensland, Australia, 2.17, 3.14, 3.15, 4.6, 4.13, 4.14, 4.15, 4.17, 5.11, 5.12, 7.15, 7.17, 7.18, 8.9; Dr T. Kobayashi, Tohoku, Japan, 6.15, 6.16, 6.17; Dr K.J. Lessman, New Mexico State University, USA, 8.2, 8.4, 8.5, 8.6, 8.8; Dr H.S. Mann, CAZRI, Jodhpur, India, 9.1; Dr J.A. Martin, Clemson University, USA, 5.10; Dr B. Mazzani, Maracay, Venezuela, 5.7, 5.9; Meggitt Ltd, Sydney, Australia, 9.7; Dr N. Mehrotra, Haryana Agricultural University, India, 2.3, 2.9; Mr Peter Milthorpe, Department of Primary Industry, Queensland, Australia, 8.21, 8.22; Dr C.B. Patel, ICAR, Igatpuri, India, 8.10, 8.11; Dr T.L. Rost, University of California, USA, 8.16, 8.17; Professor J.B. Sinclair, University of Illinois, USA, 6.18, 6.19; South African Department of Agriculture, Pretoria, 6.9; Dr P. Subrahmanyan, ICRISAT, Hyderabad, India, 2.9, 3.16; U.S. Department of Agriculture, Washington DC, 2.6, 2.14, 3.7, 4.5, 4.11, 5.1, 6.5, 6.6, 6.13, 7,9; Dr D.M. Yermanos, University of California, USA, 8.15.

The author also wishes to express his appreciation to all those people he has met or with whom he has corresponded, who contributed valuable information or experiences growing oilseeds. Some have been mentioned herein as authors of published works, and private communications from some are directly quoted. Others ranged from smallholders to those with large-scale farms, from operators of ghanis to managers of huge processing plants. To all these go my most grateful thanks, not only for the information gained but also for the opportunity to meet so many, most interesting people. Finally my thanks go to my wife, who word-processed the new text.

Introduction

Oilseeds are a very important component of semi-tropical and tropical agriculture, providing easily available and highly nutritious human and animal food. Many also have industrial uses and are relatively easy to incorporate into locally manufactured products. In selecting those oilseeds to be included in this book, castor, groundnut, safflower, sesame, soya and sunflower are obvious choices. Crambe and niger were chosen because these two crops are particularly suitable for specific regions or markets. Jojoba is a crop for arid and semi-arid areas, since it can withstand a shade temperature of 40°C and grows in areas with an annual rainfall of 300 mm. It can thus be a valuable income producer for countries with few alternative crops, or for people with a semi-nomadic existence. Other oilseeds have been excluded because they are of local importance only, and unlikely to expand significantly. Rape has been omitted from this edition since the author's subsequent experience showed it to be unprofitable in the high-altitude tropics, and the oil is easily and cheaply available on the international market.

Oilseeds have been treated as warm region crops and examples and research quoted are mainly from work carried out in these areas. Lack of mention of essentially temperate research and crop response does not indicate lack of importance, rather that the author decided it was not directly relevant. Where such information has been included it is because tropical data are lacking, and results obtained elsewhere are valuable as a guide. Some important developments, especially in plant breeding and genetics, are omitted owing to publishers' copyright restrictions, whilst certain production methods are industrial secrets.

Some of the oilseeds choosen have a restricted range of cultivars suitable mainly for their natural range. However, the author considers that biotechnology and the wide acceptance of Plant Breeders Rights will rapidly produce suitable cultivars, or perhaps virtually new species, to greatly extend their range. However, field research is also most important; for example a plant that grows naturally in a specific region may grow just as well or better elsewhere, as demonstrated with niger and jojoba.

Genetically modified plants are important in crop improvement, and the techniques are a useful tool in the analysis of basic plant processes, leading to significant advances in crop development. The success of biological techniques in modifying plants has resulted in a mass of information outside the scope of this book, which is reported in specialist journals. Where this research has produced field applications, it will be mentioned. Industry prefers oils with long chain saturated fatty acids, and the industrial uses of these oils and their derivatives could well be greatly extended in the future, partially at the expense of petrochemicals, if the full potential of these developments is utilised.

Certain genes that control the synthesis of fatty acids have been cloned, manipulated and introduced into oilseeds to change their composition. For example linseed oil contains 50% linolenic acid which renders it highly susceptible to oxidation and, together with a rather unpleasant odour, unsuitable for human consumption. CSIRO scientists in Australia have successfully modified linseed to produce linola oil, with less than 2% linolenic acid and suitable for cooking or margarine production; have inserted a sunflower gene into lupin to double the methionine content of the seed and make it more suitable for stockfeed; while a canola cultivar now contains lauric acid not normally present in canola, but a valuable component of coconut and palm oil.

The marketing of genetically altered oilseeds for use in human and animal feed stuffs has become controversial, and is discussed in the chapter on processing. More important to growers is the probability that genetically altered plants may become pesticide-resistant weeds of other crops, while volunteer plants may also act as reservoirs of pests and disease.

Since the first edition of this book was published, the massive improvement in research equipment, especially in computing power, has made the evaluation of data more easy and much faster. Laptop and similar portable computers plus mobile phones have greatly reduced transcription errors and the time required to collect and transmit field observations. Using computers to access databases has facilitated literature searches and reference retrieval.

Increased mechanisation, specialised machinery and equipment, and large-scale production have achieved a dramatic increase in crop production per unit area, coupled with a reduction in price and ease of transportation. Higher yields from current cultivars make intensive production of oilseeds on smaller areas or by smallholders increasingly attractive, not only to individuals but also to many governments confronted with an expanding population. It is as import substitutes that crops such as castor, groundnuts, sesame, sunflower and their derivatives are of great importance, not only for humans but also as feeds in intensive livestock production, especially pigs and poultry.

The crops discussed are generally well known to farmers in most developing countries, and more intensive cultivation poses fewer problems than the introduction of a new crop. Marketing and processing are also simplified because these crops are already familiar to local merchants, processors, and equally importantly individual financiers (moneylenders) or institutions who might be unwilling to support relatively unknown crops. In developed countries there is a growing consumer preference for what are perceived to be natural ingredients in foods and some industrial products. This trend presents many opportunities greatly to increase the range and usage of vegetable oils and their derivatives.

An increase in seed yield can also lead to more stable and cheaper supplies and thus stimulate demand, as processors require large, regular supplies to allow more efficient factory utilisation and a wider variety of products. Protein concentrates for example, can directly substitute for meat or fish protein in diets lacking this component; protein concentrates are also a valuable supplement in cereal-based children's food.

Yield of oilseeds per unit area in developed countries is usually high, and often quoted as the target for farmers in less developed countries. That smallholder yields are generally low cannot be denied, but the potential yield is much greater. For instance in India, where the average smallholder yield of niger is around 400 kg/ha, up to 2000 kg/ha has been obtained on a field scale with little additional input. Similar results were achieved on groundnut, sesame, and sunflower in Africa and South America.

Planting during the optimum period and timely weeding are generally far more important in raising yields at smallholder level than ploughing a larger area or adding plant nutrients, as the author has frequently demonstrated. On a local or regional basis, selection for a desired characteristic can increase local production without the necessity to establish a major breeding programme. Selection may not primarily be concerned with higher yield, but to increase tolerance of salt, drought or high-temperature, which can be selected visually by relatively unskilled staff. Simple procedures have proved valuable with groundnut in Africa, sesame in South America, niger in India, and are described in the relevant chapters. The oilseeds discussed are generally rain-grown, and although irrigation will probably become more profitable, water will become less readily available or more expensive; thus research into the optimum water use of oilseed crops is of increasing importance.

The price of meat directly influences sales of vegetable protein in markets where both are competitors. Where meat and fish are absent, or meat on the hoof is capital rather than income, this market factor is also absent. Cost of oilseed production, either free market or government subsidised, will then be the major pricing factor, and generally the greater the volume processed the lower the unit cost. While processed vegetable proteins, including meat analogues, are generally too expensive for underdeveloped markets, the simple formulations discussed in the sesame and processing chapters are more appropriate. Since many rural societies are very conservative in their diet, through ignorance, social or religious taboos, education in the use of vegetable proteins is essential. Provided such programmes are carried out realistically and by local people, problems of acceptance can be substantially reduced; free school meals are an excellent method.

The chemical control of weeds, insect pests and plant diseases is a constantly evolving process, and recommendations change as rapidly as new formulations appear. For this reason active ingredients only have been mentioned, since relevant information is freely available from manufacturers and officials, whose assistance should always be sought to avoid misuse. In this respect it would be an asset if manufacturers either resisted the temptation to describe ideal safety measures, or modified their products' toxicity to ensure that safety precautions are practical. The use of chemicals with high toxic hazards to operators of spraying equipment should be avoided in hot climates, however effective these may be against particular pests or diseases. To insist on full observance in tropical conditions of recommendations regarding protective clothing, becomes sadism! A frequent result is that few or no precautions are taken, and this must be recognised by local advisers when making recommendations for pest and disease control.

Conventional insecticides will remain for some time the main method of controlling insects, and be replaced only where resistance develops or cheap alternative methods become available. Many of the newer insecticides have lower mammalian toxicity, shorter persistence, and quicker degradation into harmless materials. Integrated pest and disease control, which includes biological methods, plus increasing a plant's own resistance, will also reduce the danger inherent in chemical control methods. Dangers not only include direct damage to plants and humans, but chemical residues that remain in seeds and their derivatives. An important example of biological pest control is a commercially produced nuclear polyhedrosis virus (NPV), which is fatal to larvae of the widespread and damaging *Helicoverpa* (*Heliothis*) spp., and the use of mycoherbicides in weed control.

Demand by industrial nations also influences the level of production or profitability of a particular oilseed in a specific country or region. For example, changes in consumer demand have greatly influenced production of polyunsaturated oil for inclusion in a wide variety of foods, and resulted in the rapid expansion of sunflower growing in the USA and Australia. Similarly the use of groundnut products in industry and stockfeeds greatly increased demand, and the subsequent rise in prices to producers substituted this crop for sesame over wide areas of Burma during the period 1900–20, a trend since reversed.

The technical knowledge available to assist growers at all levels of production and development and its rate of accumulation, is far greater than the level and rate of application. In many regions advanced techniques may also be unnecessary initially, since tropical oilseed production could be greatly expanded with practically no additional input other than manual labour, whose opportunity cost in many countries is virtually nil. Timely planting and weeding can substantially increase the yield of almost any annual tropical crop. When combined with optimum plant population and improvements to traditional harvesting/storage techniques, total availability can be dramatically expanded at little or no monetary cost. Subsidiary activities have been mentioned where they can be integrated into a cropping programme. Honey production is one which costs very little to introduce or expand, and bees increase pollination especially in sunflowers.

Processing of primary products must of necessity become of greater importance to developing countries, as flawed economic policies that supported ever greater industrialisation proved incapable of increasing national revenues, per capita income or general living standards. Equally important must be the realisation by governments and international agencies of the growing threat modern cultivars pose to indigenous species or land races. The establishment of gene pools is vital to ensure the greatest possible amount of material is, and remains, available to plant breeders.

Crash programmes, especially in agricultural development and research, invariably produce crashes! The author's experience has been gained at many levels of tropical agriculture, and has influenced his approach to commercial crop production. What should ideally be done, and what it is profitable or practicable to do, are very different. The author has therefore attempted to relate agricultural research and development to profitable production, an objective not always appreciated by agricultural scientists and economists, although the basic aim of most producers. Thus an apparent lack of interest by farmers especially in developing countries, in new methods or crops may be due to a realistic assessment of the limits of their environment (physical or social), and not to any lack of appreciation of the value of such improvements. However, profitability at a national level should not be evaluated solely on financial reward, but also by the socio-economic benefits which are not accurately measurable in money terms. Maintaining viable farming communities to provide employment, or a rural population on the land, will reduce the ever-growing drift to towns, a worldwide problem.

An extremely important aspect of oilseed production is the effect that processing plants can have on a country's economy, and virtually all developing countries exporting oilseeds could increase the value-added component by expanding processing capacity. In many such countries domestic oil requirements are often largely supplied from cottage industries, and there remains a tendency to assume that crushing plants in these countries would be at a significant disadvantage in world markets. However, there appears to be no basic reason in terms of technical efficiency and operating costs why such plants should not operate profitably.

Local processing of oilseeds has many advantages, not only are the products of greater value, this value-added component remains with the producing country. Development of practical methods of using solar and wind energy will allow construction of modest processing plants in many previously remote areas, and once established, they then become available for increasingly sophisticated products. More technically advanced units however, although desirable are often impracticable. The smallest viable solvent extraction plant was estimated in 1997 to cost about US$20 million, and many developing countries have no local supplies of food quality hexane. Thus new, or small-scale, processing plants in developing economies must be based on pressure extraction.

Since many processing plants will be built in regions with long sunny periods, utilisation of solar energy should be included in factory design. Outback meatworks and telegraphic installations in Australia for example, are already substituting solar energy for a significant proportion of their energy requirements, while small solar driers are operating successfully in Africa.

Technical and mechanical problems likely to be encountered when establishing a small-scale oilseed processing plant have been reviewed (Neumunz, 1976) and remain relevant, while there has been little change in the administrative and political problems of adjacent settlement projects (Weiss, 1980). It may be desirable to include machinery to produce flour (meal) for human consumption, thus overcoming a local lack and reducing imports.

Establishment of a plant processing local crops will encourage allied or subsidiary industries and become a focal point for development, including education, social services, banking and supply of consumer goods. This may be a surer way to overall rural development than attempting to increase production of export crops, introducing new crops, or attempting to establish alternative industries unfamiliar to the local population, which often actually reduce local employment opportunities.

Introducing a crop to an underdeveloped area or encouraging increased production of a particular crop, should be carried out with caution (Weiss, 1966b). New crops or new methods of production introduced too quickly into a traditional system of agriculture, which over the years has become delicately adjusted to local environments, can result in considerable disruption to a local society. Even if the new crop is itself successful, the consequent imbalance in the local agricultural system may negate the advantage occurring from its increased production, as has occurred with the so-called miracle wheats and rice.

Establishing a factory in a developing area can also provide many useful services for local producers, facilities for co-operatives and marketing offices, machinery for hire, or operate a machinery and farm equipment pool, supply and dress high-quality seed, and offer a guaranteed price for produce of an acceptable grade. Credit facilities may be introduced, but their extent and range should be determined after detailed and careful study of local conditions; their improper use can be disastrous, as the author well knows. Integration of marketing and processing, preferably not as an official monopoly, creates a system most conducive to agricultural progress. In India, and elsewhere, establishment of a processing unit in a district usually stimulates a larger volume of production than the independent traders or cooperative marketing organization that preceded it. For this reason the Indian government supported the establishment of thousands of small oil crushing plants in its Rural Development Plans.

Providing specialised items of farm equipment is often one of the most important services rendered to small farmers; for example the hulling of castor seed on a large scale is notoriously difficult, but factory-based or mobile hullers can reduce not only the labour involved but the proportion of damaged and unsaleable seed. Hulling can be carried out by an individual to whom the machine is rented, or directly by the factory and paid for by a levy on users. Threshing of oilseeds other than castor is relatively easy, and is probably best left to producers. Too much assistance is often more demoralising or misunderstood by recipients than too little, and self-help should be the maxim.

When a processing unit is established and financed by government to stimulate production, it may also be necessary to encourage or establish local cooperatives. Every attempt should be made to encourage local people actively to participate in such operations, and wherever practicable to share in the equity of processing plants in order to secure higher returns to their members. This ensures the enterprise is regarded as directly contributing to the local community.

Establishing or extending, a marketing system into more remote districts, will also encourage production of a particular crop. In isolated areas it is often the long journey to a trader who will buy produce that limits production. Provision of recognised buying centres, with public cleaning and weighing of produce and published prices can increase production in particular areas out of all proportion to the investment required in their establishment. The resultant increase in cash income will also probably result in additional social services, and certainly more frequent visits of local tax officials, a very persuasive argument to ensure official support! Such buying centres may also provide primary storage, and ensure a steady flow of produce to the factory. By reporting on the state of the crop and estimating probable seasonal production, supervisors of such centres will assist in the efficient planning of factory working time. Dissemination of information and advice to producers is also simplified.

Improved agricultural techniques, however, are often not the only or the most important factor influencing increased oilseed production; political decisions frequently have priority. Government controlled marketing boards in many countries are not usually an asset to producers, since the political bias of a country's rulers invariably results in the appointment of party officials with a marketing policy based on ideology and not economics. This is particularly so in one-party states or centrally administered economies, where the operations of such boards were so appallingly incompetent or corrupt that they virtually destroyed the industry they were established to encourage! This was a major reason why the production of oilseed crops including cotton, but also coffee and tobacco, fell dramatically in Africa and elsewhere.

The common claim in developing countries that government marketing organisations will increase profits to producers by eliminating the middleman has proved fallacious. The author's experience of many such organisations is that when they are successful in reducing marketing costs, their savings are then more often diverted to other official undertakings operating at a loss, or dissipated in the employment of unnecessary staff rather than returned to producers. A most pernicious use of such funds was to retain the price paid to producers at an artificially high level compared with world prices; the subsequent drastic fall in price when the support ceased disrupted the whole industry. The effect was often most severe in those localities that could least afford it, for alternative crops are generally lacking. Similarly the maintenance of an artificially high international exchange rate for the local currency had the same depressing effect by keeping producer prices low.

Establishing a factory will usually allow development of related enterprises and when residues from oil production are suitable for stockfeed, a fattening or dairy unit can be established – or a bull-breeding herd to upgrade local stock. The factory may initially require an area in its immediate vicinity to be directly farmed, to give added impetus to local production. In these circumstances the factory-owned plantation must be capable of producing the volume required for efficient factory operation. Smallholder crops would be processed as they become available, the factory being independent of minor fluctuations in supply. When local production is large enough to support the factory, the latter's nucleus plantation could be either rented to individual farmers or partially used as a training ground for non-agriculturalists. The latter could be of great importance in many areas of Africa for instance, where areas suitable for oilseed crops are often located in primarily non-agricultural, stock-raising regions with very low income.

A small experimental or trial farm should be accepted as part of the scheme's contribution to local development. Research into methods of improving local production should have high priority, and in dry-farming areas should be directly applicable locally. Programmes must also be so designed that results can be locally incorporated with little or no additional cash expenditure by farmers. The author has usually aimed to increase returns to producers by improving traditional methods, which often have a very sound basis for their continued use.

Notes on the text

The author's many decades of experience working with tropical crops will probably be apparent in some statements that may be considered as not well supported by research results. This is an inevitable and valid criticism, but the answer can only be that such observations are of value since there is a vast difference between experimental plots and 1000 ha of crop. The author has supervised experimental farms and commercial units of many thousands of hectares, and well knows the pitfalls of extrapolation. Other statements may appear to be obvious, or overemphasised, but personal experience has shown the obvious is often ignored!

The text of this edition has been completely revised, but most illustrations and tables have been retained as the information remains relevant, and to keep down the cost of producing this book. Many original references have been removed as the data has since been confirmed or new information obtained. The first edition can be consulted for these references. Cultivar is used in its normal definition, but for brevity to also include what the literature describes as selections, strains and local varieties. All values are in United States dollars (USD or $) unless otherwise specified, and all units are metric.

1
World Production and Trade

The first edition of this book included detailed statistics on world production and trade in oilseeds and oilseed products. In this edition they are replaced by a general review, which can be easily updated by referring to numerous databases and the Internet. The introduction to each chapter has been expanded to include statistics on the crop concerned. The following tables and graphs indicate the status and production trends of major oilseeds.

Caution is necessary when comparing oilseed production figures for the world or a country. The reliability of national statistics is somewhat variable, as is the period to which they refer. Overlapping of harvests between years, delays in reporting crop movement, differing accounting periods, varying data collection methods between government departments, all act as modifiers. The following tables of world production of selected oilseeds, oils and meals were compiled by the author, and include data supplied by Oils & Fats International (Oils Fats International, 1999), Mintec Ltd (Mintec, 1999, personal communication. Mintec Ltd 46–48, The Green, Wooburn Green, Bucks, UK) and Soya Bluebook (Soya Bluebook, 1998 & 1999).

Total production of oilseeds and oilseed products should be basically construed to mean world trade in oilseeds, since there is significant production for growers' domestic requirements which never appears in official statistics. World trade in oilseeds and oilseed products rose steadily in the years following the Second World War; in the decade 1980–90 the rate of growth increased and continued to rise over the next decade (Table 1.1 and Fig. 1.1). Rape and soya have recorded the greatest proportional volume increase since 1960, the former by ten times and the latter by five times; palm oil is included in oil production.

The great increase in rape and soya production is a major factor affecting international trade in oilseeds, although the dramatic and continuing rise in palm oil production is as significant. Availability of soya or its products previously dominated the world oilseed trade and virtually fixed the price level of competitive crops to 1980. However, this dominance was eroded with the growth in palm oil production, and introduction of oilseed rape producing an edible oil, widely marketed as canola oil. Meal and oil from double zero oilseed rape cultivars can now directly substitute for soya bean products.

Despite this the world's major soya producers substantially increased their production; in 1999 Brazil's crop topped 31 million tonnes (Mt), Argentina's reached 18 Mt, and the USA nearly 80 Mt. India's production of all oilseeds and their derivatives would benefit from an increase in the generally low yields, as unlike South American countries India has little additional land available.

World production of vegetable oils steadily increased to 35 Mt in 1970, rose to 60 Mt in 1980, 80 Mt by 1990, and could reach 100 Mt by the year 2000 (Table 1.2 and Fig. 1.2). The United States Department of Agriculture (USDA) predicts that should the rate of growth in oil consumption between 1990 and 1998 continue into the next century, world

Table 1.1 Calculated world production of selected oilseeds (million tonnes).

	1960	1970	1980	1990	2000*
Castor	0.7	0.9	1.0	1.3	1.5
Copra	4.0	4.0	5.0	5.0	5.0
Cottonseed	20.0	21.0	24.0	34.0	35.0
Groundnut†	12.0	12.0	14.0	17.0	20.0
Linseed	4.0	4.0	3.0	2.5	2.5
Rape	4.0	7.0	12.0	25.0	40.0
Safflower	0.5	0.6	1.0	1.0	0.5
Sesame	2.0	2.0	3.0	3.0	3.0
Soya	27.0	45.0	93.0	104.0	180.0
Sunflower	7.0	10.0	15.5	23.0	28.0
Subtotal	81.2	106.5	171.5	215.8	315.5
World total (all oilseeds)	100.0	120.0	185.0	220.0	290.0

* Estimate; † Shelled equivalent.

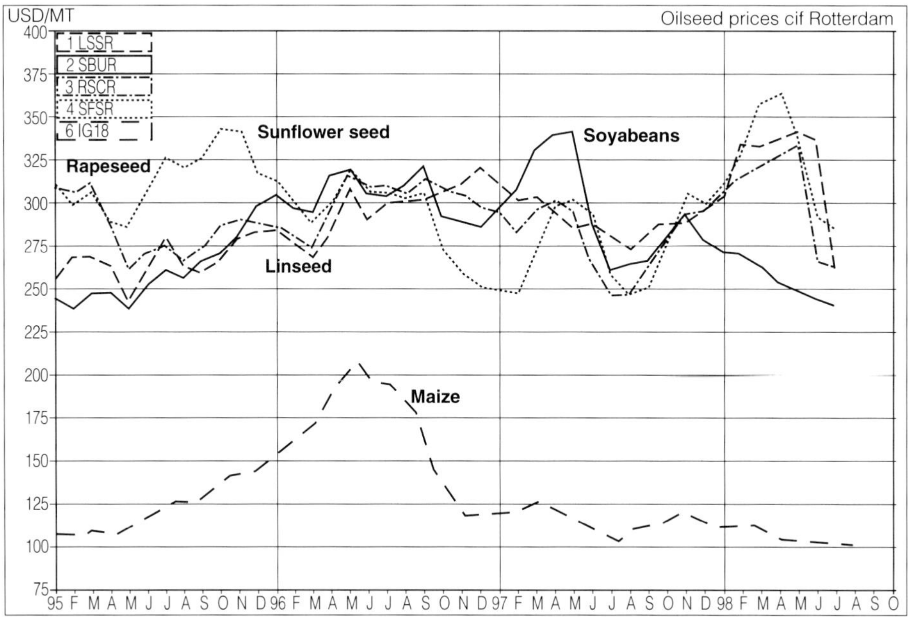

Fig. 1.1 Prices of major oilseeds CIF Rotterdam 1995–98.

oil consumption would reach 14 kg per capita by the year 2008, a calculated total of almost 110 Mt. The greatest proportional increases over the 1970–98 period were palm, rape, soya, and sunflower oils. Brazilian soya oil exports rose from 0.6 Mt in 1988 to 1.65 Mt in 1997 despite an increase in domestic demand of 0.8 Mt over the same period.

Since many vegetable oils are direct competitors, the relative importance may temporarily change as supplies vary (for example Indonesia imposed a temporary ban on exports of crude palm oil in 1998 and the El Nino weather reduced South American soyabean production in 1996/97), new products are introduced, or new uses discovered.

Table 1.2 Calculated world production of selected vegetable oils (million tonnes).

	1970	1980	1990	2000*
Castor	0.4	0.5	0.5	0.8
Coconut	2.4	3.2	3.0	4.0
Cottonseed	2.5	3.2	4.0	4.0
Groundnut	3.0	3.2	4.0	4.0
Linseed	1.0	1.0	1.0	1.0
Olive	2.0	2.0	2.0	2.5
Palm†	2.0	4.5	13.0	22.0
Rape	2.0	4.0	9.0	15.0
Safflower	0.2	0.3	0.3	0.3
Sesame	1.0	1.0	1.0	1.0
Soya	6.0	15.0	16.0	30.0
Sunflower	2.0	4.0	8.0	12.0
Subtotal	24.5	41.9	61.8	96.6
World total (all vegetable oils)	35.00	60.00	80.00	102.00

* Estimate; † Palm and palm kernel oil.

Sources also change; in the 1960s Africa supplied about 65% of world palm oil, but this has fallen to around 10% and unlikely to increase as the current continental situation is unfavourable to the long-term investment required. Approximately US$ 2500/ha was necessary in 1998 to establish a palm oil plantation which would not reach maturity for ten years, and few companies or banks considered this an acceptable risk in a politically and socially unstable region. The economic turmoil in South East Asian countries that occurred in mid-1997, temporarily affected palm oil supplies especially from Indonesia, and also the regional trade in oilseeds, meal and oil. Major palm oil producers are Malaysia and Indonesia, with 1998 production of 9.5 Mt and 4.8 Mt respectively.

The world's major oil importers in the three fiscal years to 1996/97 were China at 3.45 Mt, India at 1.5 Mt (up from 0.5 Mt in 1994/95) and Pakistan at 1.4 Mt. Indian domestic oil production rose sharply from 4.4 Mt in the 1988/89 fiscal year to 8 Mt in the 1997/98 year, with consumption rising steadily over that period to 8.5 kg/capita. Continuation of this trend will

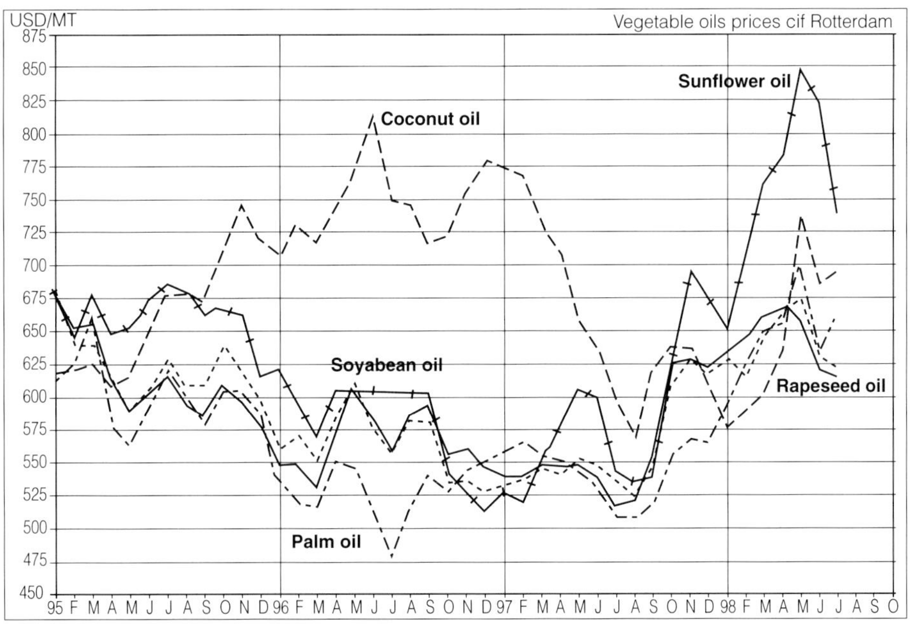

Fig. 1.2 Prices of major vegetable oils CIF Rotterdam 1995–98.

require an additional 0.35 Mt of oil annually, far in excess of any possible increase in total Indian oilseed production. However, oilseeds are mainly processed in small and inefficient plants, and consolidation into larger more modern units would substantially increase oil recovery. Demand for industrial oils and fats in the USA is forecast to rise to 2.0 Mt in 2000 from 1.5 Mt in 1990, with about one third satisfied by imports. Coconut oil and inedible tallow each account for about 20% of total usage, with tall oil in strong demand. In the decade to 1998, USA internal demand for soya oil increased by 2.2 Mt, equal to 12 Mt of soya beans.

Table 1.3 Calculated world production of selected oilseed meals (million tonnes).

	1970	1980	1990	2000*
Coconut	1.3	1.8	1.7	1.7
Cottonseed	8.0	10.0	15.0	16.0
Groundnut	4.0	4.0	6.0	7.0
Linseed	1.8	1.4	1.4	1.3
Palm kernel	0.5	0.6	2.0	2.5
Rape	3.0	6.0	14.0	20.0
Sesame	0.7	1.0	1.0	1.0
Soya	32.0	60.0	70.0	95.0
Sunflower	2.0	5.0	10.0	13.0
Total	53.3	89.8	121.1	157.5
Fishmeal	4.0	5.0	6.0	6.0

* Forecast.

World demand for oilseed meals since 1970 has been relatively greater than for vegetable oils, owing to a rise in demand by the intensive livestock production sector especially pigs, poultry and later aquaculture. Since the price received for meal often directly influences the price at which the oil can be sold, there is thus a relationship between demand for a specific oilmeal and selling price of the oil. Conversely, demand for a specific oilmeal requires that the oil be sold to keep down cost of meal production, and this may be difficult. Soya is an example where seed products are generally more important than the oil, while the reverse is true of castor.

The use of oilseed meals for stockfeeds is controlled by price and nutritional value (digestibility) of their protein content, as calculated for

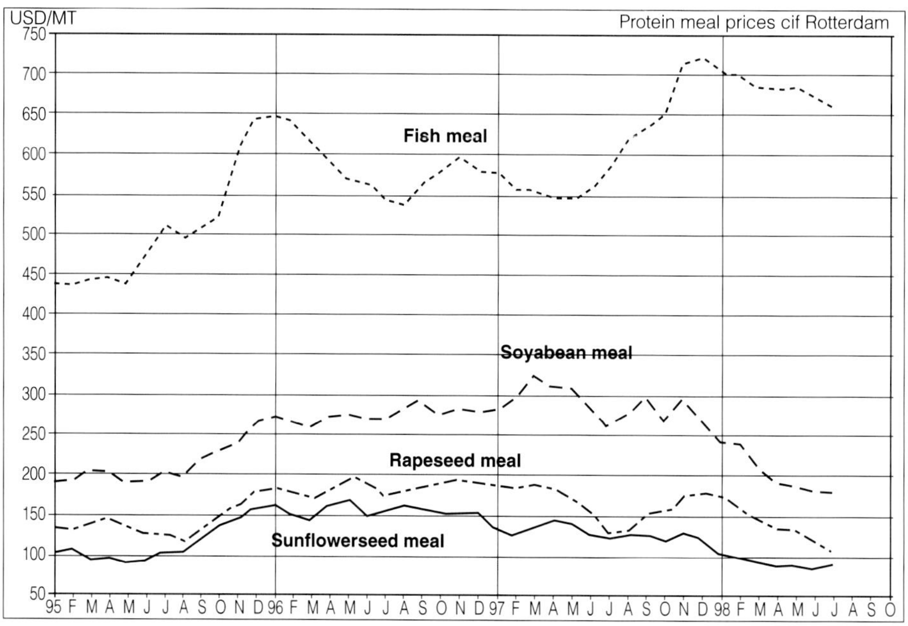

Fig. 1.3 Prices of major oilmeals CIF Rotterdam 1995–98.

individual species. Soya meal can be used for cattle, pigs and poultry, and normally only price determines the percentage of soya meal in a feed. Oilseed meal usage is thus directly linked to intensive livestock production, demand for whose products steadily rises worldwide, and this will continue to drive increased meal production and crushing capacity (Table 1.3 and Fig. 1.3).

The major producers of soya meal are the USA, Brazil, Argentina and the European Union (EU); of rapeseed meal China, India, Canada, the EU and Japan; of cottonseed meal China, India, Russia and the USA; of groundnut meal India and China; of sunflower meal the EU, Russia and Argentina; of copra meal the Philippines and Indonesia: of linseed meal Argentina, China, the EU and the USA.

Major exporters of oilseed meals are the USA, Brazil, Argentina, China and India who together usually account for about 75% of world trade. China was a major importer of soya meal in the years 1995 to 1998 and imported 10 Mt in the 1996/97 and 11.5 Mt in the 1997/98 seasons. Main rape seed importing countries are the USA, South Korea, and Japan. The major sunflower meal exporter is Argentina, and the major buyer is the EU; the latter is also the major importer of palm kernel meal.

2
Castor

The *Euphorbiaceae*, or spurge family, contains some 280 genera, in which there are a vast number of imprecisely related plants mostly native to the tropics. Many of these plants contain an irritant white latex, sometimes used as a fish poison or bird lime. The genus *Ricinus* is considered monotypic and *R. communis*, (the castor plant), includes many polymorphic types.

Dioscorides named castor *kroton* or *kiki* in Greek, and castor seed and the dog tick are known in Latin as *Ricinus*, which Linnaeus retained. The common name castor was apparently coined by English traders who confused the oil with that from *Vitex agnus-castus*, known to the Spanish and Portuguese in Jamaica as *agno-casto*. Although widely known as the castorbean plant and the seed as castor beans it is not a legume, and the names should be discarded in favour of castor plant and castor seed.

Castor's contemporary distribution in the warmer regions is worldwide, although its origin is obscured by wide dissemination in ancient times and the ease and rapidity with which it becomes established. Castor is indigenous to eastern Africa, and most probably originated in Ethiopia (Weiss, 1971). Russian plant breeders consider there are four large centres of variability, each with its own specific plant characteristics to support a polyphyletic origin; most important is the Irano-Afghanistan region, then Palestine-SW Asia, India-China, and the Arabian Peninsula (Moshkin, 1986). While this argument is acceptable to distinguish centres of diversity, as occurs with numerous plants long used or domesticated by man, there exists in East Africa a far greater range of wild types ancestral to all others. Chemotaxonomy or gene mapping should allow a more accurate assessment.

Castor seeds were an important item of commerce in Ancient Egypt, have been found in tombs dating from about 4000 BC, and sections of hollow castor stems were used to store the cosmetic kohl in the eighteenth dynasty, 1567–320 BC (Hayes, 1953). Dioscorides, Herodotus, Strabo and Pliny all mention the use of castor oil in Egypt for lighting, and the method of oil extraction described by Herodotus continued in use until modern times. Strabo stated poor people and labourers used the oil to anoint their bodies, and the odour was acrid and unpleasant. Hebrews knew castor and its oil as *qiqayon*, and it was probably the plant referred to in the Book of Jonah, which provided divine shelter for Jonah outside Nineveh. Israel is a major source of high-yielding castor cultivars.

The castor plant is naturalised throughout Africa south of the Sahara, where it is generally a fairly tall, much branched perennial, often referred to as giant castor, and seldom cultivated. The plant and its seed were described by many early travellers and explorers. Tribesmen used it mixed with local pigments to paint their bodies on ceremonial occasions and their womenfolk as hairdressing or skin conditioner. The ripe seeds, many varieties of which have most attractive coloured testa, are commonly used as counters in the universal and interminable game of *bao*, and by witchdoctors and soothsayers as part of their impedimenta (Weiss, 1966c; 1979). The seeds and leaves were also used in poison ordeal trials (Irvine, 1930). The West African *ogiri*, a popular condiment and soup mix, is made from fermented castor seeds (Aniche & Nwokedi, 1994).

Castor growing and use of its seed and oil are so well established in India that the theory of continental pairing has been suggested by local botanists to account for its ancient occurrence in India and the Ethiopian region. The *Susruta Atharvaveda*, circa 1000 BC, refers to castor as indigenous and describes two main types. A tall perennial with large, shining, reddish, high oil content seeds, but producing a dark oil unsuitable for medicines. The other a short annual, with pale grey or mottled light brown seeds, lower in oil content but producing a light coloured oil of superior medicinal quality. The *Arthasastra* of Kautilya, probably written about 300 BC, records castor as *eranda* and explained how the seed was crushed to provide oil for lamps. The earliest castor seeds were found during excavations at Ter, Maharashtra State, and dated to around 125 BC (Achaya, 1990). However Watt (1966) comments, 'Oddly, it was not mentioned by any of the early European explorers'.

The oil was used extensively in local medicines mainly as a laxative, but also to ameliorate dry and rough skin, and bruised leaves as a poultice to relieve the pain of sprains and torn muscles. In Assam and Burma the foliage is food for the local Eri silkworm, leaves being collected daily from the extremely numerous wild plants. In Malaysia a mixture of unslaked lime, resin and castor oil, *chan*, is used to caulk small boats.

The Chinese have traditionally used the oil in medicines, and the plant has been cultivated, or wild plants protected, for centuries. Castor was not, however, mentioned in literature before the T'ang period, AD 618–906, and only as a cultivated plant. In the Great Herbal, the *Pen-Ts'ao kang-mu* begun in AD 1552, castor oil is quoted extensively for internal use and in poultices (Wong & Lien-Teh, 1936). It was probably introduced to Japan from China, and although horticultural varieties are widely grown for use in flower arrangement, *ikebana*, it has never been cultivated commercially for its oil, and domestic demand is satisfied by imports.

In southern Russia castor cultivation, or more accurately preservation, has been common for centuries, and three 'species' are officially recognised – *R. communis*, *R. macrocarpus* and *R. microcarpus*. It was first cultivated in the Caucasus in the middle of the nineteenth century and grown on a commercial scale in the former USSR in the late 1930s. From 1945 the area greatly increased but from 1965 began to decline, and following the break-up of the USSR virtually ceased on a commercial scale.

Castor is a fairly recent introduction to the New World, probably transferred with slaves from Africa, but has become widely naturalised, hastened by large-scale commercial planting, especially in Brazil. Plants freely produce viable seed and large indigenous populations are quickly established. In Idaho, USA, the Rocky Mountain wood-tick, *Dermoncentor andersoni*, vector of several dangerous human and animal diseases, is known as the castor-bean tick, a reversal of castor's original name!

Castor oil's medicinal use was the main reason wild castor was everywhere tolerated, but plants were also grown for shade or as wind-breaks. Perennial, almost evergreen types which grew into small trees, were planted and much appreciated in arid, often somewhat saline areas where few other trees could exist, as in parts of Iran and Pakistan.

The root is traditionally believed to have medicinal properties, probably due to the presence of ellagic acid (gallogen). In East Africa persons infected with guinea worm, *Dracunculus medinensis*, chewed the pounded root and continuous use reportedly killed the worms. Although whole castor seeds are extremely poisonous, a method of preparing them for human food is used in West Africa. The seeds are cracked open when dry, the testa removed and the cotyledons, wrapped in small banana-leaf bundles, are boiled for 1 hour. They are then fermented in baskets lined and covered with banana leaves for two days, removed and pounded into a paste. During pounding, ashes from burnt oil palm kernels are added to discourage flies. The mass is scooped out, tied into banana leaves, dried for a day or so, and taken to market. The finished product is called *ogili-isi* (Onitsha) or *oglili-ugba* (Awka).

Castor oil and its derivatives are widely used in industry, and are a component of human and veterinary medicines. It is thus interesting to speculate on the methods used to establish that castor oil was a purge, but the whole bean a fatal poison. The classic remark of Aulus Celsus, AD 14–37, is relevant:

> In this way medicine arose from the experience of the recovery of some and the death of others, differentiating the harmful from the salutory things.

Strict safeguards are necessary to ensure that expressed oil for medical use is not contaminated by toxic elements. Frequent fatalities result from children eating the attractive seeds, and records of such tragedies are numerous and continuing.

The production of dwarf hybrid castors for mechanised agriculture has radically altered the economics of the crop, and made it increasingly profitable. Additionally, the cost of liquid fossil fuels makes castor growing even more attractive in countries with no indigenous oil reserves.

The two largest producers of castor seed are India, 750, 800 and 900 kt in 1997 to 1999, and China 300 and 350 kt in 1997 and 1998. Main producing areas in India in order of importance are Andra Pradesh, Gujarat, Mysore and Orissa, with the first normally accounting for one-third of total production. Indian exports of castor oil normally account for about 85% of world trade, and oil production rose steadily from 1980 to average about 80 000 t since 1990. More efficient expelling techniques should increase oil obtained by 10–15%. In 1998 the Chinese government stated there was an additional 7 Mha suitable for the crop in 20 provinces, and plans annually to expand the area planted and build modern processing plants. Brazil was the third largest producer until 1990 with about 250 000 t, although this was half the 1975 total, but the area under castor has since declined. Main producing areas are, in order of importance, Bahia, Sao Paulo, Pernambuco, Ceara and Parana. Bahia can produce two crops annually and normally accounts for half of Brazil's production.

Castor seed production in the former USSR was 98 000 t in 1940 from 400 000 ha, fell to 70 000 t in 1970 from 280 000 ha, to 50 000 t in 1980, and to an estimated 25 000 t in 1990, but no accurate figures from what is now the Russian Federation are available. The crop relied almost wholly on official support which it no longer receives. Main producing areas were in the Ukraine and north Caucasus, with large irrigated areas in the now independent southern Russian republics, especially Kazakhstan.

Thailand's production steadily decreased from 40 000 t in 1970 to less than 10 000 t by 1990, but since then has been given official support with a target of 50 000 t annually, which could be processed by a local Thai–German joint venture plant. In Africa a private company has encouraged local Zimbabweans to grow castor and built a processing plant with an annual capacity of 40 000 t; unfortunately in 1998 domestic production had reached only 200 t, and the company planed to import seed from neighbouring countries.

BOTANY

Ricinus communis L is known in English as the castor plant; Spanish and Portuguese, *ricino*; Arabic *kirwa*; East Africa (Kiswahili), plant *mbarika*, seeds *mbono*; Ethiopia, *gulo*; Farsi *bedanjir;* India (Sanskrit), *ruvuka*, or *eranda* now used generally; Thai, *mahung*; Chinese (Mandarin), *pi-ma*, tick hemp.

All types of castor from giant perennials to short internode dwarfs have the same chromosome number $x = 10$, $2n = 20$, and natural polyploidy is probably absent from the monotypic genus. However, cross-incompatibility occurs between strains from ecological extremes and may be the beginning of subdivision of the

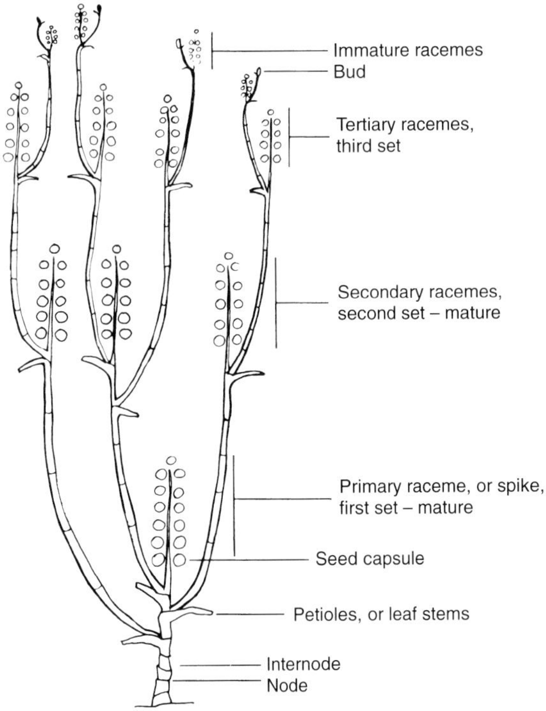

Fig. 2.1 Schematic drawing of castor plant showing sequence of branch and raceme development.

Fig. 2.2 Typical dwarf castor oil plant, USA.

species. Contemporary species with $2n = 20$ may have arisen by autoploidy from a diploid progenitor with $2n = 10$, and it has been suggested that the basic chromosome number should be $x = 5$ and not $x = 10$.

Classifications of the different castor types into subspecies (subsp.) have been published, but none are true subsp. as botanically defined and their continued use is inaccurate. Most of these so-called subsp. are merely dominant local types whose characteristics reflect adaptation to local environmental pressure, or human selection. The most commonly quoted are *Ricinus communis* subsp. *persicus* (Persian species), *R.c.* subsp. *chinensis* (Chinese species), *R.c.* subsp. *zanzibarensis* (Zanzibar species), *R.c.* subsp. *sanguineus* (Crimson species), *R.c.* subsp. *Gibsoni*, and *R.c.* subsp. *cambogensis* (Purple species). Three separate 'species', *R. communis*, *R. macrocarpus* and *R. microcarpus* are described by Zukovskyij (1962) the last occurring wild in the Middle East and southern Russia. Six subsp. were recognised by Popova (1930), the first four most commonly quoted plus *R.c.* subsp. *africanus* and *R.c.* subsp. *mexicanus*.

The castor plant, hereafter for brevity designated plant and seed, varies greatly in its growth; colour of foliage and stems; seed size, colour and oil content. Most wild types are large perennials often developing into small trees, cultivated types are usually short annuals. Colour differences in leaves, stems and inflorescence have allowed the selection of very attractive horticultural and ornamental variants. Classifying such selections as subsp. is also botanically incorrect, although they may be given names to establish a separate identity by their breeders.

A diagrammatic representation of a plant is shown in Fig. 2.1. Variation in the rate of raceme development is also typical, and a disadvantage for mechanized production. Plants can be divided into two basic types commonly known as giant and dwarf castor, Figs 2.2 and 2.3, and to avoid unnecessary repetition, these terms will be used hereafter.

Fig. 2.3 Interplanted giant castor in India.

The period after emergence to maturity is greatly influenced by environment. Giant perennials in East Africa produce the first harvestable seed in approximately 140 days, imported dwarfs after 160 days. This compares with a total life of 140–160 days for the latter in the USA, where frost often limits growth (Table 2.1). Quick maturity can be an asset, even when related to lower potential yield, as in India where an early maturing cultivar allows double cropping as the period available is too short for higher-yielding, longer-season cultivars (Yadava *et al.*, 1976). Irradiation of castor seed and seedlings produced plants with varying flower-sexuality ratios, height, maturation, and increased insect resistance.

Giant castor has a root system similar to typical woody perennials, with a large, well-developed tap-root which penetrates to several metres and has substantial laterals. The well-developed secondary root system is much branched and often deeply penetrating to take maximum advantage of soil moisture, a major factor in the plant's resistance to drought. In arid areas with only rainfall for subsistence, aerial growth tends to be slower in relation to root growth than under more favourable conditions, and the rate of root growth of such plants is often faster than in cultivated types.

An added advantage of quick root growth is increased ability to follow the moisture front into the subsoil, which in many dry areas is often better supplied with nutrients than in more temperate regions. Good yields can be obtained at a lower level of added nutrients than would be required in neighbouring areas of higher rainfall, and cultivars have been developed in India whose root growth is substantially faster and greater to take advantage of this ability. Dwarf cultivars generally have more compact roots and are thus suitable for irrigation.

Table 2.1 Average growth periods of castor (days).

Country	To emergence	To flowering	To harvest
Australia	11	50	160
E. Africa (giant)	10	40	140
Italy	14	53	166
Poland	20	70	153
USA	13	47	150

The stem is round, glabrous, frequently glaucous, and covered with a waxy bloom which gives red or green stems a bluish appearance. In India, plants with a heavy bloom are considered to be more resistant to jassids or leafhoppers, but those without bloom more resistant to mites. Data from Japan showed a highly significant increase in 100-seed weight of seed from local red-stemmed compared with green-stemmed strains (Ueda *et al.*, 1990), but this has not been noted elsewhere. The stem of dwarf cultivars generally becomes hollow with age, but in giant types it is usually solid to a considerable height. Nodes are well developed and from each arises a leaf. The node at which the first raceme appears is an important agronomic characteristic, since it is associated with quick maturity. In dwarf-internode hybrids it usually occurs after the sixth to twelfth node. Knowledge of the rate of absorption, assimilation, and transport of water within castor plants is important in relation to irrigation methods. Nuclear magnetic resonance is a useful non-invasive method of measuring water flow in stems, and a microimaging technique determined its extent and directions (Kockenberger *et al.*, 1997).

Castor is naturally multi-branched, primary branches giving rise to secondary branches, a sequence that continues over the life of the plant. This is an undesirable characteristic for commercial production and mechanical harvesting, since the fewer the branches the lower the number of racemes, and hence a more even and shorter period to overall maturity. Pruning to reduce height or number of major branches is usually ineffective on giant types, although it may slightly increase yield. The cost of the operation, which must be carried out with care if it is not to cause damage to the relatively brittle plant, is normally greater than the value of the yield increase. Topping shorter cultivars at 30–60 cm generally increased branching but usually decreased yield in India, while removal of all auxillary buds increased average yield by a third and also hastened maturity. Such operations are labour intensive and restricted to smallholder crops.

The very large leaves on long, stout petioles are alternate except for two opposite leaves at the node immediately above the cotyledons. Leaves are palmate, with five to eleven lobes, and prominent veins on the under surface; usually a dark glossy green above, duller below. Ornamental varieties have leaves and stems varying from light green to dark red, depending on level of anthocyanin pigmentation. Young leaves are mildly toxic to animals and some insects, but in central Tanzania during the dry season when green fodder is almost non-existent, herds of eland (*Taurotragus oryx*) browsed giant castor with no apparent ill effects. Mature leaves are harvested in Korea, dried and stored until winter, when they are cooked in boiling water and eaten. Leaf extracts have minor bactericidal activity, and fresh or pulped leaves are frequently used by local people as poultices on wounds.

Growth and expansion of leaves is little affected by prolonged sunlight, provided there is ample moisture for transpiration, and leaf-growth curves obtained in California indicate that temperature and water stress are more important than sunlight in control of growth. Irrigation can also increase absorption of photosynthetically active radiation due to an increase in total leaf area.

The inflorescence is a pyramidal raceme (spike or candle), borne terminally on main and lateral branches, and the node at which the first raceme originates is a cultivar characteristic. The inflorescence can reach 100 cm, but since there is wide variation in distance between flowers, yield is not necessarily correlated with length.

Castor is normally monoecious, with pistillate flowers on the upper and staminate on the lower portion of the raceme, but both can also be interspersed along the raceme. Inheritance analysis of progeny from various crosses in China showed that pistillate characteristics may be controlled by a combination of a large number of different genes (Zhang & Liang, 1993). The proportions vary naturally or can be influenced by environment; female tendency is highest with moderate temperature, maleness with higher temperature. Day length also affects the ratio of male to female flowers, and long days in Russia promoted female flowers at the expense of male, and short days reversed the process. Spraying with growth regulators can also alter the sex ratio of flowers.

Plants with wholly female flowers were first found in a field in Manchuria in 1941 and is a rare mutant in castor, the progeny segregating in a 1:1 ratio for female and monoecious plants (Katayama, 1948). Selection of strains with only male or female flowers allowed commercial production of specific hybrids. Gamma-irradiation induces pollen sterility and can also be used to produce male-sterile plants.

Pollen is normally shed from 2–3 hours before sunrise until late afternoon, with a mid-morning peak; a temperature of 26–29°C and relative humidity of 60% is the optimum; 15°C delays shedding and lower temperature can injure grains. Pollen contains an allergenic compound similar to that in the seed, but the reaction by susceptible persons is not so severe. Prolonged periods of high humidity adversely affect pollen viability and shedding, and in Dahomey, West Africa, so adversley affected normal pollen production that it significantly reduced yield. Castor is normally cross-pollinated, wind being the major agent in the tropics, but self-pollination of monoecious plants can occur.

The plant has a long flowering period and in the wild giant types may flower year-round; the lowest flowering raceme on the plant is usually the first to mature, the others following sequentially up the stem, with a distinct variation in yield of seed and seed oil content; primary racemes generally producing more and larger seeds. Table 2.2 shows the variation between cultivars and also between racemes, related to the various periods involved. Average seed size was less affected, although oil content and kernel: husk ratio differed significantly between varieties (Sichman *et al.*, 1978).

The fruit is a globular capsule, usually spiny, becoming hard and brittle when ripe; giant types and some cultivars shatter at maturity, but dwarf types generally do not. Ripening of fruits along the raceme is uneven, the lower maturing before

Table 2.2 Differences in development and seeds of sequential racemes – Brazil.

Component	Variety		
	IAC 38	Campinas	Guarani
Days to flowering			
PR	90	69	72
SR	118	82	85
TR	140	104	106
Days to maturity			
PR	146	139	145
SR	175	155	158
TR	198	175	178
100 seed wt (g)			
PR	46	44	50
SR	44	40	43
TR	40	39	35
Germination (%)			
PR	100	100	100
SR	70	70	35
TR	70	56	42
Av. seed dimensions (mm)			
Length	15	15	15
Breadth	10	9	9
Thickness	7	7	6
Kernel/husk ratio (%)	61	69	68
Average oil content (%)	40	45	48

PR, Primary; SR, Secondary; TR, Tertiary racemes.
IAC 38 is a local selection; Campinas is a cross between IAC 38 and Cimarron; Guarani is a cross between Campinas and Preta.

the upper, and in wild varieties the period to maturity between first and last fruits may be several weeks.

Non-spiny (thornless) cultivars have been developed, in others the capsule spines are soft, flexible and non-irritant. Capsule spines have been studied in detail (Ueda *et al.*, 1989). Length, thickness, rigidity, presence or absence of spines is generally a cultivar characteristic, but considerable variation can exist even on individual plants. One plant growing in a field of giant castor in Tanzania, had capsules on adjoining branches that ranged from practically spineless to fully spined.

The method of capsule dehiscence varies, and thickening of the wall, endocarp and region of fusion of the loculi appear closely related to a low degree of shattering. In some wild varieties the whole capsule falls from the desiccated raceme, the seeds remaining enclosed. This capsule may be blown by the wind or washed by heavy rainfall a considerable distance from the parent plant, thus distributing the seeds over a wide area. In other varieties, capsules remain *in situ*, and split to release seeds only after successive showers and sunshine cause capsules to fracture. These seeds often remain in the vicinity of the parent plant, and under suitable conditions dense thickets of castor occur, which are a feature of some African areas, and provide a valuable cash crop to local people.

Indehiscence is a valuable characteristic in dwarf types as it greatly reduces pre-harvest capsule loss, especially when crops are machine harvested (Fig. 2.4). The strength of the capsule wall is a cultivar characteristic, and in some it is so strong it precludes mechanical hulling. Conversely, very soft capsules may be equally difficult to hull without damaging the seeds.

Fig. 2.4 Indehiscent capsules on raceme.

A capsule normally contains three seeds, each a flattened oval, with a shiny brittle testa enclosing a white, highly oleaginous kernel. A diagram of a castor seed, compared with bean and maize, is shown in Fig. 2.5. Seeds may be coloured white, brown, buff, black or red, but usually several colours occur as very attractive mottling on the testa (Fig. 2.6). Local people use the seeds for necklaces and belts, or as ornaments on domestic utensils. Seeds vary greatly in size, from 5 to 250 mm long in giant types, in breadth from 5 to 16 mm. Seed size varies not only between cultivars, but from different racemes on the same plant. Germination of seed from different racemes varies substantially, the first to flower having greater viability than the last.

The 100-seed weight varies from 10 to 100 g, but most dwarf-internode varieties average 30 g. Weight of individual seeds generally increases as the total number of seeds produced per plant decreases. However, the increase in seed weight does not fully compensate for the decrease in number, and total yield is less. This applies particularly to the yield obtained from different times of planting, since yields normally decrease after the optimum sowing period. There appears to be a correlation between seed size and oil content; when the range of seed size is large, kernel percentage is more closely correlated with oil content than other factors.

The testa is thin and often brittle, depending to some extent on age and oil content. Seed of some cultivars may have a dormancy period of several months, but freshly harvested seed of these varieties can be made to germinate by removing the caruncle and piercing the testa at this site, leaving the endosperm intact. However, the majority of modern dwarf hybrids are not dormant, and freshly harvested seed germinates without special treatment. Some factors controlling dormancy have been investigated (Bewley & Downie, 1996). Germination is epigeal, and the cotyledons expand as green leaves, Fig. 2.7.

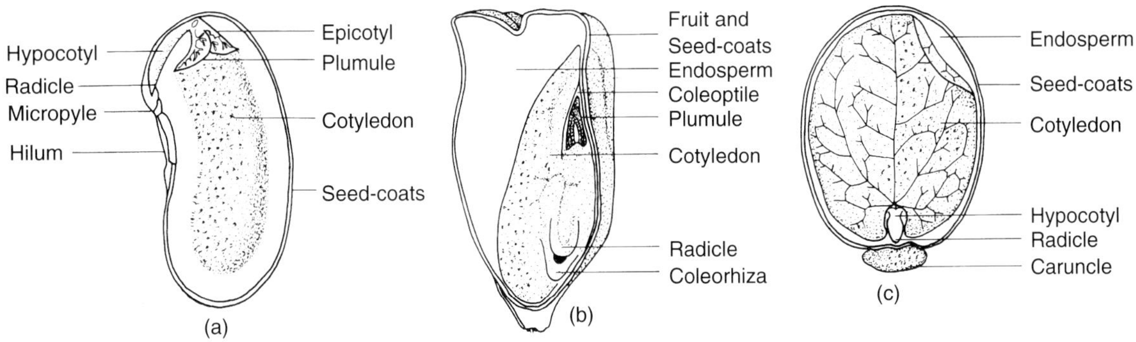

Fig. 2.5 Section through seed of (a) bean (b) maize and (c) castor; for comparison of main components.

Fig. 2.6 Castor oil seed showing mottled testa.

The most important seed constituent is the oil, usually 40–60% in commercial cultivars, high in triglycerides principally ricinolein, the fatty acid component being ricinoleic acid whose three hydroxyl groups confer on the oil the unique property of solubility in alcohol. Analyses of seed from selected countries are shown in Table 2.3. Factors that influence formation of seed oil, the period over which oil is formed and the maximum reached, are important. Such data are essential for assessment of nutrient and water requirements, optimum amount and times of application, and in selecting the most suitable time of planting and harvest.

Oil formation began 20 days after flowering on a dwarf hybrid in Canada, and two-thirds was synthesised in the next 20 days. Protein synthesis occurred over a longer period, but was accelerated during the middle 20-day period. Ricinoleic acid was absent in very young seed, but appeared at 12 days, and at 36 days it represented 90% of fatty acids; fatty acid composition of oil then remained constant. In the intervening 24-day period the amount of ricinoleic acid was lower than normal. The amount of each fatty acid component expressed on a percentage basis increased over the entire period of development, and there was no evidence of substantial conversion of oleic and linoleic to ricinoleic acids (Canvin, 1963); see also Tables 2.4, 2.5.

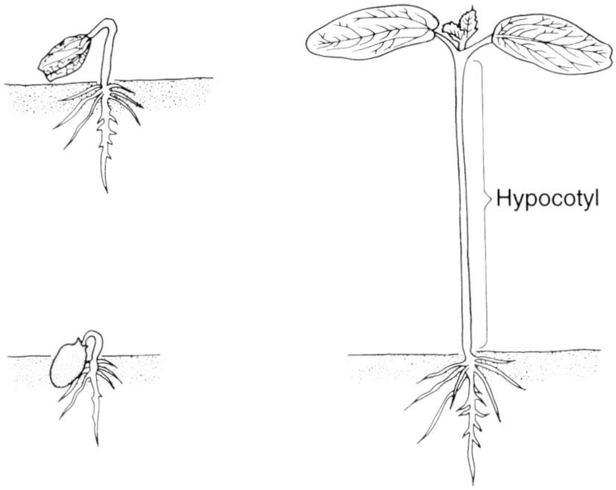

Fig. 2.7 Epigeal germination of castor with cotyledons carried above soil on long, brittle hypocotyl.

Annual variation occurs in seed oil content from the same cultivar, and was confirmed by the author in successive seasons in Kenya and Tanzania. While this suggests that climatic differences were responsible, these may also have affected the sequence of agricultural operations, which had a greater effect. For instance, too early harvesting with a high proportion of immature or still green capsules can drastically reduce oil yield/ha. Comparison of oil content from completely ripe seed of a Tanzanian giant castor i.e. with capsule brown and dry, and physiologically ripe seeds contained in a green capsule, showed the latter to have 20% less oil. In Kenya, percentage oil content from a dwarf hybrid was; ripe dry seed 44.3%, ripe but not fully dry 41.0%, green but full-sized 32.2%, green and immature 27.8%. Thus premature applica-

Table 2.3 Analyses of castor seed.

Characteristic (%)	India	Brazil	Cyprus	Philippines	Argentina	Mexico	Italy	Ethiopia
Kernel	66.8–74.4	—	60–65	—	65–67	—	—	66.0
Hull	25.6–33.2	—	35–48	—	33–35	—	—	—
Oil	45.0–51.8	48.6	53.7	53.7	43–45	46.0	52.6	47.0
Moisture	3.1–5.8	5.5	—	5.1	—	7.1	8.0	6.5
Protein	12.0–16.0	17.9	—	30.6	—	—	20.5	22.1
Carbohydrate	3.1–7.0	13.0	—	6.7	—	20.0	19.9	—
Crude fibre	23.1–27.2	12.5	—	—	—	20.0	15.9	—
Ash	2.0–2.2	2.5	—	3.8	—	1.9	2.9	2.9
100 seed weight (g)	19.0–26.0	—	—	—	—	—	—	—

Partial analyses of African castor seed*.

Characteristic	Aden: local	Kenya: Kitui local	Nigeria	
			Benue local	Kabba local
Moisture content (loss at 105°C)	4.9	5.2	5.2	5.0
Oil content per cent (MFB)	49.3	49.2	49.0	52.3
Acid value of oil (KOH/g)	1.9	1.3	1.8	1.6
100 seed weight (g)	48.9	59.2	61.3	65.2

* Carried out by the Tropical Products Institute, London.

tion of pre-harvest defoliants could have an appreciable effect on the total yield.

Whole seeds are poisonous to animals and humans, and continue to be dangerous to the unwary; human symptoms are vomiting, diarrhoea, dizziness, and a high temperature. Symptoms of poisoning in animals are noted in the literature. The seed contains the toxic protein ricin and alkaloid ricinine. Ricin was first isolated by Dixon (*Australian Medical Gazette* 4, 137–155, 1886–87). Ricin is one of a small group of phytotoxins which also includes abrin, circin, crotin and robin, all are extremely toxic, antigenic, thermolabile, and they agglutinate red cells. Oil-free meal contains about 1.5% ricine, in its purified state a white powder. It differs from many bacterial and snake venoms in the relative ease with which it is absorbed through the abdominal wall. It induces no symptoms when given by the mouth, since it is digested in the stomach, but injection of 0.001 mg/kg body-weight leads to general haemorrhage and death. Acute (lethal) oral dosage of ricin to mammals is 150–200 mg/kg, but 0.035 mg/kg may prove fatal to man. Symptoms of poisoning occur with a subcutaneous injection of only 0.001 mg/kg body-weight, and small particles in abrasions and eyes may also prove fatal.

Ricinine is a white crystalline alkaloid usually extracted from seeds, but also contained in leaves. It is synthesised mainly in growing tissues of young, and developing organs of older plants. Another group of proteins and polypeptides in seeds is the carbohydrate free allergens, which can be resolved into four bands by electrophoresis, and the various fractions may have different antigenic specificity. A highly potent allergenic fraction (CB-1A) isolated from shelled defatted seeds, is composed of low-molecular weight protein and polysaccharidic proteins. It induced positive cutaneous reactions in seed-sensitive patients when diluted 1 in 106. One of the allergens present in castor is chlorogenic acid. Ricine was also said to have been used to murder a Bulgarian in London in 1978, by injecting him in the leg with a pellet inserted from a specially modified umbrella!

Under natural conditions castor exhibits a diversity in growth, and at smallholder level visual selection for desired characteristics is a useful method of producing a locally superior type. Under more intensive production the ultimate object must be profitably to maximise oil yield per hectare. To do this it was necessary to breed castor types suited to fully mechanised agriculture. The two main obstacles were the plant's great height and extended flowering period, which made mechanical harvesting and systematic pest control very difficult. Finding short-internode plants in Brazil gave breeders the key to height reduction (Krug, 1945), and dramatically reduced plant height; since 1956 short-internode high-yielding varieties have been commercially available (Fig. 2.8).

The various types of dwarf castor were defined by Prof. R.D. Brigham, Texas A.M. University, as;

Tall normal internode, and high node number (late in initial bloom)
Intermediate normal internode, and low node number (13 or less)
Dwarf genetic dwarf (usually with low node number).

Another factor then became important – the varying thickness and strength of capsule walls which complicates mechanical hulling, since accurate and standard machinery adjustment is essential to produce undamaged seed; cultivars are now available with very uniform fruit characteristics. Resistance to specific pests or disease has been located in strains of wild castor and some Indian types; unfortunately insect resistance may be combined with some undesirable features such as small seed, and dehiscing capsules. Resistance can also be related to bloom on leaves and stems, location of flowers on racemes, and leaf secretions, as noted in the section on pests and diseases. However, no correlation exists between vegetative colour and other agronomically superior characteristics.

Castor breeding programmes in developed countries naturally focused on problems associ-

Fig. 2.8 Prof. R.D. Brigham with dwarf castor plant, USA.

Fig. 2.9 Spineless-capsule castor variety, India.

ated with mechanical production, but other factors can be more important to smallholders. Extremely spiny capsules are a deterrent to manual harvesting and thornless varieties desirable (Fig. 2.9). Cultivars with indehiscent capsules prevent loss from shattering in the intervals between collection, while those on which capsules dried and remain *in situ* allow irregular harvesting. Where the cultivated plot is some distance from the dwelling, a common occurrence, this would reduce the necessity for frequent visits to collect seed. Reducing dehiscence may be related to high wall-strength, increasing hulling time. However, this additional time would be offset against the greater return from the increased number of seeds harvested. Salinity tolerance is assuming greater importance worldwide, especially on the Indian subcontinent and Africa, where unwise irrigation decisions and techniques have greatly exacerbated the degree of salt damage to agricultural land.

The very wide range of plant types and characteristics that can be seen in any large indigenous castor population indicates the wealth of material available to breeders. Substantial increases in seed yield per plant can be achieved by a relatively modest selection programme, thus increasing castor's attractiveness to smallholders. This diversity also exists in hybrid breeding programmes, and can be either genetic or environmentally influenced. In Italy for example, significant differences existed among cultivars for yield, 1000-seed weight, number of racemes/plant, plant height and emergence to flowering period. Genotype × environment interactions were significant, indicating that each trait was very variable across environments (Laureti, 1995).

ECOLOGY

Castor is a semi-tropical perennial now found throughout the warm-temperate and tropical regions. It flourishes under such a variety of environments that its range cannot easily be defined. As a smallholder crop it can be grown on almost any available land, and this is one of castor's greatest virtues, since a saleable crop can be produced by the most primitive methods. In such circumstances it grows from sea level to high on inland mountains, often depending at the limits of its range on the degree of official support given to local farmers. Castor is commercially cultivated from 40°S to 52°N, and from sea level to 2000 m, occasionally to 3000 m, but generally between 300–1500 m, with the number of frost-free days the main limiting factor.

Extending the growing area northward was important in Russia and China, and research concentrated on developing very short-season or partially frost tolerant cultivars. The first has been more successful, although both have lower seed or oil yield/ha. Increased cold-tolerance would also be of importance in high, dry, tropical regions, where frost limits growth of existing cultivars.

Despite the agronomic and genetical advances of the last two decades, not all attempts to introduce castor as a commercial mechanised crop have been successful. In Italy the evergreen, perennial character of the plant made mechanical harvesting difficult, and an 'ideal type' for the local semi-arid conditions was described as an annual plant which should mature in 120–150 days (May–September), have a one-metre high stem completely branch-free, with 15–18 nodes to the main raceme, and a calculated seed yield of 7–8 t/ha (Baldanzi & Benvenuti, 1995).

Castor is basically a long-day plant, but adapts with some loss of yield to a fairly wide range of day length. Research showed plants developed normally on a 12-hour and 18-hour day, but growth and synthesising processes were severely retarded at 9 hours. Plants also varied in their reactions when grown in special chambers with similar photoperiods, but light of different spectral composition; under red light, plant development and chlorophyll formation were severely retarded, whereas both were accelerated with blue light of short wavelength. Very short photoperiods accelerate development of woody plants that did not flower, and plants are most sensitive to varying photoperiods between emergence and appearance of racemes. Additionally, long-established local cultivars are often well adapted to local conditions and behave differently when grown elsewhere, although the environment may appear similar.

Plants flourish between 20 and 26°C, and clear, sunny days with low humidity throughout the growing season produce high yield; cloudy or humid days irrespective of temperature will reduce yield. An interaction may exist between temperature and light intensity, with the latter having a greater effect than temperature on seed size and oil content. Very high temperature at flowering, above 40°C, results in blasting of flowers and poor seed set. High temperature can also adversely affect seed composition; temperatures above 35°C reduce oil and protein content, those below 15°C reduce oil content and alter oil characteristics. Climate can influence free fatty acid (acid value) of seed from plants grown in a particular area; seed harvested in November in Nebraska had an acid value of 0.5, but a value 10.0 the following April. Seasonal climatic variation can substantially influence oil components, as an analysis of oil obtained from local mills in India indicates (Table 2.4).

Published data on oil composition from different world regions varies substantially, and could be due to climate, cultivar, cultural methods, or processing, but seed crushers state this regional variation is fairly constant. Gas-liquid chromatographic analysis of oil from different countries is shown in Table 2.5. (Binder *et al.*, 1962).

High soil temperature decreases time from sowing to emergence, and low soil temperature extends it, and although the two are closely related the factors influencing both vary regionally. In more temperate regions the soil is not warm enough early in the year, whereas in more dry areas onset of the rains may depress soil temperature. Availability of irrigation water may also involve reassessment of data acquired from rain-grown crops, and modify agricultural operations, since time of planting can become independent of rainfall. The effect of soil temperature, as influenced by planting dates, is shown in Table 2.6. Similar results have been obtained in many other countries, and to plant

Table 2.4 Seasonal variation in Indian oil.

Year	Saturated	Oleic	Linoleic	Ricinoleic	Dihydroxystearic
1947	3.0	5.4	3.4	85.5	2.4
1949	3.5	5.1	3.5	86.0	1.9
1950	3.4	5.8	3.5	85.7	1.6

Table 2.5 Analysis of oil from selected countries.

Source	C_{16}	C_{18}	Oleic	Linoleic	Linolenic	Eicosenoic	Ricinoleic	Dihydroxystearic
Angola	1.0	0.7	2.9	4.3	0.6	0.5	89.4	0.6
Brazil	1.0	0.8	3.3	4.2	0.5	0.4	88.9	0.9
India	1.1	1.0	3.1	4.7	0.5	0.5	88.4	0.7
USA	0.9	0.8	2.0	4.3	0.6	0.3	90.3	0.8

Table 2.6 Effect of time of planting on emergence and yield – USA.

Date of planting	Number of days to 50% emergence	Mean soil temp. (°C)*	Adjusted mean yields (kg/ha)	Adjusted mean yield (kg/ha)†
18 Apr.	23	10	1776	1201
1 May	19	14	2290	1370
16 May	17	14	2220	1272
30 May	12	18	1998	1325
12 June	10	19	1826	936

* Mean soil temperature was taken at 10 cm depth at 8.00 a.m. daily for 14 days after sowing.
† From non-irrigated trials carried out at the same time.

outside the optimum period generally reduces yield. There are occasions, however, when it may be necessary, to avoid high temperature at flowering or the period when a major pest is most active, as noted later.

A temperature of –2°C for 4 hours will normally kill dwarf or annual castor at any growth stage and a frost-free growing period of 140–190 days is necessary. Giant types appear to be resistant when well grown, although succulent parts may be affected. Resistance to low temperature can be increased by selection, and there is some evidence of varietal difference in susceptibility (see Table 2.13). Freezing of fully developed but immature seed awaiting harvest can reduce viability, but ripe seed is little affected. Short-season, quick maturing cultivars allow castor to be grown in areas where the number of frost-free days is the main constraint, and this is probably easier than developing frost-resistance.

Hail can seriously damage seedlings, and if a severe storm occurs shortly after emergence most will be destroyed. Losses caused through shattering of mature capsules by late-season storms can also be serious. Hail damage to foliage can be visually severe with little effect on final yield. Giant varieties are able to compensate for damage caused even fairly late in the season by prolonging flowering. This extension has a secondary effect also tending to reduce loss, since later racemes are less liable to insect attack. However, late-flowering racemes normally bear fewer capsules, and total yield is generally less than from the same number of earlier flowering spikes. Fairly extensive defoliation of dwarf varieties up to flowering of the first raceme has little effect, but after this yields may be depressed.

Castor produces highest yield with an annual rainfall of 600–700 mm if this falls mainly in the early growing season, although 375–500 mm will produce a reasonable yield with good water conservation. There should be no moisture shortage during flowering, but protracted heavy rain at this period will reduce yield. Prolonged rain before planting will lower soil temperature, and sowing should be delayed until the soil becomes warmer. While young giant castor plants are relatively tolerant of high soil moisture levels for up to 14 days, seedlings and young plants of newer hybrids are often very susceptible, and high levels for over 7 days can cause a serious reduction in subsequent growth (Poltanee, 1995).

A close correlation generally exists between rainfall and time of planting to achieve maximum yields. The rapid fall in seed yield after the optimum sowing time is illustrated in Table 2.7. (Brazil – Banzatto *et al.*, 1975; East Africa – Weiss, 1967a; Sudan – Heipko *et al.*, 1966; USA – Kittock & Williams, 1967) and similar results have been obtained in most castor growing countries. Castor's ability to withstand rainless periods may make the relationship between rainfall and planting date less important than that between rainfall and subsequent insect pest activity, discussed later, or between planting date and maturity.

Table 2.7 Effect of planting date on yield (kg/ha).

Brazil*		East Africa*	
Date	Yield	Date	Yield
1 Nov.	2420	1 June	1269
15 Nov.	1958	22 June	807
30 Nov.	1711	12 July	583
15 Dec.	790	2 Aug.	185

Sudan†			USA	
Date	Yield		Date	Yield
6 June	924 (a)	1120 (b)	25 Apr	1139
23 June	722	1089	9 May	1208
13 July	864	1108	23 May	1262
8 Aug.	480	970	6 June	1255
23 Aug.	365	971	19 June	800

* Average of several years † Sudan yields from north (a) and south (b) of Khartoum.

Where the expected rainfall approaches the minimum required it is essential to plant following the first major rain, when soil-moisture is adequate for germination and soil temperature has not been depressed. Where rainfall is above this minimum, planting early in the rainy season may result in poor emergence owing to low seed-bed temperature and fungal attack. More important, flowering will then frequently coincide with a major insect build-up, since onset of the rainy season with its periods of high humidity acts as a trigger mechanism on insect eggs and larvae to complete their life cycle. In such areas planting should be delayed for 2 or even 3 months after the rains break. In the Sepik area of New Guinea crops grown in the rainy season failed due to insect attacks, but yielded well during the dry season. In heavy rainfall areas, quick-maturing varieties planted late are not a solution, as in the USA capsule mould and not insect attack is the main factor limiting castor growing.

In arid regions where rainfall is barely adequate to sustain a crop or where most rain falls as short, intense storms, moisture conservation is essential. Cropping in alternate years may be possible, the non-crop year being used to store moisture. To be effective, lands should be under bare fallow, or if liable to erosion, protected by a cover crop. The effect of various crops on soil moisture in sandy red loams in Tanzania is shown in Table 2.8.

An approximate guide for regions where average rainfall usually exceeds 750 mm is to allow 400–500 mm rainfall up to main flowering, subtract this from the total and plant accordingly. Where there are clearly defined rainy seasons, time of planting is critical in respect of available moisture and temperature, and formal trials essential to determine the optimum period. It is also necessary accurately to determine the cultivar most suited to a particular area, since the very wide range of yield possible can materially

Table 2.8 Effect of cropping on soil moisture.

	Bare fallow	Teff cover	Volunteer cover	Sorghum crop	Maize crop
Year 1	323	312	234	239	254
Year 2	350	295	259	267	292

Figures are equivalent to rainfall in mm at 1.8 m soil depth. Castor has been successfully grown in this region for many years, averaging two crops in 3 years.

Table 2.9 Normal planting dates in selected countries.

Africa		India	
North	November–December	Kharif	May–June
West	February–March	Rabi (i)	September–November
Central	October–November		
South	October–November	Russia	
		South	May–June
America		Australia	
North	March–May	North	September–December
South	September–October	West	January–February
	January–March		

Rain-grown crops except (i) = irrigated.

affect profitability. This is well demonstrated from a trial with 15 dwarf varieties in Kenya, where the average seed yield ranged between 33 kg/ha and 1680 kg/ha (Weiss, 1967a). Normal planting dates for selected countries are shown in Table 2.9.

SOILS

Castor will grow and produce a crop in almost any soil, with the exception of very heavy clay and poorly drained soils. It flourishes on the light sandy soils of North and South Africa, the Middle East and Australia, the more sandy loams of the USA and central Russia, lateritic soils in India, Central and East Africa and Brazil, and the alluvial clayey sands of many irrigated or seasonally flooded areas. Since castor is often selected for areas marginal for other crops, this adaptability is of considerable economic value. Highly fertile soils favour excessive vegetative growth, prolong the period to maturity and greatly extend flowering, all adversely affecting seed yield or cultural practices.

Castor's great adaptability should be exploited, since selection within local varieties for a strain suited to a particular soil type can greatly increase yield. In Uttar Pradesh, India, cultivar Kalpi 6 is suitable for the basaltic (trap) soil of the main producing districts of Bundelkand.

Within a given region, topography also exerts a considerable influence on cropping programmes independently of climate and other factors. An excellent example is the succession of soil types which occurs on the rolling hills in East Africa, the catena. The higher red soils are generally used for maize or groundnuts, with sorghum, sesame and tobacco planted on the lower slopes not subject to waterlogging. The lower grey sandy soils tend to retain subsoil moisture into the dry season, which castor sown well after the optimum date for other crops, is able to utilise.

These low-lying areas are usually too wet to plant early in the rainy season when other crops are being sown, and frequently left under indigenous grasses of low productivity offering poor quality grazing. Their usefulness can be further restricted by the presence of tsetse fly and high tick populations, which require expensive control. Planting these areas to castor is economically sound, since machinery suitable for other crops may be used for castor, thus extending its use. Alternative cash crops that persist into the dry season are limited, and utilising these often large areas can be a major factor influencing the viability of a regional development plan or settlement scheme.

Soils of slightly acid reaction, pH 5.0–6.5 are preferred, but a pH up to 8.0 is tolerated, but there is little data on the plant's reaction to high soil pH values. The influence of local conditions on castor growth is therefore best determined by comprehensive variety trials. Saline soils are unsuitable, for dwarf castor is ranked below maize and considerably lower than cotton in salt tolerance. Growth and yield in Romania were directly related to salts present, i.e. on strongly saline soil plants died; on non-saline soil plants grew to 50 cm and produced nearly 1000 kg/ha seed. Chemicals can increase salt tolerance, and there are indications that where salt tolerance exists in a cultivar, this may be increased by selection.

The salt tolerance of some castor hybrids in the USA is between 4 and 8 mmhos/cm (in SI units 1 mmhos/cm = 1 × siemens/m), but the majority of dwarf cultivars are salt sensitive. In India and Pakistan where the area affected by salinity is increasing, breeding salt-tolerant cultivars is important; the situation is similar is in Egypt, where cultivar differences in salt tolerance exist. In general those cultivars with pronounced tolerance of strongly acid or alkaline soils tend to be tree-like, and although of little commercial value as oil producers, are appreciated where few other plants will grow. Sea spray containing natural salts is tolerated by castor provided the soil moisture is not saline. On the sea coast near El Arish, in the Sinai, the author saw perennial castor growing on the seaward side of dune slopes, utilising underground fresh water overlying sea water.

Castor plants have little soil binding ability, a fairly open canopy, require efficient weed control resulting in a relatively bare soil surface, and are often grown in dry areas with high winds and high-intensity rainstorms. On soil liable to erosion these factors can result in major damage, and adequate conservation measures are essential on any large planting. In Tanzania for instance, on a tenant farming scheme supervised by the author, attempts to reduce erosion in widely spaced giant castor by interplanting with low-growing cover crops were unsuccessful. Castor occupied land for at least 2 years, this being the minimum period from which profitable returns could be obtained. The dry season in this region is severe, with no rain falling for some 5–7 months, the majority of annual herbs and grasses die, and castor defoliates. Thus at the beginning of the rains when the soil surface requires maximum protection against frequent high-intensity rainstorms, little or no cover is available.

Where pruning of giant varieties was practised, prunings laid in lines between rows planted on the contour, assisted retention of soil and water during rainstorms. Unfortunately large numbers of insect eggs, pupae and diapause larvae in prunings emerged with the onset of the rains, and their attacks on adjacent castor plants were devastating. A secondary effect was increased weed growth in the lines of prunings, as normal inter-row cultivations were impossible, and large numbers of weed seeds were washed into the prunings where they found ideal growing conditions. Contour planting on terraces was successful, with adequate weed control obtained by cutting once or twice during the rainy season.

FERTILISERS

Castor will, as noted, grow on a wide variety of soils, but it cannot be too strongly emphasised that where for reasons of natural infertility general crop yields are low, yields of unfertilised castor will also be low. The plant is tolerant of low rainfall, not low fertility, and these are often confused. Data indicates that a 1700 kg/ha seed yield removes the equivalent of 50 kg N, 20 kg P and 16 kg K, ignoring that contained in the plants themselves; while 2000 kg/ha seed and 1333 kg/ha hulls removed; 80 kg N, 18 kg P, 32 kg K, 12 kg Ca and 10 kg Mg per hectare.

Crops selected to be grown at a specific nutritional level must be adapted to produce the maximum required product at that level; thus castor is seldom grown on highly fertile soils, since these tend to produce a mass of vegetative growth at the expense of seed. Castor is most often grown on soils of moderate to low fertility, and it is on these soils when their nutrient status is increased, that cultivar reactions become apparent. This is increasingly important as the number of cultivars bred for specific regions or with special characteristics multiplies.

In west Kenya, following top-dressing with ammonium sulphate at 112 kg/ha of a trial comparing 11 dwarf hybrids, those with the highest potential yields made proportionally less vegetative growth than types with lower potential yields. These latter, however, were selected initially for East African areas of very low rainfall where the potentially higher yielding cultivars with larger foliage area and higher moisture requirements would have been at a disadvantage. Height of plants was also affected by fertiliser applications, and a variation of some 0.5–1.25 m was noted between cultivars when phosphate was added to the seed-bed (Weiss, 1967a). An interaction between variety and nitrogen application also occurred in the Sudan. There may also be a cultivar interaction between the rate and type of fertiliser and insect attack, noted later.

High levels of nitrogen often reduce seed oil content, but fertilisers in general have little effect on oil content of seeds at levels normally applied. Oil formation is most active 20–70 days after flowering, and throughout this period nutrient supply must be adequate. The effect of varying amounts of nitrogen and phosphate on bushel weight and shelling percentage in Zimbabwe is shown in Table 2.10. The full data however, showed that within the 15 cultivars tested there were significant differences. High levels of phosphate reportedly increased capsule indehiscence in Brazil, but this effect has not been noted elsewhere. The effect of varying levels of NPK on raceme and seed characteristics is shown in Table 2.11. Although seed oil content was reduced, both total seed and oil yield per hectare were highest at the maximum fertiliser application (Choi, 1967). However, it is doubtful if the increased oil yield between levels 2:2:2 and 4:4:4 would have been profitable.

When castor has not previously been grown on any scale, fertiliser levels necessary to produce high maize yields will generally be suitable for high-yielding dwarf hybrids. Where indigenous castor types are to be grown, the fertiliser type and levels must be determined by formal trials, since the author's experience is that response from dwarf types tends to be unpredictable. It has been remarked that succeeding castor crops sometimes produce a higher yield from the same application rates, and in India the average yield increased when castor was grown on the same site for several years, which received the same annual amounts of fertilizer. However, more than two succeeding crops of castor are usually inadvisable, for any increase due to such an effect is more often negated by increased loss from pests and diseases.

Castor's natural adaptability can ensure adequate returns to peasant growers in those areas where climate or lifestyle offer few opportunities

Table 2.10 Effect of fertiliser on seed weight and shelling percentage – Zimbabwe.

Component	Nitrogen (kg/ha)			Phosphate (kg/ha)		
	0	33	66	0	33	66
Av. bushel wt	46.0	46.3	46.3	46.0	46.3	45.7
Av. shelling (%)	60.6	60.6	61.3	61.3	61.0	61.0

Table 2.11 Effect of fertilizer on yield – Korea.

Nutrient levels*			Av. number of:		Av. seed:		
N	P	K	Racemes/ plant	Capsules/ raceme	Yield (kg/ha)	Wt/100 (g)	Oil content (%)
0	0	0	12	16	1500	25	51
1	1	1	13	13	1760	27	50
2	2	2	14	13	1920	27	49
4	4	4	14	12	2000	27	48

*Nutrient levels applied (kg/ha): N = 120, P = 160, K = 100.

for cash crop production, or where the existing agricultural system is such that little cash is available to purchase fertilisers. The correct time of planting and plant population then become more important than fertiliser to ensure maximum yields. Increased interest in a crop often precedes better standards of management and fertiliser use, and the level which maximises growers' return must be established. Castor grown in pure stands, instead of being intercropped or preserved where found growing, makes different demands on soil nutrients or moisture, and it may well be that the soil-nutrient status is insufficient to sustain a higher population. In such underdeveloped areas the use of imported (from whatever distance) fertilisers should be rigourously evaluated before recommending their use, although animal and plant wastes can be a locally acceptable substitute; rice husks in Nigeria for example (Ugbaja, 1996).

The thin and, often brittle testa of castor seeds, especially from dwarf hybrids, is easily damaged, and when fertiliser is placed in the seedbed at sowing, it is essential that seeds suffer neither mechanical damage during the planting operation, nor damage from too close proximity to fertiliser. Both will cause poor or uneven stands. Deep fertiliser placement also tends to produce a deep-rooted plant better able to withstand drought. Distance between seed and fertiliser is also important, since few nutrients move laterally in the soil, and banding fertiliser below and to the side of seed is recommended. However, the often costly equipment required does not always result in a profitable yield increase, and the standard maize planter, which places fertiliser directly below, but not in contact with seed, produces acceptable results. Combine drills can be used provided seed and fertiliser are sown down different spouts.

Where green manure or other bulky organic residue is ploughed-in, the carbon : nitrogen ratio becomes important. When this ratio is greater than 15:1, it may prolong the vegetative phase thus delaying harvest and reducing yield in areas where the growing season is limited. Decomposition of the herbage may also induce a soil moisture shortage.

NITROGEN

Nitrogen, (N), promotes vegetative growth, and substantial applications produce leaves with large, thin-walled cells, reduce resistance to certain insect and fungal attacks, and increase susceptibility to damage by drought and frost. Insufficient nitrogen frequently results in small, harsh and fibrous leaves with small cells and thick walls. Firing of older foliage is a characteristic symptom of a major nitrogen deficiency, and should this be general in a crop, yield will be reduced; in younger plants light-green or yellow leaves indicate a nitrogen shortage. However, visual symptoms of nitrogen deficiency are not as general in castor as in maize, and it is safer to determine by fertiliser trials what amount is necessary. Nitrogen application to crops grown in marginal areas is both unnecessary and unwise, since the reason for castor's selection initially was that it would produce a crop under existing conditions.

An important aspect of nitrogen fertilisation requiring further study is the often low N recovery. This is improved by using slow-release fertilisers, which may also be more efficient in terms of available plant nutrient in areas of intermittent rainfall. A drawback is that some cannot be mixed with strongly acidic fertilisers; an advantage is they can be broadcast before ploughing with little loss of nutrient.

Response by castor to nitrogen application in the seedbed or as top dressing tends to be erratic, and yield increases are frequently due more to the cultivar selected and soil type than the fertiliser. In general it is unprofitable to top dress giant types, and in East Africa ammonium sulphate depressed yields as frequently as it increased them; similar results were obtained in other countries. In the same areas, dwarf types have usually responded well to nitrogen in the presence of adequate soil moisture, phosphate and potash. It is doubtful if more than 100 kg N/ha in one application is profitable, and where yield response has been obtained to larger amounts, split application is recommended. Not only will plants be able to make greater use of the nutrient over a longer period, less nutrient is lost through rainstorms leaching or washing away the fertiliser.

Placing nitrogen in the seed-bed produces healthy seedlings, but often no significant increase in yield. However on irrigated crops of dwarf castor, field experience indicates that seedbed application of NPK compounds is profitable, and increases response to later N top dressing. Castor is often planted well after rains begin in tropical countries with a distinct dry season, and does not benefit from release of mineralised nitrogen. A profitable response to seedbed nitrogen may occur, but it is necessary to establish this by formal trials.

The method of top dressing is of secondary importance, provided coverage is even and timing accurate, with the last application no later than the appearance of first racemes. In the field, the main top dressing is usually applied midway between emergence and flowering and often combined with a cultivation, but for mechanical operations the height of plants will determine the timing. If manual application is used the author found approximately 20–30 days after emergence is suitable for dwarf types; in Russia floral initiation occurs in dwarf types between 13 and 27 days after emergence.

Nitrogenous fertilisers can also affect plant resistance to pests and disease. In the USA it reduced the incidence of *Alternaria* leaf spot, and in the Gezira Irrigation Area of the Sudan, those plants receiving nitrogen were less attractive to leaf eating caterpillars of *Achaea janata*, an effect the author has noticed elsewhere in East and Central Africa, not only on castor. Data on the effect of nitrogen on seed composition is lacking, but high levels can depress oil content, as occurs with other oilseeds. A high level may also adversely affect germination.

There appears to be little difference between various nitrogenous fertilisers in terms of return per unit N when applied as top dressings. When applied in the seed-bed, a compound containing nitrogen, phosphate and potash is most frequently used, and when potash is not required, diammonium phosphate or similar compounds often give the most profitable response. In Kenya, diammonium phosphate to supply seedbed nitrogen also reduced the total phosphate required, while a mixture of superphosphate and diammonium phosphate gave maximum yield increase (Weiss, 1969).

Phosphate

Root development of castor is of prime importance as a major factor in the plant's ability to withstand drought, and phosphate (P), applied in bands at sowing generally increases root growth. In tropical red earths often high in oxidized iron and aluminium and low in bases, phosphorus availability is often low. However, high residual phosphate levels occur in other tropical soils, particularly those derived from granite, and a substantial P application to a preceding crop make unnecessary additional amounts to a following castor crop. This directly affects profitability in areas where transport expense is a significant factor in total production cost. On granitic soils in Kenya, an initial application of 66 kg P/ha to maize and castor was necessary for optimum yields, but this could be reduced to 22 kg P/ha for the following crop, a significant saving (Weiss, 1967a). Higher rates depressed castor yield, an effect that has also been recorded in other African countries, India and the USA.

An assessment of results from many countries where castor is grown commercially in pure stands, indicates that 50–75 kg P/ha is normally necessary, unless castor follows heavily fertilised crops. Substantial amounts are officially recommended in many South American countries, 200 kg P/ha for instance in Mexico, but from personal observations it is doubtful if these recommendations are followed by growers. In Brazil most state recommendations are between 50 and 75 kg P/ha, and locally compounded fertilisers are usually high in phosphate, moderate in potash and low in nitrogen.

Seedbed application of nitrogen/phosphate compounds has produced conflicting results as noted. Since nitrogen content is often critical in respect of the response to the phosphate component, trials to establish the correct proportions are essential. Compound fertilisers are invariably more expensive than superphosphate, and must justify their extra cost. Seedbed placement is the most efficient method, but rock, fused tricalcium, highly ammoniated, and calcium metaphosphate can be broadcast and ploughed in. Broadcasting is fast, cheap, and by allowing removal of fertiliser hoppers from planting units, it speeds up this operation during the often very short optimum planting period. Granulated fertilisers are recommended in all areas where relatively unskilled labour is used, as they flow more uniformly and do not cake or bridge in hoppers.

Potassium

Plant roots readily absorb soluble and exchangeable potassium (K), and it is easily translocated. Potassium moves from leaves to maturing flowers, fruit and seeds, and a large deficiency must occur before seed yield is seriously affected. Potassium most commonly occurs in compound fertilisers, and is thus often applied in unnecessarily high quantities. Too much potassium is money wasted, but occasionally an excess can limit uptake of minor and trace elements, particularly magnesium. Unless there is a known potassium deficiency limiting general crop growth, no application to castor is required.

When castor is grown in a rotation with other crops receiving potash fertiliser, especially legumes and tobacco, it can reasonably be assumed that no further application is necessary. Should it be required, small annual amounts are to be preferred to a large single application, as this frequently results in plants taking up luxury amounts. Potash fertilisers are usually placed in the seedbed as part of a compound, but when potassium alone is required it is better broadcast before ploughing. Potassium does not appear to have any effect on seed constituents at the levels normally applied.

Minor nutrients

Sulphur is deficient in many tropical soils in areas suitable for castor, but the effect of a deficiency on castor is unknown. It is probable that, as with other oilseed crops, the amounts of this element present in single superphosphate is sufficient. In general, application of minor elements is unnecessary unless a known local deficiency limits plant growth. Three are worthy of mention: zinc, since castor is a poor extractor of soil zinc, and a deficiency should be corrected if it limits growth; copper, because the amounts used to control fungal diseases of tobacco which can precede castor in a rotation, may lead to toxic amounts being accumulated in the soil; and aluminium, since an aluminium toxicity has been reported from South Africa. Deficiency symptoms for nitrogen, phosphorus, potassium, iron, calcium, magnesium and molybdenum have been demonstrated under controlled conditions on dwarf hybrid castor.

CULTIVATION

When castor is grown by smallholders ploughing is usually shallow and kills few weeds. Planting may be carried out at the same time by dropping seeds into the furrow and covering them by treading down earth. In Africa, ridges on which the previous crop was grown are split and remade in the furrows, into which weeds and crop residues had been placed. Castor is sown into the new ridges usually with groundnuts but also with sweet potatoes. Weeding operations help maintain the ridge, the furrow retaining a certain amount of rainfall, and the ridges are tied where rainfall is low.

A system introduced into isolated regions of Africa involves minimum cultivation. Lines or circular areas are cleared in the indigenous grasses, and castor is planted in the cleared areas, two or three seeds/hill. The area is slashed when grass growth becomes excessive and this forms a dense mulch. The sites selected are often flood plains or seasonally waterlogged areas, and are cropped in the dry season. The mulch retains moisture and reduces soil temperatures, but can become a fire hazard. Data collected by the author in the Sudan showed seed yields of 300–1200 kg/ha were obtained with minimal effort.

For rain-grown mechanised cropping initial cultivations must be deep enough to break up any compacted layers. This will allow adequate root development, for castor's drought-resistance depends on the ability of roots to penetrate soil and obtain moisture from depth. Rainwater percolation is also enhanced, and a secondary effect is the much greater volume of soil scavenged for plant nutrients, reducing the crop's dependence on added fertiliser.

Where disc harrowing is sufficient, a seed-box may be placed above the discs and seeding carried out at the same time. Accurate control of sowing is not possible, and a higher seed rate is necessary with later thinning of unwanted plants. The operation is extremely fast especially when using a 6 m wide single line disc harrow, and was successful when planting co-operative farms under the author's supervision, where all subsequent operations were manual. Irregular plant spacing was of little importance compared with the considerable saving in mechanical costs, a weighty consideration to farmers with very slender financial resources. The operation also relieved pressure on the farmer's family by reducing labour requirement at planting, for at this time of year all available labour was required to weed maize and other food crops.

In dry regions where total rainfall is barely adequate to produce a castor crop, ridge cultivation is recommended. Ridges can accommodate one or more rows of plants, should follow the contour, and ideally be tied at intervals to reduce run-off. By using a system of contour-aligned tied ridges, the author successfully grew locally profitable crops of maize and castor on an alluvial sandy clay with a total rainfall of 400 mm somewhat unevenly distributed.

The seedbed necessary for dwarf castor is similar to that for maize or cotton, but for even germination castor requires a moist seedbed for a longer period than either, since emergence may take 3 weeks. Soil should be moist to 15–20 cm, and seed planted at least 5 cm deep. On light or sandy soils, the use of press wheels is recommended. Too deep planting of some quick-maturing dwarf cultivars however, may result in the first flowering spike being set very low, below the range of mechanical harvesters. Seedlings will emerge from a depth of 30 cm, but are usually shorter and less vigorous than those planted less than 10 cm deep. The effect of varying planting depth on field emergence in the USA is shown in Table 2.12 (Smith, 1972). Even germination and emergence are important to large-scale commercial growers, as plants then grow more uniformly to maturity, simplifying weed control and harvesting. Speed of germination, emergence and attainment of the four-leaf stage are also significantly correlated with yield. The effect of chemical and other pre-sowing treatment of seed has been investigated (Patane *et al.*, 1996), and also noted in the botany section.

Castor is often grown in agriculturally marginal areas where the relationships between rainfall, time of planting and plant population are

Table 2.12 Effect of planting depth on emergence.

Planting depth (cm)	Sown:	16 May*		24 May*	18 May†		18 June†	
	Days:	28	35	21	14	28	7	28
2.5		23	53	78	8	11	7	22
5.0		45	59	71	52	43	16	45
10.0		74	69	62	63	64	37	84
20.0		29	37	29	34	52	41	84

* Loam; † Sandy loam. Figures in table are percentage emergence at day shown.

very important in achieving optimum yields. Production cost will probably be higher than for similar crops grown in more fertile or productive regions, and the necessity of ensuring maximum return per hectare more important. The correlation between rainfall and plant population that produces the highest yield, is most important where the first approaches the minimum necessary to produce a crop. Where dwarf castor is to be grown on a commercial scale in pure stands, it is essential to establish these relationships by preliminary trials.

These trials will also give an indication of the severity of insect attack, and planting may have to be varied to avoid flowering during such times, or during a period of very high temperature as both can drastically reduce yield. Trials should be as comprehensive as possible and include a range of the most suitable cultivars. Such trials will also allow selection of several cultivars, each to be planted at a different time, or with different periods to maturity. This will permit more efficient use of expensive equipment and harvesters, by extending their working periods, or reduce pressure on labour if weeding or harvesting is manual.

A seedbed temperature of 17°C ensures uniform germination, and in the USA planting after soil temperature averages 16°C at 20 cm at 8.00 a.m. for 10 days is officially recommended (Thomas, 1960) (Table 2.13). However, there are cultivar differences in susceptibility to low seedbed temperature and subsequent emergence; in Russia a seedbed temperature of 8–12°C did not affect emergence of local cultivars. This variation can be important when related to time of planting of a particular cultivar, and where several have to be grown, staggered planting dates may be necessary. Alternatively, soil conditions may be such that the optimum period for sowing a particular cultivar may be too short to plant the whole area to that cultivar. To achieve maximum yield per hectare it will be necessary to use other cultivars selected to suit the conditions of the extended planting period. A soil moisture range of 15–20% at normal planting depth usually ensures good germination at the optimum temperature, below 11% impairs and above 25% inhibits. Heavy rain just after planting, unless followed by warm sunny days, can seriously reduce emergence.

Standard maize or cotton planters are commonly used for dwarf castor grown in an existing rotation, but where it is a major component or the sole crop, specialised equipment is necessary. Electronically controlled air-seeder units or planters ensure even depth of planting and seed rates, but are not always suitable for less-developed regions, since maintenance and repairs require skilled technicians. Many maize and cotton planters have horizontal seeder plates with relatively small holes, rotating fairly rapidly, and metal cut-off pawls. Unless modified they cause considerable seed damage unless planting speeds are reduced below 5 km/hour, substantially increasing costs.

Special castor planters have substantial seed hoppers, large horizontal plates with up to 32 holes, nylon guides and brushes, and are geared to turn slowly to avoid splitting or crushing seed.

Table 2.13 Effect of seedbed temperature* on percentage stand – USA.

Variety	30°C	20°C	15°C	10°C
Cimarron	76	71	81	79
Baker 43	65	41	85	85
Baker 195	82	38	83	89
Baker 296	90	42	28	14
Custer	69	64	62	59
Dawn	51	28	13	11
Nebraska 145	81	68	61	52

* Held at these temperatures for 14 days, then at 30°C for a further 14 days.

Fig. 2.10 Sowing castor on pre-irrigated ridges – USA.

The addition of some maize flour, talc or similar inexpensive product to seed hoppers at filling will absorb oil from cracked seeds and prevent gumming of plates. Attachments that place fertiliser deeper than for maize or cotton or to one side are desirable, since castor seed is easily damaged by too close proximity.

Castor grown under irrigation requires special planting techniques; water is applied to ridged or bedded fields some time before planting to allow a thorough wetting; planters fitted with sweeps ahead of the seeder units remove enough dry soil from the ridge top to permit placing seed into 5–7.5 cm of undisturbed moist soil (Fig. 2.10).

Seed rates are of little significance with giant types since these are generally planted by hand, with several seeds per hill and thinned as desired. Occasionally mechanical and manual operations can be successfully and profitably combined. In Tanzania the author mechanically constructed large ridges on the contour to conserve rainfall, and seed was manually planted. Fertiliser was placed in the ridge during construction.

Plant population is much more critical for dwarf hybrids, since there is considerable variation within cultivars in rate of growth and ability to compensate for seasonal variation. For a particular cultivar at a given row width, between-plant spacing may vary considerably since individual plants are capable of utilising the additional space. Other factors can have a greater influence on yield; too close in-row spacing will produce spindly plants susceptible to wind damage, too wide spacing encourages weed growth and much-branched plants with an extended flowering period.

The 1-m width adopted for many tropical row crops can be used for dwarf castor, as closer spacing can damage branches and shallow lateral roots during cultivation. A standard row width also involves the minimum of implement adjustment between crops, but for mechanical harvesting row-width may be predetermined by machinery design. If there are no local data, a guide to in-row spacing would be 25–30 cm for dwarf, 30–40 cm for larger hybrids, or about 25 000–30 000 plants/ha for crops grown in the 750–900 mm rainfall range. A guide the author established for use in low rainfall regions, 760 mm or less, proved reasonably accurate. Taking as standard a 100 cm row width and 30-cm spacing, the in-row spacing was increased by 30 cm for every 120 mm drop in average rainfall; i.e. with a rainfall of 500 mm spacing would be 100 × 90 cm. A graph easily shows intermediate spacings. At the wider spacing, leaving two plants together produced a reduction in individual plant yield, but higher total yield/ha.

For irrigated crops, row width may be predetermined by the system of watering used, and where water is not the limiting factor, a population of 30000–40000 plants/ha is suitable, again dependent on cultivar. Very high seed rates have been used in Russia and Romania, up to 60000 plants/ha, but this level should not be an initial aim since dwarf hybrid seed is expensive and often difficult to obtain.

Grading seed is normally unnecessary and usually has no effect on final yield, although a positive correlation exists between seed weight and percentage germination, while weathered or frozen seed has lower viability. Planting unhulled seed is unwise. Seed is very susceptible to seedbed fungal disease, and should be dressed with a combined fungicide/insectcide, particularly seed of dwarf cultivars grown under irrigation. Seed dressings used on many tropical crops can also be used on castor, and special preparations are usually unnecessary. In less developed areas where seed of a local cultivar is readily available and cheap, a higher seed rate is preferable and safer than chemical seed dressings. Although commonly known as a bean, castor is not a legume and bacterial inoculation of the seed unecessary. Transplanting castor is unusual and should only be undertaken when there is no alternative, as seedlings are very fragile.

WEED CONTROL

Castor seedlings, especially the dwarf cultivars, are poor competitors to the often rampant weed growth in tropical countries. A plough is often the first item of machinery to be purchased by smallholders, but in general it is not the amount of land available for planting that is the major factor limiting crop production; controlling weed growth accounts for a very high proportion of total labour. Inadequate weeding is one of the most important reasons for the failure of castor growing in many African areas suitable for its production.

Mechanical weeding is common but, because the seedlings have long brittle hypocotyls and spreading shallow root systems, it can be extremely damaging (Fig. 2.11). Plant height is also a limiting factor and high-clearance equipment may be necessary to reduce crop damage. Supervised hand weeding is effective, especially where there are patches of persistent perennial grasses. The author noted that hand weeding also produced the highest yields in replicated trials comparing various methods of weed control, an effect also noted in India and the USA. Where practical, a combination of pre-emergence herbicide followed by hand weeding is recommended.

Fig. 2.11 Seedling stages of castor plant, showing long, brittle hypocotyl susceptible to damage by tilling machinery. Seedling on right shows root damage caused by cultivator.

Chemical control is efficient, only slightly damaging to plants when correctly used, and band-spraying or granules at planting are effective for pre-emergence weed control, requiring little additional capital expenditure or equipment. A slightly deeper planting depth is recommended when applying some compounds to allow a higher rate of application if this increases weed control. Whatever type of chemical control is selected it is essential to follow the maker's recommendation for application levels and protective clothing; the latter is most important in isolated areas far from medical attention.

The following have been used at the pre-emergence stage in castor: diuron, glyphosate, linuron, propachlor, simazine and trifluralin. When used post-emergence, most herbicides have caused varying amounts of damage, but the following are considered least dangerous: alachlor, diuron, glyphosate, linuron, norea (noruron), simazine and trifluralin. For post-emergence use, the correct level of application for a particular locality and cultivar should be established by field trials before wider usage. Castor has a lower tolerance to a particular compound or active ingredient level than many other crops, and even when vegetative damage is slight, seed composition and germination can be affected. Castor plants are generally very easily, and often severely damaged by hormone type herbicides.

IRRIGATION

Giant castor is drought resistant and will produce seed on a very low rainfall, but dwarf hybrids require adequate moisture. The ability to extract water from considerable depth makes castor a frequent choice for irrigation schemes where the amount of water is insufficient to allow an economically viable area of more demanding crops. Alternatively, castor can be used to extend the area of an existing scheme by utilising water available outside the normal planting date for other crops; for instance, it can be sown several months after the optimum period for cotton, or it can utilise soil moisture remaining from the previous year. In the USA dwarf castor was able to draw on such subsoil moisture plus rainfall not usually sufficient to sustain a high-yielding crop, to produce a seed yield of 2500 kg/ha, twice that of non-irrigated crops and little inferior to one fully irrigated.

Maximum water availability in the tropics frequently does not correspond with the optimum planting date of many crops, whereas castor can utilise moisture over a long period. The ability to extract subsoil moisture was demonstrated in Zimbabwe where castor utilised this moisture more efficiently than sorghum Table 2.14. (Anon, 1960). Thus it is well suited for water-spreading techniques, such as the Gash delta in the Sudan, wadis in Arabia and Iran, or in those regions with very heavy rainfall of short duration.

Table 2.14 Percentage moisture use by Central African crops.

	Depth (cm)			
Crop	20	45	65	85
Castor	35	52	61	53
Maize	42	52	55	40
Sorghum (dwarf, grain)	29	45	48	48
Sorghum (tall, forage)	28	39	48	40

Moisture use = 100% available minus mean of available readings on Bouyoucos meter.

Although irrigation of dwarf castor requires considerable skill in water use to achieve maximum yields, the degree of land preparation necessary is less than for many other crops. Castor is deep rooted and individual irrigations can thus be relatively large and less frequent, the reduction in transit losses will increase the irrigable area. Rows can be longer, ridges higher and follow the contour, which substantially reduces the amount of levelling necessary. Large ridges also allow a greater head in supply channels with less danger of breaks or topping, and facilitate the use of temporary dams or weirs. This is especially useful where rows are irrigated by plastic syphons, a system the author successfully introduced in East Africa as it stopped the traditional system of breaching banks and virtually eliminated over-watering.

Based on development cost per hectare, planting hybrid castor on smaller irrigation projects can be economically more attractive than other crops, especially cotton. The author believes that crops on many of these smaller projects were frequently chosen by agricultural staff because of their specialised knowledge or preference, rather than a critical assessment of the most suitable long-term cropping programme or future development.

Indigenous castors are seldom irrigated, but yield increases can be spectacular. In one area of Tanzania yield of hulled seed from a giant type rose from 200 kg/ha when rain-grown to 1600 kg/ha under irrigation; when treated as a perennial yield totalled 2400 kg/ha. Rainfall was

355 mm, with irrigation supplying a further 200 mm. Alternate furrows only were watered at each irrigation, the interval varying from 14 to 28 days, dependent on rainfall. The average annual yield of dwarf hybrids was 1300 kg/ha. Dwarf cultivars normally require a total of 750–1250 mm of water, depending on local climatic conditions. Extremely windy, dry areas with high evapotranspiration rates are in the higher range. When calculating total moisture requirement, one-third of measured precipitation is considered available to plants.

Some castor cultivars, especially giant and some taller hybrids, do not wilt to any great extent, and to the inexperienced may not appear to be under water stress. Thus visual estimation of irrigation need can be very inaccurate, and the use of soil moisture indicators is recommended as they not only ensure adequate soil moisture, but more importantly preclude over-watering. This not only damages plants, it wastes water which could be used to increase the cropped area. Most dwarf types are also sensitive to quite low levels of salinity; in Egypt irrigation water with a salt content of 0.10 mol NaCl was sufficient to inhibit development.

Timing of irrigation is important, for there should be no water stress once the primary raceme has flowered or during flowering, especially where temperatures are likely to exceed 30°C. Moisture stress at such periods results in a high proportion of light seed and pops. In general, watering should cease once the majority of capsules have been formed, and 21–28 days before harvest.

When considering the introduction of castor to irrigation projects, an important factor is cultivar reaction. Locally adapted types which remain small when rain-grown may grow much taller, take longer to mature and flower over an extended period when irrigated. Thus cultivar trials are most important where castor is to be a major component of an irrigated cropping programme. Adequate water may also emphasise cultivar differences in ability to utilise added nutrients, or increased resistance or susceptibility to pests and disease. Castor grown in the summer monsoon in Rajasthan, India, for instance, used water more efficiently when given NPK fertilisers; groundnut produced the opposite result.

INTERCROPPING AND ROTATIONS

Castor is frequently interplanted with other crops, sown around margins of fields, planted on areas unsuitable for other crops, or merely protected when found growing, but is seldom deliberately intercropped as the giant types are unsuitable and dwarf hybrids usually grown as monocrops. In general the total return from intercropped or strip-cropped castor is less than from pure stands of the crops concerned (Weiss, 1971). A secondary factor for smallholders is that the spiny capsules make harvesting unpleasant, and hulling tedious. More commonly the giant types are grown round field margins, the less tall types on levees or strip-cropped.

In many regions where the number of suitable cash crops is limited and the period between succeeding crops is shortened, damage from pests and disease invariably increases. In these circumstances castor can provide a break in the rotation, and reduce expenditure on expensive control chemicals. Although it may not be as profitable as the crops it displaces, increasing the period between them raises their yield and thus the total return over a period.

Castor itself may also suffer greater damage from pests and diseases when planted adjacent to certain other crops, or when it overlaps them in a rotation. In more developed countries, dwarf castor is seldom planted for more than two succeeding years in rotations, as pests and diseases quickly build up to levels that make protection unprofitable.

An important advantage of including castor in a rotation is that it induces germination of *Striga* spp. but is itself resistant. *Striga* spp. are widespread in Africa and responsible for a major reduction in cereal yields. Crops following castor benefit from the lower *Striga* seed-bank and reduced levels of infestation. Castor can also be used to expand the cropped area of irrigation schemes through its ability to utilise sub-soil moisture, as noted in the relevant sections.

HARVESTING

Commercial plantings of dwarf hybrids are usually fully mechanised, and although generally successful in respect of field operations, hulling may still present problems. Mechanical harvesting consists basically of removing seed capsules from standing plants by a beater mechanism that strikes plants directly under the lowest raceme. Capsules fall on to a conveyer, thence to the huller. Hulls are removed in a drum, which rolls capsules between two rubber surfaces moving at different speeds. The biggest problems are uneven ripening, or varying thickness of capsule wall, either producing a large proportion of unhulled or broken seeds.

Capsules must be completely dry when harvesting commences and hulling rollers accurately adjusted. The importance of maintaining correct working speed and frequent adjustments to machinery was shown in the USA, with the seeds remaining on the ground ranging from 52 000 to 2.25 million/ha. More seed remained at field edges as harvesters slowed at row ends, a loss that can be reduced by beginning harvesting with several runs around the margins. Field harvesting of dwarf castor in the USA showed the loss grew as the forward speed of machines increased; from 4.8% at 4.5 km/hour, 8% at 9.0 km/hour to 9.7% at 13.5 km/hour. Harvesting loss can be kept below 5% by correct machine adjustment, a skilled operator, and working when conditions are favourable.

Most harvesters are designed to operate with a relative humidity below 45%, as moist capsules may remain attached to racemes and do not easily hull. When plants do not desiccate naturally, chemical defoliants may be applied 10–15 days before harvest. Correct timing is critical, and when chemical defoliation was carried out correctly in Russia and Romania yield increased by 50–70%, and the number of green capsules harvested was greatly reduced; a major aid to harvesting and hulling efficiency. In Russia also the slow death of plants sprayed with magnesium chlorate increased the flow of assimilates to seeds, and thus total weight and oil yield.

Capsules can be hand harvested by breaking or cutting off whole racemes or by some form of stripping device (Fig. 2.12). However hand harvesting is generally unsuitable for large areas, since the spiny capsules make the work unpleasant and disliked, and their bulk involves frequent transport and emptying of containers. It may however be politically desirable to employ people who would otherwise lack work. The stripping cup, shown in Fig. 2.13, has been used successfully in East, Central and South Africa, and a worker should harvest 300–500 kg of capsules daily.

Production of castor harvester-hullers is limited, but specialised attachments for standard harvesters are available. Around 500 ha of high yielding castor is required to justify the use of a harvester-huller, but half this for harvesters only, and the latter can also be profitably used on smaller fields. Two-row machines operate most efficiently at 5 km/hour, harvesting 0.75–1.5 ha/hour in fields yielding 3000 kg/ha shelled seed (Fig. 2.14). Some Russian two-row machines harvested up to 200 ha/season, and four-row types 300 ha/season. Harvesting seed crops requires great care and special skill in operating combines, but where plots are small, hand harvesting is recommended.

The author considers harvester-hullers should be used only where skilled operators, mechanics and spares are easily available. In less-developed regions eminently suitable for large-scale plantings, harvesting and hulling should be separate operations. The extra working time involved is usually offset by reduction in time lost through mechanical breakdowns, neglect of necessary adjustments during work, and higher yield of whole seed. Separation of these operations also allows hullers to be used for smallholder production, or hired to local co-operatives.

Since hulling is a most laborious operation, availability of hullers frequently results in a major increase in local production (Figs 2.15 and 2.16). The advantage of introducing a mobile huller to local producers in Africa and South America has been well documented. An added

Fig. 2.12 Harvesting giant castor, Tanzania.

Fig. 2.13 Harvesting dwarf castor with stripping-cup, Kenya.

advantage is that it stimulates interest in castor growing as an alternative crop in areas where choice is often limited. Stationary and mobile hullers with throughputs varying from a few hundred to 3000 kg/hour are available commercially, or can be manufactured locally (Fig. 2.15). Mechanical hullers should be operated at slow speeds, with a constant feed rate and dry capsules, to reduce seed damage. Whole seeds should be separated immediately from damaged seed, dried of oil from broken seed and dusted with an insecticide.

Also influencing the necessity for partial mechanisation of harvesting and hulling is the effect on a local agricultural population of large numbers of young men being absent on work elsewhere, as occurs in predominantly agricultural regions adjoining mining or oil-producing areas in Africa. Thus the available workforce consists only of women and children.

Fig. 2.14 Harvesting dwarf castor, USA.

Clusters reaped slightly green or with wet capsules must be dried before hulling, either in the sun or artificially, avoiding overlong exposure which may reduce seed oil content. Artificial drying is expensive and although most modern crop driers are suitable, castor capsules require more attention when loading than soyabeans or cereals. Drying capsules of high moisture content requires a uniform density as packing produces high static pressures, while air movement through the material should be continuous. The maximum rate of moisture removal from the first 30 cm of unhulled seed occurs after approximately 1 hour, with moisture reduced from 30% to 5% after 6 hours. Moisture loss at beginning of drying is greatest in the first 60 cm depth of seed, with moisture accumulating in the 90 cm layer. Drying capsules with high moisture content requires accurate temperature control, above 40°C may affect oil characteristics. Where one cultivar predominates, trials to determine the optimum temperature and drying time at various capsule moisture content is recommended.

Annual castor should be destroyed after harvesting preferably by burning, which substantially reduces subsequent insect and disease damage. However, where it is necessary to control erosion, crop residues should be left *in situ*, but preferably broken up to limit pest survival. Heavy harrows with discs set parallel to direction of movement, or heavy-duty rotary mowers can all operate at fairly high speeds. By corrugating the soil surface, the former will also

Fig. 2.15 East African design of hand-operated castor huller.

Fig. 2.16 Mobile huller developed in the USA.

assist rainfall penetration. Wind erosion or control of encroaching sand, can be a major problem in arid regions especially in the Middle East, Russia and China, and contour windrowing of crop residues is desirable. Where this is carried out the lines of castor stalks, etc. should be made as compact as possible, and windrows burnt before planting. If windrows are fired upwind, wildlife that shelters in them receives adequate warning and the burn is slower and more complete. The ash is a good fertiliser and soil ameliorator.

The yield of castor in selected countries is shown in Table 2.15 and is average seed yield of rain-grown crops unless otherwise stated.

Seeds are large and occupy considerable space in relation to their weight, thus on-farm storage of large quantities should be avoided, but as rail

Table 2.15 Castor yields in selected countries (shelled seed in kg/ha).

	Average	High
Africa		
West	200	600
Central	200	800
South	1000	1500
South (i)	1500	2000
Americas		
USA (i)	1100	2000
Brazil	800	1200
India	200	500
Former USSR(i)	400	650
	600	1400
China	400	1500

i, Irrigated.

or truck carriage is usually less expensive in full loads, a certain amount of storage may be unavoidable. Bagged seed can be stored in the open only for short periods, as heat and sunlight reduce oil content and quality. Low temperature storage either artificial or natural, as in Russia affects viability, and at –5 to –7°C viability was almost zero at 6 months.

Seed for bagging or transport should be handled using wooden scoops or shovels, and rubber conveyor belts and handling should be kept to aminimum. Seed for export must be clean and whole, and severe penalties or rejection of a consignment may result if the proportion of broken seed exceeds 2–5%. Most countries have regulations regarding transport of castor seed, including size and colour of identifying marks, and restrictions on carriage with food or feedstuffs. On shipboard, seeds should be stowed away from boilers. Damaged seed and hulls can be used as fuel, or composted, and are then approximately equal to farmyard manure and superior to local kraal manure.

PESTS

Castor is attacked by a multitude of insect pests, and even those which would probably never encounter the plant naturally, eat it with avidity! Thus noting of reported occurrences or listing species found on plants is of little value, since many are of no importance. However, the damage caused by insects to castor may be greater than the small amount of sap lost or leaves eaten. For instance, the action of piercing a developing inflorescence during feeding by sucking bugs, especially the Miridae, is itself sufficient to cause die-back or premature abscission of capsules. Following a severe infestation few healthy flowers remain. Simulating the action with a needle produces the same effect.

Many of the major pests of castor also damage other common tropical crops, and when grown in the vicinity of large plantings of these crops, castor becomes liable to greatly increased levels of attack. In Tanzania and Kenya castor planted close to areas of maize or sorghum suffered a higher level of insect attack than when surrounded by grassland. The proportion of mirids also increased in relation to other pests, and damage was thus more severe. Where this occurs, an integrated spraying programme should be implemented, varying the pesticides to suit seasonal variation in the insect species and infestation levels.

Intercropping also increases mirid damage to plants, although the severity will depend on the variety of castor, type of crops and extent of cultivation. The intercropping of young cashew nut plantations in southern Tanzania with castor to reduce the cost of establishing the former proved unsuccessful, as both are severely attacked by mirids, and interplanting exacerbated the overall damage. There can however, be circumstances when pesticides should not be used, as in the Gash delta of the Sudan, where wild castor plants provided a reservoir of parasites sufficient to control a major pest, *Achaea janata*, the castor looper. Loss of yield in the cultivated crop was a fraction of estimated spraying cost.

Control of insect pests may also affect other agricultural operations, an aspect of pest control the author considers has been little investigated. For instance in Tanzania, weekly dusting with insecticide of both giant and a dwarf hybrid modified the harvesting period. Dusted plants produced the greater proportion of their crop before the normal harvesting period for unprotected plants (Table 2.16) (Cole, 1958). Unprotected plants that lost a major portion of their first racemes from insect attack were able to compensate to some extent by fruiting later in the season, when insect numbers had naturally decreased. The giant type was thus able to produce a crop from undusted plants equal to that from protected plants of dwarf castor. This was due to the latter being a short-term type which died off at harvest.

Where insect attack limits seed production, selection of low-growing perennial types might be the most rewarding approach. A minimum of protection would be required, and directly related to expected increase in yield or benefit of an advanced harvesting date. Low yields shown in the table reflect a low level of inputs rather than potential yield of varieties grown. Timing of an application is also important in relation to sowing date and rainfall, for in the same area of Tanzania, one modifies response to the others (Table 2.17). The effect of correct planting time in relation to rainfall on total yield obtained from both types is of more importance than crop protection. Yields obtained at the optimum planting date were substantially increased by applying the standard local cotton pesticide. Thus, timing of crop protection applications in relation to crop growth and potential yield is important in ensuring maximum returns from the chemical used. Due to height of plants, ground spraying equipment must be specialised and used with care to minimise mechanical damage to the crop. Aerial spraying is to be recommended where possible.

Attacks by soil insects up to emergence can usually be limited by seed dressing, and thinning should be delayed where such attacks are frequent. The efficiency of a systemic insecticide is improved by placement, for increased root growth in the vicinity of fertiliser bands aids uptake of the associated insecticide.

Cutworms, *Agrotis* spp. are a common pest; *A. ipsilon* is widespread, *A. repleta* is important in the northern South America, and *Capitassia con-*

Table 2.16 Effect of pest control on harvesting dates (all yields in kg/ha) – Tanzania.

Harvested	Dwarf hybrid		Local giant	
	Unprotected	Protected	Unprotected	Protected
Before 14 Oct.	170	411	47	289
After 14 Oct.	228	125	474	319
Total yield	398	536	521	608
Harvested after 14 Oct. (%)	57	23	91	53

Table 2.17 Interaction of pest control with sowing date (all yields in kg/ha).

Planting date	Dwarf hybrid		Local giant	
	Unprotected	Protected	Unprotected	Protected
Jan.	242	417	476	835
Feb.	131	353	276	547
Mar.	96	161	243	324
Apr.	74	117	34	65
May	66	111	5	4

Rain falls in the period Nov.–Feb. and most crops are sown in Nov.–Dec.

sueta from Mexico south to Argentina; *A. segetum* is common locally in parts of Africa, the Middle East and Asia, and can become a major pest in Russia. Seedlings are eaten off just above surface level, and digging near the roots will expose greasy, dark-brownish green or grey, sometimes yellowish larvae up to 4 cm long curled up about 1 cm below the surface (Fig. 2.17a). Mole crickets, mainly *Gryllotalpa*, *Brachytrupes* and *Gryllus* spp. can cause damage to both seed and seedlings. Seed dressings give some control except when attacks are severe. Soil insecticides are recommended where these insects are common, and baiting may be necessary locally at the beginning of the rains or on irrigation schemes.

a

b

Fig. 2.17 (a) cutworm *Agrotis* spp., (b) woodmoth, *Xylentes* spp.

Small chrysomelids occasionally cause serious damage to young seedlings in East Africa and South America (see Fig. 5.12). *Aphthona whitfieldi* has been recorded on young castor plants in East Africa and the Sudan, but in Israel, *Hermaeophaga ruficollis* is the most important. Grasshoppers can be troublesome in seedlings, but seldom damage mature plants; in East and West Africa seedling are attacked and sometimes destroyed by bands of nymphs or adults of *Zonocerus variegatus*. Beating these bands with bushy branches is a most effective method of control for smallholders. *Chrotogonus*, *Cyrtacanthacris*, and *Phymateus* spp. all cause damage to young plants, becoming serious pests locally in Africa, the Middle East, southern Russia and India. *Aiolopus* spp. becomes more important in China, Korea and Japan, and *Eyprepoenemis shirakii* in China.

Stem-borers are of economic importance only on perennial giant castor and damage may so weaken plants they are destroyed by high winds. Little control of an infestation is possible, but damage can be reduced by burning stems of fallen plants and the previous crop when replanting. Since many borer species are not specific to castor, wide area control is not possible. Where persistent insecticides are economic, spraying plants from ground level to the first branches will considerably reduce the damage; systemic compounds also give substantial control. Resistance to *Ostrinia nubilalis* borer exists in some dwarf varieties. In the USA greatest damage occurred to castor growing close to maize fields, since borers were not found in castor fields at least 800 km from the nearest maize. This association of borer damage with maize indicates that the greater damage often noted on tertiary racemes may be due to second-generation borers migrating from maize. Irrigated plants are much more susceptible than non-irrigated. Damage due to this borer is generally important only in temperate regions, but it also occurs in many countries of tropical Asia as a major pest of maize.

In East and Central Africa the main stem-borer is the larvae of a moth, *Xyleutes capensis* (Fig. 2.17b). Damage usually begins in the stem near ground level. Larvae are creamy white, up to 5 cm long. The pupa formed protrudes outside the stem for most of its length. Several buprestids are common borers of the large perennial castor in the Middle East; *Sphenoptera arabica* and *S. ardens* in Egypt, and *S. fulgens* in the Sudan. *Xyleborus fornicatus* bores mature plants of perennial varieties in Ceylon, but has not been recorded on dwarf annuals. It is the most common borer in Indonesia, where a longicorn beetle *Monochamus* spp. also bores mature castor.

Caterpillars of a number of moths and butterflies can cause serious defoliation; generally these larvae are present in low numbers in most years, and at this level specific control measures are unnecessary. Most common are *Achaea janata*, the castor looper, in Asia, and *A. catella* in Africa. Control is recommended in South Africa when two caterpillars of *A. catella* are found on one plant. *Achaea janata* occurs annually during August in India and Thailand, and attacks 4–8 week-old-seedlings. Infestations can be extremely heavy, and capable of defoliating a crop in an amazingly short time. Wide spacing to discourage caterpillars moving from plant to plant and hand-picking leaves with clusters of young caterpillars on their undersides are successful in smallholder crops. On larger plantings spraying is the only practical solution, except where the crop is grown for rearing silkworm. An interesting interaction between nitrogen applications and *A. catella* was observed at Wad Medani, Sudan, where in a fertiliser trial plants which had not received nitrogen suffered greatest damage.

The bollworms, *Helicoverpa (Heliothis) zea* in the Americas and *H. armigera* and *H. punctigera* elsewhere (see Fig. 4.13), the widespread leafworms *Spodoptera litura* and *S. littoralis* and to a lesser extent *Pectinophora gossypiella*, are also damaging in specific seasons. Lymantriid larvae can cause severe defoliation; in Africa and Madagascar the tussock moths, *Euproctis rubricosta*, *E. producta* and *E. convergens* may become major local pests, and *E. fractens* similarly in South-East Asia; in India and Pakistan *E. lunata* can completely defoliate the kharif-season crop; it is also a minor pest in many parts of South-East Asia and China. The moth lays its eggs in clusters on the underside of leaves and covers them with pale brown hairs; in India eggs are often multi-parasitised. Another tussock moth, *Dasychira georgiana*, occurs in East Africa where severe attack completely skeletonises leaves; *D. mendosa* occurs in Kampuchea and Indonesia, and *D. inclusa* is found generally in Asia.

In China, *Porthesia scintillans* is of economic importance, and as it has a wide host range, an infestation can build up with great rapidity. Elsewhere in Asia and southern CIS, it is of widespread occurrence but of minor importance. *Orgyia* spp. defoliate plants to some extent in tropical Africa, and in southern Africa individual plants were stripped bare by *O. mixta*; in south China, South-East Asia and Indonesia, *O. postica* can be locally damaging especially near rubber plantations. Since many of these moths also attack other crops, castor planted in their vicinity is usually more severely attacked. This is especially so when castor is planted later than other crops, and continues to be attractive to pests after their main host has matured.

Jassids, leafhoppers, particularly *Empoasca flavescens*, attack castor and a wide range of other tropical plants, but as noted later there is cultivar resistance to this pest, and also to the capsule borer *Dichocrocis punctiferalis*. The latter is of special importance in India where local farmers have little cash for pesticides, and significant differences in resistance are recorded. Although scales and mites occur frequently on castor, they do not appear to be of economic importance. Varying resistance to mites has been observed in India and correlated with bloom on leaves and stem.

Probably the most damaging pests are those attacking the inflorescence, especially mirid and pentatomid bugs, which can be devastating when present in large numbers. Various species of *Helopeltis* cause considerable damage in India and Africa, particularly south of the Sahara. The taxonomy of this group is confused, and identification difficult. In Tanzania, *Helopeltis schoutedeni*, a red, orange or yellow bug with black-winged adults, occasionally appears in enormous numbers, and not only causes severe damage to the flowering spike but also seriously retards growth of the whole plant. It is common in the countries of the Sahel. *Lygus* spp., particularly the small greenish-brown *L. vosseleri*, are commonly noted especially where cotton is also a major crop; *L. ricini* is locally damaging in East Africa. Pesticides applied to cotton are normally successful on castor. *Eurystylus* spp. occur generally in Africa, *E. capensis* especially in South Africa where it can cause severe damage to capsules; *Adelphocoris* spp. in Africa, *A. suturalis* and *A. lineolatus* in Asia and the Far East.

Dichocrocis punctiferalis, the peach moth, is a most important pest in India and throughout South-East Asia; in India some cultivars showed resistance to this pest and also to *Empoasca flavescens*, others to one only. Plants with purple stems and small seed were reportedly more resistant than those with large seed and green stems (Fig. 2.18). Throughout East, Central and South Africa, larvae of the false codling moth (*Cryptophlebia leucotreta*) feed on panicles and bore maturing capsules and seeds and once larvae are in the capsule little control is possible. Delay in harvesting exacerbates the damage and hulling problems, due to oil from damaged seed clogging shelling plates. Control should aim at killing young caterpillars when they emerge by a persistent contact insecticide applied after flowering at a suitable time determined by local experience. Aphis and whiteflies frequently occur in large numbers, but damage is usually confined to a specific district or season; *A. leyrodidae* in Asia; *Trialeurodes ricine* in the Middle East; *Asphondylia ricini* in South Africa. Nematodes are not a serious pest of castor, which is regarded as resistant.

Few pests attack stored whole seed, but damaged seed and press cake can become infested by *Ephestia cautella* the tropical warehouse beetle, *Lasioderma serricorne* the cigarette beetle and *Tribolium castaneum*, the red flour beetle. For additional comments on the control of storage pests and pest control in general see this section in Chapter 3 on groundnuts.

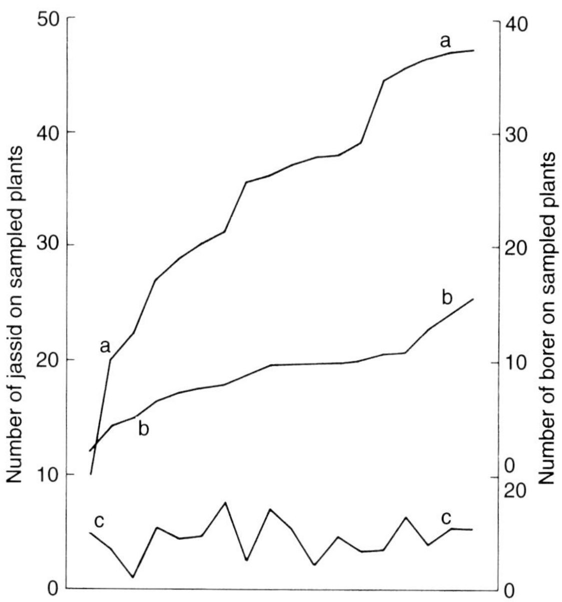

Fig. 2.18 Varietal susceptibility to insect attack: (a) attractiveness of 18 castor varieties to jassid, *Empoasca flavescens*; (b) attractiveness of 18 castor varieties to capsule borer, *Dichocrocis punctiferalis*; (c) attractiveness of same 18 varieties as (a) to capsule borer.

DISEASES

A large range of pathogenic fungi has been recorded from castor, some fungi are seedborne, but few are of economic importance. Records are difficult to correlate, since there has been considerable misidentification in the past, and many reclassifications. Common names are no guide, the same name often being applied to a range of infections with generally similar symptoms. Foliar diseases appear to have little effect on final yield unless they defoliate the plant to an extent that grossly affects growth, or are transmitted to capsules. Capsule diseases are more important, and in some areas limit commercial castor growing. Chemical control measures are generally neither profitable nor necessary, and where a particular disease causes economic loss, growing resistant cultivars and good cultural practices will materially assist in producing strong, healthy plants able to withstand infection.

The most damaging diseases attacking emerging seedlings are frequently caused by *Fusarium*, *Rhizoctonia* and *Sclerotium* spp., with *Phytophthora parasitica* most important in India and Africa; high resistance to *Fusarium* wilt exists in some Indian varieties. The incidence of seedling wilt is related to temperature, maturity of seed planted and cultivar, and local trials are desirable to determine their relationship. Plants infected as seedlings are considered susceptible to later fungal infection. Foot, or crown rot, caused by *P. parasitica* is widespread in Africa and Asia, thrives under wet conditions and is spread by heavy tropical storms, whose large raindrops carry zoospores from one plant to the next. Prolonged cool, wet conditions cause extensive damage, particularly among immature plants.

The most common foliar disease is *Melampsora ricini* (long known as *Melampsorella ricini*), now of worldwide occurrence. The author has seen it on the majority of plants growing in Africa, and many in Asia. Since it is most severe on mature leaves which are no longer contributing significantly to plant growth, its effect on yield is generally slight. From personal observation of giant types in Africa, local strains exist with resistance to the disease, which could be of value to plant breeders. Symptoms are masses of orange spores on the undersides of leaves.

Alternaria ricini, leaf spot, is more damaging, also attacks seedlings and can affect capsules. It is cosmopolitan and becoming of greater economic importance. Nitrogen application can reduce severity of the disease, as can rate and timing of irrigation. Symptoms are light brown, generally circular spots on leaves, larger than *Cercospora*, which tend to become angular with age. A grey-green spore mass may sometimes be seen. In Africa south of the Sahara, *Cercospora ricinella* has a similar effect on plants as *Alternaria*, but does not damage capsules as badly. In Somalia the disease is considered to be most serious, and can defoliate plants completely in years of heavy infection; in Togo (West Africa) it is usually associated with *Capnodium* spp. and *Nosmeniella* spp.; it is widespread in the USA, South America and the Caribbean; in India it is of little importance, but can become locally damaging in Indonesia. Symptoms of *Cercospora* attack are light brown, generally circular interveinal spots with margins of concentric rings, the outer broad and dark. With age the centre may change to grey. The spore mass is blackish.

Pseudomonas spp. cause bacterial leaf spot or rust, are seed-borne, and occur widely in the tropics. Some giant and indigenous types appear much less susceptible than dwarf hybrids. Within the latter there is also cultivar resistance, and in the USA, where this and bacterial leaf wilt are major diseases, only varieties with known resistance are grown. Damage from leaf diseases can be substantially reduced by cleaning fields of plant residues after harvest, and use of crop rotation. Symptoms are drying of leaves which turn black and fall. Branches also turn black and stems may be affected, when plants usually die.

Xanthomonas ricini causes severe defoliation in the USA, Africa, India and Japan, and many other Asian countries. In all regions it is of economic importance due to the difficulty of control, and in areas of very heavy infestations it may be the limiting factor in plant growth. Symptoms are small, rounded water-soaked lesions, most numerous towards the leaf tips. It usually occurs after periods of heavy rain; or heavy dew in the USA. The pathogen travels through the xylem vessels and veins, the leaves becoming dark brown and dying. In severe infection the organism may move down the leaf petiole and into the stem.

Capsule diseases may limit commercial production, especially the moulds *Alternaria* spp. and *Botrytis* spp. particularly in the south-eastern USA and the CIS, where they restrict the area suitable for commercial production of dwarf and hybrid castors to those regions where little or no rain falls just before harvest. The most obvious symptom of capsule mould is a dense woolly growth on flowers and capsules which varies in colour from pale to olive-grey. Leaves and stems can be infected from racemes. First symptoms are small blackish spots from which drops of yellow liquid may exude. Fungal threads, which grow from these spots, spread infection and produce the characteristic appearance of the affected raceme. The damage is exacerbated by weakening of capsule peduncles, which subsequently break. Susceptible cultivars can lose 80% of capsules, while others show considerable resistance. *Sclerotinia ricini* is known as castor oil plant ash in Colombia, South America, because of the brownish-grey mould that covers and shrivels capsules. In Brazil, close spacing of plants in the row or several plants in hill planting, increases incidence of the disease. Chemical control is not practical, and wide spacing of plants is recommended. *Sclerotinia ricini* is particularly damaging in the Gulf Coast region of the USA, where in wet periods the fungus destroys the inflorescence at all stages of development. There is no effective control. In the more humid areas of West Africa, *S. ricini* is considered the most important disease of the inflorescence. Mildews are an irritating and unsightly disease of ornamental plants, and also occur on breeding and experimental plants in greenhouses.

3 Groundnut

The Leguminosae is one of the three largest families of flowering plants, with about 700 genera containing some 18 000 species ranging from herbs to trees. The distinctive structure of the flower makes members easy to identify. A most important characteristic of legumes is their ability to maintain a symbiotic relationship with specific bacteria which fix atmospheric nitrogen, and make it available to the host plant. The Leguminosae is composed of the Caesalpinioideae, Mimosoideae, and the largest and most important the Papilionoideae, containing about 480 genera and 12 000 species. The Papilionoideae provides many important crop plants, including the groundnut (*Arachis hypogaea*), and soya bean (*Glycine max*). When used without qualification in the text, groundnut refers to annual cultivars of *A. hypogaea*, their seed, oil and derivatives.

The oldest indications of groundnut cultivation are from pre-Columbian Peruvian societies of 2000–3000 BC, well to the north-west of its native Bolivia. The plant is believed to have been originally domesticated by predecessors of the Arawak-speaking peoples who now live in its homeland. Thus groundnut probably had a much longer history of human use, for the Spanish found groundnuts widely distributed in South and Central America and the Caribbean.

The groundnut was brought to West Africa from Brazil in the sixteenth century by the Portuguese, and thence to the African east coast and so to India. Another route to Asia was from the west coast of South America, and thence to India and Africa. Thus Africa received stocks from widely separated regions of the New World, which have come together and given rise to a new and distinct pattern of variation, and Africa can now be regarded as a substantial centre of diversity. In fact, some widely grown varieties in the USA may well have reached there from Africa, since they do not occur naturally in the Americas. The routes of introduction of groundnut to Asia and South-East Asia can be deduced from the many local names (Watt, 1908; Burkill, 1966). There may also have been a direct transfer from Peru to China via the Portuguese city of Macao, since the Peruvian three-seeded type occurs in China (Fig. 3.1).

Groundnuts have become so integrated into African lifestyles that many Africans are unaware they are an introduction, and traditional customs have arisen which are quite unexplainable. In Zambia for instance, it is customary for a man to ask a friend for a handful from a basket of newly dug pods with no possibility of refusal. A custom that did not apply to other local crops.

The groundnut and its derivatives are major international commodities but it is also an important food crop, and a substantial proportion of total world production is consumed, unrecorded, by growers.

India is the world's largest producer of groundnuts, and with approximately 8 million hectares (Mha) under groundnuts in the crop years 1996/97 and 1997/98, about 40% of the world total crop area. Production was 3.7 Mt in 1997/98, but fell slightly in 1998/99, nearly 33% of world production. Main producing states are

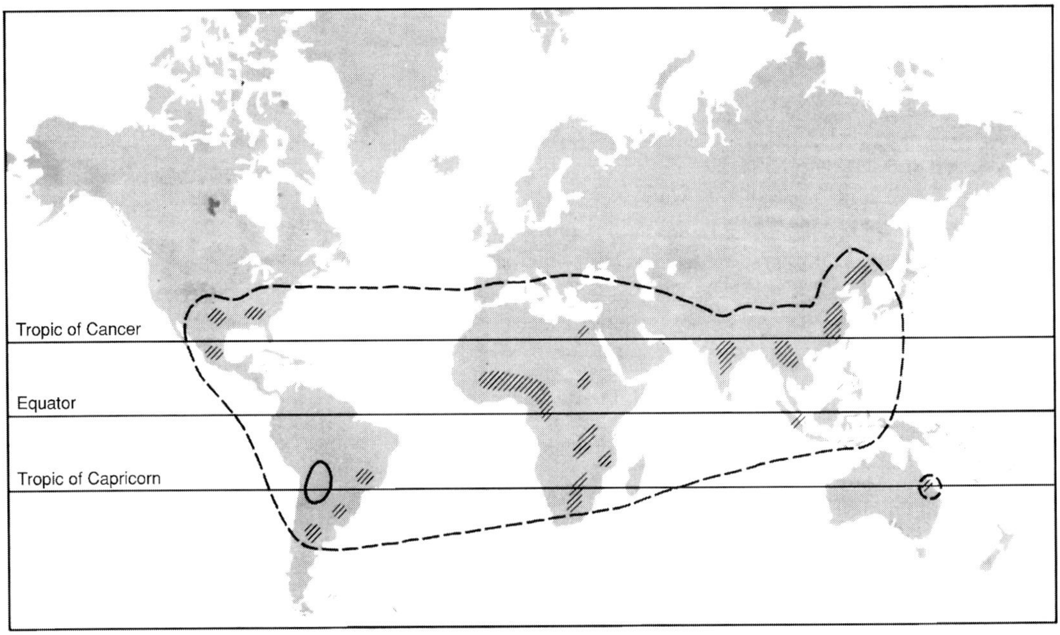

Fig. 3.1 Groundnut centre of origin (solid line), area of intensive cultivation (dotted line) and areas of maximum cultivation (shaded).

Andhra Pradesh, Gujarat, Mysore and Tamil Nadu, together accounting for about 40% of domestic production. In 1997 groundnut accounted for 33% of total oilseeds produced, but the crop is facing growing competition from soyabean and sunflower. However, groundnut production could be substantially increased or current production maintained in a smaller area if modern or efficient agricultural methods were more widely used.

India is also the world's largest producer of groundnut oil but this is insufficient to meet the local annual demand of around 2 Mt, and the shortfall is filled by rapeseed and palm oils, which are now direct competitors in the edible oil sector. Palm oil imports rose to 1 Mt in 1998; a trend likely to continue. Domestic groundnut meal demand has remained at 2.5 Mt since 1990 and all the meal is produced domestically, as imports of oil and meal are banned.

The next largest single country producer is China with 3.5 to 4 Mha planted in the last 5 years. Production has risen steadily from 2 Mt in 1970 to about 10 Mt (unshelled) in 1997, but is insufficient to supply domestic demand for oil and meal, which is increasingly satisfied by imports of alternatives, especially soya beans, soya and palm oils. In the years 1992 to 1997 domestic consumption of groundnut meal rose from 1 Mt to 2.5 Mt and that of oil from 0.8 Mt to 1.5 Mt. However external trade in both depends on political decisions rather than market forces, and tends to be unpredictable.

The USA is third with production in the last two decades relatively static at about 1.5 Mt. Main producing states are Georgia, Alabama, Texas and North Carolina in that order, together normally accounting for about 90% of the total. A high proportion is used as edible nuts or in processed foods of all kinds. The USA is also a major exporter in years of a global or particular country shortfall.

Groundnut, indeed all oilseed production in most African countries has suffered from the wars that have raged almost continually since 1970, and the continent now has an overall edible oil shortfall and this trend is unlikely to be reversed; Kenya for example imported an average of 250 000 t of edible oil annually between 1994 and 1997. Despite these continental problems individual countries continue to produce significant quantities; Senegal exports substantial but irregular amounts of groundnut oil to the European Union.

World groundnut production for crushing is expected to decline as competition from rapeseed, sunflower and oil palm rises and their oils fall in price. Edible nut production will probably increase as food products containing groundnuts increase in popularity, especially peanut butter and a huge range of snacks and confectionery.

BOTANY

The genus *Arachis* L. contains a large and somewhat ill-defined number of tropical and subtropical members, occurring naturally from Brazil to northern Argentina. The basic chromosome number is $x = 10$, most wild species being diploids with $2n = 20$, and the cultivated groundnut an allotetraploid, $2n = 40$ (Smartt *et al.*, 1978; Stalker 1997). *Arachis hypogaea* is known only under cultivation, its nearest wild relative being *A. monticola* (K. & R.), regarded as a truly wild conspecific; *A. hypogaea* has stronger pods and shorter carpophores than *A. monticola*, probably resulting from unconscious selection during domestication. It was originally considered that *A. hypogaea* may have originated from a hybrid between *A. cardenasii* (nn) and *A. batizocoi* (K. & G.), and although both putative parents occur in reasonable proximity in Bolivia, *A. batizocoi* is no longer considered a progenitor of the cultivated groundnut. Numerous wild species are known to exist in the South American centre of diversity, many are not fully described (Grosso *et al.*, 1997; Stalker *et al.*, 1994 and Stalker, 1997).

Arachis glabrata, a wild, perennial relative of the cultivated groundnut, has been under development as a forage plant for the subtropics and tropics since the early 1960s, and a significant germplasm collection has been assembled. Little is known about genetic diversity in this species, since research has focused mainly on agronomy (Maass & Ocampo, 1995).

The main subdivisions of *A. hypogea* have been defined as follows:

Arachis hypogaea	
subsp. *hypogaea*	No floral axes on main axis: alternating pairs of vegetative and floral axes along lateral branches.
subsp. *hypogaea* cv Virginia	Less hairy, branches short.
subsp. *hirsuta* cv Peruvian runner	More hairy, branches long.
subsp. *fastigiata*	Floral axes on the main axis: continuous runs of multi-floral axes along lateral branches.
subsp. *fastigiata* cv Valencia	Little branched.
subsp. *vulgaris* cv Spanish	More branched.

These four basic types are broadly characteristic of the secondary centres of diversity shown in Fig. 3.2. Centre 1, and to a slightly lesser extent Centre 2 which contains mostly fastigiate types and is the centre of Valencia-Spanish types. Centre 3 contains a unique form of subsp. *hypogaea*. Centre 4 is mainly subsp. *hypogaea* but also some subsp. *fastigiata*, and shows evidence of interspecific hybridization. Centre 5 contains subsp. *hypogaea* and *hirsuta*, and a special fastigiate form not found in 1 and 2. Centre 6 is regarded as a secondary centre.

Interspecific crosses within the genus are claimed, particularly at the tetraploid level, but it is considered doubtful if crosses outside the section *Arachis* are possible, although diploid members of this section can be crossed with *A. hypogaea* (Li *et al.*, 1994; Shen *et al.*, 1995). This cross-incompatability between wild and cultivated groundnuts impeded utilisation of desirable qualities such as disease and drought resistance, but genetic manipulation is now overcoming this problem.

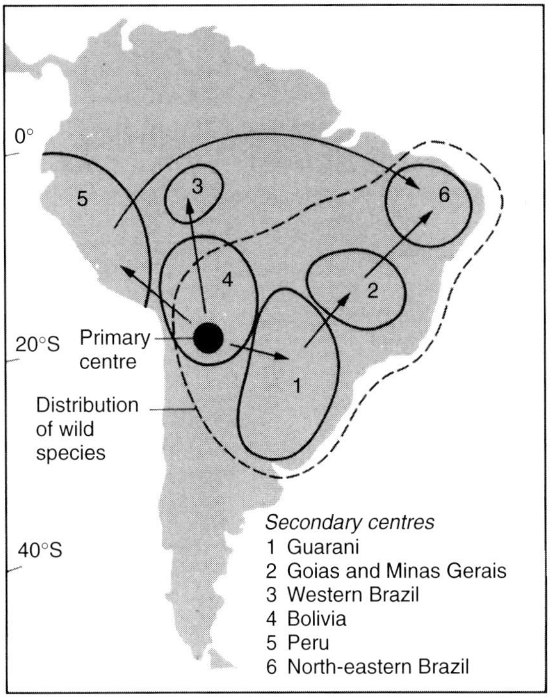

Fig. 3.2 Groundnut: centres of origin and diversity.

Artificial hybridisation is a common method of obtaining genetic variability although there are morphological aspects that make it difficult. Comparison of a conventional hybridisation method (emasculation 1 day before anthesis with manual pollination the next day) with that of a modified method (emasculation and pollination on the same day) showed the latter required 50% less work and reduced injury to flower structures (Santos *et al.*, 1994). Groundnut may also be vegetatively propagated using several techniques, including tissue culture and multiple shoots from seed, as described in the literature. All require specialised equipment and skilled operators.

The groundnut is an annual legume, with wide variation in the types cultivated in particular localities, but in general the two main types grown commercially are distinct in appearance. One is upright with an erect central stem and vertical branches, the other recumbent with numerous creeping laterals. The first is more commonly grown for mechanised production, the second under peasant farming systems. Erect types often have lower individual nut yields per plant than recumbent types under similar conditions, but because of mechanical cultivation and harvesting problems with the latter, final in-store yields tend to be lower. However, erect types tend to have slightly higher seed-oil and seed-protein content than recumbent varieties, and on an oil yield/hectare basis there is often little difference. Fig. 3.3 is a sketch of a typical groundnut plant.

The plant normally has a well-developed tap root to 2 m in uncompacted soils, with numerous laterals which themselves may turn downwards to a similar depth. The rootlets carry few root hairs, which appear to be generally absent in this species. Roots proliferate in the 0–20 cm zone, which normally contains at least half of the total root weight.

Extent of the root system, total weight of roots, and rate of penetration have been established in several countries, but an interesting study in Malawi on individual root longevity (root turnover) showed that at maturity, between 75% and 85% of the cumulative length of root systems had died. Death of root laterals progressed down the soil profile with time, while new roots were produced successively deeper from the extending tap root. Despite continued new root production, the maximum length of living roots of the cultivars grown was 2–4 weeks after sowing (Krauss & Deacon, 1994).

Root growth can be rapid under favourable conditions – an 11-day-old seedling can have roots 30 cm in length and a month-old plant a tap root 50 cm long. In the Sudan a root growth-rate of 3 cm/day has been recorded for the first 50 days of growth, and roots normally follow the moisture front downwards, stopping when dry soil is reached (Lea, 1961) (Table 3.1). Breeding for deeper rooting should develop cultivars suitable for environments where mainly subsoil water is available. However, effective techniques for screening groundnut cultivars for differences in root growth under field conditions are few; one is a soil-injection test using the herbicide metribuzin (Khalfaoui & Havard, 1993).

It is essential for maximum plant growth that there is no check to root elongation, since a large

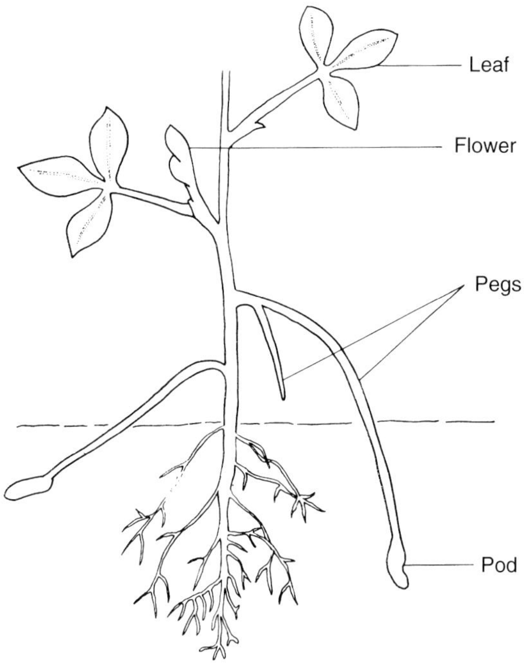

Fig. 3.3 Diagram of groundnut plant (not to scale).

Table 3.1 Rate of root extension, bunch type – Sudan.

Days from planting (cm)	Root depth (cm)	Mean daily extension in preceding period
10	31.5	3.15
20	58.2	2.67
30	88.3	3.02
40	103.1	1.47
50	141.7	3.86
60	157.5	1.57
80	165.1	0.38
100	173.5	0.41

and proliferous system is a major factor in the plant's ability to withstand the dry periods that frequently occur in such areas (Fig. 3.4). The correct time of planting in relation to available soil moisture is important in regions where there are distinct wet and dry seasons.

Nodules are small, globular, often dark brown, and abundant on main and lateral roots. Many serologically distinct strains of *Rhizobium* have been isolated from cultivated and wild *Arachis*, but in general they all appear to be equally symbiotically effective. However, some significant *Rhizobium*–cultivar interaction may occur, and collections of root nodules from South America have been established further to investigate the role and efficiency of nitrogen fixation. Mycorrhizal infection can also be beneficial, but a high soil nematode population reduces this effect.

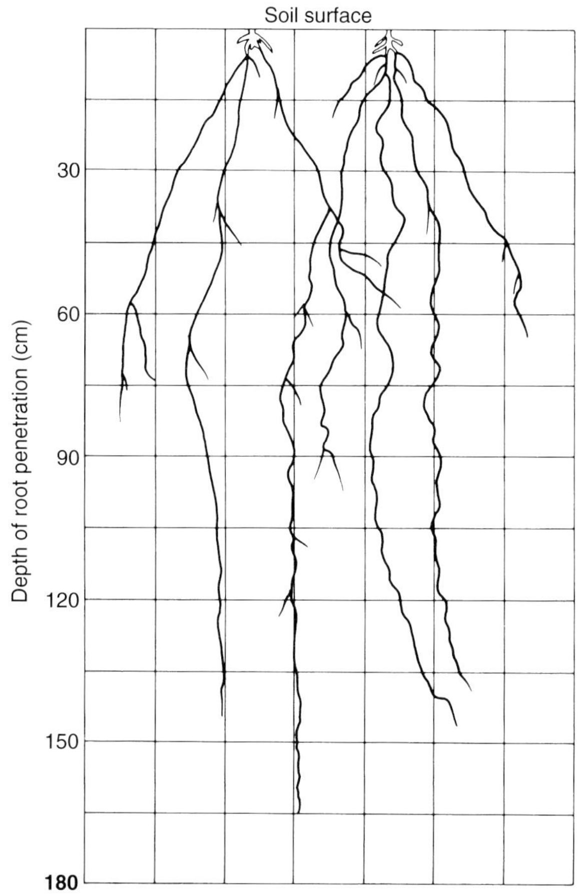

Fig. 3.4 Root system of Natal Common groundnut plants, Tanzania (rainfall 50 cm).

Nitrogen, even at relatively low application rates, appears to depress or inhibit nodulation, while phosphate has a stimulating effect. The sulphur content of some phosphatic and nitrogenous fertilisers can have a beneficial effect on nodulation, since many tropical soils have a slight to moderate sulphur deficiency. Nodulation is generally inhibited in soils below pH 5.5, and although acidity can be partially corrected by liming and nodulation increased, substantial amounts can adversely affect kernel quality. Adverse reactions to certain soil insecticides, nematicides and herbicides are also recorded in the literature, generally however, at higher levels than normally applied.

The stem is usually circular in section, covered with fine hairs, composed of short internodes sometimes tinged with red, and may become hollow with age. In erect types the central stem is vertical, to 30 cm, with numerous lateral branches, which also tend to upright growth, giving the mature plant a tightly bunched, bushy appearance. Hence the designation 'bunch' for varieties with this type of growth. Reducing internode length would shorten the distance between flower and soil surface, and by reducing carpophore length assist quicker soil penetration. Growth regulators including chlormequat, GA, IAA, PP 333, inabenfide, and TIBA, affect plant growth to some extent with little effect on yield. However, if a growth regulator were able to increase peg strength or pod attachment it would undoubtedly increase yield by reducing pod loss at harvest.

Recumbent types have a central stem which is generally less than 20 cm high, and the lateral branches remain close to the ground producing a spreading plant with many 'runners', hence the popular type name. The length and size of runners is a cultivar characteristic, and can be substantial on some African types. Runners of some cultivars grow little once fruiting has commenced, while others continue to grow to plant maturity. There are also intermediate forms, known as bunch-runner or runner-bunch, depending on which characteristic is dominant.

Leaves are alternate, pinnate, with three or four leaflets carried on long petioles; generally a light green, but colour varies considerably between cultivars from yellowish to very dark green. Leaflets are obovate, softly hairy, 3–5 cm long and show characteristic daily sleep movements, in which the petiole folds downward and leaves upward until they touch. A mutant, crumbled-leaf type, is widely used as a marker to detect outcrossing in breeding lines. Leaf growth is normally rapid for approximately 6 weeks after emergence, or until plants meet across the row. Canopy growth-rate is then usually more constant until fruit development assumes increasing importance at about 8 weeks. The rate of early leaf development varies between cultivars, and significantly affects the degree of radiation interception (Bennett *et al.*, 1993), which can also vary with leaf movement thus affecting transpiration (Isoda *et al.*, 1996).

A most obvious feature of groundnut cultivar trials nearing harvest is the great difference in number and colour of leaves remaining on plants. Some cultivars appear green and healthy, others brown and defoliated, but on lifting, the poorest-looking plants often have the highest yield of pods and kernels. The variation in foliar retention is considered to be due to greater foliar disease resistance, and the variation in yield to a basic difference in assimilate partitioning. Where leaf loss from disease significantly reduces yield, prophylactic spraying to maintain a healthy canopy could be profitable.

The small flowers, to 12 mm, are commonly yellow, but can vary from nearly white to deep orange, occasionally with red markings, and are carried on the stem and branches, singly or in small groups; a proportion may be produced underground (Fig. 3.5) (McGregor, 1976). Flowers normally open between 6 and 8 am, the anthers dehiscing from 30 minutes to 2 hours earlier and is a cultivar characteristic; thus groundnut is essentially self-pollinating. However outcrossing of 2% occurred under experimental conditions in Africa, and in commercial crops in the USA some 80% of flowers were visited by bees. Temperature at flowering affects the number produced and their fertility, which varies substantially between cultivars. The optimum appears to be 25–30°C; above 35°C adversely affects numbers and fertility more severely than proportionally lower temperatures. Sunlight can modify the effect of low temperature, and can also affect total flowering.

A groundnut plant normally produces far more flowers than fruits started, but the number

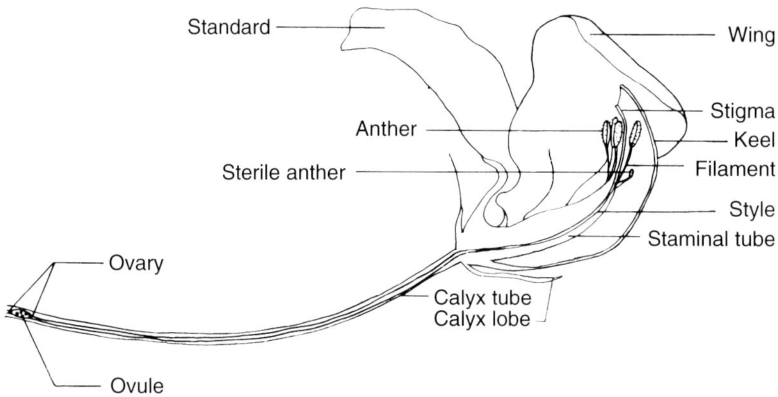

Fig. 3.5 Longitudinal section of groundnut flower.

of fruit started daily is nearly constant, irrespective of the number of flowers. This surplus flower production may explain the relatively small influence on yield caused by heavy infestations of pollen- or flower-eating beetles which commonly occur in the tropics. Flower overproduction is considered a survival mechanism from more primitive times, and plants are unable to sustain fruit production at the same level.

Flowering in some long-season cultivars may continue for 60 days, and fertilisers can extend or reduce this period, and also affect the period from emergence to flowering. Both nitrogen and phosphate can increase the number of flowers and lengthen the flowering period, although the nutrient level at which this occurs is extremely variable, and may also adversely affect the number of fruits set. A severe boron deficiency limits flower formation, and also inhibits pegging. One reason in bunch types is that flowering continues higher up the stems, becoming too high for soil penetration by the carpophore.

About 7–10 days after fertilisation the receptacle thickens, elongates, turns downward and forces the ovary into the ground (Fig. 3.6) (Smith, 1950). The long carpophore bearing the fertilised ovule is commonly known as a peg, the action of burying the immature pod as pegging, and groundnut growers describe a crop as 'pegging well' (Fig. 3.7). Pod formation can only take place in the dark, and pods terminating pegs not strong enough to enter the soil invariably die, although above-ground development is possible when humidity is high. Gynophores shortly after penetration into the soil (white gynophores) released twice the amount of ethylene as above ground green ones or gynophores bearing pods (Shlamovitz *et al.*, 1995).

The ovary develops parallel to the soil surface, into a 4–6 cm long fibrous oblong pod, smaller in bunch than runner types whose pods can reach 8 cm and 2.5 cm in diameter. Mature pods have a pattern of prominent longitudinal ridges, joined by less prominent transverse ridges. The period to kernel initiation and final maturity is influenced by time of planting. Sowing later than the optimum, and this can be as little as 1 week, reduces both and also rapidly reduces seed yield. The firmness with which pods are attached to pegs is a cultivar characteristic, and is of importance in mechanical harvesting; pods strongly attached are lifted with little loss.

The depth at which pod development takes place is basically determined by soil type and conditions, with some cultivar variation; majority of pods are usually 3–10 cm below the soil surface. Final pod dimensions may effectively limit kernel size, since pods generally reach maximum size while kernels are still developing, and as pegs may also deteriorate the developing kernel is thus deprived of nutrients. These may well be major yield-limiting factors for specific cultivars and be exacerbated by premature harvesting. Empty, moisture-free pods contain approximately 65–80% fibre, 10–21% nitrogen

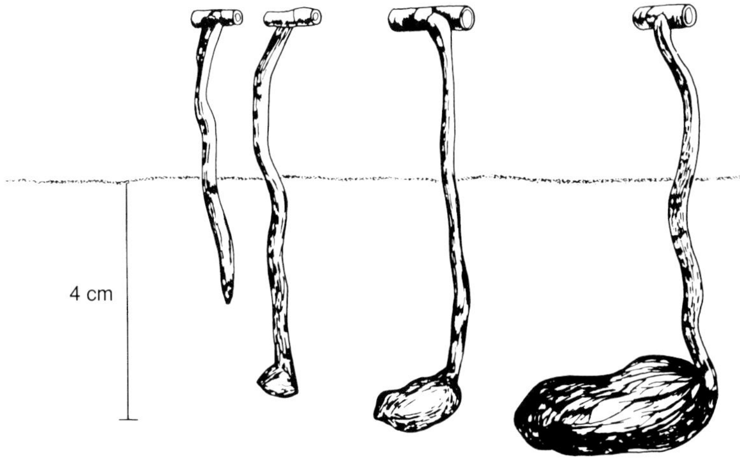

Fig. 3.6 Successive stages in groundnut fruit development, left to right.

Fig. 3.7 Groundnut plant showing 'pegs' entering soil.

free extract, 5–7% protein, 1–2% fat and 3–5% ash.

Pods usually contain two to four ovoid seeds, each enclosed in a papery testa, usually some shade of brown, but colour can vary from creamy white or yellow to dark red, almost purple. Seeds vary in shape from roughly cylindrical to almost spherical, 10–25 mm long and 5–10 mm in diameter, circular to oval in cross-section. Some newer cultivars produce very uniform seed; Florunner frequently has over 90% of its kernels within a diameter range of 1.0 mm. The 100-seed weight can vary from 30 to 150 g, and considerable variation in seed weight may exist within a genotype. Uniformity of both seed size and weight is of major importance to commercial growers using modern machinery. Seeds comprise 55–75% by weight of unhulled pods, a ratio reduced by premature harvesting, peg disease, lack of moisture or nutrients.

Seeds are often diagonally flattened at the end where they are in proximity in the pod, and each consists of two elongated, cream-coloured cotyledons with a short radicle and plumule. The seeds or kernels, commonly known as peanuts or monkeynuts, are without endosperm and can contain 50% oil high in oleic acid and about 40% protein, although the average is 40–45% and 25–30% respectively. In some cultivars significant differences occur in oil content between kernels in the same pod, the basal usually has the highest. Seed composition of a range of wild *Arachis* spp. in Peru showed great variation in all components (Grosso *et al.*, 1995).

The rate and composition of fatty acids in developing seeds has been studied in detail, and show that in general total fatty acid content was greatest at physiological maturity and remained so until maturity (harvest). During seed development, oleic content increased, linoleic acid decreased, while palmitic and eicosenoic decreased slowly from seed formation to maturity. Oleic and linoleic acids account for about 80% of the fatty acids in seed oil, and increasing the ratio of the former to the latter will improve the oil's storage capacity. Genetic studies show a single recessive gene controlled the ratio, and this simple inheritance, lack of background genetic effects and apparent undesirable linkages, indicates relatively simple incorporation of high oleic acid into groundnut cultivars (Knauft *et al.*, 1993). Organoleptic and agronomic characteristics do not appear to be affected by the change in fatty acid composition. The effect of various enzymes on seed maturation has also been investigated (Chung *et al.*, 1995).

Dormancy exists to some extent in most cultivars, and a long dormancy period is an asset in seed destined for the confectionery and edible markets. Seed in pods left unharvested in the field often have a shorter dormancy than if harvested and stored, and on germinating provide a locus for diseases and pest carryover. Dormancy can be broken with various chemical treatments, including hydrogen cyanamide, 6-benzylaminopurine (BA), gibberellic acid (GA), and ethephon (ethrel), but many have adverse effects on germination or seedling development.

Increased yield and oil content of seed has been the main object of plant breeders, and breading has been concentrated on bunch types which are grown commercially on a larger scale than runner. Selection for yield has produced cultivars that partition more of their daily assimilate to fruit, and is largely responsible for yield gains. Duncan (Duncan & McCloud, 1978) commented that:

> One can speculate that in the evolutionary history of peanut species the rate of partitioning to fruit was related to survival in different environments. Where conditions were such that insects, animals and diseases destroyed leaves rapidly, the more successful plant might have been the one that filled its few seeds only by expending most of its assimilate to maintain its leaf canopy. Under less severe conditions the more successful plant would be the one that made more seed by partitioning more of its energy to fruit. Selection for a crop plant would be toward a type that would make more seed by a higher partitioning factor.

Groundnuts may be destined for two very different markets, edible and industrial, and the qualities required are markedly different. For edible purposes a cultivar producing a relatively low-oil nut with high protein and sugar is favoured, combined in a bold, clean seed; factors often related to lower seed yield, but offset by higher prices. The only acceptable testa colours in the US edible market are tan, pink and red. The inheritance of these important colours has been investigated (Branch, 1995). Industrial nuts should have a high oil content, high shelling percentage, long dormancy, and store well.

Other factors important to producers are early maturity, flower fertility and seed dormancy. Quick growth is important where available moisture is limited, and can also assist in reducing damage caused by insects and disease, as many are most harmful to seedlings. Plants that grow quickly and uniformly also flower together, and set seed over a short period. Quick uniform growth also reduces the number of immature and aborted pods, as maturity of the earliest pods affects the growth of later. The ideal groundnut plant should set all its fruit on the same day, and spend the rest of the growing season filling them!

A great deal of genetic variability in cultivars and wild species remains unexploited due initially to problems of interspecific sterility, which biotechnology is solving. Collection and maintenance of wild and cultivated *Arachis* spp. to provide a broadly based gene pool is thus of great importance, instituted by the International Board for Plant Genetic Research and the International Crops Research Institute for Semi-Arid Tropics (ICRISAT). Breeding and selection programmes require seed of known parentage and quality, thus field multiplication techniques must ensure no seed is wasted, harvested seed is of good quality, correctly identified and properly stored.

OTHER GROUNDNUTS

Voandzeia Thou. a genus limited to Africa with the somatic number $x = 11$, with only one member, *V. subterranea* (L.) Thou. has been incorporated into the genus *Vigna*, and is now *Vigna subterranea*, the bambara groundnut, groundbean, earthpea or Madagascar peanut.

Bambara groundnut was probably first domesticated in West Africa, since the only wild form was collected in the Nigeria/Cameroons area. A number of named cultivars are recognised in West and East Africa. Although the wild and cultivated forms have been differentiated for convenience and are agronomically dissimilar, both have $2n = 22$. In contrast to the groundnut it was transported from Africa to Central and South America and the Caribbean in the seventeenth century, and later to the Philippines and Indonesia. Little research has been carried out on the crop since it is of minor importance.

The bambara groundnut is widely but locally cultivated in tropical African environments not suited to groundnuts. It is also planted in seasons when rainfall is too little for groundnut, or planted after groundnut to extend the season. The cultivated bambara groundnut is a small, annual, darkish-green, compact plant to 15–20 cm tall, with recumbent, much-branched stems rooting at nodes. Leaves are carried on long, erect, grooved petioles, and are pinnately trifoliate, larger terminal leaflet 5–10 × 1.5–3 cm, apex rounded, marginate. The inflorescence is axillary, the peduncle short and thick; flowers are usually in pairs, small, pale yellow. After fertilisation the peduncle elongates and penetrates the soil, and the generally single-seeded pods mature underground.

Seeds are spherical, to 2 cm, smooth, and very hard when mature. Seed size and testa colour are variable, and seed may be red, black or whitish, mottled or blotched. Seed weight varies with the cultivar between 1250 and 2000/kg. The highest yield obtained by the author in Tanzania from an irrigated pure stand was almost 1500 kg/ha. However, Bambara groundnut is usually grown on ridges as an intercrop. Plants mature about 4 months from planting. Seeds are usually too hard to be eaten raw, and are ground into flour or boiled. The Kiswahili name, *njugu mawe*, means literally a nut of stone! Seeds are lower in fat and protein, and higher in cellulose than groundnuts. It is grown for food, not as an oilseed.

Kirstingiella is another monotypic genus limited naturally to West Africa. Kersting's groundnut, *Kirstingiella geocarpa* Harms., the Hausa groundnut, is an uncommon cultigen whose chromosome number is believed to be $2n = 22$. It is an annual herb, with prostrate branches (runners) and long internodes, rooting at the nodes. Flowers are axillary, small, subsessile and yellowish; the calyx is deeply divided, with narrow lobes and glabrous style. After fertilisation the carpophore behaves as the groundnut. The underground pods are up to 2 cm long, usually contain two oval seeds, approximately

9 × 6 mm, resembling Bambara nuts. It is also grown only for food.

ECOLOGY

The groundnut is probably native to a region in eastern South America bisected by the 20°S line of latitude, where a large number of wild species are known to exist (see Fig. 3.1). When cultivated its main range is between 35°S and 40°N, but extends to 45°N in Central Asia and North America where there are hot summers.

Like other annual oilseeds with a short growing season, groundnut can adapt to a wide range of environments. However, some modern cultivars may exhibit a more highly developed local acclimatisation. Groundnut is normally grown commercially below 1250 m, although specific varieties can be found at much higher elevations, especially in its South American homeland, but seeds tend to be smaller and their oil content lower.

Groundnut is little affected by day length being basically day-neutral, but plants can be adversely affected by low light intensity during early growth and flowering. Cloudy weather during the flowering period reduces the number of flowers, but as far more flowers are usually produced than fruit started, there is little effect on field yield. However, an interaction between photoperiod and temperature has been noted, with a cultivar variation (Bell & Wright, 1998).

Bunch types are generally more severely affected by climatic variation within their normal range, the runner types least. While ambient temperature affects general plant growth, and above 35°C adversely affects flower production, temperature in the soil zone in which pods develop, the geocarposphere, has the greatest effect on yield. Temperatures between 25 and 30°C appear to be the optimum, below 20°C retards development. Drought and high-temperature resistant strains exist in many local runner cultivars, and a soil temperature of 60°C may be tolerated for a very short period following sowing.

Low temperatures initially retard, then prevent growth, and the plant is generally killed by frost at any stage, while freshly dug and windrowed nuts that become frosted have reduced viability. However low-temperature tolerance exists and could be further developed. Hail on newly emerged seedlings and young plants causes severe damage, especially if the hypocotyl or terminal shoot is broken or bruised.

Groundnut is drought resistant once established, more so than cotton but less than sorghum, and can also tolerate flooding for up to 7 days provided the water quickly drains away. In mist-belt areas of South Africa which are otherwise suitable, the crop is recommended for northern slopes to minimise development of foliar disease. A rainfall of 500–1000 mm will allow commercial groundnut production, but a crop can be produced on as little as 300–400 mm, when the local soil type then has a major influence. About 300 mm should fall between emergence and maximum flowering to ensure good vegetative growth, since there is a direct relationship between number of stems, flowers and subsequent pod formation. Moist soil also allows pegs to penetrate easily.

Ideally no rain should fall once pods are mature, for seed of some cultivars has a very brief dormancy and germinates quickly under suitable conditions. Seed dormancy however, is a genetic factor in many bunch types, and thus does not depend primarily on soil conditions.

Time of planting is important since the majority of groundnut crops are rain-grown, and yield falls rapidly outside the optimum planting period, which usually corresponds with the onset of a rainy season. Table 3.2 is compiled from trials in selected countries, and shows the effect on yield of various sowing dates.

In areas marginal for groundnut growing the period available for growth is of critical importance, and cultivars should be selected whose yield potential can be achieved despite erratic rainfall or temperature patterns. For instance, selection within local runner types in East and West Africa has produced strains that are resistant to a soil moisture shortage, and suitable for areas with a high-intensity short-duration rainfall season.

In the state of Sao Paulo, Brazil, where it is possible to grow two annual crops of groundnuts,

Table 3.2 Effect of sowing date on yield.

USA		Tanzania		Sudan	
Date	Yield pods	Date	Yield kernels	Date	Yield pods
20 Apr.	1810	15 Dec.	1250	1 June	3299
10 May	2018	30 Dec.	788	1 July	2838
1 June	1850	25 Jan.	280	1 Aug.	2351
20 June	1570	10 Feb.	150		
10 July	810	25 Feb.	90		

Figures in table are mean yields in kg/ha over several years at approximately the dates shown.

Table 3.3 Time of planting in selected countries.

Africa		India	
The Gambia	Mid-July	Monsoon crop	June–July
Mozambique	Early Sept	Irrigated crop	Jan–Feb
Nigeria	Early June		
South Africa	Mid-November	USA	
Sudan	Mid-June	Main growing areas:	Mid-April to mid-May
Tanzania	Nov–Dec		

there are distinct differences between the same varieties grown in the two seasons. Bunch varieties produce similar yields in either 'wet' or 'dry' seasons, but runner types are more obviously affected. Reactions within the latter group varied considerably, some were affected by high temperatures, others by high moisture. Bunch types had a consistently higher shelling percentage over runner types in both seasons, and shelling percentage was also higher in the dry than in the wet for both types. Thus there appears to be considerable scope for selecting desirable types for either season.

Local factors other than rainfall can modify time of planting, important though this may be. In India for example, a relationship between time of planting and insect damage can significantly affect seed yield. The recommended planting period in major groundnut growing regions of selected countries is shown in Table 3.3.

SOILS

The ideal soil for groundnut is a light-coloured, well-drained, friable, sandy loam, well supplied with calcium and a moderate amount of organic matter. An opinion expressed to the author in West Africa was that groundnut will grow on soil too light for rice, too poor for cassava and too dry for maize!

Soil type and composition can have a significant effect on fertiliser application and levels, since groundnut frequently gives a greater yield from fertiliser applied to a preceding crop than directly to groundnut itself. This may be so on loamy and clay soils, but is much less obvious on the more sandy soil frequently selected. At the lower limits of the rainfall range, the moisture-retaining properties of the soil are probably the biggest factor influencing its suitability and crop performance. Knowledge of local soil and water relationships is thus extremely important when considering the introduction of groundnut into marginal areas.

Groundnut prefers a slightly acid soil, pH of 6.0–6.5, between 5.5 and 7.0 is acceptable but pH 5.0 severely limits growth. Local runner cultivars are often adapted to a specific level. Saline soils are unsuitable for groundnut, which has a very low salt tolerance. Trials indicated that seed may germinate at a salinity level of ECe 13 mmhos/cm but seedlings died before emergence; at ECe 4 mmhos/cm they remained smaller than plants growing on non-saline soil. (In SI units 1 mmhos/cm = 1 × 105 siemens/m).

So popular is groundnut in peasant agriculture for its ease of production and high nutritive value, that it can be found growing on a wide range of soil types. Sandy and loam soils are most suitable as they are easily penetrated by roots and pegs, and allow easier percolation of rainfall. A further reason for utilising light sandy soils is the ease of lifting, since the operation is difficult from a loam with moderate to high clay content, and many pods break off and are left in the ground. Such loss will be increased if harvest is delayed after the crop is mature, or rain does not fall to soften the ground a few days before digging.

In Senegal, West Africa, highest yields were obtained on soils with a clay fraction below 7%, while personal experience in East Africa with soil which varied from below 5% to above 20% clay content, favoured the former for yield and ease of production. The more clayey soils can produce very high seed yields, but require greater skill in management especially for mechanised production. An additional reason for choosing light-coloured soils is that they do not stain pods, of importance when unshelled nuts are sold and eye-appeal of the product is critical.

Under smallholder production, where the work is often wholly manual, more easily worked soils are automatically chosen. Thus in Africa and India groundnut is commonly found on these soil types, and local opinion frequently rejects the more fertile clays as unsuitable when in fact the energy input may be the limiting factor in selection. For instance in the Nuba region of the Sudan, 1140 kcal/day were required for 4 hours of hoeing, plus 5 hours walking and resting. It was calculated that only 940 kcal were available in the diet for this labour, and the mounting deficiency reduced both the ability of the farmer and final crop yield (Engel, 1966). In South America, especially in the highlands, the crop is sown wherever space is available. For mechanised production in the USA light soils are favoured for ease of lifting, and crops grown on the flat or ridges as the producer chooses.

FERTILISERS

Groundnut is a major food and oilseed crop worldwide and there is thus a substantial literature available on all aspects of its nutrition. The following section reviews this information with an emphasis on commercial crop production.

There are many published estimates of nutrient removal, varying as widely as the conditions under which the crop is grown. The following values are a guide to nutrient levels which should be available or supplied to ensure profitable production. The nutrients removed by an average groundnut crop in Niger, West Africa, are shown in Table 3.4, where 1 tonne unshelled seed removed 5 kg nitrogen, 3 kg phosphate, 12 kg potassium, 7 kg calcium and 5 kg magnesium (Bertrand *et al.*, 1972). In Nigeria an average groundnut crop removes 45 kg nitrogen, 12 kg phosphate and 34 kg potassium per/ha.

Groundnut plants can scavenge a considerable volume of soil and this ability is probably a major reason for the very irregular yield responses obtained from added nutrients in many countries where groundnut is widely grown. Analyses in Brazil on plants from seedlings to maturity showed high concentrations of nitrogen and phosphate in seeds, sulphur in roots, and potassium, calcium and magnesium in stems and leaves. At harvest fruits weighed half of the total plant dry weight. In Nigeria the proportion of mean nutrient uptake by plants and which remained in the kernels, was 63% nitrogen, 68% phosphate, 23% potassium, 24% magnesium and 4% calcium. In East Africa 70% total plant nitro-

Table 3.4 Nutrient removal by groundnut and millet (kg/ha) – Niger.

Groundnut	N	P	K	Ca	Mg	SO_3
Kernels	87	6.6	12.8	0.6	3.1	10.5
Shells	5	0.1	3.3	0.4	0.3	1.3
Foliage	66	4.2	65.0	31.5	19.0	18.4
Total	158	10.9	86.0	32.5	23.4	30.2

Six consecutive groundnut crops averaged (kg/ha): kernels 1830, shells 520 and foliage 3875, which in terms of nutrient removal per 100 kg of unshelled nuts, equalled (kg) N 6.72, P_2O_5 1.08, K_2O 4.17, CaO 1.95, MgO 1.58, SO_3 1.28.

Millet	N	P	K	Ca	Mg	SO_3
Grain	46.5	5.3	18.7	1.2	2.9	12.4
Straw	42.0	2.6	91.0	14.0	15.0	37.4
Total	88.5	7.9	109.7	15.2	17.9	49.8

Six consecutive millet crops averaged 1785 kg/ha grain and 4285 kg/ha straw.

gen and 80% of phosphate was found in kernels at harvest and analyses from other countries gave similar results. In general, plants absorb approximately 10% of their total requirements of nitrogen, phosphate, calcium and magnesium during the vegetative phase, 40–50% during reproduction, the remainder to maturity.

Groundnut should preferably be grown in a rotation with crops that require substantial fertiliser applications, since as noted, high seed yields are frequently obtained when nutrients are applied to a previous, not necessarily the preceding crop. There is considerable evidence to indicate that even where a particular nutrient is deficient, higher application to a preceding crop increases the following groundnut yield above that obtained by smaller, similar amounts to both. It is probable that the additional cultivations necessary for the two crops distribute fertiliser through a greater volume of soil, to the advantage of groundnut. This is particularly so with potassium and to a lesser extent phosphate, although the preceding crop can modify this; groundnuts following wheat may require more added nutrient than when following cotton or maize.

Since the residual effect of added nutrients can be substantial, fertiliser trials should precede any new large-scale groundnut plantings, especially in less-developed regions. Fertiliser transport and storage costs are invariably high, and savings contribute directly to profits. The final arbiter as to level, time of application, type of nutrient and whether it is applied to groundnut or preceding crops, must be the return obtained.

Germinating groundnuts are easily damaged by fertilisers, and band placement 10 cm on either side of the row, preferably 15–20 cm deep, is recommended. At this depth there is also less chance of soil becoming too dry for plants to take up nutrients. This placement also encourages root penetration, which enables plants to tap a greater volume of soil moisture, and a direct correlation between high total root weight and final yield exists. Deeper than normal placement of fertiliser in Indonesia not only improved groundnut yield, it also increased the seed yield of following crops.

Rock phosphates, basic slag and some slow-release compounds can be broadcast and ploughed in, but many tropical red earths are high in aluminium and iron, which rapidly fix phosphate in unavailable forms. In these areas annual application of more concentrated phosphatic types is usually necessary, and attempts to build up the soil phosphate level unsuccessful. Wide distribution of fertiliser within the root zone is particularly successful with runner types grown on large ridges, a common practice in Africa.

Groundnut plants flower profusely, and although the total number of flowers is increased by the use of fertilisers the percentage of additional flowers that set pods is low, and the number of additional pods that reach maturity even lower. Fertilisers can also affect size of seed and number of seed per pod, but generally have little effect on oil content or composition. These characteristics are highly heritable, and environmental conditions have comparatively little effect.

Farmyard manure (FYM), poultry and pig farm effluents, compost or plant residues are commonly used by smallholders, but on few commercial plantings. When groundnut follows a crop receiving large amounts of organic fertilisers, it seldom requires any additional nutrients. In many parts of Africa and India where animal manure is applied, the addition of single superphosphate or potash has profitably increased seed yield (Table 3.5a and b). Newly broken land high in organic matter is considered unsuitable for groundnut in South Africa, nor should the crop follow cereals which have left a large amount of partially decomposed stubble or straw.

Deficiency symptoms for major and minor elements have been described, and various methods of correlating foliar analyses with nutrient deficiency published. However, in the author's opinion there is usually no substitute over wide areas of the world for precise field trials accurately to determine groundnuts' response to added nutrients. With all high-yielding commercial crops including groundnuts, fertiliser use must accompany optimum plant populations, good quality seed and cultural practices, since the indiscriminate use of artificial fertilisers is no substitute for good management.

Groundnuts can also accumulate heavy metals in foliage and seeds, and in Bulgaria the accumulation of lead, zinc and cadmium in seeds and plants grown in polluted soils reached dangerous levels, especially if used as forage or food (Stefanov *et al.*, 1995). Cadmium uptake and translocation have been investigated (Popelka *et al.*, 1996).

Table 3.5 (a) Effect of farm yard manure and single superphosphate on groundnut yield – Africa (yields in kg/ha pods).

Farm yard manure	Single superphosphate	
	Nil	100 kg/ha
Nil	1260	1820
2 t/ha	1470	2090
Mean	1360	1960
LSD 5%	197	260

Nitrogen

Groundnut is a legume, and *Rhizobium* nodules on roots can provide a substantial portion of its nitrogen (N) requirement. Well-nodulated plants also tend to be more drought resistant and produce higher-quality nuts.

An assessment of data available indicates the rate and amount of nodulation and nitrogen made available for plant growth appears to be extremely variable, as does the response to applied nitrogen. Because of the economic importance of the nitrogen supplied by *Rhizobia,* research continues into the selection of more effective strains and interactions between particular strains and cultivars. Where groundnuts have not previously been grown it is necessary to inoculate the first crop with the specific bacterial strain. Large scale commercial growers often inoculate each crop, since the cost of so doing is far less than the alternative nitrogen application, or possible reduction in yield.

A review of the data available indicates that even relatively low N application can suppress nodule proliferation, and appears to be more severe on high-yielding cultivars. However inoculated or uninoculated seed plus nitrogen may give similar yields, as occurred in Vietnam (Nguyen, 1994). Herbicides can also affect nodule numbers and activity.

Table 3.5 (b) Effect of farm yard manure and potash on groundnut yield – India (average of 4 years, kg/ha pods).

Farm yard manure	Potassium levels (K_2O kg/ha)				
	0	16	33	50	Mean
Nil	426	671	806	842	686
6.25 t/ha	761	858	889	915	855

Groundnuts may also be destined for two distinct markets, and end use has a basic influence on nitrogen fertiliser policy. Excess N usually depresses oil content and increases protein content of kernels; thus kernels required for oil should therefore receive less N than those destined for eating. On soils of low inherent fertility, nitrogen in the seed-bed or a small top-dressing to young plants can be beneficial, and increase the uptake of other nutrients to improve growth. However, the economics of so doing must be established, since in many instances the visual improvement in growth is not reflected in higher seed yield. Should growing plants show obvious N deficiency symptoms or respond to nitrogen applications, it is advisable to check plant roots to determine if the specific strain of *Rhizobium* is present and active.

The most common method of supplying N is as part of a compound fertiliser placed in the seedbed, and NPK ratios of 5:20:5 or 5:10:15 are often used. When K is not required, a NP combination such as diammonium phosphate, or a mixture of ammonium sulphate nitrate and single superphosphate is effective. The latter also supplies some sulphur, an advantage in many African and Asian countries. Applications above 20 kg N/ha are unusual, unless part of farmyard manure. There appears to be a relationship between nitrogen-sulphur levels and plant growth and yield, but this has not been well defined. In the dry tropics the release of mineralised nitrogen at commencement of the rains is well documented, and when possible groundnuts should be planted to make use of this phenomenon. N deficient plants are often stunted with chlorotic foliage, and reddish tinged stems. These symptoms should not be confused with those due to rosette virus (see section on diseases).

One aspect of nodule nitrogen often overlooked is the contribution groundnut residues can make to a following cereal crop, by reducing the amount of N necessary and increasing profitability. After one to three successive crops in Malaysia, an equivalent of 77–105 kg N/ha was returned to the soil as legume-N through groundnut residue. An estimated 20–56 kg N/ha was contributed to the subsequent maize crop (Anuar *et al.*, 1995). A similar high residual effect was noted in South Africa (Bloem & Barnard, 1995).

Phosphate

Phosphate (P) is considered essential for root and kernel development, increases the number, density and efficiency of nodules, and can also significantly increase the uptake of other nutrients. However, root development response to added P appears to vary between cultivars, and may not be predictable (Marubayashi *et al.*, 1994). Phosphate uptake is dependent on soil moisture availability, and an erratic yield response by groundnut to added P is frequently due to lack of moisture restricting uptake, rather than an inability to make use of the nutrient. The P requirement of groundnut is fairly high and seldom met from normal soil levels. Wide dispersal within the root zone is as important as the total amount, and thus higher application to preceding crops can improve the following groundnut yield, as noted. Groundnut in the rotation can make very profitable use of this residual P with a very small seedbed application to assist initial growth. Phosphate uptake by plants tends to decrease with age, and the nutrient can migrate rapidly from foliage to developing fruit.

Although very high phosphate levels are commonly applied in developed agricultural systems, as little as 6 kg P/ha has given profitable yield increases under smallholder conditions in Africa and India. In general the P available to a groundnut crop must equal that applied to high-yielding maize or cotton crops in the same area. It is seldom less than 20–30 kg P/ha and can well be twice this. Local rock phosphate can supply a major portion of the requirement, especially when broadcast annually in a rotation.

The type of phosphate can be important, for in many regions of Africa and India there is a sulphur deficiency, and single superphosphate or basic slag, which contain sulphur, are more efficient for this reason as noted in the section on sulphur. A mixture of two types can also increase yield, and in East Africa on a number of crops, single superphosphate + diammonium phosphate gave higher yields per unit of P than either used singly (Weiss, 1969). In East Africa the sulphur content of single superphosphate was usually around 11–12%, but only 1–2% in double or triple superphosphate. There is thus a strong probability that the erratic response to phosphatic fertilisers in many groundnut growing areas of the tropics may be partially due to the calcium or sulphur content of the compound used. In India however, ammonium polyphosphate gave the greatest yield increase per unit P than single superphosphate, triple superphosphate or diammonium phosphate, emphasising the importance of local trials on local cultivars for greatest profitability (Shinde *et al.*, 1993).

Phosphate deficiency symptoms occur most often in young plants, which appear stunted, with a reddish or purple tinge to leaves and stems. Symptoms often disappear as the root system expands, but in marginal areas the initial growth retardation may critically shorten the overall period available for plants to mature.

Potassium

Crop response to potassium (K) is directly related to availability of other nutrients in particular magnesium and calcium, and may also increase the uptake of nitrogen. A profitable response to potassic fertilisers has generally occurred only in those areas where there is a severe K deficiency, and an annual application may be necessary (Prasad, 1993). Potassium uptake is most active in the first 6 weeks of growth, and gradually decreases with age. The nutrient also moves readily in the plant, and is transported from older to younger plant parts and developing fruit. In a deficiency the terminal shoots and leaves are thus the last to be affected, and pod development continues at the expense of vegetative growth. Potassium apparently increases the number of pods per plant, but calcium and magnesium increase the number and size of the kernels. From an assessment of data available it appears that the balance between K:Ca:Mg may be more important than any one or the total of all.

The response to potassium may also vary with the variety grown, since this element can prolong the vegetative period; thus yields may be lower from quick-maturing types from amounts that increase the yield of longer-term cultivars. Potassium is often best applied to a preceding crop, or ploughed in some months before planting, since high levels in the pegging zone can inhibit calcium uptake by the developing pods. This interaction may be a major reason why groundnuts are considered a crop that responds better to residual rather than direct fertiliser applications. When potassium is necessary, lime, gypsum or a suitable substitute may also be required. In India the by-product of local salt works, schoenite, which contains potassium, magnesium and sulphur, was as effective as potassic fertilisers at the same nutrient level (Natarajan *et al.*, 1973). Similarly to phosphate, potassium sulphate gives a greater yield response than non-sulphur compounds where sulphur is also deficient.

The amount of K required must be determined by trial results or soil analyses; in their absence the amount of available P and K required should be considered similar until shown to be otherwise, with potassium preferably applied to the preceding crop. Unnecessary application of potassium will at best allow plants to take up luxury amounts, at worst result in a growth-limiting imbalance with other nutrients. K deficiency appears as yellowing of older leaf margins, followed by interveinal chlorosis and finally defoliation. Before these more obvious symptoms, plants may be less vigorous and smaller than normal. By the time visual symptoms appear yield potential will have been reduced, and there is little value in then applying potassic fertiliser.

Calcium

Maximum use of calcium (Ca) by plants occurs from the appearance of pegs until pods are mature, while high levels of soil calcium may reduce the incidence of pod and root rots. On some low calcium-content soils, gypsum can also increase seed-oil content. Visual symptoms of a severe calcium deficiency are small plants with a withered growing point, yellow upper leaves, and diminished and deformed root growth. The characteristic symptom of a lower calcium deficiency are less obvious and frequently confined to wilting and subsequent death of shoots. Underground symptoms include unfilled or partially filled pods with shrivelled kernels (pops) or kernels with brown discolouration of the embryo and/or a depression at the centre of cotyledons; cultivars with small seeds generally produce fewer pops.

Calcium can also increase the number of kernels per pod, not by raising the number of available ovular cavities, but because of the increased number of kernels reaching maturity (Piggott, 1960) (Table 3.6). Growers of seed crops in the USA use higher levels of calcium than commercial growers, since they consider this gives higher viability and seedling vigour.

Calcium is relatively immobile within the groundnut plant, and its concentration in a specific part is apparently dependent upon availability at the time the particular part was formed. Calcium must be in fairly close proximity to developing fruit, and placement can affect depth at which pods form. Additionally, in high-rainfall areas, gypsum can be easily leached from the fruiting zone before pods mature. Studies with radioactive calcium showed it was absorbed primarily through the shells, some being translocated to kernels and pegs, and that calcium absorbed by roots went to stem, foliage and aerial pegs, but little to the kernels. Growth retardants, including daminozide (Alar), may also increase calcium translocation to kernels and improved their quality. A Ca deficiency may not be the only cause of pops, although in practice applying calcium is the most simple method of reducing their occurrence. A common method of application in the USA is by band dressing at flowering with gypsum or any similar soluble Ca alternative (Fig. 3.8).

The type of commercial calcium compound applied is generally not important, but may be so in particular circumstances, when a neutral salt such as calcium sulphate may be preferable to calcium carbonate. Up to 1000 kg gypsum/ha may be necessary but 250–500 kg/ha more common, since Ca should be considered a plant nutrient rather than a soil ameliorator. When large amounts are required two applications are

Table 3.6 Effect of calcium and minor nutrients on groundnut yield – Sierra Leone.

Treatment	Pods (kg/ha)	Shelling (%)	Kernels/pod	Ovular cavities filled (%)
Control	560	57	1.0	40
Ca	930	70	1.9	76
K	580	41	0.6	21
CaK	1030	64	1.5	71
Mg	538	47	0.6	29
CaMg	1235	66	1.6	75
KMg	700	36	0.4	18
CaKMg	1680	44	1.6	72
P	925	70	1.4	68
CaP	897	70	1.7	81
KP	1300	60	1.1	50
CaKP	1150	68	1.5	77
MgP	1060	68	1.4	67
CaMgP	1010	69	1.8	83
KMgP	1250	55	1.1	51
CaKMgP	1570	67	1.7	76

Treatment P (18% P_2O_3); K (K_2SO_4); Mg (16% MO); at 250 kg/ha. Ca (73% CaO) at 2500 kg/ha.

Fig. 3.8 Calcium applied to groundnut rows.

preferable; one when the first flowers appear, the second 20–25 days later. Ordinary agricultural lime is not always an efficient substitute, owing to its variable composition.

Bunch types generally give a lower response to Ca than runner types, and some newly developed cultivars respond differently to earlier types. A considerable difference occurs between cultivars in their ability to produce high yields of good quality nuts on soils formerly considered as being too low in calcium. On these cultivars additional Ca may increase oil content, but have little effect on yield or soundness of seed. Where gypsum is not readily available alternatives include ground sea shells, wattle bark ash, bonemeal and anhydrite. Alternatively, phosphate with a high Ca content can significantly increase seed yield on soils with an acute Ca deficiency.

Sulphur

Sulphur (S) is deficient in many soils used to grow groundnuts and there is a growing amount of data showing plant response to this important nutrient. The increased seed yield from applying high S content phosphates by many crops in Africa and India, is well documented and the substantial response to single superphosphate is frequently due to its sulphur content (Pasrichuia & Aulakh, 1991; Mishra, 1996; Sahu & Das, 1997). There may also be an interaction between sulphur levels and irrigation (Tiwari *et al.*, 1994). It is also difficult to distinguish visual deficiency symptoms of these two nutrients; generally S deficiency symptoms first occur as chlorosis of terminal leaves, while a P deficiency shows initially on older leaves. A sulphur deficiency is usually more easily alleviated by applying a sulphur-containing fertiliser than elemental sulphur.

The effect of different S containing fertilisers on nutrient uptake in India is shown in Table 3.7a (Pathak & Pathak, 1972), and in Africa, Table 3.7b. (Laurence *et al.*, 1976). It was noted that plants receiving S produced higher quality, larger kernels.

Foliar sprays containing S can materially assist in alleviating a deficiency, and fungicides containing S have also produced yield increases, not only by controlling disease but by promoting uptake of N and P. Sulphur may affect nodulation; a deficiency reduces and a surplus inhibits, and it can also affect seed-oil percentage. The effect of fertilisers containing S on these two plant characters is shown in Table 3.8. (Pathak & Pathak, 1972).

Minor nutrients

Foliar analysis can assist in identifying regional or district deficiencies before field trials, and in Punjab, India, the most important were, in order, B > S > Zn > Mn > Cu. Where several nutrients are required it is often more practical to use a proprietary mixture rather than attempting to mix or apply each separately. Mixtures soluble in water can be sprayed with compatible insecticides or fungicides, and some of the latter may in fact contain sufficient of a particular element, i.e. copper or zinc. Chelates are a most efficient method of applying many trace elements, as their structure resists conversion to an insoluble form,

Table 3.7 (a) Effect of sulphur-containing fertiliser on nutrient uptake – India.

Treatment	20 days				60 days				120 days			
	S	N	P	Ca	S	N	P	Ca	S	N	P	Ca
No fertiliser	0.27	3.16	0.63	1.32	0.25	3.0	0.88	1.28	0.23	2.60	0.97	1.20
Non-S fertiliser	0.41	3.45	0.97	1.65	0.29	3.0	0.86	1.42	0.25	2.70	1.10	1.11
Plus-S fertiliser	0.46	3.60	1.04	1.66	0.34	3.4	1.15	1.60	0.27	2.90	1.17	1.41

Figures in body of table are percentages of element in whole plant, and treatments are means of sulphur or non-sulphur containing fertilisers, i.e. ammonium sulphate and chloride or calcium sulphate or chloride.

Table 3.7 (b) Effect of sulphur on yield and seed contents – Malawi.

Treatment	Kernel yield (t/ha)	Total oil (%)	Total protein (%)
No fertiliser	1.77	45.6	26.5
Gypsum (95% $CaSO_4$)	2.22	46.1	26.5
Sulphur (as dust)	2.08	45.1	27.3
Sulphur + phosphate (48% P_2O_5)	2.36	45.9	26.5

Note: Figures in body of table are mean of experimental sites and treatment levels. AMI = Arginine Maturity Index.

Table 3.8 Effect of sulphur on nodules and oil percentage – India.

Treatment (mean of levels)	Av. number of root nodules	Seed-oil (%)
No fertiliser	306	48.2
Amm. sulphate	376	50.4
Atom. chloride + S	380	50.7
Amm. chloride	389	47.8
Cal. sulphate	386	51.5
Cal. sulphate + S	360	51.2
Cal. chloride	319	48.6

Note: Elemental sulphur was added to fertilisers at equivalent rate to that contained in the sulphate compound.

and they are readily absorbed by plant roots or foliage.

The obvious sign of a magnesium (Mg), deficiency is interveinal chlorosis of terminal leaves and stunted growth; with a severe shortage plants become almost completely decolorised and die (Fig.3.9). Since a Mg deficiency is seldom noted in the field, this suggests that groundnut can exploit nutrients in acid soils more efficiently than other crops. A deficiency can be rectified by mixing with a seedbed fertiliser at 2.5–10 kg Mg/ha, or as a spray at 1–2.5 kg/ha. A relationship between potassium, calcium and magnesium exists in some regions of Indonesia (Taufiq & Sudaryono, 1997). When lime must be applied to acid soils, dolomitic limestone is to be preferred since it contains sufficient magnesia to alleviate a local deficiency. The relationship between magnesium, calcium and potassium is shown in Table 3.6.

Manganese (Mn), as a toxicity or deficiency is unusual, but the latter can occur on more acid soils where large amounts of lime have been applied. The usual deficiency symptom is inter-

Fig. 3.9 Magnesium deficiency symptoms; healthy leaf on right.

veinal chlorosis of leaves with veins remaining green (Fig. 3.9). Its effects on yield are slight and 0.25 kg Mn/ha spray usually sufficient.

Molybdenum (Mo) deficiency in the field is very difficult to determine, and so small are the amounts required to alleviate a deficiency it is often applied as a seed dressing. In India, N + Mo significantly increased seed oil content (Krishnappa *et al.*, 1994). Molybdenum can also affect flowering and peg development, and high soil content reduce the percentage of pegs producing mature pods.

A boron (B) deficiency adversely affects pod production and kernel quality and is often difficult visually to diagnose until it is too late! Typically, kernels have discoloured and hollow sections of cotyledons. It can be alleviated by applying 0.25–1.0 kg/ha boracic acid as a spray, or borax at 12–15 kg/ha. In Hebei Province, China, application of B promoted absorption of N, increased plant height, plant dry weight and total number of pods (Jiang *et al.*, 1994). The type of B containing compound appears to be unimportant, but the application must be made as soon as pegs appear. Boron is routinely applied as part of compound fertilisers in major groundnut growing areas, since the cost of so doing is negligible compared with a loss sustained by having kernels rejected for failing to conform to quality standards. However, continual application of boron may result in a toxic build-up in the soil.

Zinc (Zn) toxicity symptoms usually become apparent in the second month after emergence, commencing with a stunted appearance, followed by abnormal leaf growth and stem-splitting. Since there is no practical method of correction, groundnut should not be planted on soils with a high zinc content. Application of zinc sulphate and gypsum in India increased seed yield, and seed oil and protein content (Ingole *et al.*, 1998). Groundnut is highly tolerant of aluminium (Al) ion activity, and is unaffected by soils that limit growth of cotton, a crop often sown in rotation with groundnut. However, root growth, nodulation and yield are progressively reduced as the soil Al content increases. Cobalt applied as part of a compound fertiliser has increased yield in India (Jana *et al.*, 1994; Lakshmamma *et al.*, 1996). The uptake of selenium and its distribution in plants of local cultivars, has been studied in China (Wang *et al.*, 1997).

Other trace elements are unlikely to limit groundnut production except in very local areas. An iron deficiency reduces the chlorophyll content of leaves, but is difficult to see in the field as the symptoms are usually masked by a more serious deficiency.

CULTIVATION

Smallholder production is usually mainly manual, but with the increase in co-operatives more opportunities exist for mechanising operations that are either unpopular or have a limited period for optimum results. Implements include animal drawn planters, weeders, digger/lifters, and small hand or power operated strippers. Specialised or large and expensive equipment may require adaptation of traditional agricultural methods for their most profitable use; a change from runner to bunch types, or extension of the planting period to ensure greater use of lifting or threshing machines.

A fully mechanised groundnut crop in the USA in 1991, cost US cents 33.5/kg in variable cash expenses, with individual farm costs between 6.6 cents/kg to 158.5 cents/kg, with operator characteristics contributing to the distinction between low and high cost producers (Brooks, 1994). The average cost of producing an irrigated crop in Queensland, Australia in 1996 when nuts sold at Australian dollars (A$) 800/tonne was in A$/ha: seed 200, fertiliser 52.5, gypsum 40, weed control 130, pest control 15, disease control 150, irrigation 240, digging 40, threshing and cleaning 312, transport to sheller 80, drying 80, to total A$ 1611.5/ha, against a income of A$ 3200/ha.

The operations necessary to produce a seedbed for maize or cotton are suitable for groundnut. Clay soils may require additional working to produce a tilth, but the seedbed should not be too fine as it may crust after rain. Commercial growers in the USA advocate deep, 30–35 cm, ploughing to bury the topsoil and rubbish to reduce the incidence of disease. Since the fertilised ovaries elongate into long pegs which must enter the ground to develop, it is essential that the top 10 cm of soil remain loose. Minimum or no-tillage systems can be successful in specific circumstances, seldom in the tropics, and should only be adopted after field-scale trials.

For mechanised production of bunch-type groundnuts, especially in more clayey soils, a ridge is an asset to lifting and can be progressively constructed during mechanical weeding operations. Weeding/ridging must be accurate, and each run move just enough soil to clear between the rows without burying lower branches or piling earth against main stems. The first causes loss by smothering potential flowers and fruits, the second renders plants more susceptible to stem and collar rots. Size of ridge has very little effect on pod yield, but timing of the operation is more critical. In Brazil, pod number/plant and pod yield/ha were significantly and steadily reduced as ridging was delayed, but weight of seeds contained in 100 g of pods increased (Kasai & Paulo, 1993).

Well-constructed ridges have a flat, or slightly concave top on which plants grow. Sowing on shallow beds, some 7.5–10 cm high whose width suits diggers or harvesters, is of value where soil is liable to compact at harvest. In drier regions, planting on the contour will allow rainfall to be impounded between the ridges which can be automatically tie-ridged, thus acting as an efficient soil conservation measure. Runner types are often sown directly into prepared ridges, since inter-row cultivation is difficult. Smallholder crops are often grown on broad-based ridges, which are thus well prepared and easily penetrated by pegs.

Seed should be sound, clean, with unbroken testa, as they are very susceptible to soil-bourne fungal diseases. When using herbicides the selection of sound seed is more important as quick-growing, vigorous seedlings suffer less damage from these compounds. Seed should preferably be even in size and weight for sowing by the latest electronically controlled seeders, and obtained from specialised growers, who will also dress seed including any small amounts of micronutrients required. If it is necessary to save a proportion of a crop for sowing the following season, much greater care in harvesting and decortication is necessary, since even minor damage causes high seedbed loss. Where damaged seed must be used, a 50% higher seeding rate may be necessary, since seed dressing cannot fully compensate.

However, saving farm-grown nuts for sowing the next season is not to be recommended. The author has often noticed the number of rogues occurring in commercial bunch-type fields from sowing such seed, whose presence materially reduces the overall quality and value of the final crop. The sowing of two or more varieties in a mixture, a blend or composite, to reduce loss from pests or disease has been suggested since this technique has successfully increased yield in other crops. This has little to recommend it for groundnut characteristics vary widely, and components of a blend are frequently mismatched for rate of growth, time to maturity etc.

Seed dressing is recommended, combined with a locally appropriate insecticide. In many regions where cotton is also a major cash crop, the same seed dressing can normally be used for groundnuts. Dressed seed must be secured against theft, since in many rural areas sowing time often coincides with a period of food shortage. Where large quantities of nuts are involved dressing requires special equipment; for small quantities a 200 litre drum mounted and revolved on a suitable frame, or a sealed drum rolled along the ground, is quite suitable.

Pods are occasionally sown by smallholders, but this is not recommended. Apart from transporting unnecessary bulk, unshelled seed is liable to rot in the pod before germinating, and percentage emergence compared with shelled seed is low. The time smallholders spend shelling seed is usually fully recouped by the increased yields obtained, and in the Sudan, yields from shelled seed were consistently and substantially higher than from unshelled.

Many bunch varieties produce seed with very short dormancy, some germinating at or shortly after maturity. This greatly reduces the harvesting period, when delay causes severe losses, and increased dormancy is thus a valuable objective. Quick germination also produces volunteer plants, which can act as a reservoir for pests and diseases. Strength of attachment of pod to peg is of importance to mechanised producers in particular, since weakness at this point results in many pods becoming detached and remaining in the ground at lifting.

A seedbed temperature of 25–30°C is the optimum, and emergence percentage is normally reduced below 20°C and above 35°C. Emergence is unusual in that the cotyledons neither remain underground nor fully emerge, but usually just break the soil surface. Seed should be sown 5–10 cm deep into moist soil, deeper if necessary to preclude drying out. The author has sown seed of a local runner cultivar in drills 15 cm in depth, which had been opened up with disc coulters. The seed was then covered with a press wheel in the bottom of the small furrow which was left open to be filled naturally. This was in a region of Central Africa where the annual rainfall seldom exceeded 500 mm, and occurred as high-intensity storms at irregular intervals. This method produced good yields in an area where groundnuts were not traditionally grown. However, in the major groundnut growing regions of Australia and the USA, depth of planting is considered critical and the local optimum seldom exceeded. Planting too deep usually results in a lower number of mature pods (Fig. 3.10).

Groundnuts are sown preferably using special seeders or modified maize planters with groundnut plates and extra large seed hoppers, but grain and similar type drills are not usually suitable. Irrespective of the type of sowing machinery used, fertiliser must not be placed in contact with seed. When using planters on more clayey soils, it is essential to check that the seed-shoe does not become blocked during work. Semi-skilled drivers have a tendency to drop the planter unit too quickly after turning which rams soil into the spout and blocks it. Since it is difficult for the driver to see if the seed is flowing down the spouts, long gaps can result. It is also important that the recommended forward speed for a particular type of planter is not exceeded; too high a speed invariably results in gappy stands caused either by blocked spouts, or by damage to seeds moving too quickly through the planter units. Efficient organisation and supervision of the planting operation invariably ensures a higher area planted/seeder than attempts to increase the actual speed of planting. The introduction of air-flow, electronically controlled seeder units has greatly increased planting efficiency (McPhee & Norris, 1991).

Where row-width is optional, closer rows produce higher yields than closer in-row spacing, but must be related to the method of harvesting, i.e. row-widths of plants to be dug mechanically must suit the digger blade/blades, and weed control implements. Row-width can vary, from 40 to 90 cm for a single row, twin rows on large ridges, or four 25 cm rows on irrigated beds. Optimum plant population will be determined by soil type, expected rainfall, etc., but as a guide a population of 100 000–120 000 plants/ha for

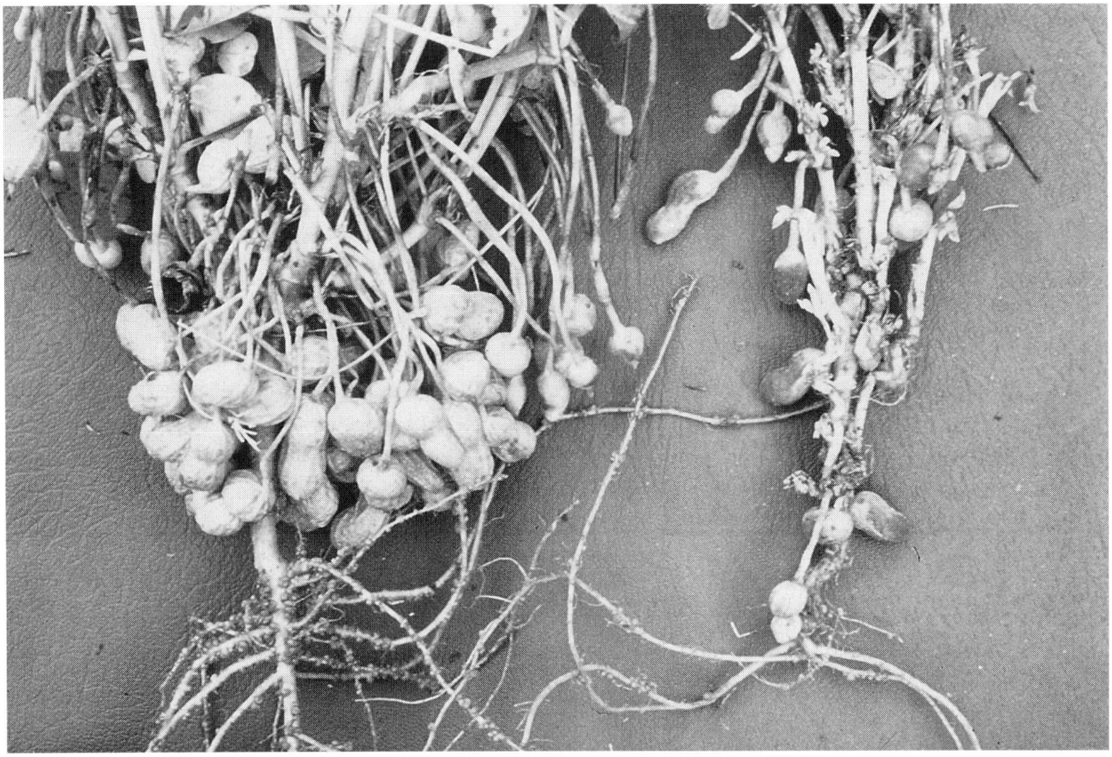

Fig. 3.10 Plant on right shows effects of too-deep planting – Australia.

bunch types (80–100 kg seed) and half this for runner is acceptable in areas of adequate rainfall, 750–1000 mm. Population should be reduced as the expected rainfall decreases; for instance at 600 mm, a population of 50 000–60 000 plants/ha would be more appropriate. Irrigated bunch types can be sown up to 150 000 plants/ha.

The population for bunch types is not critical within reasonable limits, for plants are able to compensate to a considerable extent if there is sufficient soil moisture. When runner cultivars are grown the optimum plant population must be accurately determined, since these types may use the increased space caused by low numbers to produce excess herbage. Thirty plants per square metre is a US Department of Agriculture (USDA) recommendation.

Plant population must be related to available soil moisture or expected rainfall, and populations in excess of water available to support them produce drought conditions in the planted area. The author commonly found this situation in Africa, with plant numbers too high for even the most optimistic rainfall forecasts. Although pod yields may have appeared satisfactory, shelling percentage, seed quality and oil content was poor.

WEED CONTROL

Weed control should be a continuous and integrated programme, and not treated on a seasonal or annual basis, as this can allow the build-up of persistent perennial weeds and grasses very difficult to remove from groundnut crops. Some weed species also produce growth-suppressing root extracts or may be parasitic, for example *Alectra vogelii* in West Africa. Weed-free crops also materially reduce harvester operating costs, and improve seed quality.

Groundnut seedlings are very susceptible to weed competition, and efficient early weed control is critical. Where pre-emergence herbicides have not been used, mechanical weeding is possible some 7 days after planting. A light harrow, finger weeder or rotary hoe is effective used across the rows, and this operation also assists emergence by breaking any soil crust. Specialised, inexpensive animal drawn equipment has been developed for smallholders in many countries (Jadhav & Turbatmath, 1991).

Spike-toothed harrows can be used along the rows once plants are well rooted, and this operation is least damaging if carried out during the warmer time of day when plants have wilted. Fields so harrowed look extremely attractive, with the harrowed rows alternately light and dark green. Subsequent weedings can be combined with ridging, and two operations are usually sufficient. The first is done 14–20 days after planting, and the second, dependent on weed growth, between 45–60 days after planting. Additional cultivation after main flowering can cause carpophore and root damage, and assists development of *Sclerotium* root rot.

On more sandy soils where the crop is sown on the flat, shallow ridges can be constructed by setting the rotary or disc weeders to throw soil away from plants. This will also assist in conserving rainfall and reduce wind erosion. To maintain an even surface for harvesters, the small ridges can be split during alternate weedings. Whatever method is used, care must be taken to ensure that soil is not thrown over young seedlings, or their roots damaged. Tractor tyres are another cause of crop damage and aerial infrared photographs clearly show the extent of such damage and subsequent disease infection.

Three hand-weedings produced highest yield in many areas of Africa (Fig. 3.11), but in India weeding at 21 and 42 days after planting gave acceptable control and higher yields than increasing the interval. In the Sudan, groundnut could tolerate a maximum of 10% ground cover by weeds before yields were reduced; cultivators' plots normally had over 50% cover and yield was reduced by up to 70%. Yields increased by 50–70% if main grass weeds were eliminated, and by 90% if weed cover was reduced to about 10% during the critical early to mid-season period of weed competition. The local practice of an early weeding followed by a second at the beginning of flowering produced the highest yield per unit of labour (Osman & Elamin, 1996). In Tanzania, yield of irrigated bunch-type was increased by 50% though timely weeding, without the use of fertilisers (Weiss, 1961).

Groundnut is relatively tolerant of herbicides, probably due to its quick, vigorous emergence and rapid growth of the tap-root, which reduce the time sensitive growing points are in the

Fig. 3.11 Weeding smallholder groundnut, Tanzania.

danger zone. The following compounds have been used pre-emergence either singly or in combinations: alachlor, benfluralin, chloramben, chlorimuron, diuron, ethalfluralin, fluorodifen, fluchloralin, glyphosate, imazapyr, linuron, metsulfuron, metolachlor, nitrofen, oxyfluorfen, oxadiazon, pendimethalin, profluralin, prometryn, disul (sesone), trifluralin and vernolate. Certain herbicides, for instance trifluralin, benfluralin and vernolate, require to be soil incorporated before planting for maximum effect. The effect of certain herbicides on the yield and kernel weight of rain-grown groundnut (rainfall on crop 800 mm) is shown in Table 3.9 (Hegde *et al.*, 1977). For post-emergence weed control, trials are essential to determine local and cultivar reactions, and should be applied as directed band-sprays between rows.

Combinations can also be more effective on weed complexes, and annual variation of the herbicide can increase overall weed control. Some compounds have different effects on the groundnut plant when used in wet or dry conditions, and cultivar differences exist in reaction to a particular chemical (Richburg *et al.*, 1996). Herbicides may also increase resistance to disease as in India where dinitramine and trifluralin reduced the incidence of tikka disease, *Cercospora arachidicola.* Herbicides can also affect seed and seed oil composition, alachlor for example may depress protein content, while other compounds leave residues in seeds or oil (Hashim *et al.*, 1993). Some may also affect nodulation, as noted. Where labour is available it can be profitably used to remove persistent grasses or semiresistant weeds; in Taiwan spot weeding by hand after herbicide application increased yield by a further 25%.

IRRIGATION

Modern cultivars with their higher yield potential make irrigated production increasingly attractive, as noted at the beginning of the section on cultivation, although the majority of groundnut crops are rain-grown. Erect types are generally the most suitable for irrigation, and most efficient in terms of seed produced/unit of water used. Runner types with their greater vegetative growth become more susceptible to disease, and increase harvesting cost. An important aspect of irrigation is that soil can retain sufficient moisture at harvest to allow easy lifting and retention of pods on roots, whereas a hard, dry soil causes pods to be pulled from roots.

Although the rate of root growth is relatively fast compared with many other crops, it ceases once dry soil is reached, and if this horizon should be at only moderate depth the plant's

Table 3.9 Effect of pre-emergence herbicides on rain-grown groundnut – India.

Herbicide	Rate[†]	Ranking for yield	Pod yield (kg/ha)	100 kernel wt (g)
Alachlor	2	6	1992	44.5
Benfluralin*	1	2	2175	47.1
Chloramben	2	8	1838	45.6
Dinitramine*	1	4	2054	45.0
Fluchloralin	1	3	2092	46.2
Nitrofen	2	5	2050	45.8
Profluralin	2	1	2225	50.4
Trifluralin*	1	9	1700	49.5
Unweeded control	—	10	467	40.8
Weed-free control	—	7	1854	46.1

* soil incorporated prior to sowing groundnut;
[†] amount in kg/ha active ingredient.

major root development is limited. Unless the soil reservoir is quickly replenished the plant's ability to withstand moisture stress will be severely restricted. The effect of water stress on root growth may vary with the horizon and watering interval (Meisner & Karnok, 1992)

Plants can extract water from 100–200 cm under irrigation, to around 60 cm at flowering and after this period are not greatly affected by a soil moisture deficit in the 0–25 cm zone. Thus a substantial pre-planting irrigation, or rainfall which wets the soil to depth, is necessary for high yields. Small applications at frequent intervals are not a suitable alternative, since this encourages root, stem and pod rots, and inhibits nodulation.

Groundnut fruits mature underground and the soil surface must allow penetration of the pegs, therefore a hard, dry soil is obviously detrimental. Secondly, turgescence of elongating cells in the carpophore provides the necessary force to ensure soil penetration, and since moisture must come from roots, a lack directly affects elongation and penetration ability. Developing fruits exposed above ground invariably wither.

Bunch and runner cultivars can have different water consumption patterns when grown in the same region. In arid areas bunch types have higher water consumption than runner type in the first 8 weeks of growth, but the consumption rate is then reversed. Early maturing cultivars also have higher initial consumption rates than late maturing cultivars.

Highest yields usually result from an irrigation programme that maintains soil moisture depletion in the 0–50 cm level below 25% during the vegetative period. Maximum water requirement is at peak flowering, 6–8 weeks after sowing. Where water is limited, or ceases to be readily available, a final irrigation at the beginning of pod filling is the least damaging. Results from Brazil showed that an irrigation at this time gave a similar yield, 100-seed weight, germination percentage and seed oil content to that obtained by further irrigations (Galbiatti *et al.*, 1995). The effects of various watering intervals and a growth regulator are shown in Table 3.10 (Gorbet & Rhoads, 1975).

Sprinkler irrigation of groundnut is widely practised where the system already exists for other crops, but it is seldom profitable in less-developed regions. Bed or furrow irrigation is very suitable for groundnuts, since it renders unnecessary the extensive, and expensive, earth moving required to produce level lands. Additionally, it avoids the considerable topsoil dis-

Table 3.10 Effect of irrigation and growth regulator on yield – USA.

Treatment	Harvested pod yield (kg/ha)	Total pod* yield (kg/ha)	Value of crops per tonne (US$)
1. Control No irrigation	4430	4850	307
2. Irrigated at soil moisture tension of 40 cbar†	4380	5830	316
3. Irrigated at soil moisture tension of 60 cbar†	4520	6430	323
4. Irrigated when plants wilted at midday	4540	5700	317
5. No irrigation. Split application of Kylar‡	4400	4850	295
6. Treat. 2 + Kylar as 5	5200	6810	317

* Including pods left in ground at harvest and collected later;
† at 20 cm depth;
‡ 1.12 and 0.56 kg/ha Kylar (85°70 succinic acid 2,2-dimethyl hydrazide).

placement which often causes nutritional problems. Furrows that follow the contour cause least soil disturbance, require the minimum surface levelling and resist erosion, all cogent reasons for their use in underdeveloped regions, and equally valid elsewhere (Fig. 3.12).

Construction of the ridge, either partially or wholly, can precede planting, and be increased in size during weeding operations. Tie-ridging at intervals to create basins to impound rain or sprinkler water increases efficiency of water use and reduces run-off in dry regions, as in Israel (Ben-Hur *et al.*, 1995).

Seed can be dry planted then irrigated, or rows irrigated then sown with special planters that place the seeds into ridge tops. Watering intervals must be established locally, together with reasonable estimates of crop water use, since these are the main parameters determining the area that can be watered, or cropped. As a guide, the following regime has been used by the author where no data was available: a pre-planting irrigation to ensure the soil is wet to at least 90 cm, part of this is usually supplied by rainfall; moderate irrigations at 14 and 28 days; with a substantial watering at 52 days after planting (or earlier if peak flowering is reached). The amounts at each irrigation are modified by rainfall falling since the previous application. A further small irrigation is given 14 days later if no rain falls.

Fig. 3.12 Bunch-type groundnut on contour ridges, Tanzania.

Plant population is important to obtain optimum water use in terms of seed yield per unit of water consumed, and rows should be narrow rather than widely spaced. By decreasing row-width from a standard 90 cm to 60 cm and irrigating alternate rows the author significantly increased yields in Tanzania with the same amount of water; similar results have been obtained elsewhere Table 3.11 (Oklahoma State University data).

Fully irrigated groundnuts will probably require the equivalent of 1000–1200 mm of rain to produce high seed yields. Supplementary irrigation to bring rainfall up to the equivalent of 1000 mm is generally profitable in established groundnut-growing areas, particularly when a dry period coincides with peak flowering. The amount of water required must be ascertained locally, and evaporation pans and tensiometers for soil moisture testing are essential. Environmental conditions are infinitely variable, and there is no substitute for accurate local data to ensure most profitable water use.

The quality of irrigation water is as important as application rate since groundnuts have low salt tolerance, well below that of cotton and safflower, and although selection has produced cultivars tolerant of moderate salinity levels, yield is generally lower. Salinity affects groundnut yield by reducing size and number of pods, size and number of kernels, in addition to a general reduction in emergence, plant growth and vigour.

Another aspect of salinity, which is not always obvious initially, is that caused by a high level of dissolved salts in subsoil water. Irrigation may bring these to the surface, where they accumulate and reach toxic levels, as has occured on irrigated groundnut crops in Asia and Australia. The addition of gypsum can ameliorate a slightly saline soil sufficiently to allow groundnuts to be grown, or it can be added to irrigation water. A salt concentration in the subsoil may present no immediate threat to other crops, but the rapid

Table 3.11 Effect of plant population on irrigated groundnuts – USA.

System	Row spacing	Plant/30 cm of row	Yield dried nuts (kg/ha)		
			Year 1	Year 2	Mean yield
Irrigated	40	2	2000	1360	1680
	40	8	1450	1270	1360
	100	2	1250	900	1075
	100	8	1400	816	1108
Non-irrigated	40	2	650	900	775
	40	8	700	1000	850
	100	2	450	600	525
	100	8	550	700	625

root growth of groundnut plants quickly brings them into contact with subsurface salts, and is then reflected in a general reduction in plant growth.

Mulching to conserve soil moisture is effective but seldom used on commercial groundnut crops for economic reasons. It can be profitable for smallholders, and a variety of local materials are used; vinyl sheet was successful in Korea (Choi *et al.*, 1997; Pae *et al.*, 1998).

INTERCROPPING AND ROTATIONS

Smallholders frequently intercrop or strip-crop groundnut with other local crops, as in China (Wan *et al.*, 1997); Indonesia (Harsong, 1997); and India (Khistaria *et al.*, 1998). Both systems usually produce lower seed yield than groundnut alone, although the combined crop may locally be more profitable or produce more food. Undercropping groundnut in tree plantations, alley cropping, is generally unprofitable although it can produce a small crop for domestic use. The adverse effect of seven tree species was demonstrated in India (Nadagouda *et al.*, 1996).

Crops following groundnuts in a rotation generally benefit from residual nitrogen, and usually have fewer weeds as a result of the digging operation, while groundnuts benefit from fertilisers applied to preceding crops. There can be other soil changes as occured under continuous groundnut cropping in Korea, where pH, exchangeable Ca and the Ca : Mg ratio increased, but the number of actinomycetes decreased (Kang *et al.*, 1992).

A major advantage of a rotation is the interval between groundnut crops, which substantially reduces pest and disease damage (Brenneman *et al.*, 1995). While successive crops can be grown for several years on the same lands, crop protection generally becomes increasingly expensive as the period lengthens, and is accompanied by a deterioration in seed quality. The cost and damage can be partially alleviated by sowing different cultivars in each season, whose disease resistance varies.

HARVESTING

Harvesting is usually fully mechanised on large-scale groundnut crops, but elsewhere can vary from wholly manual to semi-mechanical. Harvesting includes digging, lifting, wind-rowing, stooking and threshing. For fully mechanised crops about 300 ha of a high yielding cultivar is the minimum necessary, including a drying plant.

Bunch-types are ready for harvesting 110–130 days after planting, runner types after 130–150 days, and when the majority of pods are mature any delay in harvesting results in substantial loss. However, if it is necessary to harvest immature pods for breeding purposes, pods lifted 30–40 days after pegging had an acceptable germination and seedling survival rate in Korea (Kim *et al.*, 1997). Maturity can be easily established in the field by uprooting and inspecting plants. Pods are mature when the kernels are fully developed, the testa assumes the cultivar colour, and inside of pods darken to brown. In some runner types pods become brittle and crack open when pressed. Defoliation is no accurate guide to pod maturity since some bunch types rapidly lose a proportion of their leaves as pods mature while in others, and many runner types, foliage may remain green, and regular inspection is necessary to decide when to begin lifting.

The major reason for harvesting loss is digging outside the relatively short optimum period; too early is as damaging as too late, and the loss can be exacerbated if the crop is also sown outside the optimum time of planting (Weiss, 1961) (Table 3.12). If the soil becomes hard, as occurs very quickly in the tropics when rains cease, pegs break on lifting, leaving pods *in situ*. Mature pods are also attacked by termites and other soil pests. Should the soil remain moist when plants are mature, seed of bunch cultivars may germinate.

Groundnut production on a commercial scale depends primarily on the soil type since harvesting must be mechanised. Mechanical digging and separation of pods from plants, although possible on a range of soil types, is in fact limited to those conditions where mature plants are not held too firmly and pods separate easily from the soil. Harvesters that dig, lift and separate pods from plants and soil are available in a range of types and models, with some specially developed for use in less developed areas or smallholder co-operatives (Bindir & Kilgour, 1995). A local field test should precede any purchase, as harvesters differ and none is ideal!

Table 3.12 Effect of harvesting date on yield – Tanzania.

Bunch type		Runner type	
Days to kernels harvest	Yield kernels (kg/ha)	Days to harvest	Yield (kg/ha)
99	107	130	1310
104	1060	140	1495
109	2130	150	875
114	1270		
119	1001		

Lifting is the most easily mechanised operation, and there is a range of equipment available, animal or mechanically powered (Fig. 3.13). It is essential that the blade or share be set deep enough to cut below pods, but not so deep as to unnecessarily increase draught. Accurate setting of depth-control wheels or hydraulic controls will materially assist in maintaining the correct working position, which will usually vary several times during the day. Sweeps or fingers may be necessary on the digging blades to ensure plants are left well to one side of the opened furrow and not covered with soil.

Slow speeds and more implements are preferable to higher speeds with fewer, especially when seed is for human consumption. The gain in yield, quality and therefore price received offsets the additional digging cost. Harvesting techniques can also affect milling quality of nuts. Up to 10 ha/day can usually be lifted in the right soil and weather conditions. Where it is necessary to windrow several rows of plants this must be completed as soon after lifting as practicable, or pod loss can occur. Raking early in the day when plants are moist reduces this danger.

Lifted plants can be stooked by hand to dry, often on some kind of simple frame or pole, to be threshed by a mobile machine or transported to a stationary thresher. Stooking groundnuts normally produces excellent quality hay. When plants must be left in the field to dry after lifting, some method of keeping pods off the ground is necessary. This will reduce losses from rotting, termites, etc. and allow air to circulate more freely. The simplest method is to invert the plants in lines, pods upward and resting on the foliage.

Machines that dig, shake and produce inverted windrows are available, and consist of digging blades which span two rows, loosen the soil and cut tap-roots, a pick-up and shaker elevator to lift and clean plants of soil, and a mechanism to invert the plants and place them in windrows (Fig. 3.14). Some cultivars produce substantial herbage which remains green when pods are mature, and to reduce the volume passing through lifters and threshers, foliage must be

Fig. 3.13 Author adjusting digger blades, Tanzania 1950.

trimmed before digging. The cut herbage can be elevated into trailers and used for stockfeed. To prevent blockages when digging runner types, cutting discs may be used between rows to sever runners and enable plants to be lifted cleanly.

The windrows can be threshed when pod moisture content has fallen to 20–25%. In hot climates leaving plants in windrows for more than a few days will reduce seed quality, and can increase threshing damage. Conversely freshly dug and windrowed pods that become frosted will have kernels with impaired viability and reduced seedling vigour.

Pick-up threshing of groundnuts presents few problems provided the maker's instructions on operating speeds are followed and drum settings altered to suit varying field conditions; commonly 12–15 ha, or 20–25 t pods/day is achieved. Most machines are of the bulk tank type, and transport of pods to store should be so arranged that machines are not kept idle in the field because of full tanks. Where plants are also required for fodder, they should be left in the field for the minimum period required to dry pods. In arid regions, stems and leaves rapidly desiccate, become brittle and are shattered during threshing. Bunch cultivars can yield half as much hay as pods by weight.

Freshly lifted pods have a moisture content of 30–40% which must be reduced to 20–25% for threshing, and it is necessary to maintain a balance between digging, windrowing or stooking, and threshing. Whatever system of harvesting is employed, strict supervision of labour or accurate adjustment of machinery is essential. When using hand labour the author estimated on many occasions that 10 per cent of the pods were left in the field after harvesting. This was intentional in some instances so that family members could glean pods remaining, but was as frequently due to indolence. In the USA, harvesting losses were officially estimated as 3% when

Fig. 3.14 Combined digger, shaker and windrower.

digging, and 5% when lifting and combining due to faulty drum, fan and shaker settings.

Seed crops must be handled with much greater care than normal commercial operations allow, and when groundnut is a major cash crop in an area, special seed crops should be grown. The extent of mechanical damage and its effect on germination and seedling vigour are not always appreciated, since damage to kernels may not necessarily be visually obvious. In India hand-harvested seed normally has double the yield of seed mechanically harvested. Many countries now have central crop marketing boards which use the production of pure and improved seed as a method of increasing yields or maintaining standards.

Seed yield varies as widely as production methods and the following are from commercial crops in regions where groundnut is a major cash crop. All figures are tonnes/ha, unshelled, from rain-grown crops with average yield outside and high inside brackets; Australia 1.0 (2.5), West Africa 2.0 (3.0), Mexico 2.0 (3.0), India 2.0 (3.0), Sudan 1.5 (2.5), former USSR 1.0 (3.0), USA 2.5 (4.0). According to the USDA, 1 t of groundnuts as harvested by a domestic commercial grower consists of 55–60% edible nuts, 15–17% crushing nuts, the remainder being hulls.

In the majority of groundnut-producing countries pods are usually left standing in full sacks, tops open, until dry enough to be shelled, stacked or transported. In developed regions, on-farm storage is reduced to the minimum with pods delivered to local processors when moisture content reaches designated levels. In the USA, recommended drying conditions are: air temperature range 35–38°C or 10°C above ambient, an air flow of 8 cubic m^3/min, and a pod depth of 1.25–1.5 m at 30% moisture. In South Africa the recommended maximum is 35°C with 40% humidity. Cultivars react differently to drying temperatures, and these differences must be determined to maintain quality or viability.

Moisture passes quickly from seed to pods during drying, and since this process continues after drying has ceased, overdrying of pods must be avoided, since very dry seed is badly damaged during decortication. Overheated seed is extremely brittle, with detrimental colour and oil changes reducing its value as snack food. Slow drying is preferable to fast in terms of final quality, and a drying rate of 0.5%/hour reasonable. Portable trailer bins, radial drying bins and vertical flow bins are all suitable for drying pods. Continuous flow dryers are normally only suitable for very large producers or co-operatives, since they require constant, skilled attention.

STORAGE

Pods store easily in bags or bulk, but should not be left in direct sunlight as seed becomes very dry with consequent damage as noted. Tobacco is commonly grown in the same areas as groundnuts, and tobacco curing barns can easily be adapted to dry and store pods. Vermin are probably the greatest cause of loss to smallholders generally, especially in India (Ishwar & Ghosh, 1992), but sound, unbroken pods are less susceptible to insect attacks than nuts. Pods should have a moisture content of 8–10% for short-term storage, and below this level many nuts are split during decortication.

Decortication is a major operation and commercial growers are advised to delegate the operation to their co-operative or specialist processors. In countries where most groundnuts are grown by smallholders, a marketing authority is usually responsible for this operation. Decorticated seed (nuts) store well if sound and dry, but are very susceptible to insect and vermin damage. Nuts are easily transported in bag or bulk, but where oil content is important, they should preferably be processed locally. Nuts for edible or culinary use should be kept separate, and since they are usually more valuable, greater care is necessary in handling and storage.

The life cycle of many storage pests can be less than 4 weeks, numbers increase with great rapidity, and the importance of cleanliness and proper hygiene in stores and godowns cannot be overemphasised. Permanent stores and bins were often made of metal or wood, but are now increasingly replaced by permanent or semi-permanent plastic structures. These can vary from what is basically a large plastic bag to well-constructed buildings holding thousands of tonnes. Since they can often be more completely sealed than other materials, these buildings are easily fumigated thus reducing storage loss.

For storage of shelled seed and breeding stocks, an air-conditioned store is almost essential if seed is to retain its viability for more than a season. A temperature of 2–4°C with a relative humidity of 6.5% allows storage for at least 2 years with little loss of viability. For longer term storage of genetically valuable material, seed should have a moisture content of 5.5% and be kept at a constant temperature of –18°C or less. Low oxygen or nitrogen atmospheres in cool stores are adequate for shorter periods and less valuable seeds. Small amounts of shelled dry seed treated with an insecticide/fungicide, can generally be stored in laminated polyvinyl bags for one year without significant loss of viability. However, cultivar differences exist between storage period, viability and seedling vigour, and these must be determined.

Germination of groundnuts in India was highest when pods were dried by heaping and turning for 5–6 days, and gradually decreased when dried in the shade followed by drying in the sun (Nautiyal *et al.*, 1993). The causes of seed deterioration following natural and accelerated aging have been studied in detail, and the changes analysed (Zhu *et al.*, 1994; Perez & Arguello, 1995). It is considered that percentage germination does not indicate the degree or type of change.

A specific problem of groundnut storage is the production of aflatoxin in groundnut meal. Although this is primarily a problem for seed crushers, initial infection of pods can occur in the field or farm storage. The toxin is produced by a fungus, *Aspergillus flavus*, which is of worldwide occurrence and commonly known as yellow seed rot. Normally the mould is killed by hot, dry weather following lifting, but periods of rain and delays in harvesting or drying pods produce conditions suitable for its continued growth. Poor storage which allows pods or nuts to become damp or contaminated by vermin faeces, also increases the possibility of aflatoxin formation. A relationship between calcium content and a reduction in fungal infection occured in Argentina (Fernandez *et al.*, 1997).

Aflatoxin cannot be detected in the field, but any crop which had a high incidence of fungal infection, especially if associated with moisture stress, is suspect. Infection can be partially controlled by good cultural methods, quick harvesting and drying, and good storage. Development of cultivars with resistance to fungal disease is probably the best method of reducing the danger. Aflatoxin is further discussed in Chapter 9.

PESTS

Chemical methods of pest control will probably remain common, but breeding of pest-resistant cultivars or development of biological control is preferable. It is also possible to reduce the severity of insect infestation or damage by simple procedures; sowing several rows of tall plants around fields can provide a virtual barrier to some insects, and the use of biological preparations such as *Bacillus thuringiensis* sprayed on lepidoptera larvae can be most effective. Where chemical spraying is necessary, the amount used and area covered can be substantially reduced, for example, by spraying a prophylactic band around the edges of the cropped areas. This is particularly effective against those insects which progressively invade a crop from field margins; or sowing in advance around field margins rows of a plant that is more attractive to locally important pests, which are then sprayed at an appropriate time. This may not stop the main crop from becoming infested, but can allow it to reach a stage when it is less susceptible to damage. Intercropping, adjustment of row or in-row plant spacing, and similar procedures can also reduce the damage caused by a specific pest in a particular locality.

Insect pests of groundnuts, apart from locusts and similar plagues which can decimate crops in outbreak years, are generally considered of local importance, although the total loss in a particular region or country may be high. An extensive survey of central African countries established the high level of loss due to insect damage, and that chemical control would be profitable given the substantial increase in yield obtained (Wightman & Wightman, 1994). This supports the author's opinion gained from examining groundnut crops in many countries, that damage caused by insect pests on a world scale is greatly underestimated, especially in smallholder crops. An indication of the extent of losses caused by insects and disease in oilseed crops is contained in Cramer (1967). He calculated that in 1965, loss of yield in groundnut crops worldwide totalled 11.4 million tonnes (mt) valued at US$ 15 million, including 5 Mt due to insects, nematodes etc., 3 Mt from diseases, and 3.5 Mt for all other reasons.

The USDA estimates losses from insect pests and nematodes to be 8–12% annually, including direct field loss and commercial downgrading of affected seed. The total loss of yield from all causes was estimated to be at least 25% in an average year, but could reach 30% in certain years. In South Africa the loss of yield from unprotected crops can reach 40–50%. Chemical pest control remains the most common method of limiting the damage, and although generally effective can have side-effects; one causing increasing public concern is residues in seeds and oil (Hashim *et al.*, 1993).

The extent of loss due to insect and allied pests in many countries as noted, is not well documented, partly because of the difficulty in assessment. Loss or damage to aerial inflorescences is obvious, in contrast groundnuts can present a healthy, vigorous appearance but produce few pods. Pests that attack the vegetative parts or maturing fruits cause the major loss, since a large number of flowers can be destroyed without a drastic fall in yield. This is in direct contrast to castor, where attack on inflorescences is the most damaging and most obvious.

It is as carriers of virus and other diseases that insects may become limiting factors in groundnut production. Probably the most well documented example is transmission of rosette virus complex by the groundnut aphis, *Aphis craccivora*, to crops planted on the Overseas Food Corporation's huge projects in Tanganyika (Tanzania). So damaging and widespread was the disease that it effectively precluded the growing of bunch-type groundnut on a commercial scale, the *raison d'etre* for the project. Viruses are discussed more fully in the section on disease.

Seedlings may be attacked just before and at emergence by *Agrotis* spp. cutworms, especially the widespread *A. ipsilon* (see Fig. 2.17a). A survey of groundnut fields in central Africa found some 40 species of Scarabaeid larvae were present (Wightman & Wightman, 1994). Scarabaeids are most damaging in areas of intensive agriculture on sandy or loamy soils with a rainfall of at least 1000 mm/year. Underground damage also occurs pre-harvest, as puncturing and chipping of ripening pods by *Agrotis segetum*, elaterid adults and larvae occured in South Africa, although pod scarification by Tenebrionidae to mature pods was the most frequent type of injury observed. Insect numbers 110 days after planting had the most pronounced effect on damage to mature pods, with the tenebrionids being the most abundant (Eeden *et al.*, 1994). Crickets, especially *Brachytrupes magacephalus* in Africa (see Fig. 7.17) and grasshoppers can also cause extensive local damage, but are easily controlled by baiting in smallholder crops. Other ground living larvae cause damage to developing pods (Fig. 3.15).

Empoasca spp., leafhoppers/jassids, are widespread and polyphagous, and can be found in most groundnut crops. Both adults and larvae feed on the plant, but the amount of damage caused, or degree of infestation that requires control, has not been defined. It is also possible that some species can transmit the rosette virus. *Empoasca fabae* occurs in the USA, *E. dolichi* in Africa and *E. flavescens* in Asia. A leafhopper, *Hilda patruelis*, feeds on roots and underground portions of pegs in East Africa, and can cause serious damage in particular regions in a specific year. It may in fact be more damaging than is presently realised.

Fig. 3.15 Larvae of white fringed weevil, *Graphognathus* spp., attacking pods.

Thrips are of widespread occurrence; *Frankliniella fusca* causes severe damage to young plants in North America; *Taeniothrips sjostedii* to a lesser extent in Africa; *Mylabris amplectens* and *Scirtothrips dorsalis* in Asia and *Spilosoma strigatrila* in Indonesia. Elsewhere thrips are not considered important, but this may be because their effect is masked by the more obvious damage caused by other insects. Various species of spider mites, *Tetranychus* spp., commonly occur on groundnuts worldwide, the most obvious symptom being foliage covered by fine webbing. Mites are seldom of economic importance, but numbers can build up rapidly following spraying with pesticides to which they are resistant. A mealy bug commonly found on groundnuts in Africa, India and parts of North and South America, but apparently causing little damage, is *Ferrisia virgata*, which is a major pest of cocoa and coffee.

Flea beetles commonly occur in fields of young plants, and damage is directly related to numbers, since a small population has no apparent lasting effect. *Systena* spp., *Paraluperodes* spp. and *Ootheca mutabilis*, the brown leaf beetle, are most often noted, the first in the Americas, the others in Africa. In Australia *Monolepta australis*, the red-shouldered leaf beetle, is most important.

The leaf miner, *Biloba subsecivella*, groundnut surul, is an important pest in India, and also occurs widely from Africa to the Far East. Also in India *Aproaerema modicella* can be damaging in specific seasons. Larvae of the weevils *Alcidas arcuastus*, *Alcidodes dentipes* and *Systates alticallis* mine stems in Africa; in India larvae of the beetle *Sphenoptera perotetti* mine the stem and bore down into the roots. The lesser cornstalk borer, *Elasmopalpus lignosellus* occurs mainly on crops grown in the vicinity of more favoured hosts in the USA. Resistance to this borer exists in wild *Arachis* spp. (Kamal, 1976). The shield bug, *Nezara viridula* (Fig. 5.12), may also cause stem damage locally in many tropical areas, particularly in Africa, but is not of major importance, and the white-fringed beetle, *Graphognathus leucoloma* causes damage to a lesser extent in the Americas and Australia (Fig. 3.15).

A blister beetle, *Coryna apicicornis* feeds on flowers and is persistent and troublesome in eastern Africa, where it can occur locally in large numbers. The scarab *Schizonycha africana* is a similar problem. In Brazil, *Slegasta basquella* can cause local damage by feeding on terminal buds, and infestations may be more severe where groundnut crops overlap seasonally, or are sown near other legumes. Elateridae larvae, wireworms, attack roots and lower stems in Africa and many other countries, but are usually of minor importance. By contrast in the USA, *Diabrotica undecimpunctata* larvae are a major source of crop loss.

The well-known bollworm/earworm, *Helicoverpa* (*Heliothis*) spp., is a common and widespread pest, and when present in large numbers can cause serious defoliation, see Fig. 4.13. Damage is greater when groundnut is grown in rotation or in proximity to more attractive host plants. A similar situation occurs with *Spodoptera* spp., the armyworm, which may sometimes attack groundnuts but often ignores the plants, preferring grass and weeds within the crop. It is a major pest in Brazil. Records of the cottonworms *S. littoralis* and *S. litura* causing major damage are rare, but it is commonly found on crops growing near cotton in Africa. Resistance to *S. litura* has been identified in wild *Arachis* spp. including *A. appresipila*, *A. batizogaea*, *A. chacoensis*, *A. glabrata*, *A. kemphmercadoi*, *A. paraguariensis*, and *A. pseudovillosa.* Larvae fed diets containing the chemical extracts of dried leaves of these species either died or had lower weight gains, and preferred not to eat their leaves. The physical characteristics of the leaves and foliar chemicals were considered responsible (Stevenson *et al.*, 1993).

Serious pests are the mung moth, *Maruca testalis*, and the red hairy caterpillar, *Amsacta albistriga*. Both are of widespread occurrence in India and neighbouring countries, where they can cause severe local damage. Since there can be several generations in a season, numbers rapidly increase unless controlled. In North America the alfalfa caterpillar, *Colias eurythme*, causes locally severe defoliation on irrigated crops, and could become more important as the irrigated area increases. Also in North America, and to a lesser extent elsewhere, the velvet bean caterpillar, *Anticarsia gemmatalis*, becomes a local and sporadic pest of groundnut, the severity of the attack often depending on the availability of alternative host plants.

Nematodes, usually the groundnut nematode *Meloidogyne arenaria, Pratylenchus* and *Scutellonema* spp., are widespread and sometimes damaging, especially when crops of groundnuts are grown in succession. A survey in India found *Tylenchorhynchus brevilineatus, T. nudus, Meloidogyne arenaria* and *Helicotylenchus dihystera* in large numbers associated with Kalahasti disease of groundnut, which blackens pods. *Pratylenchus brachyurus, Rotylenchulus reniformis* and *Hoplolaimus indicus* were also found

(Prasad & Rangappa, 1994). Nematodes can be controlled by soil fumigants, but control is usually unprofitable except in seed or special crops.

In many dry regions or when plants and pods are mature, termites, commonly *Microtermes* and *Hodotermes* spp., attack stem, pegs and roots and render harvesting difficult with many pods left in the ground. Pods are also eaten, and even if the kernels are not also damaged, they become infected by soil-borne disease. Losses can reach 30% of pods when harvesting is delayed. This may be relatively unimportant in peasant agriculture where nuts are grown mainly for food, and displaced pods are easily gleaned during hand digging, but it is a major factor influencing the efficiency of mechanised harvesting. The author commonly found in Africa that the first pods to mature had already been attacked by termites while the majority of pods were still unripe. This was particularly noticeable in dry years, when pods quickly dried out and were attacked in succession as they matured.

Storage pests include *Aphanus sordidus*; the groundnut beetle *Carydon gonagra*; the rice moth *Cocyra cephalonica*; the warehouse moth *Ephestia cautella*; the grain beetle *Oryzaephilus mercator*; the Indian meal moth *Plodia intertpunctella*; the cadella beetle *Tenebriodes mauretanicus*; and the flour beetle *Tribolium castaneum*. Most are cosmopolitan and attack a range of stored products (Fig. 9.10).

Control of storage pests using conventional chemical treatments is becoming more limited as insect pests develop resistance and consumers demand end products free of detectable chemical residues. The number of acceptable chemicals will diminish, although certain gases remain in general use, and alternative control measures sought. One is biological control using predatory or parasitic insects and mites; protozoa, bacteria and fungi; or manipulating an insect's chemical communications system, pheromones, to disrupt mating and as attractants in traps. Biological control is most effective when combined with other control measures noted at the beginning of this section.

DISEASES

The loss caused by disease in specific groundnut crops has seldom been quantified, despite the comment in an authoritative monograph that all the major diseases of groundnut in the USA in 1950 remained so in 1970 (Garren & Jackson, 1973); the situation was similar in 1990! According to the USDA, kernel loss from all diseases is approximately 20% of potential yield. Infra-red photography from satellites is of considerable value in detecting, or assessing, the severity or onset of disease infection.

The importance of correct agronomic and control methods to reduce the loss caused by disease is widely recognised, and in many important commercial groundnut producing regions special handbooks or CD-ROMs are available to growers, as in the USA and Australia. Rotations are also recommended to reduce the incidence of disease in groundnut crops, but cultivars vary in their response to infection, and there can also be considerable local variation in the degree of reduction in foliar and soilborne diseases, as occurred in the USA (Brenneman *et al.*, 1995).

Seedling rots are a major cause of gappy stands, and two of the most common are caused by *Aspergillus niger* or *A. flavus*. *Aspergillus flavus* produces yellow spores on the seed, and *A. niger* masses of black spores on both seed and germinating seedling. Damage from the last is exacerbated by soil insect attacks. Seed dressing assists in reducing the damage but a higher seed rate and sowing sound, good quality seed is often a more practical and cheaper method. Resistance exists (Zambettakis, 1975), and it can be further developed since resistance to *Aspergillus* is located in seed-coat characters (Mixon, 1979). Mature seeds are liable to infection by *A. flavus* if they become damp in the soil or store. Metabolism of the fungus can produce a series of very toxic substances, the aflatoxins (see also Chapter 9). Very damaging is the seed rot *Rhizopus* spp. which covers the seed in a white woolly coating. Since *Rhizopus* cannot attack germinated seed, planting with optimum seedbed conditions greatly reduces its severity.

Two leaf spots very similar in appearance and damage are caused by *Cercospora arachidicola* and *Mycosphaerella berkeleyi*; for all practical purposes they can be considered the same. Both cause dark-brown, zonate, circular spots, surrounded by a yellow halo, hence the popular name of frog-eye leaf spot; *tikka* in India. The small spots usually appear 6–7 weeks after emergence, can increase rapidly until they merge, when the whole leaf dries and falls. If stems become infected, long, dark-brown lesions appear. The disease is spread by infected plant material, and can persist over a normal dry season. In regions where the disease is severe, groundnuts should not be planted in successive years. High resistance to *C. arachidicola* reportedly exists in some newer cultivars. Crop surveys in the USA showed that *Cercospora* and *Sclerotinia* infections accounted for at least half of all disease-attributed losses, including a reduction in quality. In Africa and Asia, *Phoma arachidicola* is a most important leaf spot.

Rust caused by *Puccinia arachidis* is now virtually worldwide, and limits groundnut production in the Caribbean Islands and Central America. Symptoms on leaves are small brown lesions, which spread rapidly over the whole plant. Badly infected leaves fall, and in a serious attack plants are completely defoliated and often die (Fig. 3.16). Some recent cultivars are resistant, and complete resistance reportedly exists in Brazilian *Arachis* spp. Blight caused by *Botrytis cinerea* is also now worldwide; first symptoms usually occur at soil contact points along lateral branches, and under humid conditions the infection spreads rapidly, with foliage and stems becoming progressively blackened.

A very serious bacterial wilt is caused by *Pseudomonas solanacearum*, and is a major constraint to groundnut production, especially in China, Indonesia, Vietnam and Malaysia. Details of strains of *P. solanacearum*, symptoms, disease diagnosis, causal organism, isolation and identification, disease cycle and disease management have been published (Mehan & Liao, 1994). Another wilt is caused by *Cylindrocladium crotolariae*.

Two root rots damaging in more humid areas are white root rot caused by *Sclerotium rolfsii* and fusarium wilt due to *F. oxysporum*. Both have a wide range of hosts, including many crops grown in rotation with groundnuts. Both attack the lower stem and roots, causing plants to wilt quickly and die. The widespread *Sclerotinia sclerotiorum,* is of major economic importance. Similarly to *Botrytis* blight, it first infects plant parts with soil contact, and spreads rapidly through stems. Lesions are light yellow, with a distinct demarcation zone between healthy and diseased tissue.

Pod rots due to *Macrophominia phaseolina* and *Pythium* spp. cause major loss in many regions. Severe pod discoloration, blackhull, is due to *Thielaviopsis basicola,* and seriously reduces their market value. Since this pathogen is soilborne, crops destined for this market should not be grown in infected soil. Pods may also be discoloured by *Alternaria* spp. and *Penicillium* spp. if rained on after digging and before threshing, or if irrigated crops remain in wet fields after lifting. The main diseases of stored pods and nuts are due to *Aspergillus flavus* and *Rhizoctonia* spp.

Although most groundnut diseases can be partially controlled by chemicals especially systemic fungicides, their use is seldom profitable, although ultra-low-volume spraying may be in those areas where the equipment is readily available. Suitable cultural techniques and the planting of resistant varieties are probably the most practical methods of alleviating disease effects in the majority of producer countries.

The introduction of pest and disease resistance to bunch types has had limited success, and in 1973 it was stated that cultivated groundnuts appeared to be devoid of qualitative genetic resistance to the major pests and diseases, although this has since been partially resolved by genetic manipulation. Selection for disease and insect resistance may also be introduced by diploid-level hybrids whose progeny could be directly crossed with Arachis hypogaea. Some introduced resistance is also related to undesirable characteristics such as low yield or longer maturity.

Fig. 3.16 *Puccinia arachidis* (groundnut rust), India.

The runner types offer a much greater opportunity for improvement, and simple selection procedures can achieve rapid results. In East Africa for instance, selection within the local Asirya Mwitunde cultivar produced strains with a marked resistance to aphis, and thus to the virus causing rosette disease. Other strains showed considerable tolerance to *Cercospora* leaf spots. Similar results were obtained with West African and Indian runner varieties, and to wilt disease in the Philippines and Indonesia.

It is expected that some wild South American species will be able to contribute towards disease resistance in cultivated groundnuts, since resistance to *Puccinia* rust has been reported in a Peruvian cultivar, and high resistance in Brazilian sources of *Arachis glabrata* and *A. monticola.* Resistance to *Aspergillus flavus* exists and also to *Diplodia* blight. Evaluation of 34 wild genotypes of the sections Erectoides, Triseminalae, Rhizomatosae and Extranervosae in China, revealed that most of them were resistant to *Cercospora arachidicola* (*Mycosphaerella arachidis*), *C. personata* (*M. berkeleyi)* and *Puccinia arachidis* (Li *et al.*, 1994; Tang *et al.*, 1996).

A number of virus diseases of groundnuts have been identified, including bud blight disease, rosette disease, groundnut (peanut) mottle disease, groundnut (peanut) stunt disease, groundnut mosaic disease, bunchy top complex, chlorotic spot disease, groundnut (peanut) green mosaic disease, peanut stripe disease, cowpea mild mottle virus disease, peanut chlorotic streak disease and groundnut bunchy plant disease The symptoms, transmission, host range and management of some have been described. However, lack of precise identification of the viruses concerned may have resulted in the same, or similar, diseases being given different names.

Groundnut rosette disease (caused by a complex of groundnut rosette virus and its satellite RNA, and groundnut rosette assistor luteovirus) is the most damaging disease where it occurs, and in parts of Africa and Asia limits groundnut growing. It is spread by aphis or leafhoppers, and plants can be infected as soon as shoots have emerged. Symptoms are variable, since there are a number of virus strains producing slightly different symptoms, which also differ between cultivars. The main effect is rosetting or stunting, and narrowing of leaves, which are often wholly or partially yellow. If infected early in growth, plants remain stunted and die. In later infections, plants produce small, pointed, yellowish leaves, with aborted flowers and few pods. The disease is not seed-borne and control is possible by repeated pesticide sprays to kill the aphids, but this is generally unprofitable. The elimination of volunteer plants, not growing out-of-season irrigated crops, and resistant cultivars all reduce the severity of attacks. Bunch types are extremely susceptible, but there are highly resistant strains within African and Indian runner types. It may be possible to transfer resistance to groundnut rosette disease from the wild *Arachis* spp., *A. chacoense, A. cardenasii* and *A. stenosperma*, into cultivated groundnut (Moss *et al.*, 1993).

A survey of insect pests in central Africa noted that *Aphis craccivora*, the vector of groundnut rosette virus (GRV), was apparently naturally controlled in some regions. It was suggested that the low incidence of GRV may be partially due to the general early sowing (Wightman & Wightman, 1994). In West Africa the leafhopper *Cicadulina arachridis* is believed to be a local vector of rosette virus and in India jassids, *Orosium* spp., can infect groundnut with the phyllody virus so damaging to sesame.

In Africa there may be a relationship between high incidence of rosette disease and conditions basically unfavourable to groundnut production. When a resistant local runner variety was grown in areas with high disease potential, it showed few symptoms but had a very low yield. Grown elsewhere, but on more fertile soils and higher rainfall, it produced good yield, suggesting that unhealthy plants are most seriously affected.

4
Safflower

The Compositae is one of the largest families of flowering plants, containing about 800 genera and some 20000 species, mainly annuals and perennials of worldwide distribution. Many members are cultivated as ornamentals, vegetables, spice or medicinal plants. The genus *Carthamus* L. contains about 25 valid species distributed from Spain, North Africa and West Asia to India. Many are indigenous to the Mediterranean region, but only one, *Carthamus tinctorius*, is cultivated commercially. The probable origin of cultivated safflower is an area bounded by the eastern Mediterranean and the Persian Gulf (Fig. 4.1).

It is grown in small plots, strip-cropped, on field borders, or around homesteads over the greater part of tropical Asia, Africa, Russia and China, or wherever Spaniards or Asians have settled the world over (Weiss, 1971). *Carthamus tinctorius* is known to exist only under cultivation and seldom persists for any period as an escape. Safflower is more drought resistant than other oilseeds and can produce good crops in dry regions on subsoil moisture, while its salt tolerance is a valuable asset as the area affected by some degree of salinity steadily increases worldwide. Throughout the text safflower refers to *C. tinctorius* cultivars, their seed and derivatives.

The plant has long been domesticated, initially for the orange dye obtained from the florets, and has been identified as growing in Egypt in 2000 BC, probably introduced from the Euphrates region. A bundle of single safflower flowers inserted in packets of willow leaves was found with the Eighteenth Dynasty mummy of Amenophis I, about 1600 BC (Schweinfurth, 1887). An accurately dated papyrus, Revenue-papyrus of Ptolemy II, 259–258 BC, states the king had complete monopoly of production and marketing of certain vegetable oils, including safflower, sesame, castor and linseed (Keimer, 1924).

Pliny (first century AD) wrote that the oil *oleum cnicium* was used in Egypt as a milder substitute for castor oil but not for edible purposes, and was not then known to the Romans. Safflower continued to be grown in Egypt, and the harvesting and preparation of the dye was described in detail by Hasselquist (1762). While on a visit to Alexandria he noted the dye was exported to Italy, France and England, where one use was to colour cheese. Other travellers also noted the cultivation of safflower by the Egyptians in the eighteenth and nineteenth centuries. Until quite recently the local eye cosmetic *kohl* was made of soot from charred safflower plants.

The Arabs believed safflower to have alexipharmic and diaphoretic properties, and used it to promote sweating to alleviate a fever. Mesua (c. AD 1000), who lived in Baghdad, is author of a major work on contemporary Arab and Greek medicine. He called the plant *cnicus*, noted it was both wild and cultivated in the eastern Mediterranean region, and that the large white-seeded variety was the most valuable. An illustration in the translation by Marinus (AD 1562) is quite recognisably *C. tinctorius*. Another Arab author, Abu Hanifa (quoted by thirteenth-century writer Ibn al-Athir), mentions wild and cultivated safflower as available in Arabia. The great

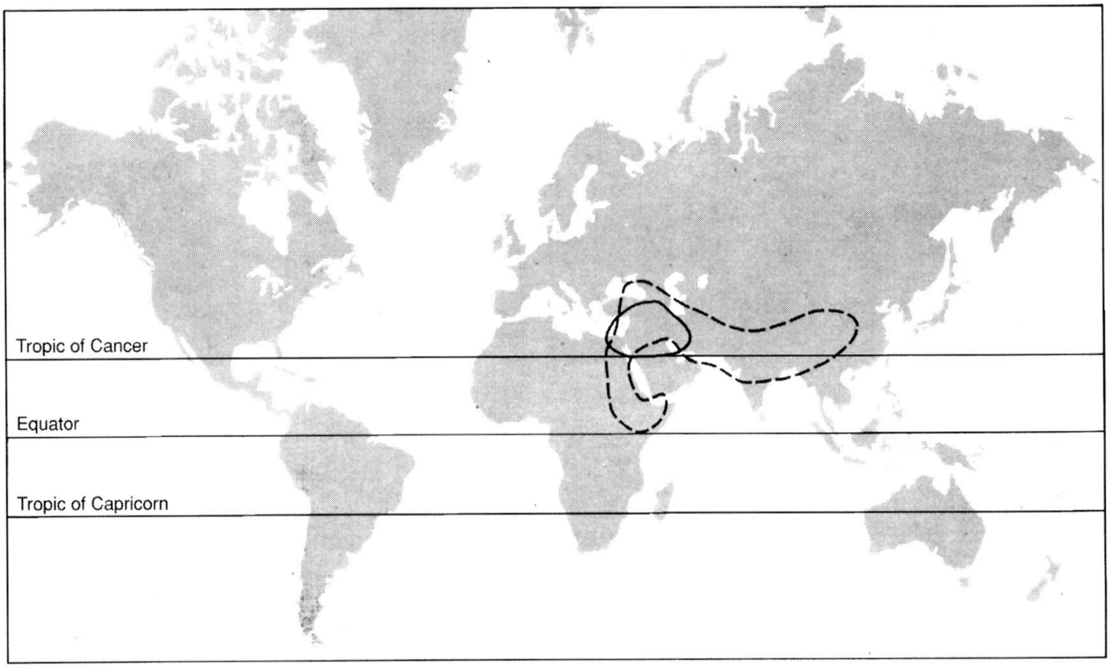

Fig. 4.1 Probable centre of origin and centre of domestication of safflower (solid line) and area of ancient cultivation (broken line).

Arab traveller Ibn Batutah described safflower during his journeys in Africa and Asia between 1325 and 1354. A Greek, Pedanius Dioscorides, served as an army doctor under Nero and assembled a wealth of material concerning traditional medicines and herbs, published in his *De Materia Medica.* In it he accurately described and illustrated about 600 medicinal plants, among them castor, safflower and sesame. The last two were used to colour and flavour various potions and unguents, and safflower was also used as a mild laxative and pot-herb.

Western expansion of the Arabs in creating the Muslim Empire of the fifth and sixth centuries probably expanded safflower cultivation along the Marghreb and into Europe via the Iberian peninsula. Introduced into Britain from Egypt in 1551 *(The Botanical Register)*, safflower was grown by monks and in private herb gardens mainly as a pot herb but also as a dye. In Gloucestershire such large quantities of flowers were used to colour sausages and cheese that it had a considerable purgative effect (Hanelt, 1961). Mixed with French chalk it formed cosmetic rouge. Retained in the European Pharmacopoeia until fairly recently it was seldom prescribed as a specific remedy.

Safflower was a source of dye from ancient times for carpet weavers of the Irano-Afghanistan area, from where it was probably introduced into southern Russia. The oil, known locally as *kusumbha*, was regarded as purgative, so that identical properties were assigned to it in Egypt, Africa and India, in a similar manner to the universal use of castor oil. However, although the oldest Sanskrit authors credit safflower as being purgative and mention a medicated oil, some early materia medica and herbals make no mention of the plant (Bower MSS: Hoernle trans. 1893–37). Watt (1966) remarks that it is curious that none of the early travellers in India mention the dye.

In Afghanistan, Pakistan and India it is used mainly to produce an edible oil, with the florets added to rice, bread and pickles to give them an attractive orange colour, and small yellow rice-balls are a common sight at the feet of stone figures in wayside shrines. A brew made from its foliage is said to prevent abortion and the barrenness dreaded by local wives. In India and Burma leaves and thinnings of less spiny varieties are used as a vegetable. The tubular florets are commonly used as an adulterant or substitute for true saffron. Safflower cultivation by smallholders for domestic use remains widespread, especially in central and south Asia, but with the introduction of cheap synthetic dyes it is now of little importance as a dye source.

Safflower was probably introduced to China around AD 200–300 initially to produce a dye, had a prominent place as a cosmetic, and to a minor extent in medicine. The oil was considered inferior to sesame oil for cooking, but included in many recipes where a less characteristic flavour was required. Its use as a substitute for saffron in colouring rice dishes, soup and sauces was common. Safflower was later grown extensively for its dye in many areas of China, particularly in Yangtze and Szechwan provinces. Its importance as dye source was greatly diminished at the beginning of this century, but costly brocades were still coloured with safflower red.

Safflower was introduced to Japan from China probably about the third century AD, but was seldom cultivated and the cosmetic *beni* was made mainly from imported Indian material. The oil was little used in cooking until the twentieth century, but is now a major import. A blend of safflower, sesame and other oils is used to cook a national dish, *tempura*, and individual restaurants owe their popularity to the flavour of a particular oil, blended to a secret recipe by the chef.

Safflower was introduced to Central and South America by the Spaniards, Portuguese or their allies, but it is nowhere common. One of the first references to safflower in the USA is a research report of the University of California at the beginning of the twentieth century, and the university is a leading centre of research on safflower. Safflower became a major oilseed crop in the USA after World War II, when private breeders produced high yielding, high oleic cultivars with greater resistance to disease. Safflower later suffered severely from competition with oilseed rape, *Brassica* spp. The USA normally exports about 80% of oil produced, about half each to Asia and the EU, as the domestic market has not expanded to meet supply.

Since 1975, world production of safflower has fallen as the crop suffered from increased plantings of sunflower and soya. While the oil has valuable specific characteristics, as an edible oil it has generally been replaced by cheaper alternatives, and this trend is likely to continue. In Australia for example, the 30 000 ha under safflower in the 1970s and 1980s has been virtually replaced by sunflower. However individual growers in Australia and elsewhere have a found niche markets producing organically grown safflower oil which sells at a premium locally and for export.

BOTANY

Classification of the genus *Carthamus* L. remains incomplete, but comprehensive reviews have been prepared by Ashri & Knowles (1960), Hanelt (1961), Weiss (1971) and Estilai & Knowles (1978), which describe many species in detail. Cultivated safflower, *C. tinctorius* L., has 12 pairs of chromosomes, $2n = 24$, together with *C. oxyacantha* M.B., *C. palaestinus* Eig. and *C. flavescens* Spreng. The first three can be readily crossed and yield fertile hybrids. Isoenzyme patterns have confirmed the close genetic relationships between *C. tinctorius* and *C. oxyacantha.* Other species have $2n = 20, 22, 44$ and 64, and some successfully crossed with safflower, which has also been used to produce interspecific and intergeneric hybrids. Tetraploids of *C. carthamus* have been induced using various chemical treatments and irradiation.

Vegetative hybridisation with grafts of safflower on sunflower, sunflower on safflower, and sunflower on *Xanthium strumarium*, were successful, and viable seeds obtained from the first and last. Grafted safflower cultivars in the USA showed no influence of stock on scion. Heterosis has been widely reported in safflower for yield, oil percentage, plant height and other characteristics, and the discovery of a gene which caused partial male sterility was important in the study of heterosis and related processes (Carapetian,

1994). A mass emasculation technique has been developed for use in breeding programmes (Singh V. *et al.*, 1995); in vitro techniques also enable large scale production of selected strains, and methods have been reviewed (Palmer & Keller, 1994).

A variety of techniques under the general heading of genetic engineering, including gene mapping, recombinant DNA etc., are now employed, and the information gained is reported in the specialist literature. When the results can be applied to commercial production especially in respect of increased oil yield, desirable oil components and disease resistance, they will enhance safflower's position as an important edible oil source. Some are noted in the text.

In considering the origin of cultivated safflower, both *C. oxyacantha* and *C. flavescens* are adapted to a habitat associated with man, and in their homelands are commonly found in crops of wheat or barley, growing on field margins and around habitation. This would suggest that both originated when the local peoples' nomadic lifestyle was replaced by a more settled existence with permanent agriculture. *Carthamus palaestinus*, however, is a wild desert plant, which suggests it is an ancient species, perhaps ancestral to the others.

The English carthamus is the latinised version of the Arab word *quartum* or *gurtum*, alluding to the colour of the dye obtained, and *usfar* is probably the origin of the English name safflower, via various written forms of usfar, affore, asfiore, saffiore, to safflower. Safflower is also known as false saffron, bastard saffron, thistle saffron and dyer's saffron. In India, Sanskrit authors described safflower as *kusumbha*, hence the common modern name kusum, although it is also known as kardiseed, kazhirak, ghurtum and suff.

Safflower was a comparative latecomer to mechanised arable cropping. There has been considerable research into its genetics and breeding, which has also included work on many related species considered a valuable source of genetic material. A gene bank is maintained by the USDA at Washington State University, a world germplasm collection at Solapur, India, with other collections in Israel and Russia. In this context it is worth noting the comment by the late Professor Knowles of California University, that in view of export from the USA of their highly productive varieties to all parts of the world, it becomes increasingly important to collect, protect and maintain pure strains of indigenous varieties as valuable genetic reservoirs.

Safflower is cultivated for a variety of uses, and growers therefore require a range of plant types. A smallholder may desire a brightly coloured floret with high dye content or leafy strains as green fodder; a producer using highly mechanised methods requires evenness of growth and maturity for ease of combine harvesting; resistance to root rot and other diseases under irrigation.

The ideal safflower plant for mechanised production has been described as of medium size, 60–80 cm high, maturing in 130–150 days, with appressed to semi-appressed branches; 6–8 primary branches bearing 12–14 large heads with 30–40 seed/head, containing bold seeds weighing about 5 g/100, a minimum seed oil content of 50%, and high protein content. Additional assets would include rapid seedling growth, hardiness, low seed hull content, and resistance to major pests and diseases. Selection for a particular characteristic is possible, since the range of variation in *Carthamus* spp. is high.

Two very important improvements were the introduction of reduced hull types in the USA, and the Australian discovery that two different oils can be produced. Thin-hulled seed cultivars performed well in the USA but were less successful in Australia. Seed oil has risen significantly from 30% but must reach at least 50%, or seed yield must be greatly increased, if safflower is successfully to compete with other oilseeds.

Cultivars with a high oleic acid level made safflower attractive as an olive oil extender, and a reason for the crop's rapid expansion in Spain. In Australia high linoleic acid content seed is preferred, and one cultivar has 76–78% linoleic and 10–12% oleic acid. High demand for such seed is generated by customer preference for products containing unsaturated fats and oils.

Increasing seed protein and lysine levels and reducing fibre content, made safflower meal more attractive to stockfeed manufacturers, as did removal of the bitter principle, matairesinol monoglucoside.

Safflower is a highly branched, herbaceous, thistle-like annual, varying in height from 30 to 150 cm, generally with yellow flowers, whose seeds contain 35–45% oil. Development of safflower from planting to maturity is shown graphically in Fig. 4.2 (Stern & Beech, 1965). Five growth stages were defined in China, based on organ development, plant growth, dry matter accumulation, phytosynthate distribution and metabolism (Li *et al.*, 1995). A comparison of the various growth periods from a number of countries is shown in Table 4.1, compiled from published data.

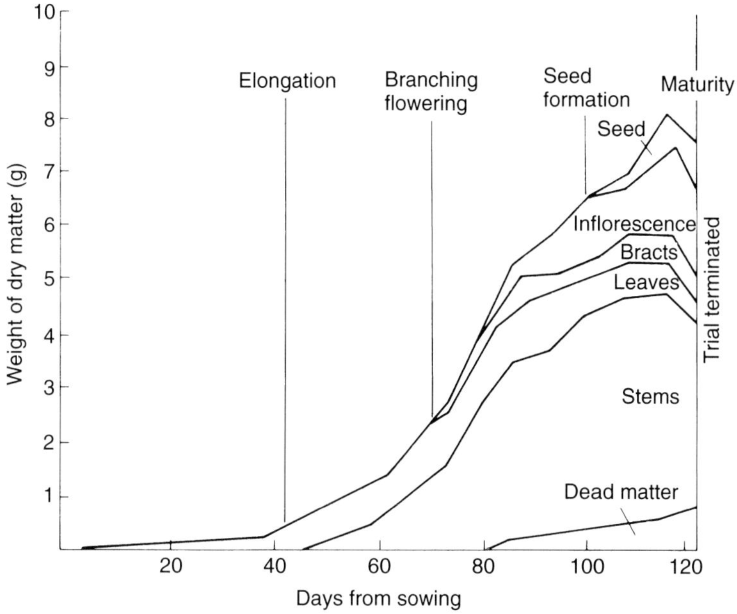

Fig. 4.2 Distribution of dry matter with age, Australia.

Table 4.1 Rate of plant development in selected countries.*

Country	To emergence	To elongation	To branching	To flowering	To harvest	Remarks
Argentina	7	30	—	85	130	Rain-grown
Australia	7	25	50	89	142	Irrigated
India	8	30	—	82	127	Rain-grown
Japan						
(autumn-sown)	20	—	—	234	280	Rain-grown
Poland	12	—	—	82	136	Rain-grown
South Africa	12	32	—	85	140	Rain-grown
Sudan	10	—	—	80	120	Rain-grown
Turkey	12	36	—	100	150	Rain-grown
USA						
(autumn-sown)	18	121	150	185	220	Rain-grown
(spring-sown)	5	28	55	90	135	Irrigated
Venezuela	8	30	—	75	120	Rain-grown
Kenya	14	30	58	95	150	Rain-grown

*Time given is the average in days from the optimum period for sowing.

The plant has a well-defined and frequently fleshy tap root, and normally produces numerous thin horizontal laterals. Although the tap root commonly penetrates to 2–3 m, the mass of roots is usually in the first 30 cm. Cultivations must allow this root development with the minimum of hindrance. Tap root elongation can be retarded by compacted or dry soil, lateritic layers as in East Africa and India, or fine-textured clay, but safflower roots generally have greater penetrating power than other crops. In Australia for example, roots of safflower penetrated to 220 cm, while wheat roots in an adjoining plot reached only 40 cm. This deep-rooting characteristic also allows plants to draw moisture and nutrients from a considerable volume of soil (Meyer & Barrs, 1991; Materechera *et al.*, 1993; Ghorpade *et al.*, 1993). Salinity may also affect the extent of root penetration and the horizons in which secondary roots proliferate.

The stem is stiff, cylindrical, fairly thick at the base, becoming thinner as branching increases, smooth, glabrous, light grey or green to white, with fine longitudinal grooves, becoming brittle when mature. Anthocyanin has been identified in a genotype, and used as a marker in breeding studies (Singh V. *et al.*, 1995). Stem girth of mature plants at ground level ranges from 3 to 12 cm, and this characteristic has been significantly correlated with yield. Following emergence, the stem apex produces a rosette of leaves on the soil surface, which rapidly develops into a true stem. Cultivated safflower has no true rosette stage as occurs in other *Carthamus* spp., but the rosette stage can be prolonged when the crop is sown in autumn in the northern hemisphere, or by inimical weather conditions. In contrast to its relatively slow start, safflower grows rapidly after the stem begins to elongate. Dry matter distribution in China was 87% to leaves at the rosette stage, 50% to stem at elongation; while dry matter was translocated to reproductive organs from the flowering stage (Li *et al.*, 1995).

Seedling growth can be affected by growth hormones, but their application may also have damaging side-effects; foliar sprays of gibberellic acid applied at the seedling stage to field plantings in California caused rapid elongation of internodes, with a subsequent considerable increase in lodging. Lodging is also higher in thin-hulled safflower cultivars since the main effect of the gene controlling this factor basically reduces secondary wall-thickening in sclerenchyma cells. Thus the fibre cells surrounding vascular bundles were weaker than in normal thick-hulled types. Gibberellin may also promote plant growth at the expense of reproductive capacity (Potter *et al.*, 1993), while chlormequat and TIBA increased seed yield in India (Deotale *et al.*, 1996); ethyl methane sulphonate (EMS) reduced plant height and time to flowering.

Height is a cultivar characteristic, between 50 and 200 cm, and is affected by climate and cultural techniques. The tallest types come from the Turko-Afghanistan area, the shortest from India. Dwarf plants some 35 cm in height when mature, are related to lack of a rosette stage. The central stem branches at 15–20 cm to form secondary stems, which themselves branch, each branch terminating in a flower head (Fig. 4.3). The angle of branches to the main stem is a cultivar characteristic, but can be considerably affected by environment.

Most commercial cultivars have a branch–stem angle of 30–70°, but the angle can range from 10° to 90°. Removing shoots just before flowering can increase branching, number of seed heads per plant and total seed yield. This can be an advantage where safflower is grown as a border crop around fields, and yield per plant important. The excised young tips are used as a vegetable. The effect of defoliation on yield is discussed in the section on pests.

Leaf size and shape varies considerably between cultivars and on individual plants, from 2.5–5 cm × 10–15 cm long, usually large and deeply serrated on the lower stem, short, stiff, ovate to obovate at the inflorescence, where they become superimposed as the involucral bracts. A relationship may exist between the total area of outer involucral bracts, seed yield, and moisture utilisation (Zope *et al.*, 1994a). The lower leaves are generally spineless, but the number of spines on upper leaves is a cultivar characteristic, and can vary from spineless to strongly spined. This spiny character is utilised by smallholders in many countries, who plant several lines of plants on bunds or field borders to repel small livestock and children. Spininess is of little importance in mechanised production, although spiny cultivars may require a higher plant density to produce the same yield as non-spiny.

Fig. 4.3 Typical safflower plant showing branching and flower heads.

Spininess is however, a major drawback to smallholders wishing to grow safflower on a larger scale with manual harvesting, as later noted. Spininess may also be associated with desirable factors such as insect resistance, or male sterility. Spineless strains have been selected, some specifically as ornamentals.

The inflorescence is typical of the Compositae, consisting of numerous florets collected closely together on a circular, somewhat flattened, receptacle. Floret number varies with the variety, between 20 and 180, and can also be affected by environment. The receptacle is surrounded by several layers of involucral bracts, the outer ring being heavily spined. Individual flowers are provided with bracts in the form of small hairs. The involucre is conical, with a small apical opening through which protrudes the corolla tubes.

Flower colour can vary from a whitish yellow to red-orange, with deep yellow the most common. Flower colour may be used as a method of classifying cultivars of *C. tinctorius*, and the colour types listed in Table 4.2 generally recognised.

Two colouring materials can be obtained from flowers, a water-soluble yellow, carthamidin, and the formerly important colouring material cartharmine, an orange-red dye insoluble in water but readily soluble in an alkaline solution. Changes in floret colour and the red compounds obtained have been identified using a spot scanning technique (Fukushima *et al.*, 1997).

Commercial extraction has virtually ceased but it is still produced on a small scale by local

Table 4.2 Flower colour types.

Type	Bud	In bloom	Wilted
1	White	White	Greyish white
2	Light yellow	Light yellow	Greyish white
3	Light yellow	Light orange base	Orange base
4	Yellow	Yellow	Yellow
5	Yellow	Yellow	Orange
6	Deep red	Reddish orange	Deep red
7	Yellow, lobe base and tips orange	As bud	Orange

farmers to supply traditional and religious occasions. In Sri Lanka flowers are collected fresh early in the morning, dried on muslin trays and then stored in tins for later extraction. Yield of florets is 70–100 kg/ha. In India, fully opened heads are collected before fading, dried and corollas removed. These are then packed for sale, or extracted by washing corollas for 3–4 days in acidulated water. The pulp remaining is dried into small cubes and sold. Commercial safflower carmine is extracted from washed floral material by a sodium carbonate solution, and precipitated by dilute acids. Bajra millet, *Pennisetum* spp., provides the carbonate. First and last pickings of florets are considered to yield an inferior dye to mid-season pickings. Dye from the Dacca area of Bangladesh is considered to be superior to that produced in India, and a normal crop yields 90–140 kg/ha of dried florets.

Capitula are borne at the end of stems or branches; flowering begins at the inflorescences that terminate the main plant axis, followed by the most mature of the main branches; secondary and tertiary branches follow sequentially. Although flowering may thus be extended, heads do not easily shatter when ripe, a distinct advantage for combine harvesting. Flower heads can vary in diameter from 1.25 to 4.00 cm and the number per plant from 5 to 50. One reason for this diversity may be evolutionary pressure; growers requiring flowers for dye would select plants with large heads and few involucral bracts, while natural selection produced small heads, tightly enclosed, with long involucral bracts for protection.

Although large head size usually indicates more or larger seeds, the relationship between head size and other plant characteristics is inconsistent. The number of seeds per capitulum is directly related to 100 seed weight, and in general as the number rises the weight falls; however, capitulum weight/plant and seed weight/capitulum are positively related to yield (Patil *et al.*, 1990).

Close spacing directly reduces the number of heads produced, but although the number of heads can be reduced to a low average of three/plant, further reduction is resisted (Fig. 4.4). Reduced yield from this point is due almost entirely to a reduction in number of seeds per head. Individual head size also depends on its location; primary and secondary heads are largest, with tertiary and subsequent heads usually smaller.

Flowering normally begins at the head margin, proceeds centripetally and requires 3–5 days to complete. Since plants are multi-flowered, total flowering may extend over 10–40 days. Time to flowering in a cultivar is basically genetically controlled, but can be greatly influenced by environment.

Flowering is initiated by approximately 14 hours of daylight in the USA, but can be modified by temperature during growth. High

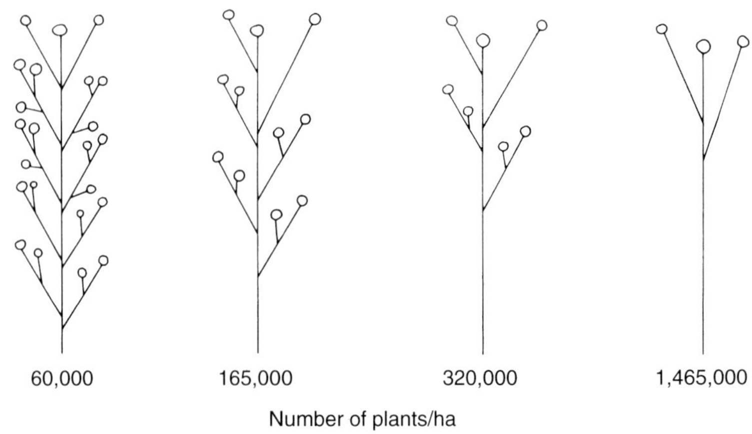

Fig. 4.4 Effect of plant population on plant structure.

temperature during early growth accelerates flowering, and when it is the result of sowing after the optimum date, time from emergence to flowering can be reduced by 30–40 days. Salinity may also accelerate the onset of flowering, but reduction of the period from emergence to flowering also varies with the cultivar and type of salinity. Salinity also reduces the number of flowering heads and seed yield per head, although the number of seeds per head may not be affected. Temperatures approaching 0°C at the late bud or early flowering stage will adversely affect fertility; a range of 24–32°C is considered most favourable, although in Australia temperatures above 40°C occur after flowering with no appreciable damage when soil moisture is adequate. Sterile but vegetatively vigorous, plants which produced heads but no seeds, and plants with closed-flowers (cleistogamic) have been reported.

Safflower is basically self- not wind-pollinated, but bees or other insects are generally necessary for optimum fertilisation and maximum yields, related to the method of flower opening. Before flowering the stigma is enclosed by the five fused anthers, which are attached by very short filaments to the tip of the corolla tube. Usually all florets that open during a given day have begun to elongate by sunrise. Anther dehiscence, which normally occurs soon after sunrise, takes place at the top of the anther column as the stigma emerges from within the anther tube. The combined elongation of the style through the corolla tube pushes the brush-like stigma through the anther tube until all the stigmatic surface of the pistil has grown well beyond the tip of the anthers. By the time this process of elongation is completed, the stigma is usually well covered with the floret's own pollen. A technique for mass emasculation of safflower is based on this flowering process (Singh V. *et al.*, 1995).

Honey bees (*Apis mellifera*) are attracted to safflower fields during flowering because of the plant's abundant pollen and nectar; and bumblebees and syrphid flies are attracted to a lesser extent. Honey-bees are most efficient pollinators, as clearly demonstrated in breeding plots of thin-hulled safflower strains, since inheritance of this character also affects several other plant characteristics including anther dehiscence. Interestingly, bees differentiate between normal and thin-hulled lines; pollen collectors worked only the normal line, nectar collectors worked both. However, high temperature can reduce the hours that bees work crops, and thus the degree of natural outcrossing (Ahmadi & Omidi, 1997). When insects are excluded from plants at flowering, yields may be reduced by up to half, depending on the cultivar, but in Victoria, Australia, exclusion of large insects mainly bees, from plots of Gila safflower had no effect on seed yield.

The degree of cross-pollination to be expected is always of importance to plant breeders, and although safflower is basically self-pollinated, a high degree of crossing can occur, particularly in thin-hulled types. Sterility, both male and female, occurs in safflower, and can complicate a breeding programme. Structural male sterility is linked with thin hull, and in India delayed anther dehiscence in this type is used to produce hybrid seed. Cytoplasmic male sterility would greatly assist breeders, as it has done with other oilseed crops. Techniques used to ensure seed purity or breeding lines normally require considerable skill or supervision, and it may be easier to grow these crops outside the normal season; for example India, where seed multiplication crops are grown in the dry season and harvested before commercial crops of the monsoon are sown.

The fruit, an achene, resembles a small, slightly rectangular sunflower seed, but with a thicker, more fibrous hull (Fig. 4.5). Pappus is normally absent, but may occur on seed from the flower centre, and on all seed of some strains. The testa is generally cream or off-white, but grey and mottled types occur. A black or brown testa invariably indicates crossing with a wild species, and a striped testa occurs in crosses to produce thin-hulled seed. The seed-coat, about 15% of the hull, consists of two brown epidermal layers, with the parenchymous layer of the seedcoat and endosperm normally only a few cells in thickness. In the thin-hulled type the outer sclerenchyma cells are very thin and the melanin layer is visible, producing greyish or brownish colour

Fig. 4.5 Safflower seed.

seeds. Variation in the thickness of the outer sclerenchyma layer through which the melanin layer appears, results in grey-striped seeds, while variation in the melanin colour produces variously striped seed.

Seed size is a cultivar characteristic, is variable, and affected by environment. The Gila variety has seed 6–9 mm long and about half as broad, with a 100 seed weight of about 4 g, but nearly 10 g has been recorded from the Indian cultivar Bagewadi-1. The average seed composition of white-hulled commercial types is in the ranges, hull 25–45%, kernel 55–65%, oil 35–60%. The high proportion of hull was a handicap to commercial production, but the introduction of thin or reduced hulls was also associated with less desirable characteristics. Very thin hulls may also cause problems during mechanical harvesting, bulk seed movement, storage and processing.

The period to anthesis and days to maturity directly affect rate of grain filling and thus yield, and in general, seeds normally reach maximum dry weight, oil and germination percentage, iodine value, and minimum hull percentage four weeks after flowering (Zope *et al.*, 1994b; Kumar, 1996). Seed from heads at different positions on the plant, and similar positions on different plants, can vary considerably, and oil content tends to be highest in later and lowest in early flowering heads. Under irrigation, increasing plant populations may reduce the dry weight of tops and number of seed heads per plant, but the relative seed and oil yield of individual primary, secondary and tertiary heads remains fairly constant.

Seed composition, oil content and quality as expressed in terms of iodine value and component fatty acids, is influenced by environment; including latitude and altitude, day and night temperatures and amount of rainfall during periods of flowering and seed setting. Some fungi also appear to have lipolytic activity, and can increase free fatty acids in seed. Oil quality may be so modified that oil with differing characteristics is obtained from the same variety grown under widely different conditions. Before 1957, for example, the iodine value of safflower oil was considered to be in the range 138–145, but in 1964 a variant with an iodine content of approximately 100 was reported from Australia which had the proportions of linoleic and oleic acids reversed. A very high stearic acid strain found in experimental lines in the USA is interesting genetically, though of limited practical value (Table 4.3).

Selection can also alter oil composition by varying the proportions of seed components, as cotyledons and embryos contain almost equal amounts of lipids, the tegument a lesser amount. Embryo oil has a greater degree of unsaturation than the other two components, thus any substantial variation in these will alter expressed oil composition. Seed hull percentage is of great importance, and current cultivars have thinner and less fibrous hulls and a higher protein content, up to 17% in whole seed and 38% in the kernel, and are thus more attractive

Table 4.3 Composition of safflower seed oil from experimental lines (%) – USA.

Type	Palmitic	Stearic	Oleic	Linoleic	Misc.	Non-sapon.
Commercial	6–7	2.4–2.8	9.7–13.1	76.9–80.5	0.2–0.8	0.9–1.6
High linoleic	6–7	2.1–2.9	9.8–12.7	77.5–81.0	0.2–0.8	0.8–1.4
Intermediate	5–6	1.2–1.4	45.1–47.3	46.5–48.2	0.8	1.3
High oleic	4–8	4–8	74–79	11–19	—	—
High oleic	4.5	1.2	86.8	8.7	0.5	2.3
High stearic	6.2	9.5	11.9	71.0	1.4	2.1

Note: The high stearic acid strain is interesting genetically, though of limited practical value, but those with high oleic acid content could be promising for commercial oil production.

to crushers and feed compounders (Rojas *et al.*, 1993).

Growth regulators, as noted, can influence the rate of development of plant and seed, and although they may promote plant growth in specific circumstances, generally have an adverse effect on seed yield and composition. Gibberellic acid in particular can reduce yield and seed-oil content, but not affect oil composition.

Safflower oil use in the industrialised countries was initially as a drying oil, but it became popular as an edible oil following medical reports on its advantages over animal fats in the human diet. However, the relationship between a diet containing safflower oil and blood cholesterol or serum lipid content, and between cholesterol content and various heart ailments, differs between individuals (Wardlaw *et al.*, 1991).

Safflower is usually considered an oilseed crop, but its drought resistance makes it valuable as green fodder or silage for drier regions. Once livestock become accustomed to the spines and unfamiliar taste, they readily graze safflower or its stubbles. Leafy strains have been selected in India, and there are high leaf-protein selections in India and Mexico (Musa, 1993; Satpute & Kothekar, 1995). A review of the chemistry and biology of safflower has been published (Monika Verma *et al.*, 1997).

Other *Carthamus* species

The most agriculturally important species closely related to *Carthamus tinctorius* are *C. lanatus* L. and *C. oxyacantha* Bieb.

Carthamus lanatus, the saffron thistle, occurs wild over the entire range of the genus. It has been introduced into the USA and Australia, and can be found growing up to 1800m in Kashmir. Seed oil content is about 16%; the oil is pale yellow, and similar to safflower oil. In Australia it was tested as a cultivated crop, and could be reaped with a grain combine. The harvesting period was more critical than with safflower, and plants were cut before completely ripe as oil content fell rapidly from seed maturity. The plant was originally introduced into Australia as seeds in feed imported with camels from India. Now classed as a noxious weed of cultivated crops, it is controlled by herbicides.

Carthamus oxyacantha is a medium-sized, branching, spiny, annual widely distributed from the Caucasus to northern India and Afghanistan, and the seed is collected where the plant occurs in sufficient quantities. Seed oil content is approximately 28%, but oil yield by local extraction is around 18%. The oil, *pohli* oil, is used in cooking, lighting, and formerly in the preparation of roghan used in the manufacture of Afridi cloth. Roghan is prepared by heating *pohli* oil in an earthen pot for some 12 hours over a low and uniform heat and then pouring the boiling oil into cold water, when it solidifies into a gelatinous mass. *Pohli* oil can be used in the manufacture of soap and varnishes, and as a glass cement. In the Punjab it is used especially for the treatment of skin diseases and dressing ulcers. The seeds are also parched and eaten, either alone or mixed with wheat or flour.

Oil characteristics between the three species differ considerably (Table 4.4). The iodine value of *C. oxyacantha* oil varies substantially and may provide genes that affect this characteristic. *Carthamus lanatus* yields a pale-yellow oil which dries to a hard, pale film, and is almost identical with safflower oil.

Saffron

Saffron is the dried stigmas of *Crocus sativus* L., for which safflower florets are used as a cheaper substitute and adulterant. *Crocus sativus* is a small bulbous plant similar to the common cultivated crocus, whose range extends throughout the Middle East, Asia and China. It was formerly widely grown for the yellow dye obtained from the stigmas. Flowers are collected daily, just as they open, the stigmas removed by hand and immediately dried. In India and Ceylon this is often carried out over a low fire, but in Europe and China special ovens are used. Yield varies from 5 to 30 kg/ha, and approximately 90 000–125 000 flowers are required to produce 1 kg of dried saffron. The immense amount of hand labour involved is seldom justified by the price received, as many inexpensive artificial substitutes are now available.

ECOLOGY

Safflower is basically a crop of the warm-temperate regions whose range has been greatly extended by selection and breeding. It can be grown as an autumn-sown crop in the northern hemisphere, or under irrigation in the dry tropics. Distribution as a smallholder crop is approximately bounded by latitudes 20°S and 40°N. Commercial production on a large scale is concentrated in fairly low-altitude, semi-arid areas where the incidence of disease is low. Safflower is normally grown below 1000 m and seed yield and oil content fall with increasing height. In the high-altitude tropics safflower will make good growth, as the author demonstrated at 1800 m in Kenya, but yields were low in comparison with those obtained below 1000 m. Since the two areas were not markedly dissimilar in terms of climate and rainfall during the plant's growing season, there would appear to be no major reason why suitable cultivars should not be bred for these very extensive highland areas. Data indicates that the most productive cultivars often show the least variation in growth and yield when introduced into an ecologically similar area.

Although safflower is considered day-neutral, cultivars are adapted to specific photoperiods, and a short photoperiod can prolong the rosette stage. Temperature can be as important as day length, as was demonstrated at the Oil Crops Institute of Yunnan, China, where some 3000 accessions were separated into four major types; photo-sensitive, thermal-sensitive, photo-thermal-sensitive and photo-thermal-insensitive (IPGRI, 1994). Good yields have been obtained in the plant's natural habitat in long-day areas of the Mediterranean region, and in Australia under a 12–hour day in the field (Fig. 4.6), and 16–hours in a phytotron. In the USA a 14–hour day is considered necessary to initiate flowering.

The most important effects of temperature on safflower are seedling tolerance of low temperature, and susceptibility to high temperature at flowering, although there are considerable culti-

Table 4.4 Comparison of oil obtained from some *Carthamus* species.*

Species	Iodine value	Sapon. value	Unsapon. value	Refract. index	Oil (%)	Fatty acids (%)			
						Linoleic	Oleic	Palmitic	Stearic
C. lanatus	139	187	1.00	1.4729	16	70	15	10	7
C. oxyacantha	140	195	0.42	1.4715	28	68	24	10	4
C. tinctorius	145	190	0.75	1.4691	40	73	21	7	3

* Average values.

Fig. 4.6 Commercial safflower crop, Queensland, Australia.

var differences. Seedlings can tolerate –7°C, with cultivar tolerance to –12°C, but become more susceptible to frost damage at cessation of the rosette stage. However, when germinating seed of a local cultivar was exposed to 0°C for 15 days in China, the plants developed normally regardless of subsequent temperature (Li *et al.*, 1995).

Cultivars with a short rosette stage are generally more susceptible to frost damage than those with a long stage. Frost resistant types in Iran can survive temperatures to –14.4°C, and with the cold-hardiness which exists in wild safflower species, breeding cultivars with greatly enhanced cold-hardiness is possible. Growing plants are generally damaged by temperatures approaching freezing, and even when there is little visual effect the yield can be drastically reduced. Frost on maturing crops affects both yield and oil content of seeds, and thus the period from stem elongation to maturity should be frost free, approximately 130–160 days.

The optimum temperature at flowering is between 24 and 32°C, but adequate soil moisture reduces the adverse effect of high temperature, and 43°C in California had no major effect on irrigated safflower. High temperature also tends to decrease seed weight and oil content, and affect oil constituents, while high temperature and high humidity can severely reduce seed yield. Plant population can also allow temperature to exacerbate these effects, as high temperature at lower populations encourages production of secondary and tertiary heads which prolongs the flowering period.

Since time of sowing basically governs the temperature range at which plants grow, the optimum must be determined locally. Delayed planting in either hemisphere that causes flowering to coincide with periods of high temperatures, or maturity with frost, steadily reduces yield. In addition to the fall in yield there is generally a fall in seed oil content, further reducing the profitability of late-sown crops, which emphasises the importance of planting at the local optimum time. There can also be a relationship with pest and disease damage, as noted later.

The effect of a range of sowing dates on plant growth is shown in Table 4.5 (Beech & Norman, 1963), and the effect on yield and seed characters in Table 4.6. (Yermanos *et al.*, 1967). The recommended sowing date for selected countries is shown in Table 4.7.

Safflower can be substantially independent of rainfall, since it is capable of obtaining moisture from levels not available to the majority of crops, and can thus tap subsoil moisture or seepage. However, if safflower's water requirement can only be met by rain falling on the crop during its life, then commercial production will be hazardous. If sufficient pre-planting soil moisture is available, i.e. approximately two-thirds of total water requirements, the remainder can occur as rain with no major depressing effect on yield, provided that humid conditions do not persist for long periods. Rainfall should also preferably fall between the rosette stage and flowering.

Safflower is tolerant of a moisture shortage at the rosette stage, and dry conditions at this time do not appear to have any major effect on subsequent growth and yield. A moisture shortage during the period of maximum growth from rosetting to flowering, will markedly reduce growth and delay maturity, while moisture stress during flowering and maturation will reduce yield and oil percentage and may hasten maturity. Two dry periods during growth frequently have a greater depressing effect on yield than one.

Pre-planting irrigation, flooding or delaying sowing until after the main rains have fallen is common practice. In dry areas where rainfall is scarce and high temperature during the growing season encourages quick maturity, planting should preferably be delayed to avoid development of a large plant that will exhaust available soil moisture and produce foliage at the expense

Table 4.5 Effect of sowing date on safflower growth – Australia.

Period	Sowing date		
	5 May	5 June	5 July
Days to emergence	7	10	12
Days to elongation	40	45	47
Duration of rosette (days)	33	35	35
Days to flowering	93	99	89
Days to seed maturation	142	136	122

Table 4.6 Effect of sowing date on seed yield and characteristics.

USA	Seed yield (kg/ha)	Oil content of seed (%)	Iodine value	1000 seed wt (g)	Hull oil content (%)
25 Jan.	2605	32.6	145	46.6	45.3
14 Feb.	2264	35.8	144	42.7	41.7
10 Mar.	2057	37.7	145	41.6	36.9
13 Apr.	1614	39.7	144	40.0	35.5

Australia	Seed yield (kg/ha)	Oil content of seed (%)	Iodine value	1000 seed wt (g)	Hull oil content (%)
5 May	3365	36.8	145	51.0	10.8
5 June	3095	37.8	143	47.8	11.9
5 July	1804	33.9	142	32.4	8.6
4 Aug.	517	29.3	141	22.2	6.4

Table 4.7 Recommended sowing dates for safflower.

USA	Arizona	Early Nov.	Winter-sown
		1 Jan.	Spring-sown
	California	1 Dec.–15 Feb.	Rain-grown plus
	Dakota	10 Apr.–20 May	Irrigation
Canada	Alberta	1 Apr.	Rain-grown
S. America	Argentina	Sept.–Oct.	Rain-grown
	Mexico	Oct.	Rain-grown
Africa	W. Kenya	June	Rain-grown
	Tanzania	May–June	Irrigated
	South Africa	Oct.–Nov.	Rain-grown
	Sudan	Mid-Sept.	Rain-grown
		Mid-Oct.	Flood-irrigated
	Egypt	Apr.–May	Rain-grown
Mediterranean	Italy	March	Rain-grown
	Turkey	Mar.–Apr. or Oct	Rain-grown
	Israel	Jan.–Mar.	Rain-grown
Asia	Iran	Dec.–Jan.	Rain-grown
	India	Oct.–Nov.	Rain-grown
		October	Irrigated
	Pakistan	Oct.–Nov.	Rain-grown
	Bangladesh	Dec.–Jan.	Rain-grown
	Sri Lanka	Oct.–Nov.	Rain-grown
Australia	North-East	June–July	Rain-grown
	West	Apr.–May	Irrigated

of seed. Thus the drier areas of the USA, Russia and Australia proved very suitable for irrigated crops, and large-scale rain-fed plantings are often relatively unprofitable. The dry summer and autumn areas of Middle Anatolia in Turkey are considered ideal for non-irrigated safflower production.

A rainfall of 600 mm is the minimum necessary to produce a commercial crop, the major portion falling before flowering. In drier, windy regions most suitable for safflower as the incidence of disease is lower, 800–1000 mm is desirable, the larger amount when irrigated. In areas where there are no hot, dry winds, and with adequate pre-planting soil moisture levels, 300 mm of rain will produce excellent yields if this falls mainly before flowering. Excessive rainfall or humidity increases damage from fungal disease at all stages of growth, and in areas where foggy or misty periods occur during flowering, head rots can be severe.

Prolonged rainfall at flowering adversely affects pollination, and after flowering causes discoloured and undersized seeds. Persistent rain or periods of high humidity when heads are mature encourages seed to germinate in the head. An indirect effect of rainfall is on the total area planted to rain-grown crops in the USA. When a long dry period follows the first rains in autumn, farmers tend to plant the major part of their land to barley. Continuous wet weather in the autumn makes cultivation and planting very difficult, and land not planted to barley is sown to safflower in the spring.

Safflower is little damaged by quite severe hailstorms early in growth, particularly the rosette stage, but once the stem elongates it becomes more susceptible. Hail immediately before formation of flowers and secondary branches can do great harm, especially where the growing season is limited, but mature plants are resistant to all but very severe storms. Fully grown plants are extremely wind resistant and even after seeds are mature, there is little loss from lodging or shattering. Winds of 24 km/hour caused little damage to mature plants in California, nor 50 km/hour in Tanzania. However, etiolated plants produced by too close spacing or incorrect irrigation practice are more susceptible to damage. In areas where persistent high winds are common, wider in-row spacing and closer rows reduce wind damage. Where possible rows should also follow the prevailing wind direction, and this can also reduce in-crop humidity and so lessen disease infection.

SOILS

Safflower is grown by smallholders on a wide range of soils, pH 5–8, but for commerical production highest yields are obtained on fairly deep, well-drained, somewhat sandy loams of neutral reaction. Highest yields are obtained in dry regions on rather sandy loams with sub-irrigation. Irrespective of their fertility, shallow soils seldom produce high yields, and this is invariably due to insufficient moisture. Dense subsoil will retard root growth, as will a layer of impermeable clay, and the yield from rain-grown or sub-irrigated crops will be substantially reduced even though the subsoil below the layer has adequate moisture. Acid soil may also increase damage from *Fusarium* root rot, since soil acidity is known to favour development of several biological forms of this fungus.

Australian commercial crops are generally located on sandy, moderately fertile soils occasionally under irrigation. Soil permeability is important, and good drainage considered essential. When grown on the relatively heavy clay soils of the Kimberley area, cultivation techniques and time of planting become more critical. Early planting following winter rains may be impossible because the soil is too wet, and late planting can reduce yield. There is a tendency for these soils to crust, and a higher seed rate is necessary to ensure even emergence and adequate stand. Well-drained, sandy or clay loams are preferred in the USA, although it has been grown as a rotation crop on rice lands.

Safflower is grown on a variety of soils in India according to season and production method; in Maharashtra, Andhra Pradesh and Karnataka it is grown on black-cotton soils, in other states loams and light alluvium are preferred for both irrigated and rain-fed crops. On heavy soils it can follow an early kharif crop and may often be the single crop of the rabi season. On lighter soils it is almost entirely irrigated, and can be strip-cropped with a cereal or legume. High yields have been achieved on river silts in the Sudan and on Tozi clays, but on the Gezira cotton scheme, crops were disappointing.

Safflower is considered to be salt tolerant ranking just below cotton, although many commercial cultivars are salt-sensitive, and it is similar to barley in its tolerance of alkalinity when rain-grown (Fig. 4.7) (Bernstein, 1964). Safflower is especially tolerant of sodium salts, but less tolerant of disalts and trisalts containing either calcium, magnesium or both (Uma *et al.*, 1993; Ashraf & Fatima, 1995). Salinity delays initial emergence which also tends to be irregular, while very high levels of salinity also reduce germination. The use of saline soils for crop production is increasingly desirable and some of the newer safflower cultivars are highly tolerant.

FERTILISERS

When grown as a smallholder crop little fertiliser is applied directly to safflower, but commercial crops are fertilised as necessary (Singh V. *et al.*, 1995). The nutrient most often required is nitrogen, phosphate requirement is moderate, and potassium is needed only where there is a major local deficiency. Newer cultivars specifically developed for irrigated cropping have a higher nutrient requirement. Fertilisers generally have little direct effect on seed composition or oil percentage at the levels normally applied, but by increasing seed yield they increase total oil yield/ha. High levels of nitrogen however, may

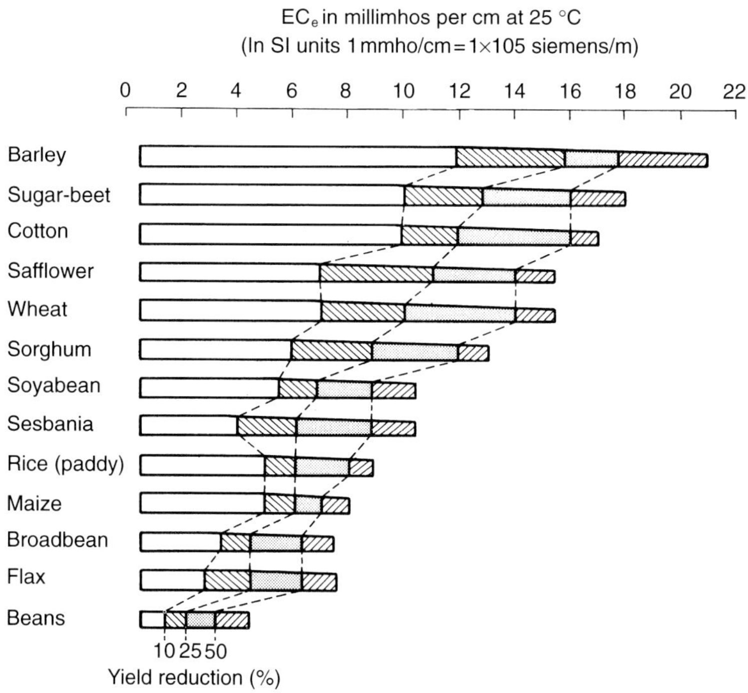

Fig. 4.7 Salt tolerance in field crops.

increase seed-protein content at the expense of oil percentage.

Placing seed and fertiliser in contact adversely affects emergence, and band placement is recommended; however as safflower is easily and efficiently sown by grain drills which are seldom equipped for band placement, ensuring that seed and fertiliser are sown down different spouts is sufficient, and blanking-off outlets in the grain or fertiliser boxes is a simple operation. Seed-bed placement of nitrogen generally increases uptake of phosphorus, but the economics of so doing must be determined.

Nitrogen

Crop response to nitrogen (N) is generally greater than to other nutrients, and affects not only total yield of seed but also seed composition as in Egypt (Ashoub, 1995) and India (Memon *et al.*, 1995). It does not, however, usually affect oil content or iodine value of the oil except at high levels. Nitrogen increases seed yield primarily through its effect on the number of heads per plant and the increase is greatest in tertiary and to a lesser extent in secondary heads. In the United Arab Emirates, for example, the number of capsules/plant and seeds/capsule, seed yield/plant, test weight and seed oil content were highest following application of 120 kg N/ha (Ibrahim, 1994).

Nitrogen recovery in safflower is very variable, and why this is so has still to be fully determined. The total N in aerial portions of safflower at different stages of maturity and fertiliser levels, are shown in Table 4.8, and as total N uptake increased by 50% between first bloom and maturity, an adaquate supply at this time is essential (Jones & Tucker, 1968).

The amount of N applied, especially as a top-dressing, should be related to local environmental conditions. To apply substantial amounts to a crop and encourage growth is unwise if the growing season is so limited that any addition to the period to harvest will reduce yield. Nitrogen also tends to promote and prolong flowering, but if this then takes place during a period of high temperature, a larger proportion of heads will be

Table 4.8 Nitrogen recovery in safflower – USA.

Determined at:	Ammonium nitrate (kg/ha)							
	0		67		135		269	
	%*	Wt†	%	Wt	%	Wt	%	Wt
First blossom	1.4	34.1	1.6	51.7	1.8	60.2	2.4	85.6
Maturity	1.0	71.5	1.2	127.0	1.2	123.4	1.3	150.5

* Nitrogen recovery = per cent per plant; † kg/ha.

Table 4.9 Effect of previous crop water and nitrogen application to irrigated safflower – USA.

Water regime		Ammonium nitrate (kg/ha)				Percentage increase
		0	67	134	269	0–269
Fully irrigated	A	4520*	4372	4543	4818	7
	B	2351	3056	3816	4245	80
Moderately irrigated	A	4095	4212	4426	4491	10
	B	2573	3233	3328	3749	46
Rain-grown	A	3649	3670	3765	3892	7
	B	2972	3167	3428	3317	12

* Yield in kg/ha. A = Following well-fertilised barley. B = Following unfertilised hay crop.

sterile or produce few seeds. Fertilisers applied to a previous crop can affect the N requirement of a following safflower crop (Table 4.9). Soil moisture also affects the availability of crop nutrients, and when N is applied to irrigated safflower, there is an optimum level for both, and a shortage of one not compensated by increased amounts of the other. The effects of varying levels of nitrogen and water application are shown in Table 4.9 (Worker *et al.*, 1965).

Commercial growers consider urea in the seedbed can affect germination and reduce stands, but as a foliar spray it increases seed yield. However, these increases are often similar to those from seedbed application of other types of nitrogen, which are usually less costly to apply. The level of soil moisture or its rate of depletion during growth also directly affects nitrogen availability and uptake, as occurred in India (Bansal & Katara, 1993; Katara & Bansal, 1995). Irrigated safflower plantings may receive drilled liquid or gaseous fertiliser, since these require adequate soil moisture for optimum results. When nitrogen is applied in the seedbed, combined NP fertilisers including diammonium phosphate are recommended as being less harmful to seedlings.

Published data comparing the various nitrogenous fertilisers commonly available indicate that type is of little importance, especially when applied as a top-dressing. The rate and timing of application are more critical in terms of yield increase/unit N. On current high yielding cultivars, up to 150 kg N/ha is applied to irrigated, and up to 56 kg N/ha to rain-grown crops. Split application is recommended at levels above 100 kg N/ha, when it is preferable accurately to determine the optimum amounts, since as noted, safflower's requirements vary with age and moisture availability. Australian growers favour 20–30 kg N/ha in one top-dressing on rain-grown, up to 100 kg N/ha on irrigated crops which also receive seedbed application. Nitrogen top-dressing can only be profitable if the plant's requirements of other nutrients are satisfied, and in their absence nitrogen frequently depresses yield.

Phosphate

When phosphate (P) is normally required by other crops, the minimum recommended rate should be applied to safflower until trials establish the optimum level. Safflower is deep-rooted and can draw on a considerable volume of soil for nutrients, and if a high level of P is applied to the preceding crop, or as broadcast rock phosphate, it is frequently not required for safflower. However, since most arable crops respond to adequate phosphate in the seedling stage, small amounts band-placed at planting promote rapid initial growth and improve general plant health. More rapid seedling growth will also enable young plants more effectively to compete with weeds. Adaquate soil moisture levels are essential to allow uptake of P by rain-grown crops.

Safflower is not usually grown in pure stands in Asia, and fertilisers are normally animal or crop residues, or rock phosphate. Application of 5–12.5 kg P/ha is recommended for smallhoder crops in India, Iran, Pakistan and Afghanistan, but seldom applied, resulting in very low average yield. This may not be the best criterion on which to assess the crop's potential, as in many instances the crop is not grown for sale but to supply a small amount for home consumption. Where it is grown commercially in India it frequently follows grain or cotton, which receive whatever fertiliser it has been possible to purchase. A small amount of rock or superphosphate may be broadcast at final ploughing, but ash and plant residues of the previous crop are usually all that safflower receives. Fertiliser trials with seedbed P frequently gave profitable yield increases, and an additional increment when combined with nitrogen.

Phosphate is not usually applied to safflower in the USA, except in small amounts at planting where this is considered desirable, and 15–30 kg P/ha the normal range. Since safflower seldom follows safflower and the majority of other crops in a rotation are well supplied with P, no field deficiency of this nutrient has been recorded. Combined NP fertiliser placed in the seedbed is recommended when safflower is spring-planted, to give strong healthy seedlings. Phosphate is deficient over huge areas of Australia, and in Western Australia on virgin lands, 35 kg P/ha is recommended for the first arable crop, reduced for succeeding crops, with 10 kg P/ha as an annual application. In the Queensland wheat belt, safflower usually follows a fallow to ensure sufficient soil moisture in this area of scanty and erratic rainfall. Since wheat is usually well supplied with phosphate, 5–10 kg P/ha in the seedbed is sufficient for safflower.

Phosphate is usually applied as superphosphate, but single superphosphate with its slightly higher sulphur content frequently gives a higher yield/unit P than other types, since many tropical soils have a minor but important sulphur deficiency. Rock phosphate, which is usually cheap when locally available, should be broadcast and ploughed in on a preceding crop, since its residual effect is high.

Potassium

Potassium (K) application to safflower is usually unnecessary except where a deficiency limits all crop growth. In general, the application of K to a preceding crop is sufficient for safflower, and few commercial growers specifically apply this nutrient. Where it is required, adequate N and P levels must exist, since application of K alone is rarely successful; in Saudi Arabia split applications of K decreased seed yield, but interestingly, seed-oil content increased. A relationship exists between sodium and potassium, and when levels of the latter are low, sodium can be substituted in the plant. Sodium can also improve growth independently of the potassium level, but information on the role of this nutrient in safflower is scanty. It may also affect the uptake and availability of magnesium.

Sulphur

Sulphur (S) is commonly deficient in many tropical soils, and often alleviated by single superphosphate, as noted. There can also be a high residual effect from application of fertilisers containing sulphur. Safflower yield increased significantly in India following application of gypsum, ammonium sulphate or single superphosphate to a preceeding crop of sunflower. Gypsum gave the highest residual effect in terms of increased seed yield (Sreemannarayana & Raju, 1993).

Minor nutrients

Data on minor element deficiency or toxicity affecting safflower yield in the field are few, and indicated no response by safflower to those applied to other crops at normal levels. Artificially induced reactions have occurred in laboratory trials and are noted in the literature. Copper for example, affected photosynthesis and transpiration (Nalini & Sharma, 1996) and zinc reacted with salinity to induce structural changes in plants (Gadallah & Ramadan, 1997).

CULTIVATION

Safflower is readily integrated into normal mechanised small grain production, but being deep-rooted, chisel ploughs and subsoilers should be used to fracture plough pans, clay or lateritic layers within the root zone. Where moisture is the limiting factor in crop production, cultivations should aim at retaining the maximum amount in the soil.

Minimum surface cultivation using shallow sweeps or chemicals, can be an advantage, and in the USA gave higher yields than disc or mouldboard ploughing. In arid regions minimum cultivation leaves crop residues on the surface, reduces erosion and conserves moisture, but chemical ploughing has disadvantages where there are persistent grasses. It may also induce a a temporary moisture deficit, since planting time in many tropical countries often coincides with low soil moisture levels following the dry season.

Current safflower cultivars are more environmentally sensitive and selection is more important, thus seed must be obtained from a reliable source. Seed from local or commercial crops can be used, but it must be carefully cleaned and graded, and usually requires a higher seed rate. Seed size is important as large seed generally has the highest percentage emergence. Large seeds are also able to compensate for uneven planting by their more vigorous initial growth, which can be important when using unskilled workers or unsophisticated equipment. Pure seed of known variety is essential for large-scale production and, if it cannot be easily obtained, should be produced in separate, isolated plots which receive special treatment and attention – the additional cost and care will be repaid many times over.

Seed dressing with a suitable fungicide/insecticide is recommended as standard practice on commercial plantings, since seedlings are highly susceptible to fungal disease and require maximum protection at this stage. Many diseases are seed borne and others locally endemic, thus neglect of this simple treatment will reduce emergence or cause gappy stands from premature death of seedlings. The extent to which poor emergence caused by disease can reduce stands is often not appreciated. The author found that a major factor contributing to low safflower yield in Eastern Africa was poor emergence, and specific trials determined that 25–35% of seeds not treated with a seed dressing failed to emerge.

Seed should preferably be dressed at least 24 hours before planting as this increases the efficiency of the fungicide by allowing it to penetrate the small cracks and crevices of the seed-coat and reach fungal spores lodged therein. If an insecticide is not applied as a seed dressing, it can be placed or sprayed during planting. Where high infestations of cutworms, wireworms or crickets are common, seedbed application of insecticide is effective, and combined insecticide–fertiliser granules are most efficient.

When broadcast with other crops in India, Pakistan and neighbouring countries, a seed rate of 5–12 kg/ha is common, the amount varying according to the percentage area safflower occupies. When grown in strips alternating with other crops, a common row spacing is 45 cm, with a seed rate of 18–22 kg/ha.

A major factor determining plant population for commercial rain-grown safflower crops is available soil moisture. As noted, rain that falls on the crop during its life is less important than the amount of soil moisture available to plants over this period. In areas where crops can be grown for only a certain number of years in any given period, selection of the correct plant population can mean the difference between profitable yield or complete loss; seed rates will also vary for different cultivars.

Safflower can compensate for spatial variation within a fairly wide range by producing more secondary and tertiary heads per plant. However, while final seed yield/ha may be little affected it will produce less oil/ha as seed from these heads are generally smaller, with lower oil content than seed from primary heads. Excessively branched plants also increase harvesting cost by slowing the operating rate.

For large commercial crops seed rates average 30 kg/ha when rain grown and 60 kg/ha when irrigated, and must be adjusted for cultivar selected. Wide rows, 35–60 cm, occasionally up to 90 cm, with close in-row spacing generally give a higher yield than the converse on commercial crops. However, in Nebraska, USA, sowing rain-grown crops in solid stands consistently produced high yield, but it is doubtful if this method would be successful where weed growth is prolific and rapid, or where there are perennial weeds that are difficult to eradicate.

Determination of the optimum planting period and plant population for a particular area under specific cropping conditions, is essential to optimise yield. However the time and resources allocated for these trials is often far too limited. The author's personal experience indicated that research data from developed countries is not always transferable to less-developed countries contemplating expansion or improvement of safflower crops.

Other factors affecting seed rate are depth of planting, soil structure and fertilisers. Where late planting is unavoidable, seed should be sown deeper and seed rate increased to ensure adequate emergence. The seedbed should be free from clods, but not so fine that it caps or crusts after rain thus hindering emergence. There is an interesting discussion on changes in germinating safflower seeds published by Chitta *et al.* (1997).

The factors affecting seed rate in order of importance can be broadly stated as:

(1) method of production, i.e. rain-grown or irrigated;
(2) time of planting;
(3) row-width and method of sowing;
(4) cultivar.

Safflower seed germinates moderately slowly, requires adequate moisture and fairly high soil temperature over this period. Thus in some years it may be necessary to postpone planting of rain-grown crops until the soil moisture has reached a suitable level and depth. Under irrigation a pre-sowing watering should ensure soil moisture is at field capacity for 1.5–3.0 m. Germination may take up to 30 days when autumn sown, and such plantings normally require higher seed rates to obtain adequate and regular stands. A seed-bed temperature of 13–18°C ensures quick, even emergence.

The influence of soil temperature is shown by data from many countries; temperature at planting of 8, 12 and 21°C resulted in times to emergence of 21, 12 and 7 days in the USA; in Poland higher temperature reduced the sowing to emergence period to 12 days in late May from 16 days in mid-April; in Australia planting on 5 May, June and July gave sowing to emergence of 7, 10 and 12 days; in Turkey sowing at 14-day intervals from 15 October to 25 November increased the time to emergence almost linearly from 18 to 56 days. Thus soil temperature at sowing, together with adequate soil moisture, are the main factors influencing emergence and to ignore either is to risk crop failure.

No specialised equipment is necessary to sow safflower, and most grain drills, small planters, and traditional seed drills such as the Indian goru (Fig. 8.13) are all suitable for their particular conditions. Safflower is often broadcast and hoed in by smallholders. Large fields are usually planted with grain drills, with special seeders for bed planting irrigated crops. Since safflower is frequently grown in rotation with grain crops, adaptability of equipment is important in the economics of safflower production.

When using grain drills for sowing safflower depth of planting must be even, though the actual depth will vary according to the amount of seedbed moisture. Even depth of planting ensures regular emergence and growth, and the resulting uniform crops allow easier and more efficient field-work and combine harvesting. Even plant growth also ensures that large plants do not dominate the weaker, which become more susceptible to pests and disease. A uniform crop also makes more efficient use of fertilisers and irrigation water.

Safflower seed flows easily through grain drills, and as seed size is a cultivar characteristic, that selected will vary the rate of flow. A grain drill set to sow 22 kg/ha of barley or 50 kg/ha of wheat will sow approximately 33 kg/ha of Gila safflower. For exact seeding rates machines should be individually calibrated and a check run on hard ground will show how many seeds are being planted per metre; 26 seeds/m in 45 cm rows is approximately 22 kg/ha of Gila. The author used combined grain drills with unskilled operators to sow safflower on a variety of soil types in East Africa. There were few problems in obtaining adequate stands on sandy soils, but in more clayey red earths there was frequent trouble with blocked spouts. Modified groundnut planters were also used on the red clays, but it was difficult to ensure an even flow of seed, and stands were irregular, allowing substantial in-row weed growth.

Seed should ideally be sown 3–5 cm deep into moist soil, but when the top 5–7.5 cm of topsoil is dry and loose, seed may have to be planted 10–15 cm deep. Special openers must be fitted to seeder units, and if necessary the small furrows so formed should not be completely filled. If later considered desirable, a chain harrow can be run over the field at right angles to rows to level off lands. Where soil is inclined to dry out quickly after planting leaving the surface fluffy, a light rolling is recommended. Timing of this operation is critical, for should a crust subsequently form, emergence can be seriously reduced. When correctly timed in autumn or spring in more cold climates, light rolling substantially reduces frost damage to plants in the rosette stage. A light rolling after planting may also assist in retaining seed-bed moisture.

Although broadcasting and discing-in seed is quick and labour-saving, yields are often disappointing, generally due to weed growth that is almost impossible to control. Minimum tillage so successful with soyabean is not widely practised with safflower for the same reason.

WEED CONTROL

Mechanical weeding of young safflower is difficult, and thus pre-planting cultivations should aim at maximum weed reduction. Sowing at a specific time to reduce the period spent in the rosette stage will assist in suppressing weed growth, and lessen crop damage from cultivating machinery. A light harrowing before emergence effectively kills young weeds, and while safflower is still small, finger weeders and rotary hoes can be used overall with little damage. Operations are most effective and least damaging to seedlings when carried out at right angles or diagonal to plant rows. On hot, sunny days these operations should be done as early in the morning as possible, when plants are more ductile.

When plants reach about 15 cm cultivations should be limited to the inter-row, and rotary weeders or duck's-foot sweeps are most effective. Mechanical weed control has frequently been compared with hand weeding in many countries, and as has been remarked elsewhere on other oilseed crops, careful hand weeding often gives the highest yield.

One or two efficient post-emergence weedings are usually sufficient in commercial crops, and additional cultivations do not significantly increase yield. In regions where quick-growing weeds can be troublesome in a mature crop at harvest, delaying the last weeding as long as possible can substantially reduce the number of weed seeds in combine samples. High-arch tractors and high-clearance equipment will allow the greatest delay, and when machinery for new projects or agricultural development is planned, these requirements should influence the choice.

Traditional hand hoes are the commonest weed-controlling implement in world use, but wherever possible these should be replaced by a tool that is less harmful to crop roots, and causes less indiscriminate damage when carelessly used. In smallholder crops planted in pure stands, the author found push-type wheeled hoes or weeders most successful, since these can be accurately adjusted for depth of working and are less tiring and faster to operate than hand hoes. Ease of working applies particularly to rotary machines which are most efficient in close-planted rows.

Herbicides are widely used in commercial safflower crops, combined with mechanical interrow weeding, and the most efficient and profitable combination must be determined locally. Equally important is that cultivars differ in their reaction to a specific compound and this must also be established for the cultivar chosen. The following have proved successful pre-emergence; alachlor, benfluralin, diuron, fluchloralin, glyphosate, isoproturon, nitrofen, pendimethalin, and trifluralin. No commercially available herbicide that gives adequate weed control can safely be used post-emergence on safflower. Safflower can also be affected by herbicides applied well before sowing, as fluometuron applied 6 months before planting severely damaged safflower in the Sudan, and an interaction occured between herbicides, water applications and nitrogen in Australia. Since herbicides are either too expensive or difficult to obtain by smallholders, methods of integrated weed control that reduce the amounts required are important, as in Egypt (El-Bially 1997).

IRRIGATION

Safflower is a very suitable crop to be grown under irrigation in dry areas, but the system to be adopted must be chosen with care. A most effective method of producing high-yielding healthy crops is by sub-irrigation, or by ensuring the soil reservoir is filled to a depth of at least 2 m by a substantial pre-planting application. In California, a depth of 2–4 m is considered desirable, in Israel 2–2.5 m, and in Russia 1.5–2.5 m. If soil at planting has sufficient moisture at 3–4 m the plant can develop roots rapidly enough to exploit this moisture to produce high yield. In this respect safflower is probably superior to any other annual crop. (See also remarks on water spreading in the introduction.) When grown on residual soil moisture following a rice crop in Pakistan and adequately fertilised, safflower yielded 1439 kg/ha seed over a 3-year period. In Australia safflower can be very valuable in a rotation with irrigated pasture/cotton, to 'mop-up' excess subsoil moisture.

Autumn sown safflower obtains moisture from snow and rain and the minimum of additional water is necessary to produce an economic crop. Such a situation exists in Iran, where above-ground growth virtually ceases during the winter, but root growth continues. These over-wintering safflower cultivars are considered to be more resistant to pests and disease than when spring-sown.

In the USA and Australia 1500–2500 mm/ha of water is necessary for high-yielding commercial crops. In Israel 600 mm of rain plus a similar amount as irrigation, and in Tanzania 400 mm plus 450 mm were the minimum necessary. In Tanzania 2250 mm applied in the dry season to a wholly irrigated crop produced twice the wet season yield, partially due to less damage from pests and disease. A rain-grown crop in India requires 650–1000 mm, but in the dry season under irrigation 1800–2100 mm/ha, less if the preceding crop was rice.

Mean consumptive water use and seasonal soil moisture depletion of irrigated safflower in the USA is shown in Fig. 4.8, and in Australia in Fig. 4.9. The first gives a very good indication of water use related to growth, and daily consumption. Although this data may be modified for a specific region or cultivar, it can be used as a basic guide when assessing water needs. Testing at greater

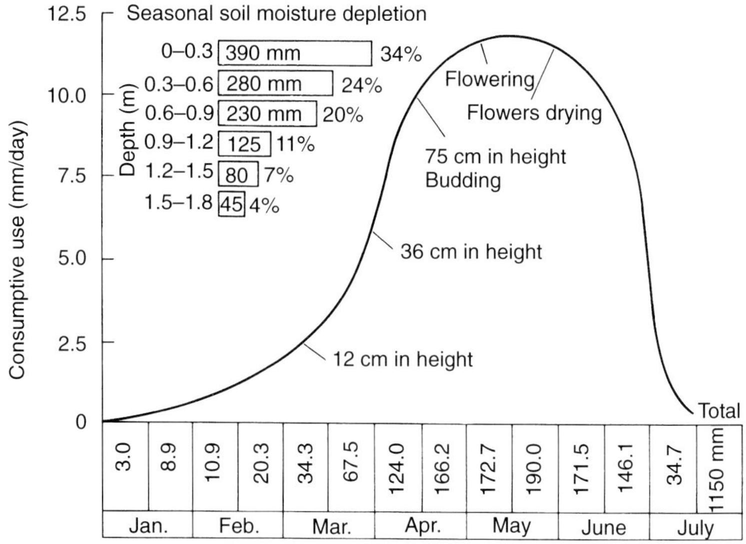

Fig. 4.8 Seasonal water loss and soil moisture depletion, safflower, USA.

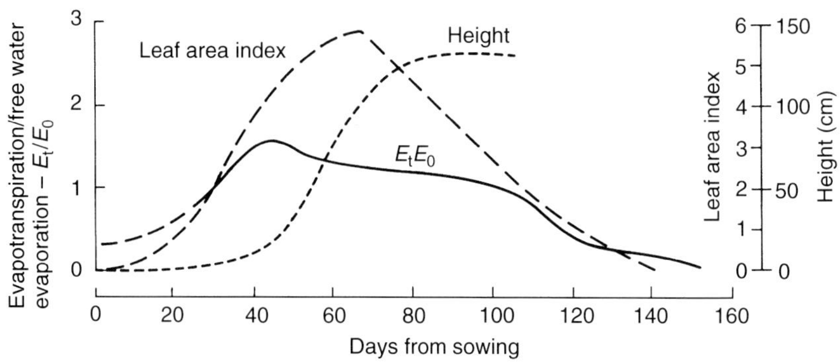

Fig. 4.9 Water use and plant growth, Australia.

soil depths than for other crops is necessary, and if the soil is moist from 30cm down, further checks at the same depth every few days will determine the rate of water use. The next irrigation should aim at refilling the soil reservoir to at least 2.0m. Irrigation trials in the USA showed that seed yield was significantly depressed when plants had used 90% of available moisture, but not when irrigation was applied at 60–70% depletion.

Other factors that modify water use are the influence of time of planting and fertilisers, and increasing the amount of either or both cannot compensate for loss of yield caused by sowing after the optimum time. Fertiliser, particularly containing nitrogen, can also interact with the amount of water applied, and significantly affect yield and profitability. A comparison of various levels of N and water application showed N can produce large increases in yield when combined with adequate water supplies, but had little effect when water was deficient; in India (Bansal & Katara, 1993; Bansal, 1995; Patil & Sabale, 1998). Similarly in the USA, where increased water application without adequate nitrogen depressed yields (Table 4.10) (Abel, 1976a).

Surface-irrigated safflower is susceptible to root disease and, as noted, has great ability to draw on subsoil moisture. The first requires very accurate control of the amount of water applied; the second allows fairly long intervals between

Table 4.10 Interaction between nitrogen and water application (yields in kg/ha).

Amm. nitrate (kg/ha)	Net water application (mm)			
	43	61	107	Mean
0	1320	1060	1040	1140
67	1410	1430	1360	1400
135	1520	1480	1690	1563
269	1470	1670	1880	1673

waterings, making application more efficient in terms of labour and water-transport loss. A common error is to apply the final irrigation too early and unless there are substantial subsoil moisture reserves the yield is reduced. The result is not readily apparent, because the thick hull gives seed a normal appearance, while it may be a pop or only partially filled. Conversely, continuing irrigation past 90% flowering can increase weed growth, thus hindering harvesting and increasing cleaning and drying costs. In East Africa and India sufficient water could be applied via every second or third furrow once plants had formed buds, with substantial savings in water and labour costs. Large furrows with a substantial water capacity are to be preferred to smaller, and allow fewer waterings.

Safflower is highly salt tolerant, and a comparison between it and other arable crops is shown in Fig. 4.7. Plants are often more severely affected by salinity at one stage of development, and safflower seedlings are only half as salt tolerant as mature plants. Methods of reducing the effect on germination and young seedlings must be incorporated in the agricultural or irrigation system. Salt levels in various soil horizons also affect growth and yield, and a higher salt level may be tolerated in a specific layer than in another. In Hungary 38 mg salt/100 g soil in the 0–20 cm layer and 10 mg in the 20–50 cm layer, gave a 40% increase in seed yield over 26 mg and 61 mg salt respectively in the same layers (Tesu *et al.*, 1975).

Salt tolerance of safflower is of considerable importance, and local methods of irrigation and cropping saline soils should be intensively investigated. The area of such soils is likely to increase rather than diminish, and their cropping and rehabilitation become of greater importance as the total area of land suitable for irrigation itself becomes limited. The most suitable fertilisers for saline soils must be determined locally using those types which will normally be available, for fertilisers can significantly change plant response to salinity. Climate also influences salt tolerance which tends to be reduced under hot, dry conditions. The type of irrigation water and the frequency of watering are important, and may even be decisive in their influence on plant response.

Safflower is less tolerant of salinity during emergence, and it is then that salt problems are often most troublesome, due to salts concentrating in the top 3 cm of surface soil. Thus germinating seed may be exposed to salt concentrations much greater than those encountered during later growth. This effect is generally not so important in flooded or basin-irrigated lands, but under furrow-irrigation salt accumulates in the vicinity of the seed (Fig. 4.10). Water permeating ridges carries salt into the centre and crown of the bed, and salt concentration there is often many times as great as in the original soil. An initial salinity of only 1 mmhos/cm ECe can become five to ten times higher after a single irrigation.

Salt accumulates less at the sides of the ridge than in the centre, and sowing on sloping beds allows the advancing water front to carry salt past the seed, and deposit it higher up the slope. Normal germination can be obtained even when the salinity of irrigation water is extremely high, for it is usually the initial salinity of soil at sowing rather than salt in the irrigation water that causes germination problems. When growing safflower on saline lands, watering should aim at maintaining an ECe below 4 mmhos/cm during the germination and early seedling stage. Salinity accelerates both flowering and maturity, the former by a few days but the latter by up to several weeks. In arid regions where temperatures are high, planting dates on saline soils should avoid plants flowering during periods of high temperature.

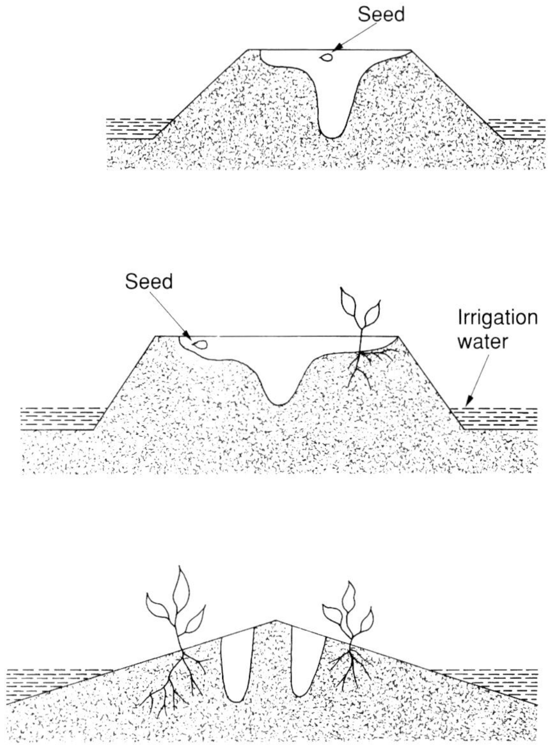

Fig. 4.10 Sowing seed on ridges in saline soils (areas of salt concentration shown unshaded).

INTERCROPPING AND ROTATIONS

Intercropping safflower is uncommon in smallholder crops as yields are generally low as safflower suffers from competition with almost all the crops involved, and the level of pest damage rises (see section on pests). Unless safflower is a major proportion of the crop, it is adversely affected by weed competition and initially more vigorous interplanted crops, as was apparent in Italy (Paolini *et al.*, 1998). In these circumstances safflower is generally less profitable than the increased yield to be obtained from the crop it displaces. However, small clumps of safflower are often grown in smallholders' fields as a source of a food colourant and for religious purposes.

Strip-cropping safflower with chickpea, pigeon pea, wheat or linseed in a ratio to suit local circumstances is profitable in India and Pakistan when the overall level of crop management is high. Safflower may be sown in closely spaced rows around field margins, when its prickly foliage also discourages trespassing by children and damage from small domestic stock.

Safflower should generally not follow safflower on the same land for at least 2 years, as it is extremely susceptible to soil-borne fungal diseases carried over on crop residues. It is usually rotated with small grains in North America and Australia, and may be grown after wheat or rice in Asian countries. The period between successive safflower crops must be locally determined, since available data indicates this varies widely, depending mainly on the predominant local disease and cropping programme.

HARVESTING

Harvesting safflower is comparatively simple since the crop does not generally lodge or shatter. For smallholders the extended harvesting period is of little importance, as the grower is often able to collect individual heads as they ripen. With commercial crops, relative immunity to damage allows the time of harvesting to be chosen with care, resulting in the highest possible yield and grain quality. A mature crop may be left standing in the field for up to one month with small loss. Light cold rain or frost does little damage. However, certain cultivars germinate in the head should there be periods of warm wet weather at maturity, and should not be grown where such conditions are common. Long delays in some regions can cause severe loss from birds, which quickly discover mature safflower fields; e.g. white-winged doves in California, parrots in Australia, love-birds and various species of weaver-birds in East Africa. The average seed yield of commercially grown safflower has increased steadily to around 1500 kg/ha, nearly twice this under irrigation.

Harvesting can usually begin 35–40 days after maximum flowering, when the plant is quite dry but not brittle, bracts on the heads turn brown, and seed has a moisture content below 8% preferably nearer 5%, easily determined by a moisture meter. Seed can also be tested for maturity by squeezing several heads between the fingers; if seed separates easily it is ready for harvesting. Safflower can be combined with a seed moisture content between 12 and 14%, but this is not recommended as drying is expensive or time-consuming. Artificial drying may also be detrimental to crops grown for seed.

Grain combines are quite suitable for harvesting safflower, but it cannot be cut as fast as wheat or barley, and the speed also depends on whether the crop is destined for crushing or seed. In a good thick stand, a reel will not be necessary and cutter-bar losses are small, even when the seed has a very low moisture content of 3%, and relative humidity at 15%. The drum will require adjustment and shoe screens and straw walkers checked during work, as they can quickly become clogged. Shaker speeds higher than those used for small grain are recommended, because of the branched nature of safflower straw and the tendency to bunch. Radiators and air inlets should be frequently checked and cleaned, as hairs from seed heads can cause overheating of engines and can also become a fire hazard. In more remote areas machines can also be out of action for long periods when they are most urgently required, awaiting spare parts to replace those broken by overloading. Tanker-type combines should be used whenever possible, since bags tend to pick-up plant spines making handling most unpleasant (Fig. 4.11).

Seed loss from the back of the combine can be estimated by counting seed in the straw row, or if a chopper and distributor are fitted, by holding a container at the back of a working combine for some 5 m. Seed loss is calculated in the USA using a standard seed-dockage pan and a correction table, and in Australia is usually 3–4%; about 1% at the cutter bar and 2–3% going through the combine, calculated by the number of seeds/square m in the straw row. In India, shattering was the largest single source of seed loss at 10–12% of total loss, and was substantially reduced by fitting a special reel to grain combines (Bhende *et al.*, 1996). Loss due to unthreshed heads can be calculated by counting the number of whole seeds that are hand-threshed from all heads found in a running metre

Fig. 4.11 Harvesting comercial safflower crop, California, USA.

of straw row; approximate loss in kg/ha is the number of seeds obtained divided by two-thirds the width of cut.

The main cause of damaged seed is incorrect adjustment of cylinder and concave, which not only reduces oil yield and quality, but can cause damage to the combine, for oil from broken seeds will collect dust and rubbish on moving parts leading to excessive wear, blockages and breakdowns. In addition to visible damage to seed, there is hidden damage to the embryo, even when hulls appear unbroken.

Cutting and wind-rowing before pick-up combining should only be carried out on commerical crops as an emergency measure, for direct harvesting costs less and normally produces a higher yield and better-quality seed. Windrowing some 3 weeks after 90% flowering, with a 25% seed moisture content, gives yields similar to combining dry plants. Windrowing can be advantageous when soil moisture remains high, soils are unusually fertile or in a cool growing season – all factors that tend to prolong the period plants remain green. Difference in the rate of maturity between varieties then becomes important, and early maturing varieties or those with a tendency to dry off quickly following flowering should be selected.

Excessive weed growth in nearly mature crops makes combine harvesting difficult and windrowing essential and timing important. Cutting and leaving in the field for too long can allow regrowth of weeds through the cut crop. Cutting should be carried out when seed moisture content is low enough for pick-up combining immediately the windrows are dry. Residue from windrowed crops is a valuable fodder, and safflower stubbles are readily grazed by stock.

STORAGE

Seed should be stored in bulk where possible, and grain bins are quite suitable, provided that the seed moisture content is between 5 and 8%, preferably the lower. Forced-air in-bin drying is not common owing to high cost of the installation; it may be justified for certified seed producers. Seed intended for crushing should be stored only for short periods, with delivery to the processor as rapidly as possible. Where seed has to be transported some distance to buyers or millers, it may be necessary to store a certain tonnage to take advantage of bulk freight rates. Safflower is both edible and saleable, thus in many countries stores must be thief-proof and this is an additional reason for swift movement of produce.

Viability loss in stored seeds is related to rate of respiration, which can be restricted by refrigeration, reducing initial seed moisture content, or available oxygen. A moisture content of 10% is too high for safe storage at temperatures above –1°C, and 7% in sealed cannisters. The type of container affects seed viability; in general bags composed of laminated plastic and paper or metallised polyester, are more suitable than either plastic or paper alone. However, during a temporary seed shortage in East Africa, the author planted seed that had been well dried and stored for at least 2 years in large double-layered paper bags in an airy shed on an experimental station. The humidity within the building varied as the outside air, and interior temperatures between 7°C and 24°C. Germination was not determined, but seedling emergence was 85% of the required stand.

Long-term storage of breeding stock requires careful control of humidity and temperature in particular, and about 5% moisture and a constant 10–12°C will maintain viability for many years, although there is considerable cultivar variation. To reduce the rate of deterioration, seed can be treated with a number of chemicals, including the halogens chlorine and iodine (Singh & Joythi, 1996; Ramamoorthy & Natarajan, 1997). The most important factor in safe long-term storage is sound seed with a low initial moisture content, irrespective of container, atmosphere or temperature.

PESTS

The majority of insects that attack safflower are of little economic importance and do not require control, especially in smallholder and interplanted crops, when an overall spray with a particular chemical may be of little use, and specific control of individual pests may involve a number of compounds and applications. However specific insects such as the fly *Acanthiophilus helianthi*, can virtually preclude safflower growing, as noted later.

Chemical control of foliage pests is seldom necessary except in large commercial fields, but may be required for pests that attack buds and flowering heads. The degree of damage that plants must suffer before the yield is affected and the value of the yield decrease in relation to the cost of protection, are major factors in the profitable use of pesticides. The leafworm, *Spodoptera litura*, initially attacks lower leaves, then progressively moves up the stem. To determine the degree of damage causing a yield decrease, safflower was artificially defoliated when plants were fully grown and first flowers visible, when *Spodoptera* attacks normally occur. Seed yield decreased in direct proportion to the degree of defoliation, and oil content was significantly reduced, as shown in Table 4.11. (Beech, 1964).

The necessity for balancing chemical control of insect pests against allowable damage is essential, for complete control is seldom possible or profitable on the majority of safflower crops. Integrated pest control including suitable cultivation, time of planting, and resistant varieties should be practised. Natural predators exert considerable control, for example 22 species of insect and one Trombiform mite in Israel. Occasionally a particular pest, such as the safflower fly, *A. helianthi*, can be so damaging that it effectively precludes profitable production.

Acanthiophilus helianthi is small, rather bristly, greyish in colour, and has light-brown legs (Fig. 4.12). Maggots feed upon florets and thalamus, infected buds rot, and an evil-smelling fluid oozes from the apical opening, giving the whole bud a bedraggled appearance. The fly can have up to three generations over the life of a crop, but an early infestation may remain in the terminal shoots before flower formation. In France it decimated plantings and was the major reason why the crop was abandoned. The fly is common in the Mediterranean region, and if safflower plantings are extended would probably become of greater importance. It has been recorded on Compositae in East Africa but the author did not record it on safflower grown in Tanzania or Kenya, although it is common in Ethiopia at all altitudes. In Israel it is important on the coastal plain and Jordan valley where it is common on wild Compositae. It

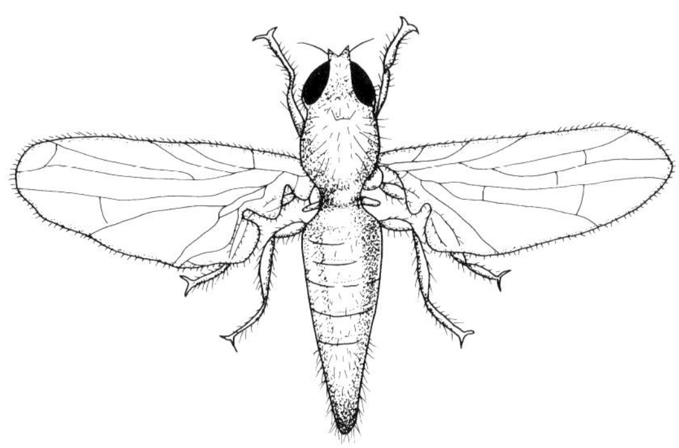

Fig. 4.12 *Acanthiophilus helianthi* (safflower fly).

Table 4.11 Effect of defoliation on safflower – Australia.

Attribute	Degree of defoliation			
	0	$\frac{1}{4}$	$\frac{1}{2}$	$\frac{3}{4}$
Seed yield (kg/ha)	2787	1911	1577	1450
Oil content of seed (%)	32.1	31.4	31.2	30.2
Oil yield (kg/ha)	877	652	490	437
Plant population (thousand/ha)	370	235	191	247
Number of heads/plant	3.13	3.47	3.42	2.83
Number of seeds/head	24.00	22.22	20.74	20.94
100 seed wt (g)	4.54	4.78	4.94	4.37

causes extensive but local damage in Turkey, Iran, southern Russia and Pakistan, but can limit safflower growing in some regions of India. In tropical climates it is often kept in check by predators, *Chrysops vingetes* in India, and parasites. Chemical control is difficult as the rate of reproduction and reinfestation is high.

Seedlings are attacked before and during emergence by the local *Agrotis* spp., cutworms, in virtually all the countries safflower is grown; see Fig. 2.17. It is sometimes possible to estimate the larval population by examining areas where birds are seen feeding following ploughing, as cutworms are a favourite food. Cutworms are 4–5 cm in length, usually greenish-brown or yellowish-grey, smooth and shiny, and normally feed at dusk by eating through stems of seedlings at ground level. Since they seldom move far, examining the ground near damaged plants will usually discover the species responsible. Lesser damage is caused by wireworms, *Agriotes* spp.; scarabs, *Anomala* spp., are more important in Africa and Asia. All can be controlled by seedbed insecticides, or baiting in smallholder crops.

Stem-borers, particularly *Melanagromyza* spp., can cause significant damage. Larvae bore stems and sometimes leaf ribs, and can be seen as very small, shiny white maggots or shiny black pupae in the stem. In India and neighbouring countries the pest is widespread, but economic damage occurs only on particular crops. In the USA, *M. virens* larvae live in stems and cause wilting of growing shoots in late spring sowings, but rain-grown crops are less susceptible than irrigated crops. In Africa *Ophiomyia phaseoli* is the most common. In the Caucasus, Cyprus and Israel, larvae of the weevil *Lixus speciosus* produce spindle-shaped galls on stems; a heavy infestation reduces yield, but normally only odd plants are attacked. Larvae of *Agapanthia dahlii* and *A. cardui* bore stems in Israel and other eastern Mediterranean countries, but damage is usually confined to odd plants, or a small group. This pest also attacks sunflowers and where these are grown adjacent to safflower, damage could be greater, especially if growing seasons overlap.

Foliage pests are generally of no great significance, but leaf worms and various species of looper caterpillar are locally damaging. The noctuids *Spodoptera litura*, *S. littoralis* and *S. exigua*, leafworms, are widespread and economically important pests of many cultivated plants. They occur in the majority of countries in which safflower is grown, and the degree of damage caused is related to the proportion of local cropping that safflower occupies. Safflower interplanted with small grains is often little affected, but when planted in the vicinity of cotton, tobacco, legumes or vegetable crops much more severe damage occurs. Larvae feed on foliage, but may also tunnel into developing buds. Since the moth is polyphagous, large areas must be sprayed if chemical control is to be effective, but hand picking of caterpillars is a useful control method at smallholder level.

Bollworms, *Helicoverpa* (*Heliothis*) spp., are common pests, and cause varying degrees of damage locally; *H. punctigera* is of importance in Australia, and control measures are necessary; *H. armigera* is the common gram pod borer in India, but also attacks safflower when the two are grown in proximity (Fig. 4.13). The degree of damage to plants and pods varies but can be severe in specific crops. *Perigea capensis* and *Trichoplusia ni* are common in India, Pakistan and South-East Asia generally.

Aphids are common and widely distributed safflower pests, and large colonies cause stunting and severe leaf distortion. They also produce

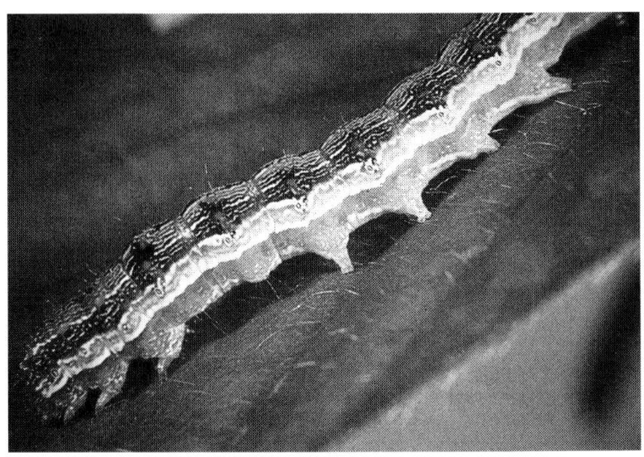

Fig. 4.13 *Helicoverpa (Heliothis)* larva on leaf.

honeydew which spreads over leaves and encourages growth of sooty mould, reducing photosynthesis, and on flowers reduces pollination. By far the most important effect of aphids is that they are able to transmit more viruses than any other insect. The green peach aphid, *Myzus persicae* is cosmopolitan, polyphagous and a major pest of safflower. In the USA it can become abundant enough in spring to seriously damage buds and seed heads. Primary buds so infected develop a spotty or mottled yellow coloration, and seed yield can be reduced by one-third. In some years East African plantings were continually reinfested from seedlings to flowering; in others there was one peak just before flowering. In Australia the aphid is often controlled by natural predators, including parasitic wasps and hover flies.

The aphid complex on safflower in India, which included *M. persicae*, was actively predated by *Coccinella septempunctata*, *Menochilus sexmaculatus* [*Cheilomenes sexmaculata*] and *Syrphus* species (Gautam *et al.*, 1995). *Macrosiphum* spp. are common in the Middle East, Asia Minor, Asia and South-East Asia, and can become major pests; the small black insects infest stems and leaves. In India, *M. compositae* and *M. carthami* are serious pests; *M. sonchi* in the Middle East, Asia Minor and southern USSR. Other aphids have also been recorded as causing economic damage in various countries, but identification is unreliable. Leafhoppers, usually *Empoasca* spp., occur widely but damage is minor.

Lygus bugs of various species attack developing flower heads and buds, and one bug to six or eight buds is sufficient to significantly reduce yield. Adults and nymphs cause injury by puncturing the buds, and at harvest results can be seen as unfilled heads or darkened, shrivelled seed. *Lygus hesperus* and *L. elirus* are important in the USA, and *Taylorilygus vosseeri* adults were found by the author on safflower growing in the vicinity of irrigated cotton in central Tanzania. As this is the most common lygus on cotton throughout Africa, it is probable that growing the two crops in proximity, especially under irrigation, could result in increased damage to both, since their main growing seasons overlap. *Nysius vinitor* may cause extensive damage in Australia (Fig. 4.14).

To determine the damage caused by *Lygus* spp. which would seriously reduce yields, plants in the USA were artificially debudded, or insects

Fig. 4.14 *Nysius vinitor* nymphs (on right) and adults.

controlled by spraying (Table 4.12) (Knowles & Miller, 1965). Removing all the buds did reduce the number of seed heads, but the weight of seed in the remaining heads was significantly increased. De-budded plants produced more but not larger seeds per head, and this increase was sufficient to compensate for reduction in head numbers. Safflower cultivars are able to compensate for such attacks to a very high degree, which explains why yields in the USA have not been materially reduced by insects such as thrips and lygus until prolonged high-level infestation occurs.

Thrips are frequently present but unless populations become very large are of little significance; *Frankliniella occidentalis* and *Thrips tabaci* are most frequently reported. Mites occur generally on safflower, and many species are now resistant to chemicals used to control other insects. Some species may be predators on other mites, and their destruction thus allows an increase in the mite population and plant damage. *Tetranychus urticae*, the two-spotted spider mite, and *T. pacificus* occasionally occur in California in very large numbers, and are capable of defoliating a late-sown crop before it matures; *T. telarius*, carmine spider mite, is widespread and recorded on safflower in many countries, but is usually of little importance. It is the most common spider mite in Egypt, and of some note in Cyprus, Jordan, Iran and the southern Russia. Although common on castor in India and South-East Asia, it has not been reported as damaging safflower.

The susceptibility of safflower to soil nematodes and the degree of damage caused is difficult to assess. Reports from the Middle East, India and South-East Asia of nematodes in association with roots of safflower indicate there was no major damage to plants. In East Africa the author carried out an extensive survey of the nematode population on lands previously under black wattle (*Acacia mearnsii*), since in other countries it is considered host to a number of parasitic nematodes. Although nematodes were found in soil samples from subsequent maize, castor, wheat and safflower crops, they were usually saprophageous and no major damage was found that could be directly attributed to nematodes.

Greenhouse trials in the USA concluded that growing safflower on root-knot nematode infected land could result in economic damage and maintain nematode populations, especially in areas where high soil temperature is common. Safflower should therefore be excluded from rotations with sugar beet when root-knot nematodes are present; however, where only the sugar beet nematode is present safflower may be useful

Table 4.12 Effect on seed yield of debudding safflower – USA.

Treatment	Number of healthy heads/plant	Seed/plant (g)	Seed/head (g)	100 seed wt (g)	Number viable seeds/100 seeds
Control	22	32	1.5	3.8	97.7
Sprayed twice*	21	29	1.6	4.0	98.9
25% debudded	18	32	1.6	3.8	97.9
25% debudded and sprayed twice*	21	31	1.7	4.1	97.9
50% debudded	17	32	1.7	4.2	97.9
40% debudded and sprayed twice*	16	29	1.9	4.1	97.9
100% debudded	13	32	2.3	4.4	98.0
100% debudded and sprayed twice*	13	31	2.3	4.3	99.1

*Dimethoate at 0.56 kg a.i./ha in two application 3 weeks apart.

because some cultivars are not suitable hosts. Gila was susceptible to *Meloidogyne javanica* and *M. incognita acrita* under glasshouse conditions, but not when planted-out in the field; the last named also attacks cotton and sorghum, two crops commonly grown in safflower-producing areas. Rotation including these crops in close succession could therefore be unwise.

Whole, sound, dry safflower seed is resistant to many storage pests, but since few bulks are composed wholly of such seed, a range of insect pests can usually be found in stores. *Lasioderma serricorne* is common in the USA, India and Israel, and has been seen by the author on stored seed in Mexico, the Sudan, Pakistan, Hong Kong and Japan. This insect, and others, were able to damage and breed on safflower seeds and meal in the USA, but only *L. serricorne*, *Stegobium paniceum*, *Trogoderma glabrum*, *T. inclusum*, *T. parabile* and *T. surplex* were able to breed on good-quality seed. Many other species are able to increase in damaged or broken seeds, which suggests that prevention measures should begin during harvest, when minimum damage to seeds will limit the development of insects during storage. Thin-hulled cultivars are more susceptible. *Lasioderma serricorne* has also been found feeding on maturing safflower heads in the field, and when plants are being grown for seed production, spraying may be necessary to prevent introduction into uninfected stores or seed stocks. For additional information on storage pests and pest control in general see the section in the Chapter 3 on groundnuts and Fig. 9.8.

Plant parasites occasionally become important; broomrape, *Orobanche* spp. has been reported in Europe, and dodder, *Cuscuta* spp. in North America. Birds can cause damage to mature plants left standing in the field, as noted in the section dealing with harvesting techniques. In tropical countries the various rodents known locally as ground squirrels dig up newly sown or freshly germinated seed around field margins adjacent to uncropped land, and small buck may also be troublesome. In Africa porcupine and wild pig can cause extensive damage, and the author once lost an irrigated seed-multiplication nursery on the banks of the Ruaha river in Tanzania when a hippopotamus grazed through it one night!

DISEASES

A range of diseases has been recorded from safflower although few are of economic importance, but some are so damaging they can limit commercial production. The USDA estimated disease damage as follows: rusts 5%, root rots 3%, bud rots 2%, leaf spots 1%, verticillium wilt 0.5%, and diseases causing damping-off 0.3%.

Safflower is extremely susceptible to root rots, or foliar fungal diseases caused by excess moisture, and irrigation practices should be aimed at minimising their incidence. Pre-planting irrigation is one of the most effective methods of ensuring adequate moisture for plant growth and reducing the possibility of root infection. Watering alternate furrows will also assist, especially where there may be warm cloudy conditions, or periods of high humidity. Excess moisture or drought conditions during rapid growth can also increase damage from root rots; disease appears when water is reapplied or rain falls.

Root rots caused by *Phytophthora* spp. including *P. cryptogea* and *P. drechsleri* are widespread and very damaging. *P. drechsleri* is a serious disease of surface irrigated safflower, and its incidence and severity is increased if irrigated safflower has suffered moisture stress. Fluctuations in dry and wet conditions are most damaging, as dryness increases plant susceptibility and wetness favours the pathogen. A symptom of infection of irrigated safflower by *P. drechsleri* is that plants wilt even in the presence of adequate soil moisture, as this pathogen causes water stress by increasing resistance to water uptake by roots and through the stem xylem. Plants are susceptible at all stages of growth, but most especially in the seedling stage (Fig. 4.15). Symptoms become most obvious at flowering; plants wilt, turn yellowish and die. In the early stages root tissues may show a red colouring, but later both roots and stems become dark and discoloured. Autumn planted safflower is often little affected, as low temperature suppresses the pathogen.

Fig. 4.15 *Phytophthora* root rot of safflower seedling.

First dead plants, which can occur individually or in patches, are usually noticed in spring when daytime temperatures reach 30–32°C. Irrigated plants killed by *Phytophthora* are most evident 4–5 days after watering and their light-green colour is characteristic.

Wilt or stem rot caused by *Sclerotinia sclerotiorum* is widespread and often seriously damaging in humid conditions which encourage moist soil at the base of plants. During mechanical weeding, wet soil should not be built up at base of young plants, as this materially assists penetration by the fungus. Above-ground symptoms resemble those of *Phytophthora* root rot. Plants become yellowish, wilt, turn brown and die. Large black sclerotia are formed on the crown, inside the stem and on adjoining roots, though the roots themselves are not generally affected. Cortical tissue of the lower stem shreds. At an advanced stage flower heads can fall from affected plants, leaving the outer involucral bracts *in situ*. This effect is considered to be caused by the development of a large pear-shaped sclerotium in the thalamus. Plants only slightly affected do not show major symptoms, but at maturity the heads often contain few or no seeds, and varying amounts of dusty-looking debris. The fungus persists in the sclerotoid form in soil and plant remains, and since it has a wide variety of hosts, control is difficult. Collection and destruction of infected material, deep ploughing and clean weeding can help prevent spread of the disease. On small areas required for breeding or seed production soil fungicides are effective. The disease also attacks sunflower, and the two crops planted in proximity of following each other in a rotation, can exacerbate the degree of damage.

Another source of seedling loss at or just before emergence is a damping-off caused by *Pythium* spp., especially in warmer regions. In Canada *P. ultimum* var. *ultimum* significantly reduced stands in the southern prairies (Mundel *et al.*, 1995). Safflower sown into warm soil is most affected by damping-off especially if the soil is wet. The damage can be substantially reduced by early sowing before soil temperature had risen.

Wilt caused by *Verticillium* spp., particularly *V. alboatrum*, is a potentially serious disease of cultivated safflower and of worldwide occurrence. Various degrees of damage occur and symptoms can also vary from plant to plant. Some may be killed before flowering, others develop sterile heads, or develop apparently healthy heads, which on inspection contain small or light-weight seeds. The usual symptoms are intervenal yellowing of lower leaves, with progressive upward discoloration. Frequently only one side of the plant is affected initially, but finally the whole plant succumbs. Plants are attacked at any growth stage during cool weather, but are seldom killed unless infection is severe. Affected plants generally mature more quickly than healthy plants. *Verticillium alboatrum* also attacks a wide range of other cultivated crops including cotton, sorghum and groundnut, and where incidence of this disease is high, safflower should not follow or precede these crops in a rotation. Interplanting or strip-cropping with these crops will also increase total damage. Resistance to *Verticillium* wilt exists in some cultivars.

Another wilt, due to *Fusarium oxysporum f. carthami*, can cause considerable damage especially to young seedlings. Symptoms are unilateral yellowing of leaves and sometimes death of branches on the affected side, frequently progressing into flowering heads on the main stem; heads become blighted and distorted. Cultivars are available which are highly resistant to *Fusarium* wilt, and others which show resistance to *Verticillium* and *Phytophthora*.

The most economically important disease attacking safflower is rust, *Puccinia carthami*, which occurs in all areas of commercial production, and is endemic over a wide area of safflower's natural range (Fig. 4.16). Plants are attacked from pre-emergence to flowering, but it is on young plants that most serious loss occurs. Symptoms on young seedlings are yellow discoloration, drooping and wilting of cotyledons and first leaves. Seedlings may also be infected in the

Fig. 4.16 *Puccinia carthami* rust on safflower leaf.

hypocotyl and die without exhibiting above-ground symptoms. On older plants stem girdling can appear, roots can be affected, and chestnut-brown pustules appear on leaves and bracts.

Safflower rust has two pathological phases; the seedling phase follows infection of hypocotyls by seed-borne or soil-borne teliospores and causes serious reduction in stand when untreated rust-infected seed is planted, or when a crop follows a badly rusted safflower crop. The foot and root phase is favoured by temperatures of 15, 20 and 25°C and hindered at 30–35°C, but soil moisture is apparently not critical. The foliar phase causes a substantial reduction in yield if rust develops early in the growing season. As *P. carthami* is an autoecious macrocyclic rust that rapidly completes the sexual cycle, maximum opportunity exists for the recombination of existing genes and development of new races; a number have been identified in the USA. Cultivars may have resistance to some but not all races, and in certain thin-hulled cultivars resistance is not linked to, or suppressed by, the gene controlling this character.

Two important and widespread leaf spots are *Alternaria* spp. and *Cercospora* spp. Leaf spot due to *Alternaria carthami* is widespread and can be particularly severe on irrigated plantings. Moisture appears essential for infection of plants, and in warmer areas heavy dews or frequent showers, or periods of coastal fog as in California, favour its spread. The disease can be seed-borne and cause seed rot and damping-off of seedlings; in Canada, *A. carthami* was recorded as being present on 95% of untreated seed (Fig. 4.17). The optimum soil temperature for its growth and spread is 25–30°C, with a pH 6.0. In more mature plants, symptoms on leaves are small brown to dark-brown spots with concentric rings 1–2 mm in diameter. Spots gradually increase in size to approximately 1 cm and may coalesce. The centre of these mature spots is usually lighter in colour and surrounded by alternate rings of light and dark brown. Fully mature spots tend to develop holes, and in severe infections irregular cracking of the leaf occurs. On stems and petioles the disease is less severe, with elongated spots. On flower heads the fungus first

Fig. 4.17 *Alternaria*-affected safflower seeds. Healthy seed left, diseased seed right.

attacks the base of the calyx and later spreads to other parts of the flower. Infected buds remain unopened, shrink and desiccate. The fungus can survive on plant debris for at least 2 years, and to reduce loss non-susceptible crops should be sown in rotation. Some *Alternaria* spp., but not apparently *A. carthami*, may also increase the free fatty acids of safflower seed.

Cercospora carthami is of worldwide occurrence, but as with many diseases of safflower, is important mainly where the crop is grown in large fields. The disease can attack plants at almost any stage of growth when there is a moist humid atmosphere with little wind, and can occur a few weeks after emergence or when plants are flowering. Symptoms are isolated, round or irregular brown spots, 3–10mm in diameter, on lower leaves. Spots are slightly depressed, with a yellowish border and sometimes concentric rings. Affected leaves become brown and distorted, with intervenal necrosis; minute black fructifications of the pathogen appear on upper and lower surfaces of the spots on affected leaves. Under moist conditions the spots have a velvety, greyish-white appearance caused by spore production. Stems and nodes may also be infected, and in a severe infection, bracts and the entire capitulum are attacked progressively. An infected bud turns brown and dies while flowering heads wither.

Leaf spots, which may cause severe local damage on commercial crops are *Macrosporium carthami*, and *Macrophomina phaseolina* in Australia, which also attacks cotton and castor in India where it is a serious disease of sesame. *Stemphylium* spp., *Phoma* spp., and *Septoria carthami* occur in the USA and North Africa, and *Gloeosporium carthami* in the USA and Australia – all cause varying degrees of damage in specific crops.

Head rot, head blight or grey mould, caused by *Botrytis cinerea* commonly occurs on heads in regions where there are long periods of high atmospheric moisture before and during flowering. Spores of the fungus are wind borne and infection occurs from budding to post-flowering. The fungus also attacks a wide range of cultivated plants. Symptoms of infection are flower heads becoming light yellow followed by complete bleaching. Badly infected heads can be readily detached from the stem. Another widespread head rot is caused by *Ramularia carthami*, which can sharply decrease yield and quality of seed. It is common in Russia and the Indian region. Some resistance to the disease has been found in India, where it is also a troublesome leaf spot.

Quick-maturing varieties first produced in Mexico show strong resistance to *Fusarium* and *Verticillium* wilts. A rapid method of determining degree of resistance to fungal infection in segregating populations has been developed in India based on leaf extract. It appears that resistant varieties exhibit polyphenol-oxydase activity, which inhibits spore germination. The test can be used to determine resistance to *Alternaria*, *Ramularia*, *Cercospora* and *Esysiphe* spp. (Karve & Deshmukh, 1977). Resistance to *Phytophthora* root rot, *Verticillium* and *Fusarium* spp. exists in the USA world collection and is being transferred to commercial cultivars.

Viruses isolated from safflower include lettuce, chilli, alfalfa and cucumber mosaic, aster yellows and prunus types A, B, E, and G. The most general symptom on infected plants is necrosis in a more or less severe form. Plants nearing maturity show necrosis of terminal apices, stems and portions of leaves, and plants that survive yield only small amounts of seed. Seed harvested from plants infected with cucumber mosaic virus produced no diseased plants, indicating the disease was not seed-borne, and transmission would appear to depend on a heavy aphid population, and appearance of infected aphids early in the growing season. *Myzus persicae* can mechanically transmit a virus in Morocco. Control of virus diseases is difficult since they can be soil-, seed- or insect-borne. The only reliable method of reducing the incidence of seed-transmitted viruses is to use seed from virus-free crops.

5
Sesame

The Pedaliaceae contains the Tubiflorae of 16 genera and some 60 species, with the genus *Sesamum* containing about 36 species including cultivated sesame, *S. indicum* L. Throughout the text, sesame without qualification refers to cultivated *Sesame indicum*, its seed and derivatives.

Sesame is probably the most ancient oilseed used by man, and its domestication lost in the mists of antiquity. Although considered to have originated in central Africa, most probably Ethiopia, Indian botanists argue there is convincing evidence that sesame originated in India (Bedigian, 1984). Sesame was widely dispersed by man both westward and eastward, reaching China and Japan, which themselves became secondary distribution centres.

Sesame was an important crop in the Mesopotamian region during the third dynasty of Ur, 2130–2000 BC, which became the main centre of distribution of the species as a domesticated plant (Burkhill, 1953). Clay tablets list oil mill components used around 2000 BC, but the method of expression is not firmly established, although mill owners were organised into a guild. King Sargon II (722–705 BC) considered the oil so important he introduced price control, which remained in place during first millennium BC. Sesame was so important and valuable that loans were negotiated in sesame seed as an alternative to silver. Sesame has been dated at 2000 BC from excavations at Harappa, Pakistan and the Medical Papyrus of Thebes, Egypt, about 1500 BC, lists a number of herbal remedies including sesame oil and seeds.

Despite its many uses in ancient Sumeria, sesame was not generally employed in early biblical times in the Palestine region, and there is no attributable mention of it in the Bible. It was apparently not used by the Jews until the second century AD, with the Hebrew name of *sumsum* or *semsem*. It was mentioned by Pliny, Herodotus, Strabo and other early historians, and in Roman times was a cultivated plant in Italy (White, 1970). Introduced into England in 1731 (*Bot. Mag.*, 1688), it remained a botanical curiosity. Commercial cultivation as a cash crop in Russia dates from the end of the seventeenth century, but in Central Asia it was known much earlier, having been introduced via Iran and northern India. In Marco Polo's description of his journey to China he noted that the sesame oil of Balashan, in the extreme north-east of Afghanistan, was 'better and has more flavour than any other'.

The earliest records from Africa south of the Sahara are in books written by explorers and travellers, and there has been considerable argument as to whether it spread from west to east or vice versa. In the 1850s sesame was grown by Arabs on the East African coast north of Mombasa, and in 1852 a French company exported the seed from Lamu, while Zanzibar exported 3628 t of sesame seed valued at $ 20 000 in 1859. *Sesame angustifolium*, *S. radiatum* and *Ceratotheca sesamoides* (false sesame) are also cultivated very locally in Africa. It was a valuable cargo along the Red Sea and southern Arabian coasts in the fourth century BC. In the Hadhramut a mixture of sesame oil and indigo was applied as a sunscreen and skin conditioner, giving the skin a blue-black appearance.

Cultivation in the Asian region is ancient if the Harappa finds in Pakistan are accepted, but as Sir George Watt (1966) remarks 'It is certainly remarkable that few, if any, of the early European travellers in India such as Garcia de Orta or J.H. van Linschoten, make mention of this plant or its oil. In the *Ain-i-Akbari* (AD 1590) frequent reference is made to both black and white seeded forms, and there is abundant evidence that it was an important crop for at least 300 years previously.' One can only speculate on reasons for these omissions for the plant, its seeds and oil have been widely noted in Indian literature from as early as the Vedic period. The *Athavaveda* of about 1000 BC calls it tila, and the *Samhitas* of two hundred years later mentions wild sesame seed being used to make a gruel, a porridge made by boiling sesame seeds, rice and milk, and young leaves and stalks as pot-herbs (Achaya, 1990).

By the first century AD sesame was essential for local religious ceremonies including *sraddha* and *pitryana*, but it is odd that its use at night should be avoided. The oil is used to promote softness of, and give a silken sheen to the skin, and is massaged into joints to keep them supple. In India jaggery (raw sugar) is sometimes added to the mix when crushing black sesame seeds to counteract the slight bitterness of the oil. India has the largest area under sesame, at about 26 Mha, equal to 40% of the world total and a quarter of world production; 6 Mt in 1998 and 5.4 Mt in 1999.

Sesame is common in Burma and South-East Asia generally, although groundnuts are tending to oust it as a cash crop. From inscriptions found in Burma, sesame was well known in the eleventh and twelfth centuries AD, and according to contemporary religious law 'Sesame was to be cultivated by slaves dedicated to the three gems, Buddha, Law and Clergy,' and pagoda lamps burnt the oil. Ash from sesame plants was employed in washing clothes before soap was generally available. Marco Polo on his return from China by sea via Ceylon, mentions the plentiful supply of sesame oil on the island.

Cultivation and use of sesame in China is ancient, and in the sixth century BC the seed apparently achieved enough prominence to be used as currency. Early Chinese writers' habit of confusing sesame with other oil crops, particularly linseed, unfortunately obscures its early history, although indicating it came from the West. When it was introduced to Japan is uncertain, the first record is AD 702 but from AD 900 it was being cultivated and by AD 1700 was a well established field crop. It remains an important domestic and commercial cooking oil, with many traditional uses. Parched sesame seeds are sprinkled on a variety of ceremonial dishes including *sekihan*, rice cooked with red beans, offered to guests in return for their congratulatory presents at weddings and similar occasions. A saying among rural Japanese in feudal times regarding the tribute exacted by their rulers stated that 'A farmer is like the sesame seed, the harder you squeeze the more you get!'

The Portuguese introduced sesame to Brazil where it is known as gengelim, a name so directly comparable with the India name gingelly as to suggest it was from their Indian colonies. Brazilian production is substantial, mainly for domestic consumption; it has flourished as a commercial crop in Venezuela where conditions are especially favourable, and local research on the crop is of the greatest international value. Slaves are considered to have introduced sesame to North America towards the end of the seventeenth century, and in the southern states it is still referred to as benne, almost identical to its West African name of beniseed. The first deliberate introduction to the USA was from Liberia in 1898.

The majority of sesame seed harvested is consumed by local producers, and it is doubtful if 10% of total production enters world trade (Weiss, 1971). Many countries have no accurate statistics of internal production from crops which are neither marketed or exported, and in these circumstances production figures must of necessity be very approximate. In 1970 world seed production was estimated at 2 Mt with exports of just over 200 000 t; in 1980 it was 2.6 Mt with a similar amount exported, and this level has been approximately maintained although total world demand has risen. Few accurate statistics of world production are now available, particularly from Africa, where sesame was grown throughout the continent, currently ravaged by endless wars in the main producing countries, including the Sudan.

World production of sesame seed remained almost static for 30 years to 1995, despite an increasing demand for sesame oil in non-European countries. Sesame seed has declined in international trade in favour of other more easily, and therefore more cheaply, produced oilseeds. In some countries however, Burma and China in particular, the area planted has increased to supply local and regional demand.

Practically all world trade in sesame is as seed, and only minor quantities of oil and cake are shipped. Although some 100 000–200 000 t of sesame seed is normally available annually for export, most movement is between neighbouring countries, for example Burma and other South-East Asian countries. Japan, Italy and China are major markets for truly international trade, with Israel, Saudi Arabia and Greece substantial consumers.

Africa with an estimated 20% of total world sesame production to 1980, previously contributed nearly 70% of exports with Sudan dominating the export trade, frequently accounting for half of world exports. This trade has now substantially diminished, and world sesame seed demand towards the end of the twentieth century was probably greater than world production.

The great diversity of sesame types, their wide environmental adaptation and considerable range of seed oil content and characteristics, make an exceptional gene pool available to plant breeders. There is thus no basic reason why cultivars should not be produced to greatly extend the range and profitability of sesame growing, as increased availability would easily be absorbed by unsatisfied world demand.

BOTANY

Sesame indicum L. (main syn *S. orientale* L. but there are numerous others – Weiss, 1971); sesame in English, but also widely known as simsim, beniseed, gingelly, till or ellu. The name sesame is believed to have been derived from the Assyrian *se-gis-i*, currently simsim in Arabic.

Sesame indicum together with *S. capense* Burm. (*S. alatum* Thonn.), *S. malabaricum* Nar. and *S. schenkii* Aschers. have the somatic number $2n = 26$; *S. laciniatum* $2n = 28$; *S. angolense* and *S. prostratum* $2n = 32$; *S. occidentale* and *S. radiatum* Schum & Thonn. $2n = 64$. *Ceratotheca sesamoides*, related to *Sesamum*, has $2n = 32$. The Indian *S. mulayanum* is very similar to *S. indicum*, but is has the valuable characteristics of being resistant to phyllody and wilt.

Polyploidy can be induced, but colchicine treated plants tend to produce a low yield even when fertility of pollen is high and seeds per capsule not reduced. The number of capsules set is often very low, but seed oil content can be extremely high. The rate of the growth and general vigour of tetraploids can exceed that of diploids, the plants are taller at maturity, with longer leaves, larger flowers, capsules and pollen grains. Sterility of pollen grains in tetraploids can vary between 20 and 40% and of diploids between 5 and 30%.

Interspecific hybridisation is possible, and crosses may produce viable seed (Prabakaran & Rangasamy, 1995). The discovery of genetic male sterility in sesame (Osman, 1981) has eased the production of hybrid seed (Tu, 1993; Ganesan, 1995; Kang *et al.*, 1995; Wang *et al.*, 1995).

The number of sesame species remains unclear but about 40 have been described, 36 in the *Index Kewensis.* Many occur in Africa (18 exclusively), 8 occur in the India-Ceylon region (5 exclusively). The Australian records are probably due to imports by Chinese immigrants in the mid-nineteenth century (Bennett, 1996). *Sesame indicum* has a number of local cultivars as noted in the literature, but it is claimed the genus *Sesamum* has only one cultivated species, which can be divided by seed colour into white sesame *S. indicum* spp. *indicum*, and variable sesame *S. indicum* spp. *orientale* (Zhang *et al.*, 1990). Based on seed colour and inheritance, it is postulated that sesame evolved from symmetric to asymmetric types, and from *S. capense* to *S. indicum* in the sequence black, brown, yellow and white seeded types (He *et al.*, 1994). Research into the fatty acid composition of seed oils from *S. indicum*, *S. alatum*, *S. radiatum* and *S. angustifolium* (Olive) Engl., indicated that the fatty acid composition of the total lipids and of the different acyl-lipid classes indicated that *S. radiatum* and *S. angustifolium* are more closely related than to the other two species (Kamal-Eldin & Appelqvist, 1994).

Substantial collections of *Sesame* spp. and cultivars exist in the Americas, India, Russia, China and to a lesser extent Japan, providing a valuable gene pool. However, a more rigourous selection and maintainence procedure may be necessary (Ashri, 1995). Reports on activities and accessions to collections are reported in a number of publications, including that of the International Plant Genetic Resources Institute (IPGRI Newsletter). A study of genetic variation in these collections identified eight major landrace groups (Bedigian, 1984).

Earlier, Hildebrandt (1931) and Vavilov (1949), concluded that each country contained a variety of types with each locality frequently represented by a distinct ectotype. Two distinct plant regions occurred in Russia; Transcaucasia and Central Asia. The former was characterized by a very uniform plant type usually single-stemmed, with broad, deeply divided leaves and only one flower in the axil, resistant to mildew (*Oidium*) and wilt (*Fusarium*). The Central Asian region had a much greater range, with ectotypes adapted to widely differing environments; the Tadzhik area had local strains resistant to *Fusarium* wilt. Seed oil content was greatest in varieties from the eastern Mediterranean especially Syria and Palestine, and lowest in the Far East particularly Japan, where plants also tended to be small. Secondary centres of diversity occur in Japan where types with tetracarpellate capsules occur.

South American collections resembled those from India and the Ethiopian–Eritrean area, and similar types occur in East Africa, where varieties are normally well branched and single-flowered. Indian regional cultivars have been well described and are too numerous to mention. They can be broadly divided into early, little-branched, few-flowered, and late, many-branched multi-flowered types. Seed colour is variable, but black-seeded cultivars tend to branch more freely and give higher seed and oil yield/ha. Seed characteristics of cultivars in the Venezuelan germplasm collection have been described (Mazzini & Layrisse, 1998)

Similar to other plants that have long been domesticated, many hundreds of cultivars of *S. indicum* exist, which differ substantially in size, form, growth, and colour of flowers; seed size, colour and composition.

A Russian expedition collected 500 specimens from which 111 types were distinguished. The numbers of types found in Pakistan, Burma and the Sudan were 62, 49 and 54, respectively, and similar wide variation exists in other parts of Africa and South America. The large number of local cultivars is considered an asset in breeding programmes, and it is also important to include a high yielding, locally adapted cultivar when breeding for high yield in sesame (Pathirana, 1995a).

The diversity of plant types and ease of crossing, together with the great adaptability of the plant as demonstrated by the range of environments in which it is grown, indicate great scope exists for improvement of current, and breeding new, cultivars. The introduction of legislation in many countries to protect the rights of plant breeders has promoted development of new and locally superior cultivars, Australia for example (Australia, 1994). Diversity also renders description difficult, and reference should be made to the literature for particulars of a specific cultivar or type. South Korean plant breeders have developed what they consider 'The Ideal Sesame Plant' for local conditions (Kang *et al.*, 1995).

Cultivated sesame is typically an erect, branched annual, occasionally perennial, to 0.5–2 m, with a well-developed root system, multi-flowered, whose fruit is a capsule containing a number of small oleaginous seeds (Fig. 5.1).

Two basic sesame types are usually distinguished; long-season, occasionally treated as

Fig. 5.1 Typical sesame plant, showing capsules in different stages of development and flowering continuing.

perennials, with an extensive and penetrating root system; and short-season, with less extensive, more shallow roots. A close relationship exists between climate and growth of both upper plant and root system. A cultivar can behave quite differently in succeeding years of climatic variation, or when planted in the *kharif* (S.W. monsoon, May–October) and *rabi* (N.E. monsoon, November–April) seasons in India for instance.

Roots of the short-season, usually single-stemmed cultivars have a more rapid rate of elongation than the longer-season branched cultivars, but the latter have a faster rate of root spread. The ratio of air-dry weight of roots to tops is greatest in monsoon and least in dry-region types; in India tap root length is directly correlated with growth. In the Sudan, dominance of red-seeded sesame on the Kordofan sandhills is considered due to its rapid and vigorous root development (Tothill, 1952); similarly with variety Yandev in Nigeria. Initial rate of root growth of rain-grown sesame is frequently less than that of groundnuts, maize, or sorghum, a factor of importance with interplanted sesame. However, sesame roots may exert an allelophatic effect on some crops (Yooprasert & Sooksathan, 1992). Rate of root penetration and extension of sesame grown on a clay soil in the Sudan is shown in Fig. 5.2 (Lea, 1961).

Soil type and available moisture influence the rate and type of root growth. This is most noticeable when local rain grown sesame is grown under irrigation, when above and below ground growth is much greater. This prolongation of the growing period can result in unfavourable harvesting conditions, or labour required at harvest not being available due to other seasonal demands. Roots develop more profusely in sandy than in clay soils, and sesame's drought resistance is partly due to this extensive system. Waterlogging is not tolerated, even for relatively short periods such as occur after heavy rain or rain following irrigation. Root growth is also inhibited by relatively low salt concentration, much lower than that tolerated by safflower.

The stem is erect, smooth, slightly to very hairy, normally square in section with definite longitudinal grooves, although rectangular and abnormally wide, flat stems occur. A correlation may exist between stem pubescence and drought resistance. Stem colour can range from light green to almost purple, most commonly a darkish-green; banded stems are uncommon. Stem height is normally 60–120 cm, but may be up to 3 m. Breeding plants resistant to lodging is now possible with the discovery of a stiff-stemmed mutant (Murty *et al.*, 1995).

Growth regulators, including gibberellic acid (GA), generally increased length of the internode(s) elongating at time of spraying, branching, and number of capsules, but had no significant effect on total oil yield. However, GA and IAA have been more extensively studied for their effects on germination, noted later. While growth hormones may affect various aspects of plant growth and seed yield, and thus the profitability of a crop (Tripathy *et al.*, 1996), they are of little commercial significance.

The extent and type of branching is a cultivar characteristic, as is height of first branch. Some

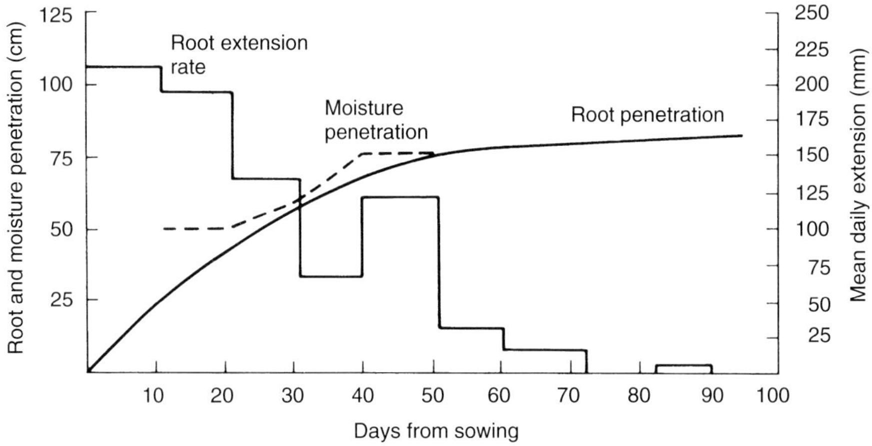

Fig. 5.2 Rate of sesame root extension and penetration, Sudan.

cultivars have few branches, or a small number rising from the axils of the lowest leaves, others form primary then secondary branches to a high level on the main stem. A relationship exists between the number of primary and secondary branches and thus capsule number per plant, and total seed yield. The degree of branching is a cultivar characteristic, but is affected by plant population, rainfall, day length, etc. and cultivars sown outside their normal habitats or season generally produce abnormal growth. In India for instance, the *rabi*, dry-season, varieties when grown as a *kharif*, monsoon, crop produce much vegetative growth but few flowers or capsules.

Short-stemmed, little-branched cultivars are generally early maturing, the taller-branched types late-maturing and more drought resistant. Long-season cultivars also elongate more slowly as seedlings, but increase their later growth rate. Single-stemmed types have thicker stems as seedlings, but there is little difference at maturity. With early cultivars short photoperiods increase stem diameter; high temperature increases it most in medium-maturing cultivars as also do long days; in late cultivars long days had a greater influence than temperature. Capsules tend to be closely appressed to the stem in single-stemmed cultivars and these produce a very compact plant. Division of cultivars into branched and unbranched is satisfactory, and in general use (Fig. 5.3).

Leaves on sesame plants are most variable in shape and size, on the same plant and between cultivars. Generally, lower leaves tend to be broad, sometimes lobed, margins often prominently toothed, with the teeth directed outwards. Intermediate leaves are entire, lanceolate, sometimes slightly serrate; upper leaves more narrow and lanceolate. Leaves may be opposite or alternate, or can be opposite below and alternate above. Leaf arrangement is important as it controls the number of flowers borne in the axils and thus number of capsules; opposite leaf arrangement encourages multiple flowering.

Leaves vary between 3.0 and 17.5 cm long and 1.0 and 1.7 cm wide; petioles are 1.0 and 5.0 cm. Generally of a dull, darkish-green, leaves can be much lighter, with occasionally a yellowish tint. Leaves are generally hairy to some degree and very hairy leaves often appear bluish, especially when grown on fertile soils. Leaves are mucilaginous. Total leaf area can be calculated using leaf length and width (Rao, 1990), while a method of photographically determining leaf area indices in Australia was effective up to 22 days after emergence, but was less reliable thereafter (Martin, 1996). Leaves of indehiscent cultivars have a much faster rate of water conductance than dehiscent varieties and are thus

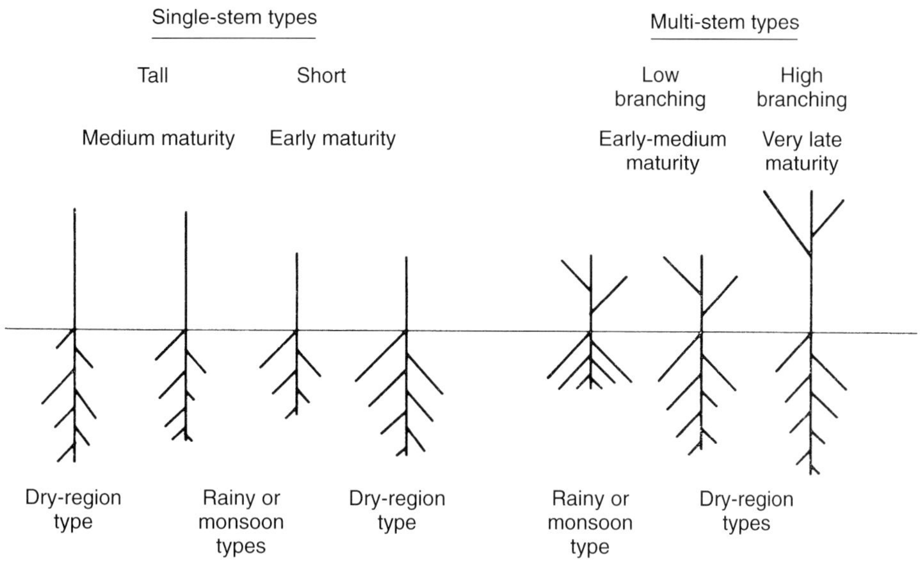

Fig. 5.3 Types of root growth, sesame.

less suited to areas with restricted water supply. Since stomata vary in number and size between cultivars, this could be a reason (Boonjan & Voorasoot, 1992).

Defoliation caused by hail or artificially in the USA, showed that recovery from defoliation was highly dependent on when it occured (Tewolde *et al.*, 1994). One third defoliation had little effect on seed yield regardless of the growth stage; at two thirds defoliation before heavy fruiting yield was reduced by up to 20% but there was a cultivar difference with some less affected. Defolitation above two thirds directly reduced yield by an amount that increased as the defoliation coincided with full bloom and fruit formation. The effect of defoliation by insects is discussed in the section on pests and diseases.

The process of floral bud differentiation has been described in detail (Liu, 1992; Liang *et al.*, 1993: Yuan *et al.*, 1996). Sepals arise first, followed by petals and stamens, finally the carpels are initiated, forming a bicarpelling, bilocular superior ovary with several dozen anatropous ovules. Flowers arise in leaf axils on the upper stem and branches, and the node number on the main shoot at which the first flower is produced is a cultivar characteristic and highly heritable. Flowers generally occur singly in lower leaf axils with multiple flowers on the upper stem or branches. Flowers are borne on very short peduncles; the corolla is five-lobed, the lower lobe being the longest, the upper the shortest. Corolla colour is generally white or very pale pink, but can be much darker to almost purple, and the inner surfaces may have red or black spots, occasionally with purple or yellow blotches.

Flowers are normally self-pollinated, but insect pollination is common and wind pollination uncommon. Natural cross-pollination up to 10% has been widely reported, but to 50% is reported in specific cultivars (Pathirana, 1995b; Manoharan *et al.*, 1995). Flowers open early in the morning, 95% between 5 and 7 a.m., wilt after midday, and usually shed between 4 and 6 p.m. Anthers open longitudinally and release pollen shortly after the flowers open; the interval is a cultivar characteristic. The stigma is receptive one day before flower opening and remains receptive for a further day. Under natural conditions pollen remains viable for approximately 24 hours, and its detailed development has been studied (Subramanian, 1995).

Development of flower and seed is shown graphically in Fig. 5.4 and photographically in

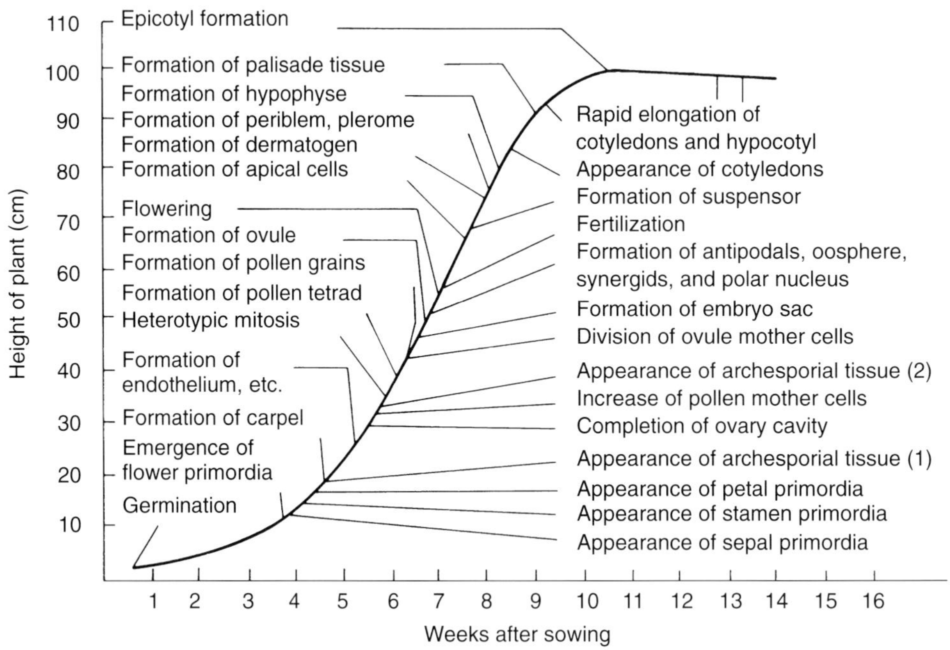

Fig. 5.4 Development of sesame flower and seed.

the original paper (Nakamura *et al.*, 1956 and 1958). The various development stages of Venezuelan cultivars have been determined by Rincon and Salazar (1997). Low temperature at flowering can result in sterile pollen or premature flower fall, while 40°C or above will seriously affect fertilization and reduce the number of capsules. A constant temperature of 24–27°C proved the optimum for early flower initiation in greenhouse plants, but a constant night temperature of either 33°C or 15°C, delayed flower initiation.

Sesame should be considered for all practical purposes as self-pollinated, with the exceptions noted. There are many easily recognisable, simply inherited characters, and in breeding plots or nurseries rogues are easily detected; cupping of leaves is linked to indehiscence and detected at the cotyledon stage. Controlled crossing is simple and, as 50 or more seeds are obtained per pollination, a single plant can produce several thousand; together with very close spacing this allows very large segregating populations. The level of cross-pollination occurring naturally or in breeding plots varies considerably, and is partially dependent on local conditions and insect population. In India the average is about 5%, but up to 14–65%; in Nigeria up to 50% occurred when plants of three cultivars were distributed in a 0.4 ha sesame field.

The literature contains numerous references to the use of seed colour as a marker; natural cross-pollination during both the summer and winter seasons in Mexico produced an average for summer of 3.6%, and for winter 15.3% when the number of potential insect pollinators was much higher. In India cross-pollination averaged 10% between wild and cultivated sesame, with 14 m between plants an effective barrier.

Inter-planting up to three lines of sorghum between sesame rows, or rows approximately 2 m apart had no effect on the extent of natural outcrossing. When honey-bees were used to produce hybrid seed in the USA cross-pollination was 60%. Under Texas field conditions, 180–360 m is considered sufficient for almost complete isolation of different sesame cultivars.

Pollen from non-members of the Pedaliaceae (intergeneric hybridization) will germinate on the stigma of sesame species but seldom sets viable seed. In India, pollen from the introduced American weed *Martynia dianalta*, germinated on the stigma and grew in the style of *S. indicum*; and similarly in Venezuela. Also in Venezuela a large number of grafts between sesame varieties proved the indehiscent character was genetic; neither indehiscence or characters associated with it were transmitted from one component of the graft to the other. Interestingly, self-grafts of *S. indicum* and *S. prostratum* grew more vigorously than ungrafted plants.

Fruit is a capsule, somewhat hairy, deeply grooved, with a short triangular beak (Fig. 5.5). Capsule size is most variable and within the basic flat-sided, cylindrical shape there is a wide range of types. Each shape tends to be a cultivar characteristic with environment a major modifying factor, and several forms can occur on one plant.

Fig. 5.5 Sesame capsules in leaf axils. Capsules dehisce from top downwards.

Capsule length varies from 2.5 to 8.0 cm, diameter varies between 0.5–2.0 cm and there are four to twelve loculi. Height to first capsule is a cultivar characteristic, and important for mechanical harvesting.

A capsule dehisces by splitting along septa from top to bottom, or by two apical pores. The degree of dehiscence is a cultivar characteristic and of great importance for mechanised harvesting. Capsules near the stem base normally ripen first, those nearest the tip last. A mature plant left standing in the field for a long period before harvest loses much seed through shattering. To encourage commercial mechanised planting, it was necessary to develop cultivars that either ripened all capsules over a very short period or were non-shattering. The discovery of an indehiscent mutant by Langham (1946) concentrated work on the second alternative. Most cultivars developed from it were difficult to thresh, but introduction of paper-shell capsules helped solve this problem. Shattering is apparently controlled by only one locus, indehiscence being recessive. Lack of correlation between yield and indehiscence would appear to indicate that yields could be further improved without losing the indehiscent feature (Delgado *et al.*, 1994). Indehiscent plants are short, erect-branching, resistant to lodging and medium early-maturing. The indehiscent capsules are bicarpellate, medium-long papershell, one per leaf axil and easily threshed. The seeds are tan, medium-sized and uniform in shape and weight.

The number of capsules per plant is directly related to the number of flowers, but climatic conditions can affect the percentage of fertilised flowers as has been noted. A short photoperiod can increase the number of capsules per plant on early and medium-late cultivars. Plant populations also directly influence the number of capsules per plant, and high population or close in-row spacing tends to reduce both the number of capsules and number of seeds per capsule. Seed yield is directly related to the number of branches, but the total number of capsules has the greatest direct effect on seed yield. The number of seeds per capsule, their weight, oil and other constituents vary with capsule position on a plant, irrespective of cultivar, and are also directly affected by environment. In India the highest oil yield is usually obtained when all capsules in the central axis are mature (Muthuswamy & Thangavelu, 1993).

Capsules ripen unevenly, as noted, with considerable seed loss from the earliest ripening capsules. A major reason is that virtually all cultivars are indeterminate with stem growth, flowering and capsule production continuing well after main flowering. Following the inducement of a determinate type in Israel by Ashri in 1985, development of determinate cultivars is currently an objective. Some of the factors involved have been evaluated (Wongyai, 1997).

Seed is small, ovate, slightly flattened and somewhat thinner at the hilum; testa may be smooth, ribbed or rough, and the last is genetically dominant (Anbanandan *et al.*, 1993). Colour can be black, white, yellow, reddish-brown or grey, but dark-grey, olive-green and very dark-brown seeds also occur. Light-coloured seeds, rather than dark seeds, are generally preferred for oil, or when roasted and eaten, and usually command a market premium. It is widely believed in India that black-seeded cultivars generally have a higher oil content than light-coloured, but research indicates that variation is high both within and between light and dark seeded cultivars (Murali *et al.*, 1996). One thousand seed weight is usually 2–4 g, and an important subsidiary trait in improving oil yield in sesame is to increase seed size and weight (Wongyai & Juttupornpong, 1992; Mishra, 1994).

Data on the development of seed components following fertilisation showed that active synthesis of oil occurs between 12 and 24 days but continues at a reduced rate to 27 days, with a a slight fall in oil content before maturity. The free fatty acid percentage is highest at the beginning of synthesis, declines rapidly around 18–22 days and then more gradually to seed maturity. In China dry-matter content increases most rapidly between 12 and 24 days, in parallel with the rate of oil synthesis, and continues to increase slowly until maturity (Chung *et al.*, 1995); in Turkey seed oil content increased rapidly in the first 40 days after 50% flowering, then more slowly to reach its maximum at 50 days, then slowly decreased (Turgut *et al.*, 1996). However, there is considerable cultivar variation, and in Venezuela seed size and oil content reached a maximum at 89 days after sowing in cultivar Aceitera, but not until day 97 in Glauca (Mazzani & Allievi, 1966).

Seed averages 44–45% oil, 19–25% protein, and USDA analysis of clean dry commercial sesame seed is shown in Table 5.1. The factor of 5.3 to convert sesame seed nitrogen to protein is considered accurate. The determination of oil content in small samples of seed is often impor-

Table 5.1 Analysis of sesame seed – USA.

	Whole	Hulled		Whole	Hulled
Moisture (%)	5.4	5.5	Ca (μg)	1160	110
Energy (calories)	563	582	P (μg)	616	592
Protein	18.6	18.2	Fe (μg)	10.5	2.4
Fat (g)	49.1	53.4	Na (μg)	60	—
Total carbohydrate (g)	21.6	17.6	K (μg)	725	—
Total fibre (g)	6.3	2.4	Vit. A (IU)	30	
Ash (g)	5.3	5.3	Thiamine (μg/100 g)	0.98	0.18
			Riboflavin (μg/100 g)	0.24	0.13
			Niacin (μg/100 g)	5.4	5.4

tant, and the nuclear magnetic resonance (NMR) method does not destroy the seed or damage viability.

Seed oil content can vary considerably between cultivars and seasons, and oil percentage tends to rise with increasing length of photoperiod. Highest oil content in Russia was in varieties from Central Asia, Afghanistan and Turkey, and were claimed to reach 61–63%. An average oil content of at least 50% is required to make sesame commercially viable at present seed yield levels.

Early planting dates increased oil content in the Caucasus and Egypt; while location also has a substantial effect. Cultivars grown at numerous sites in the USA showed a significant cultivar–location interaction affecting seed composition. Average seed content was: oil 53.4% and protein 26.3% on a moisture-free basis; iodine value 110.0; the variation in oil content due to location was 6%, that for protein content 7.3%, and iodine value 9.2 units. Data from China indicate that seed protein content falls rapidly as seed develops, from nearly 50% at 10 days to 20% at 35 days (Li & Li, 1992). Other data indicate that genetic and environmental influences affecting protein content inversely affect oil content. Thus protein is initially favoured over oil synthesis as nitrogen supply to the seed increases. However, high rates of nitrogen fertiliser reduced oil content in India, but had no influence in Israel. Red and white seeds in Turkey had different oil contents, averaging 55% compared to 51%, and protein content of defatted seeds was highest in red seeds at 44% (Kaya & Savran, 1996).

Greenhouse trials showed high levels of nitrogen can increase protein and decrease oil content, phosphorus has no effect, and potassium a similar but smaller effect than nitrogen. Laboratory trials also show that a substantial variation in day length affects oil content, as oil percentage is unaffected at 6–10 hours, but increased under longer periods up to 20 hours.

The average vitamin content of seeds has been determined as vitamin A potency (IU) 20/200 g, thiamine 1.0 mg/100 g, riboflavin 0.25 mg/100 g, niacin 5 mg/100 g and a trace of ascorbic acid. Riboflavin may range from 0.22 to 0.33 and niacin 38 to 58 mg/454 g, with an iodine value from 101 to 117, the average being 110. Seeds are also rich in calcium oxalate, source of the bitter tasting oxalic acid in unhulled seed, but between 50% and 66% of calcium is nutritionally unavailable. Sesame seed is the single readily available source of protein high in sulphur-containing amino acids, and making its derivatives more widely available would be most beneficial, as would greater removal of minor bitter constituents. Average oil composition is shown in Table 5.2, together with values obtained from sampling the USA World Collection (Yermanos *et al.*, 1972).

The effect of treating sesame seed with growth regulators has been fairly extensively studied, but results are not always in agreement. An interesting finding was that of dormancy, not previously known to be present in sesame. This was discovered in a Mexican cultivar but disappeared 6 months after harvest, and could be broken by soaking seeds in a GAA solution, while seed of an Indian cultivar C 50 was viable 20 days after fertilisation, but maximum germination of 98% was at 60 days. Sesame seed is relatively delicate and damage at threshing directly reduces viability, with even microscopic cracks in the seed-coat sufficient to render seed non-viable. The free fatty acid content of oil in damaged seed increases gradually after harvest, and can be

Table 5.2 Average composition of sesame oil – USA.

Constituent	Commercial samples (%)	World collection (%)
Oleic acid	45.3–49.4	32.7–53.9
Linoleic acid	37.7–41.2	39.3–59.0
Palmitic acid	7.8–9.1	8.3–10.9
Stearic acid	3.6–4.7	3.4–6.0
Arachidic acid	0.4–1.1	—
Hexadecenoic acid	0.0–0.5	—
Myristic acid	0.1	—
Iodine value	—	106–130

Note: Linolenic acid appears to be lacking.

directly correlated with reduced viability in storage. See also the section on storage.

Seed for sowing the following season in Africa and India is frequently kept in the ubiquitous kerosene can or grease drum at ambient temperature. Provided the seed is clean, dry and containers are sealed, this is perfectly satisfactory for the relatively short period involved. In many African villages small quantities of seed are stored in earthernware jars, sealed with dung and hung in the smoke of a fireplace to prevent insect infestation. To maintain viability under conditions of high humidity as in India, seeds can be treated with a dessicant and this may also help to control fungal infection.

ECOLOGY

Sesamum indicum is basically a crop of the tropics and subtropics, but newer cultivars have extended its range into more temperate regions (Figs 5.6 and 5.7). Sesame occurs naturally mainly between 25°S and 25°N, but up to 40°N in China, Russia and the USA, 30°S in Australia and 35°S in South America, generally below 1250m although some varieties may be locally

Fig. 5.6 Sesame plants in Central African smallholding, ready for harvest.

Fig. 5.7 Sesame crop, Ecuador, South America.

adapted to 1500 m. These high-altitude types are usually small, quick-growing and relatively unbranched, with frequently only one flower per leaf axil and low seed yield. Within cultivars seed yield and oil content invariably decrease with altitude, but there is no data that directly compares yield from a particular variety at differing elevations.

Sesame is grown in the Himalayas to 1250 m and collected in Nepal to 2000 m, has been grown experimentally in Kenya to 1800 m and in South America also experimentally at similar altitudes in rain-shadow areas. Trials in Venezuela with a number of varieties found oil yield was highest at lowest altitudes; in El Salvador there is a definite decrease above 600 m, and in Nepal plants growing at lower altitudes had a higher seed oil content.

Sesame flourishes under tropical conditions, and to produce maximum yields 2700 heat units are required in Israel during the critical 3–4 month growth period. A temperature around 30°C encourages rapid germination, initial growth, and flower formation, but up to 40°C will be tolerated by specific cultivars; temperature below 20°C normally delays germination and seedling growth, and below 10°C inhibits both.

Approximately 150 frost-free days are required, and a hard frost at maturity kills plants, reduces seed and oil quality, and also adversely affects minor seed-oil constituents, such as sesamolin and sesamin. However, at the local genotype collection at Hubei, China, seed oil content was higher in genotypes from the (colder) north than in those from the (warmer) south, and was also higher in genotypes with light coloured seed (Huang, 1994).

Sesame is a short-day plant, and with a 10-hour day will normally flower in 42–45 days, but many cultivars have become adapted to local light periods. When these cultivars are introduced into other areas which have a similar day length but dissimilar rainfall or temperature patterns, considerable variation in growth and yield frequently occurs. The effect of different photoperiods on growth and flowering of sesame has been widely studied, since this is a major factor influencing yield; for example a positive correlation exists between photoperiod and time to flowering in Japan. Differing photoperiods or light intensity can also affect seed oil and its constituents.

Adaptive reactions to shade and changes in productivity and seed oil content indicate the significant morphogenetic effect of light intensity, but the basic causes of change induced by varying light intensities and/or periods has been less remarked than the gross effects produced. In general, cultivars from relatively high latitudes are less photoperiodically sensitive than those from lower latitudes. Reducing the light period on an Indian long-season cultivar from 14 to 10 hours for example, produced a major reduction in the periods to bud and flower initiation, plant height, and drastically reduced seed yield. Local and introduced cultivars in the Sudan can be classified into groups according to length of vegetative phase (time from planting to opening of first flower) and reproductive phase (from then to dehiscence of lowest capsule). The groups are considered to represent differing response to daylight in relation to latitude. The local adaptability of sesame has been well utilized in countries such as India with two distinct seasons, *kharif* and *rabi*; sesame may be grown in either, but even in the same district two distinct cultivars must be grown.

Photoperiods are of little importance in the field compared with the greater effects of environment; for example sowing after the optimum time in terms of available rainfall directly reduces yield as the period increases. The relationship of time of planting to maximum yield is generally well appreciated in the main sesame-growing countries, where the optimum has been determined empirically by local cultivators. The effect of various sowing dates on yield is shown in Table 5.3. The rapid fall in yield is typical of most areas, and indicates how important it is to determine the optimum sowing period.

In Africa sesame is planted after the first rains where these are of short duration, but may be sown later where the crop is required to mature in the dry season. The actual date in any year is mainly dependent on the onset of the rains, and this is true of Central Africa as a whole. Recom-

Table 5.3 Effect of time of sowing on yield.

Country	Date	Yield (kg/ha)	Country	Date	Yield (kg/ha)
Sudan	Mid-June	536	India (Bengal)	Mid-May	325
	Mid-July	346		Mid-June	101
	Mid-Aug	292		Mid-July	45
Tanzania	Mid-Jan.	973	Korea	Mid-Apr.	677
	Mid-Feb.	835		Mid-May	764
	Mid-Mar.	521		Mid-June	437

Table 5.4 Recomended sowing dates for selected countries.

Africa		Burma	May
North	May		
West	April	China	
Central	Jan–April	East	April
South	January		
		Australia	
Iraq	April	North	January
		East	December
India		Americas	
North		USA	May
Rabi	November	Mexico	April
Kharif	May	Venezuela	September
South			
Rabi	February		
Kharif	June		

mended times of planting for selected countries are shown in Table 5.4.

Sesame is considered drought resistant, but good growth and yields can only be obtained with adequate rainfall. Established plants are capable of withstanding high moisture stress, but the seedling stage is extremely susceptible. Rapid wilting of young plants during a break in the early monsoon in Burma is a conspicuous feature of local sesame fields. Sesame is relatively independent of rainfall when soil moisture is adequate, and its adaptability demonstrated by the numerous successful local cultivars. Sesame can be grown in relatively arid zones and produce a high-quality oilseed, and it is in such evironments that sesame is sown over the greater part of its range, since there are very few large-scale irrigated plantings.

Selecting a cultivar with a high average yield over a fairly wide rainfall range is a major factor in ensuring a profitable return from annual sesame, as in Mexico (Santacruz-Varela *et al.*, 1996). The crop is important in many low rainfall areas because of its resistance to water stress and its adaptability, but if the conditions under which a local crop is grown are changed, it is most probable that the local cultivar will no longer be suitable. This frequently occurs when irrigation water is made available, and in expectation of higher yield, fertiliser use is increased and more advanced agricultural techniques employed.

Ideally rain should fall as follows: germination to first bud formation 35%; to main flowering 45%; to maturity not more than 20% falling as seeds are filling and ceasing as first pods begin to ripen. Heavy rain at flowering will drastically reduce yield, and if cloudy weather persists yield can be exiguous. Rainfall when plants are ready for harvest prolongs the period required for capsules to dry out, while repeated wetting and drying of capsules increases the degree of shattering.

Localities suitable for sorghum and rain-grown cotton are also suitable for sesame, which will produce an excellent crop on a rainfall of 500–650 mm, but at 300 mm or 1000 mm specialised agricultural techniques and suitable cultivars are necessary. Over wide areas of central and northern South America a rainfall of 600–800 mm is necessary, and where rainfall is only 400–500 mm, early, quick-maturing cultivars are grown. In Ecuador and Venezuela long-season or disease-resistant cultivars are sown in high-rainfall areas. From West to East Africa sesame is generally grown in areas too dry for groundnuts, about 400–750 mm, or where local dry periods make groundnut growing risky. Good rains during active growth in May–June improves yields in northern Nigeria, and in January–February in Tanzania. Sesame is grown mainly in the 300–600 mm rainfall areas of the Sudan, and at 500 mm excellent crops are produced in the Fung and Gedaref districts. Above 600 mm soils must be well drained, and in the southern regions a rainfall of 850 mm will be tolerated. The rain-grown *kharif* crop in India and Pakistan requires at least 500 mm, and local cultivars give maximum yields at 600–1000 mm.

Irrigated sesame requires the equivalent of 900–1000 mm rain to produce optimum yield from newer cultivars. The total water required varies with the district and soil, but if more than 1000–1200 mm are required, sesame is usually replaced by oilseeds better able to utilise large amounts of water. This is discussed further in the section on irrigation.

Sesame seedlings are extremely susceptible to waterlogging; one day under such conditions may substantially reduce growth and yield, while 3 days can be fatal to many cultivars. The most sensitive times in Korea were at flowering, followed by vegetative and maturing stages (Hwang *et al.*, 1992). Heavy or continuous rain will also greatly increase the incidence of fungal disease, although some new cultivars are less susceptible. Rate of infection in high-rainfall zones can be reduced by selecting free-draining soils.

Sesame is susceptible to wind damage after the main stem has elongated, and tall cultivars are not recommended for areas where high winds at harvest time are common. More sturdy but dehiscent types are also unsuitable for windy localities, as a high proportion of the seed can be lost through shattering. In the valleys of Kashmir very cold winds during early growth and flowering cause severe injury to plants. Hail causes severe damage to sesame at all growth stages. Before flowering, stems can be badly bruised, sometimes broken, and terminal shoots so damaged that distorted growth occurs; at flowering both buds and flowers may be stripped from plants, or damaged buds may produce abnormal flowers. Heavy storms can virtually strip plants of all leaves, and recovery is slow. Sesame production on a commercial scale in known hail belts such as occur in a number of tropical regions is hazardous.

SOILS

Sesame thrives on soils that are moderately fertile and free-draining, but its adaptability has been known for centuries, for the Roman Columella in the first century AD noted in his *De re Rustica,* 'Sesame usually requires a crumbly soil such as the people of Campania call *pullum,* but it thrives just as well on rich, sandy, or mixed ground.'

Sesame is sensitive to salinity, and salt concentrations that have little effect on safflower or cotton are fatal. Soils with a neutral reaction are preferred, and although good yields have been obtained on soils with pH 5.5–8.0, sesame does not thrive on acid soils.

Within these rather broad limits sesame can successfully be grown on a very wide range of soils, but this is due more to the diversity of cultivars well adapted to local conditions than to the basic adaptability of any one cultivar. In India where the range of local cutlivars is probably greatest, sesame is grown on poor upland soils, lateritic *raker* soils, on *goradu* sandy loams, heavy black cotton soils, and paddy-fields. In South-East Asia fertile loam soils give excellent yields, but the region's low average yields reflect the generally poor soils on which it is sown. Sesame is grown throughout China on light soils, but also on permeable alluvial soils; on very fertile soils special cultivars such as San-hsieh-chi and Pa-wang-pien are grown.

Sandy soils are selected in the Middle East and Africa because it is the most suitable crop for this soil type. The more fertile soils are reserved for other cash crops and only sown to sesame where irrigation water is not available. The sand-dune region east from the White Nile is the main producing area of the Sudan, but good crops are obtained on the free-draining clays of the Fung,

Gedaref and Nuba mountains. Fertile sandy loams, or well-drained heavier soils are chosen in South America, although sesame is grown as a smallholder crop on many soils, especially in the dry uplands or mountainsides. Excellent conditions exist in Venezuela, with its extensive areas of well-drained sandy loams, and large range of cultivars. Well-drained fertile soils of medium texture and neutral reaction are preferred by commercial sesame growers in the USA.

The interactions between damage caused by insects, diseases, and various soil characteristics including pH, temperature, and moisture, are discussed in the section on pests and diseases.

FERTILISERS

Data from formal fertiliser trials in many sesame growing countries are lacking, and the following section includes personal comments, observations, and unpublished information collected by the author.

Sesame is grown as a smallholder crop over the greater part of its range, and it was often stated that sesame responds poorly to fertiliser. This statement is very difficult to reconcile with the plant's performance on fertile and infertile soils in any given area. Anyone who has seen pure stands of sesame in regions where little or no fertiliser is applied will have noticed the generally uneven growth. This irregularity is invariably related to soil conditions, and where waterlogging is not the cause, differences in fertility are. To demonstrate the necessity for fertilisers, the author made a large cross with ammonium sulphate on local crops, and advised cultivators to consider the often dramatic results!

An important factor to be considered when attempting to increase smallholder yields, is that the local cultivar is often extremely well adapted to local conditions; alter these and it is usually necessary to select another cultivar able to take advantage of the new circumstances. However, the reportedly poor response of sesame to fertilisers may not necessarily be due to natural causes. On collective farms in the former USSR, fertilisers allocated to sesame were diverted to more profitable crops. The very low sesame yields which resulted were less an indication of the plant's potential in that area than of the political ambitions of the collectives' controller!

Incorporation of organic manures or plant residues into sandy soils increases their humic content and moisture-retaining ability. Their effect on subsequent sesame crops in tropical areas tends to be erratic, often depending mainly on rainfall before planting or the addition of nitrogen to assist decomposition. Clay and straw plus nitrogen applied to a sandy soil in Egypt substantially increased the seed yield of subsequent sesame crops, as did other organic wastes (Abdel-Sabour & Abo El-Seoud, 1996), but ploughing-in sunhemp at the end of the dry season in Tanzania depressed the yield of a following sesame crop by inducing a moisture shortage. Another natural product easily available to local farmers in Brazil, is an extract of cassava roots known as *manipueira*. When this was sprayed on sesame plants in the field it increased capsule numbers by 67% and seed weight by 52% (Ponte *et al.*, 1997).

An indication of the nutrients required by sesame is shown in data from Korea (Xu *et al.*, 1992). To produce 1755 kg seed/ha, plants required 6.76 kg N, 1.0 kg P, 5.25 kg K. The nutrient content in different plant parts decreased gradually during growth; N and P in leaves were higher than in stems, but stems had higher K content. At maturity, N and P were higher in seeds, followed by leaf, capsule shell and stem; K content in aboveground plant parts was in decreasing order, capsule, shell, stem, leaf then seeds. The nutrient content at harvest of sesame grown in Venezuela and compared with other local crops is shown in Table 5.5, and indicates the level of nutrient renewal necessary. Distribution of these nutrients within plants is shown in Table 5.6. (Bascones & Ritas, 1961).

In countries where sesame is grown on a commercial scale, NPK mixtures of 5:10:5, 12:12:6, and 10:14:10 at rates of 500–700 kg/ha are commonly applied at planting. For rain-grown sesame in Israel fertilisers are preferably applied to the preceding crop, often wheat, but irrigated sesame also received additional amounts. The usual practice in India is to grow sesame on the

Table 5.5 Nutrient content of sesame compared with other local crops – Venezuela (all weights in kg/ha).

Crop (%)	Yield	N	P	K	Ratio
Cotton	1700	73	28	56	100:38:77
Maize	3800	106	39	78	100:39:74
	2200	119	32	136	100:27:114
Sesame	500	25	3	25	100:13:98
Tobacco	1700	90	22	129	100:24:143

Table 5.6 Total dry matter and nutrients in sesame plants at harvest – Venezuela all weights in kg/ha.

Plant part	DM	N	P	K
Roots	779	2.84	0.95	4.33
Stems	2846	10.24	7.94	42.98
Leaves	2058	34.98	12.30	16.74
Capsules	4429	71.74	10.63	72.42
Total	10112	119.80	31.82	136.47

Note: Seed yield was 2200kg/ha.

residual fertility of the preceding crop, and apply what additional fertiliser can be spared, commonly farmyard manure (FYM). Where no formal trial results are available, a good guide is to apply the fertilisers recommended for local cotton or groundnut until the optimum levels are determined.

Fertiliser application must be related to cultivar, plant population and soil moisture availability, and the interaction between these is well documented, and shown in Fig 5.8. Fertilisers also affect plant height, branching, and number of capsules per plant, but visual effects produced by added plant nutrients are not always correlated with yield. This is particularly so with seedbed nitrogen. Fertilisers generally have little effect on seed composition or oil content except at much higher rates than normally applied.

The most common nutrient deficiency symptoms in sesame are as follows:

(1) General chlorosis, with or without withering of plants, appearing initially on older plants and associated with the following:
- Nitrogen deficiency; plants light green, stalks slender, branching absent, and deficiency leaves erect. Lower leaves become lemon yellow, turning to orange. Discoloured leaves are shed.
- Phosphate deficiency; branching suppressed, stalks slender, lower leaves dull dark greyish green. Necrosis of lower or majority of leaves is followed by defoliation.
- Magnesium deficiency; lower leaves develop intervenal chlorosis light yellow in deficiency colour becoming orange later. Green colour persists in midrib and veins giving a characteristic pattern.

(2) Chlorosis appearing as mottling, with or without necrosis:
- potassium deficiency; plants dwarfed, and margins of lower leaves become wavy and deficiency cupped upward. Light lemon-yellow chlorotic mottling appears which later turns bright orange and finally copper-coloured. No defoliation.

(3) Effects localized in the growing region. Terminals die back:
- Calcium deficiency; terminal dies out following distortion of the tips and bases deficiency of young leaves. Hooking downward of the young leaf tips followed by twisted and puckering.

The relationship between fertiliser application and insect pests or diseases is discussed in the section on pests and diseases, and between fertilisers and irrigation in the section on irrigation.

Nitrogen

Nitrogen (N) application must be related to available phosphate, or it can depress yield and may also adversely affect seed-oil content. Nitrogenous fertiliser may also increase susceptibility to fungal disease. Seedbed application of nitrogen as part of a mixture frequently increases yield, but the ratio of N to P and K should be locally ascertained. Where high rates of phosphate or potash are necessary, many proprietary mixtures contain excess nitrogen. If potash is not essential, diammonium phosphate with approximately 18% N is recommended. When sesame is planted at the beginning of the

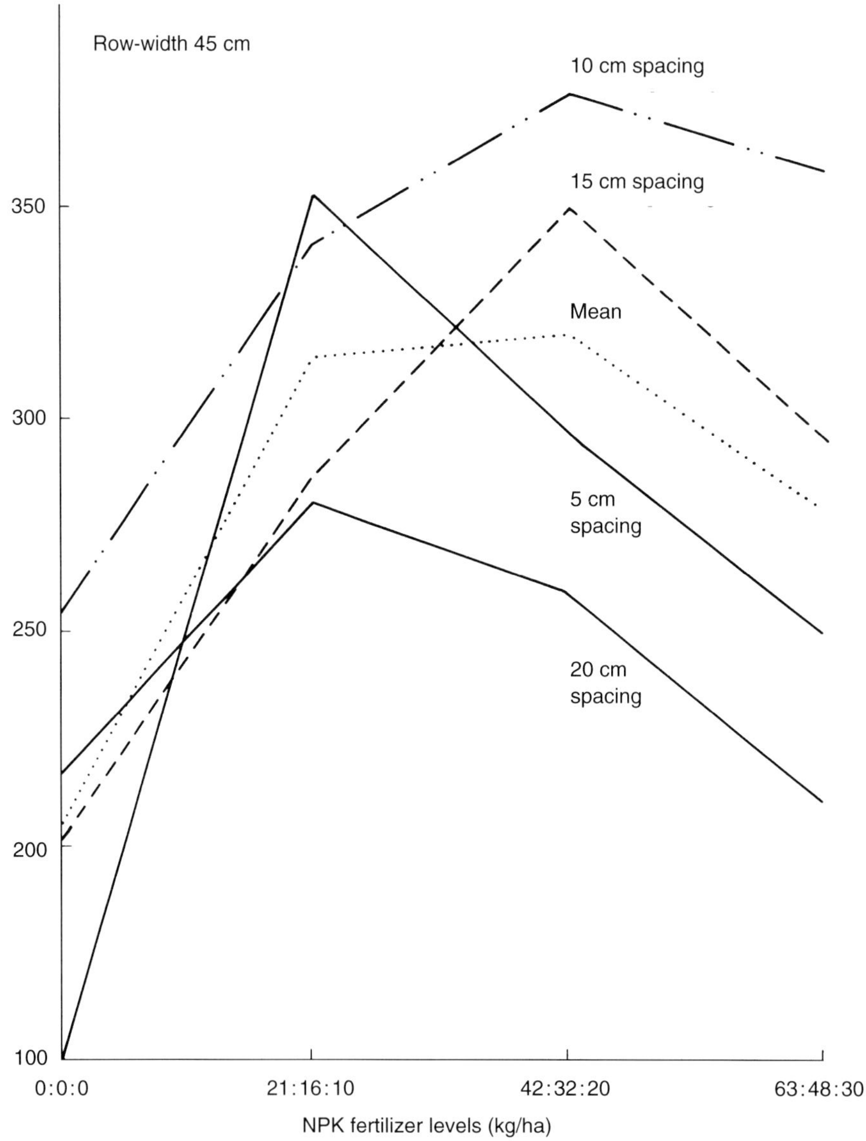

Fig. 5.8 Interaction between fertiliser, spacing and yield, Korea.

rains following a pronounced dry season the release of mineralised nitrogen, as occurs widely in Africa, may make a seedbed application unnecessary.

An interaction frequently occurs between N levels, and timing of irrigations or total water applied, with an increase in both often increasing seed yield. The optimum ratio must be established since an excess of either adversely affects yield.

A nitrogen top-dressing is advisable on commercial crops. In the presence of adequate phosphate, 20 kg N/ha can safely be applied and increased as experience indicates. It is unlikely that more than 50 kg N/ha will be profitable except on high-yielding, irrigated crops. Application should preferably be made when plants are 20 cm in height unless they are showing deficiency symptoms, and not after first buds appear. Split application is unnecessary and merely increases costs. The method of application is also of little importance provided coverage is even and timing accurate. Foliar sprays of urea or other soluble N compounds can increase yield, but in general top-dressing gives similar or better results at lower cost. There are few data comparing differing nitrogenous fertilisers, either singly or as part of a mixture. In general choose the easiest to apply or the cheapest!

Phosphate

Phosphate (P) is most important on irrigated crops. Uptake of N, P and K is related to general growth of plants to approximately 60 days, when the proportion of dry material supplied by leaves falls, and with it uptake of nitrogen and potassium, the latter to a lesser degree as it is also related to capsule number. Although the rate of phosphate uptake also declines, it continues at a higher level than the other two as the number of capsules increases. Unless there are specific recommendations to the contrary, 40 kg P/ha should be applied to commercial crops. When sesame follows crops that have been heavily fertilised with phosphate, 10–20 kg P/ha should suffice.

On high-yielding commercial crops up to 60 kg P/ha has been applied but there is little economic data to justify this amount; should it become profitable the P unit price becomes important, and comparison trials necessary. The effect of different phosphatic fertilisers on certain plant characteristics and yield is shown in Table 5.7. On these Venezuelan soils potash is seldom necessary. Residual P can be substantial in some tropical soils, when the requirement can be limited to 'pop-up' amounts. Superphosphate and similar types should be applied in the seedbed at planting, but rock phosphates and bonemeal should be broadcast before ploughing; the P content and availability must be determined to establish the ammount. Crop response to single superphosphate may also be due to its small sulphur content, since this element is often deficient in tropical soils. See also the following section on sulphur.

Potassium

Potassium (K) content is usually highest in stems and capsules of mature plants, and it was determined in Korea that for every 100 kg seed produced, 6.7 kg K was required. Uptake of K during the 46 days from emergence of the sixth leaf to cessation of apical growth accounted for 95% of total K usage by plants (Li *et al.*, 1996).

Unless there is a known local deficiency, application of K other than in small amounts is

Table 5.7 Effect of different phosphatic fertilisers and application rate on sesame growth and yield – Venezuela.

Treatment/ha	Number of capsules/plant	Height of plants (cm)	Yield (kg/h)
75 kg Superphosphate + 75 kg ammonium sulphate	71	177	1468
150 kg Superphosphate + 150 kg ammonium sulphate	87	153	1212
150 kg Superphosphate	69	141	1230
300 kg Superphosphate	77	129	1282
150 kg Hyperphosphate	76	144	1375
300 kg Hyperphosphate	71	135	1350
Control	46	114	722

seldom necessary. It is often applied as part of a compound mixture, not because it is required but because a mixture without potash is not available. On large agricultural projects standard fertiliser recommendations cover all crops grown, as it is not feasible or profitable to carry stocks of widely varying composition. Since there are no reported instances of potassium toxicity affecting sesame at levels normally applied, little harm results. When substantial amounts of nitrogen and phosphate are applied to irrigated sesame crops, potassium may be necessary to maintain a nutritional balance.

Sulphur

Many tropical soils are deficient in sulphur (S) and oilseeds grown on them generally benefit from its application, as mentioned in other chapters. This is frequently so in India, where up to 40 kg S/ha applied to local sesame was profitable in West Bengal (Mandal *et al.*, 1993). However, addition of sulphur is often not necessary, since a slight deficiency can be alleviated by choosing, as previously noted, single superphosphate to supply P, with its higher sulphur content. An interaction between S levels and number of irrigations or total water supplied, has been noted in a number of countries.

Minor nutrients

Minor elements seldom significantly affect field or commercial crops of sesame unless they also affect all local crops, and no obvious responses occured in observation plots established by the author in a number of countries. Sesame is sensitive to boron, more so than cowpeas or beans, and the typical toxicity symptom is firing of leaf margins and tips. Levels of heavy metals found in seeds following application of organic manures in Egypt, established they were neither toxic to plants nor humans (Abdel-Sabour & Abo El-Seoud, 1996). Salinity can affect the uptake of some essential minor nutrients (Yahya, 1998).

CULTIVATION

Sesame is generally grown as a smallholder crop, and a major aim was to increase yield or oil-content by selection within local cultivars in the context of sesame's place in local crop rotations. Considerable mass selection had also been carried out by producers over the long period of plant domestication, resulting in a multiplicity of cultivars or local strains with extremely specific requirements, known in India as 'tract bound'.

A major drawback of local cultivars is the mixture of colours in any seed sample, which reduced its commercial value. Selection within local sesame in the Sudan, showed that within a type producing seed of one predominant colour several strains could be isolated. These gave a highly variable yield, indicating that separation not only ensured purity of seed colour, but by eliminating low-yielding elements appreciably increased potential yield. Increased resistance to bacterial leaf spot (*Pseudomonas sesami*) was also exhibited by some of these selections.

Sesame was not normally grown as a mechanised crop and thus selection of suitable types is comparatively recent. A major problem for commercial producers was the dehiscing characteristic and uneven ripening of capsules, which precluded mechanised harvesting. This made the crop unattractive in comparison with groundnuts for instance, although there is a substantial demand for sesame seed. Introduction of indehiscent varieties removed this disadvantage.

Cultivation required for wheat, sorghum or similar small grains is suitable for sesame. Level lands are important to ensure even depth of planting, and lands may be ridged to assist drainage in areas where high intensity storms are common. Immediately before planting lands should be harrowed to kill weeds, since sesame seedlings make slow initial growth. Weed control while plants are small is difficult, and the aim should be as weed-free a seedbed as possible.

In Africa south of the Sahara sesame is grown almost wholly by manual labour, the exceptions being government-sponsored mechanised agricultural schemes. The land is broken with a hoe immediately following onset of the rains, or earlier if the ground is not too hard. It is then worked into a reasonable tilth during pre-planting weeding operations, and large ridges may be constructed. This is common over large areas of Central Africa where only the top few centimetres of soil contain any nutrients, and ridges concentrate that available. On lands with a high water-table, ridging ensures that roots remain above the temporary waterlogging that often occurs. In arid districts, ridges can be tied to produce basins in which rainwater is retained to ensure maximum percolation.

Seed quality is seldom a major factor limiting yield, and does not become important until the general agricultural level is raised. This will then most probably result in the local cultivar being replaced by a higher yielder and thus present a further opportunity for a general upgrading of agricultural standards. Mechanised crops require high-quality seed of known variety, and it is especially important to ensure there is no mixture of varieties or types, since uneven ripening or rates of growth complicate inter-row cultivations and harvesting, and an admixture of colours downgrades the seed produced.

Seed dressing is usually unnecessary in smallholder crops as a higher seed rate to compensate for seedbed losses is cheaper. Dressing with an insecticide and fungicide is desirable in commercial crops especially with non-shattering cultivars, as these tend to take longer to emerge and require greater protection against seedbed pests and disease. Seedling vigour may also be increased (Charjan & Tarar, 1993).

Fairly high seed rates are often required to compensate for loss between emergence and harvest, which is often greater than expected until specific counts are taken; in Venezuela loss per unit row between 80–100 days after sowing was 22%, and in Tanzania over 2 years losses were 25% and 34% in the local Rufiji variety.

A major agronomic factor influencing yield is plant population, and for optimum yield this must accurately be determined by trials with local cultivars, and will differ if rain-grown or irrigated. When sown as an intercrop, spacing and seed rate depend on local preferences, and in India 0.5 kg seed/ha is considered adequate. Smallholder crops are invariably sown broadcast at 5–15 kg/ha, and cultivar has little influence on seed rate within wide limits. When sown in rows the cultivar has a much greater influence on final yield, rate and type of growth (Table 5.8).

The local Tanzanian sesame is highly branched, Inamar is semi-branching, Acarigua poorly branched. The local type gave the highest yields at the widest spacing; Inamar at a wider row than in-row spacing 36 × 18 cm, and Acarigua at 18 × 18 cm. This result is similar to those obtained elsewhere, with growth directly related to spacing, irrespective of local yield level. Addi-

Table 5.8 Effect of variety and spacing on yield (kg/ha) – Tanzania.

Spacing (cm)	Population of plants/ha	Variety			
		Local	Inamar	Acarigua	Mean
18 × 8	736 000	150	90	62	101
36 × 8	368 000	188	119	92	133
18 × 18	316 000	236	138	121	165
18 × 28	200 000	250	83	82	138
71 × 8	183 000	280	120	62	153
36 × 18	158 000	279	152	87	173
36 × 28	99 000	236	115	101	151
71 × 18	79 000	237	82	52	123
71 × 28	49 000	389	64	75	176

Table 5.9 Effect of spacing on seed yield – India.

Plants per m^2	Spacing (cm)	No. of branches/plant	No. of capsules/plant	Seed yield (g/plt)	Seed yield (kg/ha)
11	30 × 30	6.7	100.3	6.78	746
11	45 × 20	7.0	96.7	7.49	824
22	45 × 10	5.1	60.9	5.53	1217
22	30 × 15	5.2	73.3	5.64	1241
33	20 × 15	3.6	33.1	3.04	1003
33	30 × 10	4.3	49.2	4.48	1478

tionally, an interaction between soil type and cultivar can be expected, as a relationship between aerial and root growth has been previously noted.

Row spacings of 30–45 cm have generally given higher yields than wider row spacings, with distances between plants depending on cultivar or season. With standard 45 cm rows, a monsoon crop in India may have an in-row spacing of 25 cm and the dry-season or irrigated crop 30 cm. This is due not only to season, but also because two different varieties will be grown. The substantial effect of population on various plant characteristics and yield in India are shown in Table 5.9 (Manivannan *et al.*, 1993), and similar results have been obtained elsewhere.

In Venezuela, where there is a basic difference between branched and unbranched varieties, high yields from both are correlated with high plant population. The branched varieties Inamar and Morada grown at densities from 36 630 to 200 000 plants/ha, varied little in seed yield and oil percentage, branching, height and height of first pod from the ground with different densities, but highest yields were obtained at highest populations.

Row spacing in the USA varies between 46 and 76 cm depending on equipment available and cultivar, and twin rows invariable depressed yield; a similar depression has occured in most countries with twin rows. For commercial sesame crops a population of at least 300 000 plants/ha is recommended, 3–4 kg/ha depending on cultivar, which can be reduced to half this when using precision seeders.

Sesame requires a warm, moist seedbed and a temperature of 25–27°C for quick and even germination, up to 34°C is tolerated, but above this emergence is erratic. Cold wet conditions following planting will seriously reduce emergence, below 18°C will retard seedling growth, and below 10°C normally inhibits germination. When grown in areas where there is a short but intense rainy season, sesame should be planted after the main rains have fallen and soil temperature has risen. However, trials in several countries indicate that within cultivars there can be strains exhibiting considerable differences in time to, and percentage emergence at marginal soil temperatures, indicating considerable scope for selection. A soil moisture level of at least 10% at and below seed level is necessary for germination, and 15% is preferred. When grown under irrigation, a substantial pre-sowing watering is to be preferred to immediate post-emergence application. Sesame germinates moderately slowly, and young seedlings also make slow initial growth.

Smallholder sesame is sown by hand and the small seeds are often mixed with sand, soil or ash to increase the volume and assist even distribution. In India seed may be mixed with wood ash when the soil is still too dry for quick germination, in order to reduce loss caused by ants carrying off seed. Broadcast sowing followed by one or two thinnings often gives highest yield in India either raingrown or irrigated. In Indonesia where sesame is sometimes sown together with rice on freshly cultivated soil, wood ash is added to the seed mixture to prevent it sticking together. In Burma a 50:1 ratio of dry sand to seed increased evenness of the stand and substantially reduced the amount of seed required per hectare.

In the West Nile province of Uganda, local cultivators have developed an ingenious method of establishing sesame on sloping land where erosion can be severe following the common heavy rainstorms. Following cultivation, grass is allowed to grow and sesame seed, at a higher rate than on a clean plot, broadcast over the area when an even stand of grass seedlings has become established. Grass prevents the sesame seed from being washed away, and is allowed to remain until young sesame has become established; then women remove as much of the grass as possible.

When planted mechanically the equipment may vary from small hand-operated seeder units or animal-drawn drills to tractor-operated, electronically controlled air seeders. Depth of planting is usually 2–5 cm, but can be 10 cm in loose soil. Irrespective of the equipment used, an efficient method of controlling planting depth is essential, since even depth ensures regular emergence and crop growth, thus facilitating subsequent tillage operations and combine harvesting. Small plants with capsules close to the ground are either left in the field, or if the cutter-bar is set low enough to harvest all, then speed of harvesting is slowed by greatly increased throughput of green material.

Rubber press wheels following seeder units to firm light soil over the seed is preferable to deeper planting, but should not be used where there is any danger of soil crusting. Alternatively, if seed must be sown more deeply into moist soil, the drill furrow should not be filled but allowed to silt up gradually after emergence.

WEED CONTROL

Smallholder sesame is seldom weeded unless it is the sole crop grown in rows, when one or two hand weedings are usual, the first within a month after emergence as sesame seedlings make slow initial growth. In India generally, only one weeding reduces the yield by up to 70% (Dharan Singh *et al.*, 1992), but in Ethiopia one hand weeding increased yield of irrigated sesame by 80% (Zewdie, 1996). A combination of a pre-emergence herbicide and one hand weeding is becoming more common where sesame is grown as a cash crop.

A weed-free seedbed is most important for a mechanised crop, since weeding is difficult as the fine, fibrous roots are easily damaged. Persistent and vigorous grasses such as *Digitaria* spp., *Imperata* spp. and *Sorghum* spp. are major weeds which must be removed before planting as later control is almost impossible. Their rapid growth also allows them to dominate in utilising soil moisture and applied fertiliser, thus compounding the damage. Some grasses also produce root secretions which suppress germination and growth of other plants. Other weeds produce seeds similar in size or weight to sesame, and if these are harvested with the crop are extremely difficult and expensive to remove. As a contaminant of commercial sesame seed they can result in downgrading or rejection of consignments; particularly so with Johnson grass (*Sorghum halepense*) in the USA and hand weeding to remove it is sometimes necessary.

A range of implements can be used to weed row crops provided they are set as shallow as possible; rotary or finger weeders are probably the most effective. Sesame plants grow rapidly from about 10 cm in height and few cultivations are then necessary, and two or three shallow weedings are usually considered adequate in pure stands. Close rows can also assist in reducing late weed growth if this is troublesome at harvest. A number of pre-emergence herbicides have been successfully used, including alachlor, diphenamid, fluchloralin, glyphosate, pendimethalin, prometryn and trifluralin, but none is safe post-emergence. A combination of directed band spraying plus inter-row cultivation frequently gives effective and low cost weed control.

IRRIGATION

Highest irrigated yields are obtained when sesame is grown in semi-arid conditions, where the sunny dry climate is most suitable, and the low humidity reduces the incidence of fungal disease. When grown in desert areas of the USA, for example, sesame produces high yield with less irrigation water than required for high cotton yield. Flood irrigation is normal for smallholder crops, and row or furrow irrigation is practised only when it is grown more extensively or as a mechanised crop. However the high cost of water in many arid areas often precludes growing sesame; alternatively the cost of production when rain-grown is so low that irrigated crops cannot compete on local markets. This is so in the Sudan and other African countries.

A common method is to impound rainwater on fields, allow it to percolate, sow sesame into dry topsoil, and plants grow on residual moisture. This can be a most effective method, and on fertile soils good yields are obtained. This system also utilises flash floods that occur in river valleys of many semi-arid regions, water which would otherwise be lost. Bunded fields can be found in Africa, Arabia, India, and the Pacific region especially China.

Irrigation of sesame and water use has thus been empirically derived and adapted to local conditions. Growing local sesame cultivars on specially constructed irrigation schemes, or continuing their use when irrigation facilities were introduced to a locality, often proved disappointing. Sesame was then replaced by another crop, but the wisdom of so doing is questionable. Improved cultivars are capable of very high oil yield per hectare, and the dietary value to people and animals of the seed, oil and cake is undeniable.

There is a need for more detailed information on the water requirements of sesame, not only the gross amount required but also the rate of water absorption at various stages of growth, interaction with fertilizers, etc. The very large number of local cultivars and their markedly different behaviour when sown elsewhere under similar ecological conditions, tends to make data obtained from a particular cultivar of little general relevance. Only recently have newer, more adaptable types been widely grown, and it is from trials with these cultivars that more accurate and more widely applicable information will be obtained. Especially those factors controlling water uptake and utilization, which are essential for the optimum use of valuable irrigation water.

Data available indicates that the total water required to produce a fully irrigated commercial sesame crop varies from 1200 to 2500 mm/ha, the very wide range is due mainly to the efficiency of water use management. Water absorption gradually increases from the beginning of growth to flowering and is often not related to air temperature. After flowering the absorption rate can be directly influenced by air temperature, especially at 30°C or less, when water absorption sharply decreases. Maximum water absorption is at flowering, when moisture stress drastically reduces yield. A marked time lag has been recorded between peak daily air temperature and peak daily water use. It is unfortunate that indehiscent cultivars require more water than shattering types, due to an inherently greater transpiration rate. An interaction may occur between total amount of water, seed weight and oil percentage, and a high application rate tended to reduce both.

Two main systems of watering are used in Russia; large applications every 25–30 days, or lesser amounts at 15–25 days; the first method is used on large, much-branched, long-season varieties, the second on the short season types. In Israel, sesame planted in late June and grown as a catch crop maturing in 70–90 days required 2250 mm/ha. A short watering interval normally gave higher yield than a larger application at longer intervals. A dry season planting of the local Rufiji variety, grown entirely under irrigation by the author in Tanzania, required 2200 mm/ha to produce a yield of 360 kg/ha. Lowest soil and air temperatures of the year occur at this period. This yield contrasts with a following wet season yield of approximately 900 kg/ha on a rainfall of 400 mm.

The first watering in India is generally immediately after sowing to aid germination and promote seedling growth; the second when plants are 15 cm high and have been weeded;

subsequent waterings are at intervals of 15–20 days, depending on water availability, soil type and cultivar. It is considered essential to water at, or just after, maximum flowering to ensure full capsule development. The great range of local cultivars is matched by the variation in the amounts of water required to produce a crop. Extent of the rooting system was a significant factor affecting water extracting ability of different sesame cultivars in Venezuela, and the ability to withstand dry periods to maintain high seed yield (Cabrera & Ricon, 1997).

The interaction between the amount of water and nitrogenous fertilizer applied to sesame is now well documented, and as newer cultivars require higher levels of both to achieve their potential, local trials to determine the correct amounts are essential. The water–fertiliser interaction is less critical with other major nutrients, although the ratio of phosphate and potassium to nitrogen remain important, as noted in a previous section.

Sesame is extremely sensitive to salinity, and salt concentrations which have little effect on safflower or cotton are fatal. Trials in India on a number of cultivars, using solutions of sea water killed all but TMV-1, which showed little reduction either in percentage germination or early seedling growth at 10 000 ppm. However, considerable cultivar variation in the period to death, indicated selection for salinity resistance is possible.

INTERCROPPING AND ROTATIONS

Smallholder sesame is frequently intercropped, but this is as often due to tradition rather than an attempt to increase the total return per hectare. Many farmers in Africa broadcast short-season sesame seed into maturing crops, especially beans, in years of adaquate moisture to ensure a supply of seed for domestic use. The converse also applies to maturing sesame, and is catch-cropping rather than intercropping.

Strip-cropping sesame is more common, and in fact the tall, long-season cultivars can benefit from the wind protection afforded by maize or tall sorghum. In general however, sesame as a cash crop produces a higher yield when grown in pure stands, as demonstrated in India (Gnanamurthy *et al.*, 1992). Yield can also be depressed by shading and the difficulty of weed control in intercrops, also some weeds are allelopathic. Strip-cropping may ensure that in a low rainfall year at least one of the crops planted will mature.

Sesame is attacked by a number of damaging root rots as noted in the section on diseases, and should be planted in a rotation with less susceptible local crops, including groundnuts, beans and sorghum.

HARVESTING

The indehiscent character, so important for mechanised crop production, was first discovered in a field of variety Venezuela 52 by Dr D.G. Langham in 1943, and was associated with other less welcome characteristics which the introduction of the character known as papershell capsule helped to alleviate. The latest indehiscent cultivars can be directly combine-harvested and are suitable for dry-land and irrigated production. Smallholder yields seldom exceed 300–500 kg seed/ha when planted in pure stands, but for profitable commercial production 2000–2500 kg seed/ha with an oil content of around 50% is necessary, and has been achieved by some efficient producers.

Sesame is harvested 80–150 days after sowing, most commonly 100–110 days, with some very short-season cultivars in 70–75 days. At maturity, leaves and stems tend to change from green to a yellowish then reddish tint. In Africa south of the Sahara, sesame is often stooked in the field and a considerable amount of seed is lost in handling from the field to threshing site. Where it is a major crop, as in the Sudan and Nigeria, special racks are built for drying to reduce seed loss. In Nigeria harvesting loss was below 2% when plants were bundled and tied upright to racks and threshed daily, compared with 10% when plants were tradionally harvested and threshed once only. Where insufficient labour is available for harvesting, sowing several cultivars with different maturity periods can markedly reduce field loss by extending the harvesting period.

Capsules ripen irregularly from the lowest upward, and it is essential plants are cut before all capsules are mature, since seed loss in the field due to late harvesting of the shattering types may reach 75%, compared with 9–20% with non-shattering cultivars (Fig. 5.9). A position/maturity interaction in India indicated that highest oil yield would be obtained when all capsules in the central axis had matured (Muthuswamy & Thangavelu, 1993).

The optimum time to harvest a cultivar should be established. Maximum yield and seed size were attained 97 days after sowing from variety Glauca in Venezuela and 89 days for Aceitera, and the reduction in yield caused by only a few days before or after the optimum period was surprisingly large (Table 5.10). (Mazzani & Allievi, 1966).

The introduction of non-shattering varieties allowed mechanised harvesting, either with a reaper-binder, cutting the plants with a mower and then using a pick-up reel on a combine, or by direct combining (Fig. 5.10). The first method is preferred by many growers, who cut the crop when it is not fully mature and combine from the windrows, as this greatly reduces seed loss, and the straw has better feeding value. Most standard combines with a pick-up reel are suitable, and forward and drum speeds may require adjustment during work. Threshing of sesame requires accurate setting of concave and cylinder as seed is easily damaged, and even microscopic cracks in the seed-coat are sufficient to affect viability and oil quality.

Since capsules ripen irregularly from the base of the stem upward, the topmost are often only half-mature at harvest, and the drying period when using a reaper-binder or pick-up combine allows the greater part of the seed to ripen without major loss. Seed harvested from shattering varieties is usually of variable quality because of this uneven ripening, and this variability may be increased by unfavourable weather at harvesting.

Direct combining of some indehiscent cultivars is difficult, as losses occur at the cutter-bar due mainly to whole plants falling off the header, or pods shattering at the cutter-bar. Such loss can

Fig. 5.9 Sesame stooked in field to dry, Venezuela.

Table 5.10 Effect of time of harvesting on seed yield and oil percentage – Venezuela.

Variety	Time to harvest (days)	Seed produced /plant		100 seed wt		Seed oil (%)
		(g)	(%)	(g)	(%)	
Aceitera	78	606	64.5	1.90	86.7	36.6
	82	703	74.8	2.03	92.7	38.5
	85	878	93.3	2.15	98.2	42.9
	89	939	100.0	2.19	100.0	47.6
	92	787	83.8	2.33	106.4	44.2
	96	937	99.7	2.36	107.8	39.1
Glauca	83	295	24.6	1.86	83.7	33.1
	87	521	43.5	1.95	87.8	37.3
	90	695	58.0	2.09	93.7	36.3
	94	735	61.4	2.13	97.9	37.8
	97	1197	100.0	2.22	100.0	37.0
	101	745	67.3	2.25	101.2	39.1

Fig. 5.10 Harvesting indehiscent sesame, USA.

be greatly reduced by fitting a six-slat reel and adjusting it to stroke plants towards the cutter-bar; or alternatively a reel fitted with sorghum attachments. Speed of rotation of the reel must be so adjusted that the slats do not hit plants hard enough to cause capsules to shatter. Seed collection trays will also reduce seed loss at the harvester front. It is also possible to cut sesame plants and leave them on the stubble to dry, to be threshed with a combine fitted with a pick-up reel and seed collection trays, similar to that used for oilseed rape.

Sesame seed is easily damaged and cylinder speeds must be slower than those required to thresh sorghum and other small grains. To keep seed loss and damage to a minimum, a setting that produces a relatively high content of extraneous material is recommended – this material is easily removed by re-cleaning. The chaffer sieve, chaffer extension, air blast and ground speed of the combine should be so balanced that neither the sieves nor the tailings elevator become overloaded. Damage to seed can be checked while the combine is working by examining 100 seeds under a low-power hand lens. If more than 20 have visible damage, cylinder speed is too high or threshed seed is returning to the cylinder with tailings.

When using a stationary thresher it is necessary to ensure seed is not being blown out with chaff or tailings, and elevators are sound with no holes, as sesame seed is small and runs easily. Tarpaulins or hessian sheets should be spread on the ground to catch seed. A stationary combine harvester should be closely inspected after a short period of working and all openings from which seed is being lost sealed. It is advisable to check why seed is being lost, for worn conveyors or other parts can rapidly cause seed loss whose sale price is far greater than the cost of replacing worn components.

Artificial methods of defoliation or desiccation are not commonly used in sesame crops, some tend to stain seed, which could be important if required for culinary use, and there may also be residues in seed but not in oil. An interesting effect of chemical defoliation was noted in Korea, where the labour required for threshing sesame crops was reduced, as plants and capsules were drier (Han *et al.*, 1993).

STORAGE

Sesame seed can generally be stored for up to 2 years with little loss of viability provided it is dried below 8% moisture content and kept in airtight containers. Seed for experimental, breeding or long-term storage should be maintained below 15°C. Insects attacking stored sesame seed are discussed in the section on pests.

Bulk storage of sesame seed presents few problems provided it is clean and dry; seed that heats or is contaminated by extraneous material produces discoloured or rancid oil. As seed is small and relatively heavy, it occupies a small volume for a considerable weight, is moved easily and efficiently by most modern conveyors, and provided these are in good condition they cause little or no damage to seeds. Remarks regarding bulk handling and storage of safflower generally apply to sesame, but precautions against theft in rural areas may have to be more stringent.

Storage methods and containers used by smallholders are varied and often ingenious, for sesame seeds are readily attacked by a number of insects, or eaten by vermin and domestic animals. A very common container nowadays is the ubiquitous kerosene can or grease drum, both having tight-fitting lids. In Africa, small quantities may be stored in earthenware jars; in East and West Africa conical mud and wattle granaries hold about 100 kg, the narrow opening sealed with a mud bung. Several may be grouped together on a raised platform, protected by a thatched roof, and seed retains its viability well when so stored.

PESTS

Sesame is attacked by many insects, but there is considerable variation in their relative importance. In some regions those attacking flower heads and young fruit assume greatest economic importance; in another foliage eaters cause the major loss. In East Africa 30 species were found feeding on sesame, but only half a dozen were of any importance; similar but longer lists have been compiled in other countries, but the proportions are similar. The reduction caused by insects can reasonably be assumed as 25% of potential yield worldwide, and the author estimated damage on smallholdings and commercial crops in Africa, Asia and Central America where no, or minimal, pesticides were used, as seldom less than one-third. The damage was calculated by estimating the loss of yield from missing plants, stunted growth, lack of flowers and capsules, damage to capsules and seed, comparing this with growth or yield of nearby healthy and relatively undamaged plants.

Chemical control of sesame pests is generally unprofitable, but the extent of potential yield loss must be realised and offset as far as possible by cultural techniques. For instance, high seed rates can compensate for seedbed losses, branching varieties produce more flowers, planting dates that avoid the main outbreak period of a major insect pest, and destruction of crop residues assist in reducing succeeding infestations, should be combined with local cultivars with resistance to a local pest. In India for example, cultivar Gouri was found to be highly resistant to *Antigastra catalaunalis*, aphids and thrips (Padmaja & Savithri, 1993).

Insect attack may influence traditional planting and cultivation methods, and control may require a change in techniques; late thinning may compensate for gaps caused by mole crickets or cutworms, but if these pests are controlled plant populations may be too high, and early thinning or lower seed rates become necessary. Losses occur not only in the field, but also during storage; in Pakistan 25% of total seed produced is saved for sowing the following season, sufficient for many times the actual area planted.

Cutworms, especially *Agrotis ipsilon* and *A. segetum* are widespread and common pests in sesame fields, causing considerable local damage to seedlings (see Fig. 2.18a). Scarabs can become troublesome to emerging plants, particularly *Holotricia helleri* in Indonesia. *Zonocerus variegatus*, a brightly coloured grasshopper, can be most damaging when heavy local outbreaks occur in many regions of Africa. *Cyrtacanthacris tatarica* has been recorded as damaging crops in Africa south of the Sahara, India and Pakistan. In Indonesia the most common is *Valanga nigricornis*. Where sesame is grown on the bunds of rice-fields, or on higher lands adjacent to paddies, the grasshopper *Hieroglyphus banian* can become a serious pest. In Japan, *Gryllotalpa africana* is often troublesome in newly emerged seedlings (see Fig. 7.17). *Chrotogenus trachypterus* is particularly damaging to young crops in northern Pakistan, India and Afghanistan. Control measures are seldom used, but baiting may become necessary and is usually successful.

Stem-borers are seldom reported as major pests of sesame, although several species have been recorded as of minor or local importance. For instance, *Melanagromyza* spp. and *Baris helleri* have been observed as occasionally attacking interplanted sesame, or sesame growing in the vicinity of more favoured hosts in Africa. *Oberea sesami* occurs locally in India, particularly in the Baroda area. The insect lays its eggs on the leaf midrib into which the larvae bore. Infected plants are usually confined to a small area or group in a field.

Herbiferous insects are of major importance, and most common are *Antigastra catalaunalis*, *Acherontia styx*, *Diacresia obliqua* and *Amsacta moorei*; the first two are of widespread occurrence, the last two more important in Asia and the Far East. The cosmopolitan leafworms, *Spodoptera* spp. can also be extremely damaging in specific years.

The adult moth of *Antigastra catalaunalis*, has a wingspan of 15–25 mm with the brownish yellow forewings decorated with indistinct zigzag lines. The hind wings are pale yellow, almost transparent, and the moth looks cream-coloured in flight. Fully grown caterpillars, 11–15 cm long, are green to yellowish green with black spots and a dark-brown head. Newly hatched larvae mine young leaves and shoot tips; at a later stage they fasten together leaves and shoots and feed inside. Pods and flowers may also be destroyed. The adult moth of *Archerontia styx* is large, stout-bodied, brown, 9 cm across the wings, the hind being yellowish with dark streaks. Larvae feed on leaves and pods and, if the infestation is large, may completely defoliate an entire crop. Fully grown caterpillars are 7 mm long, green with oblique, yellow body streaks, and a recurved horn on the posterior segment. In India maximum infestation normally occurs in September, and on smallholder crops hand-picking is recommended. In all areas ploughing of cropped fields after harvest will expose pupae and reduce the incidence of the pest in the following season. The frosted hawkmoth, *Psilogramma ineteta*, causes damage in China and Japan.

Diacresia obliqua, has a wide host range on cultivated crops, is a major pest of sesame in Pakistan, India and eastern Asia, but is less important in Africa. The moths are pale buff in colour, 25–40 mm across the wings. On hatching, the larvae feed gregariously, and if the infestation is large completely defoliate plants. Collecting the eggs, which are laid in large numbers on the underside of leaves and covered with buff-coloured hairs and are easily seen, can substantially reduce the severity of attack in smallholder crops. Another tiger moth with a more restricted range is *A. moorie* occurring mainly in the Pakistan–India area; it is polyphagous and therefore difficult to control. The adult is a heavily built moth with white, glossy wings. Eggs are laid in conspicuous yellow masses, generally on the underside of leaves. The larvae feed on the leaves and shoots and migrate from one field to the next as their food supply becomes exhausted. Cutting off the clearly visible egg masses materially assists in reducing damage in smallholder crops. Regular use of light traps from early in the season also reduces damage caused by the pest in India and Pakistan. Biological control with parasitic *Aspergillus* is possible.

Aphis can attack plants at all stages, are widespread, and important not only for the direct damage they cause but for the virus disease they transmit. There are many generations in a year, and control measures must be early and adequate to ensure that population build-up does not cause permanent plant damage. Spraying to control other insects can increase aphid infestation through destruction of their predators and parasites. Sesame cultivars vary greatly in their resistance to aphids, and in areas where the pest is prevalent and protection costs high such resistance is valuable; cultivar Morada is more resistant to aphid attack than other Venezuelan varieties. The cosmopolitan *Myzus persicae* is the aphid commonly found and the most damaging, and considered to be the most serious pest of sesame in the New World. The aphid is light-green or yellowish with indefinite strips of dark colour on the abdomen, some 2 mm in length with antennae as long as the body. Typical symptoms are young shoots deformed and leaves strongly wrinkled and curled downward. Growth of young plants may be severely checked, and if the infestation is not controlled, yields are greatly reduced. Tetraploid sesame is said to be highly resistant to *M. persicae*.

Aphis gossypii is also of widespread occurrence, but generally not as common on sesame as *Myzus persicae*. It can be more damaging locally, especially in Africa where sesame is grown adjacent to or interplanted with cotton. The aphid, some 1.5 mm long, may be yellow, green or black. Population increase is extremely rapid, with a generation every 8–10 days. Symptoms are generally similar to those caused by *M. persicae*, but initially may often be confined to the growing points. The cosmopolitan *Aphis craccivora* may heavily infest local crops, espe-

Fig. 5.11 Green shield bug, *Nazara viridula*, with eggs.

cially in an arc from the Mediterranean through southern Russia to northern India. In the latter country it can cause considerable damage.

Jassids are damaging pests when numbers become large, but *Deltocephalus* spp. is especially important as a vector of the virus disease known as phyllody; individuals becomes infective for life after a latent period varying with the season, the minimum being 11 days. Phyllody is serious in India, Africa and Asia Minor and can also be transmitted to other *Sesamum* spp.; thus control measures must also include the destruction of alternative hosts. The leafhopper *Circulifer opacipennis* occurs from southern Europe and the Mediterranean area through the Middle East to Iran, southern Russia and Afghanistan, and is of particular importance in the Khouzestan province of Iran. It causes very severe damage locally in specific seasons, and may even destroy fields of sesame seedlings. In Australia the common brown leafhopper (*Orosius argentatus*) transmits little leaf disease caused by a bacterium (Mycoplasma), but the damage is generally slight overall. Whiteflies (Aleyrodidae) can also transmit virus diseases. The species most commonly reported on sesame is *Bemisia tabaci*, which is about 1 mm long, and lightly covered with a white waxy bloom. It is common from Africa to India, where it is the vector of leaf curl of sesame caused by nicotiana 10 virus. As the insect also attacks a wide range of other cultivated and wild plants, control is difficult. Tolerance, but not resistance, exists in some cultivars, apparently related to the amount of foliage (Avila & Montilla, 1997).

A bug of widespread occurrence is *Cyrtopeltis tenuis*, recorded on sesame from the Americas, Africa, Asia and Indonesia to Japan. Nymphs and adults suck the sap, and produce characteristic symptoms; leaves and young shoots are wrinkled, often deformed, while the stems and shoots bear distinct swellings. The bug also attacks tobacco, and when the planting seasons for sesame and tobacco overlap, as occurs in Central Africa and Central America where native tobacco is grown for chewing and snuff, damage to both crops is exacerbated. No specific control measures are necessary. *Taylorilygus vosseleri* can become a major pest locally in Africa, especially in the main cotton-growing areas. Symptoms are leaves becoming mottled, slightly curled and drooping; necrotic patches form which desiccate and disintegrate, giving leaves a perforated appearance.

Pests which specifically damage the flower, fruit or seeds are often the most economically important. Although the physical damage may be proportionately much less than foliage eaters, their impact on seed yield can be much greater.

The sesame gall midge *Asphondylia sesami* occurs in the majority of pure sesame stands, or where sesame forms a substantial portion of an intercrop, and causes extensive damage in Africa and India in particular years. Eggs are laid in ovaries of flowers, and the gall begins to develop before the petals wither. Fruits may be shed after attack, or develop into larger galls inside which the maggot feeds and eventually pupates. The adult emerges through an exit-hole, leaving the empty pupal case protruding.

The Pentatomidae, stink bugs, cause extensive damage to developing capsules. They occur from the Americas through Africa and Asia to Australia and the Far East, and usually occur in almost any sesame planting. The family is characterised by having a scutellum which extends at least to the membrane of the wings, and some are brightly coloured. *Nezara viridula* (Fig. 5.11) is the most widely occurring, but it is in Africa that the greatest degree of damage probably occurs, and several other species are also noted from this continent. Populations as low as 1–2 bugs/plant can cause serious reduction in yield. *Calidea nana* is occasionally a serious pest in the Sudan, and *C. dregei* in East Africa, especially Tanzania. *Agonoscelis pubescens*, the well-known *andat* of the Sudan that is such a serious pest of sorghum, also attacks immature sesame pods.

Small chrysomelids, flea beetles, common in East and Central Africa may sometimes occur in very large numbers (Fig. 5.13). Damage caused is directly related to the degree of infestation, but as populations quickly build up, control measures must be taken before major damage occurs to young seedlings. *Aphthona fuenti* and *A.*

Fig. 5.12 Leaf-eating beetle, *Chrysomelidea.*

latipennis cause minor damage in the Sudan, Egypt and other Mediterranean countries. *Systena* spp. are the most common in the north of south America, and local cultivars are often more resistant to their attacks than exotics. The most important halticids in Russia are various *Phyllotreta* spp., which also attack a wide range of crops. *Eusarcoris* spp. may cause damage locally from Libya to India and Pakistan, but are normally of little importance; *E. inconspicuus* is common in North Africa, Egypt and the Sudan, and *E. ventralis* in India. *Acrosternurn hilare* the southern green stink bug, is occasionally damaging to specific crops in the USA.

A very small bug, *Aphanus sordidus* can cause substantial damage by sucking maturing seed in the field before harvest, when stooked, and even on the threshing floor. Damaged seed is shrivelled, stained black, and fails to germinate. It occurs in Africa, India and the Far East, and also attacks groundnuts and sorghum. It congregates in considerable numbers on threshing floors, and overall spraying of these sites and adjacent buildings is recommended.

Sesame is generally believed to be highly resistant or immune to *Meloidogyne* spp. root-knot nematodes, and also perhaps to other species in the USA. In Iran, however, *M. marioni*, locally known as *nematode risheh*, also attacks tobacco and other local crops included in rotation with sesame, and damage in specific years is severe. There are indications that infection by *Rhizoctonia* spp. and *Fusarium* spp. can be associated with high nematode populations.

Mature plants may be attacked by termites in dry regions, in seasons of below-average rainfall, or in clean-weeded crops. Many rodents, small buck and other animals find sesame attractive, and breeding or seed-production plots should be fenced. Birds can be particularly harmful to irrigated sowings grown out of the normal season in the arid tropics.

Insect damage to stored sesame seed and cake can be substantial, but there is little published information on the actual degree or amount of damage in a specific region. Almost all official publications on the crop make reference to this loss, and to the necessity of providing insect-proof storage for the following season's seed. The extent of this loss can be judged from official estimates of the Ministry of Agriculture, Pakistan, which quotes the amount of seed stored from one season to the next as 1200 tonnes, 25% of total production; in India an on-farm storage loss of 15% is common.

The most common storage pests are *Tribolium castaneum* (the red flour beetle), *Trogoderma granarium* (the khapra beetle), and *Ephestia cautella* (the tropical warehouse beetle). All are cosmopolitan and attack a range of stored products, and also the most frequently noted as infesting ships' cargoes. In India *Tribolium castaneum* is the most common pest of farm-stored sesame, causing an annual loss of 10–15%, with light seeds more susceptible than dark. Also recorded are *Corcyra cephalonica*, the rice moth, from India, Africa and Brazil; *Cryptolestes pusillus* and *Oryzaephilus mercator* in Africa; *Carpophilus* spp. in India and Pakistan; *T. confusum* in Uganda and the Central African Republic (see Fig. 9.8).

The fumigant phosphine can affect the viability of some oilseeds including sesame; and the USDA official tolerance level for post-harvest fumigation of sesame seed with hydrogen cyanide is 25 ppm. For additional information on storage pests and pest control in general see this section in Chapter 3 on groundnuts.

DISEASES

Chemical control of sesame diseases is difficult and seldom economic, and cultural methods should be used wherever possible to reduce the incidence and spread of disease. When considering chemical spraying, some sesame varieties tend to be sulphur sensitive. Seed dressings, use of resistant varieties, destruction of crop residues and alternate hosts, and rotations giving a quarantine period should all be tried.

It should be emphasised that improved agricultural practices will lose much of their value unless accompanied by research into methods of increasing disease resistance.

Many of the damaging fungal diseases that attack sesame are of restricted distribution; some of the most important are known to be seed borne, and wherever possible disinfection of seed should be routine. However, the only safe method when introducing sesame to districts where it has not previously been grown, is to obtain seed from known disease-free areas.

Pseudomonas sesami is a worldwide and damaging disease, and some resistance exists but is apparently difficult to transfer; similarly with *Fusarium* wilt and phyllody. However, a survey of *Sesamum mulayanum* in Maharastra State, India, revealed very high resistance to phyllody and wilt. Further screening under artificial conditions revealed resistance to powdery mildew, *Erysiphe cichoracearum*, but moderate susceptibility to wilt (Mehetre *et al.*, 1994). Resistance to *Sclerotium*, *Rhizoctonia*, and tolerance to leaf curl exists in various *Sesamum* species. Sesame cultivars considered resistant in their own district or country, frequently proved highly susceptible when planted elsewhere. Currently there are no cultivars fully resistant to any major disease, and few indications that this situation will quickly improve.

Probably the most damaging diseases worldwide are *Pseudomonas sesami*, bacterial leaf spot, *Alternaria sesami* and *Cercospora* leaf spots. Wilt caused by *Fusarium oxysporum* can be devastating on susceptible varieties, but some local sesame varieties have a degree of resistance. *Phytophthora nicotianae* var. *sesami* and *Helminthosporium sesami* are often locally damaging.

Bacterial leaf spot resulting from infection by *P. sesami* produces light-brown, dryish, angular spots with a darker, more purple margin, generally located between the leaf veins, but can advance along the veins and petioles, when they become dark-brown to purple shiny lesions. Spots often coalesce to form large necrotic areas on the leaves. Spots on pods are usually slightly sunken, purplish and shiny. It is known in the Sudan as *Marad ed Dum*, blood disease, owing to the red colour of the infected plant tissue.

Alternaria sesami is a seed-borne pathogen of worldwide distribution and attacks stems, leaves and green capsules, and may occasionally be severe enough to kill seedlings and young plants. Main symptoms are dark-brown, spreading, water-soaked lesions, which often occur on the whole length of stems. The lesions also occur on the midribs and leaf veins, without the typical leaf spots. Crop loss from this disease can exceed 50%. An efficient and simple test to determine if seeds carry the pathogen is the oatmeal agar test (Ojiambo *et al.*, 1997).

Cercospora leaf spots are usually irregular brown spots some 2.5 mm in diameter which can increase to 10 mm; they first appear on the lower leaves, and can coalesce to form larger patches which do not cross the midrib or larger veins. *Cercospora* spots can be distinguished from *Pseudomonas* spots, which have a pale grey centre and are generally longer and less angular. *Phytophthora* blight or rot has characteristic water-soaked spots on the leaves and stems, which are invariably brown but become black with age. The blackening of the stem is often most marked at ground level. Affected branches produce poorly formed capsules with shrivelled seeds. In humid weather the white woolly growth of the fungus can be seen on the surface of affected pods. Incidence of the disease is higher on more clayey soils.

Fusarium wilt can be devastating to susceptible varieties but local strains appear to have some degree of resistance to the local race of the pathogen. Terminal leaves become yellowish, droop and desiccate, the symptoms progress down the stem. Severe infection can cause the entire plant to defoliate and desiccate. In less severe infections, or when more mature plants are attacked, only one side of the plant will show symptoms. If infected plants are uprooted, roots will be either wholly or partially rotten on the affected side of the plant.

Blight caused by *Helminthosporium sesami* could become of greater economic importance; high humidity favours its spread and young plants are most susceptible. Charcoal rot caused by *Macrophomina phaseolina*, the sterile stage is *Sclerotium* or *Rhizoctonia bataticola*, attacks stems causing them to become characteristically black and rotten; initially infected plants make poor growth, remain stunted, and many die. Capsules are also affected, and contain seeds with small black sclerotia of the fungus. Increased sowing of sesame monocrops is favouring spread of the disease. High soil temperature and moisture increase the spread of this fungus and its severity; it is also dispersed in irrigation water. Periods of drought between heavy rainstorms are also thought to favour the spread of the disease in Africa. The very wide range of host plants of this fungus makes control almost impossible, and as infection is soil- and seed-borne, only by avoiding lands known to be infested can the severe losses be reduced. The fungus can survive tropical dry seasons in the soil and reinfect subsequent crops.

Stem anthracnose caused by *Colletotrichum* spp. is important locally in Asia and Africa; symptoms are unmistakable. The stem becomes discoloured on the affected side, initially a dull-green turning later to brown or black; cortical tissue of the infected stem and branches cracks and exposes the inner tissues; brown streaks or black spots may be continuous along the length of the stem, or may be interrupted by areas of healthy tissue. The leaves are not attacked, but those on the infected side of the plant become light-coloured and fall. Infected plants produce few pods. In India early-sown crops occasionally escape infection, but in northern Nigeria few plantings are completely free.

Mildew, usually caused by *Oidium* spp., is commonly found on sesame in the more humid areas, or following prolonged rains as in Assam. Symptoms are masses of white mildew spores covering the leaves, which turn yellow and wither. It is generally of little economic importance but may become so locally.

Phyllody virus is a most destructive disease, the characteristic symptom is deformation of flowers, which do not properly develop and remain green, with the calyx and corolla sometimes stiff, forming a half-open hood. Tops of plants so deformed give the impression of witches-broom or bunchy top, usually have shorter internodes and are much-branched. Phyllody is of considerable economic importance in India and Burma and has been recorded from other countries which have imported material from these sources. Nicotina 10 virus, which can be transmitted by the whitefly *Bemisia tabaci*, also attacks sesame and may be more common than is recorded. Little leaf disease which causes greening of floral parts and leaf reduction, is spread by *Orosius argentatus* in Australia. A non-parasitic malformation of plants, the specific cause of which is unknown, is faciation; affected plants have flattened stems much longer than normal, often with numerous longitudinal ridges; leaves are small and crowded towards the stem apex.

6
Soya

The Papilionaceae includes some 40 species of frequently twining shrubs, distributed generally in the Asia/Australasia region. It includes *Glycine max*, soya or soybean, found only under cultivation, although there is a closely related wild species *G. soja.*

The original home of soya is in dispute, and although the cultigen is generally considered to have originated in north-eastern China, the genus *Glycine* has two major gene centres. One in eastern Africa, the other in Australasia with a secondary centre in China. Russian botanists argue that cultivated soya resulted from centuries of breeding/selection from an ancestral form similar to *G. soja,* and that African and Australian species were not involved. When soya is used in the text, it refers to the cultigen *Glycine max*, its seed and derivatives.

Soya is frequently stated to be one of the oldest cultivated crops, whose domestication was long attributed to the Chinese Emperor Shen Nung, 'The Divine Farmer', who is credited with compiling around BC 2800 a treatise on medicines and herbal remedies covering 365 plant species including soya. In fact he was a mythical person, and no records of soya exist predating The 'Book Of Odes' written during the Chou dynasty in the eleventh century BC. Soya beans were originally known as *shu*, but now generally called '*ta tou*', great bean.

It is remarkable that Marco Polo, who spent some two decades in China in the thirteenth century, makes no direct reference to soya; adding to the confusion is the habit of early Chinese writers of referring to several oilseeds by the same character, the meaning apparently changing over the centuries according to the relative importance of the various crops. A similar confusion exists with safflower. On the basis of historical and geographical evidence it is reasonable to assume soya was originally domesticated in north-eastern China (Fig. 6.1). From China soya spread to the neighbouring countries of Korea, Japan, South-East Asia and finally around the world.

The peoples of many Pacific Rim countries did not drink animal milk in the quantities considered normal in Africa and Europe, but substituted various liquid soya bean preparations, and it became known as the 'cow of China' so varied were the types available. The first Europeans to visit Japan in the sixteenth century falsely assumed the very popular bean curd 'tofu' to be cheese. However, Indonesia claims to have developed the equally popular fermented soya product 'tempe', subsequently introduced into neighbouring countries. It was not until the German botanist Englebert Kaempfer visited Japan in 1691, and published *Amoenitatum Exoticum* in 1712, that the Western world fully appreciated the significance and importance of soya in Eastern diets.

Soya beans were known to Europeans in the seventeenth century 'as a curious food from the Orient', and beans sent from China in 1740 by missionaries were planted in the Jardin des Plantes, Paris. Brought to England in 1790, soya was cultivated at the Royal Botanic Gardens, Kew (Aiton, 1814). Professor Haberlandt of Vienna attempted to expand soya production in

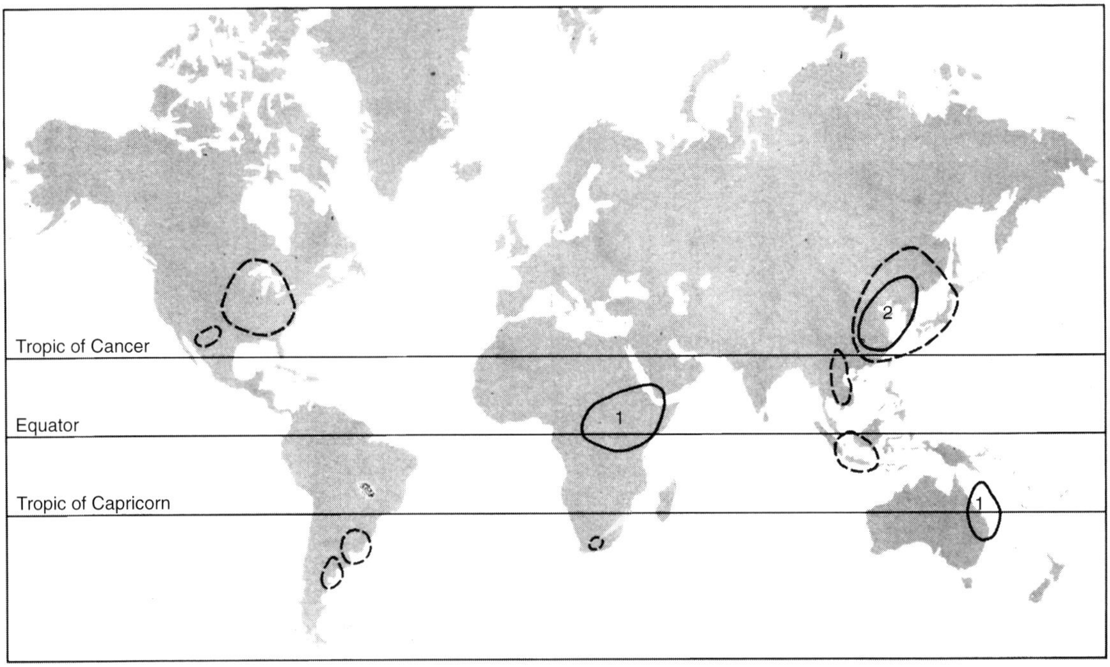

Fig. 6.1 Probable centres of origin of genus *Glycine* (1) and cultivated soya (2) in solid rings. Areas of intensive cultivation in broken rings.

Europe between 1875 and 1880 with little success (Morse, 1950).

Soya is traditionally believed to have been introduced into the USA as ships' ballast from the Far East, and first mentioned in USA agricultural literature in 1804 (Mease, 1804). Captain Perry's famous expedition to Japan in 1854 brought back many local soya cultivars, some flourishing when planted in the USA, where the seed became known as the 'Japan pea'. One of the first specially developed cultivars was Mammoth Yellow, produced by the Carolina Experimental Agricultural Station in 1882, and soya was initially used almost exclusively for forage. Oil was first extracted from imported seed in 1911, also in Carolina, and from local seed in 1915. Dr W.J. Morse is credited with founding the modern US soya industry, and in the 1920s he collected, introduced and tested thousands of soya types mainly from China. Intensive breeding in the USA from the 1940s on has greatly increased average yield, and cultivars suitable for a range of environments; as a result the USA overtook China in the 1950s as the world's largest producer of soya, a position it has retained.

Soya was introduced to Brazil in 1882 (Daffert, 1892) but did not become important for almost 100 years, (Seguy & Bouzinac, 1996), and later introduced into Argentina, and both countries are currently major producers and exporters. The area sown to soya in South America has increased dramatically in the last decade, and could expand further as other countries boost production, including Paraguay (3.0 Mt in 1999) and Bolivia (1.6 Mt in 1999).

Soya is mainly cultivated for its seeds (beans) used commercially as human and stock food, and oil extraction. Until the advent of oilseed rape, *Brassica campestris* var. *oleifera*, producing the widely known canola edible oil, soya was the world's leading grain legume. Fodder production is of little importance, and is not discussed in the text. A comprehensive review of soya products and usage is contained in The Soya Bluebook (1999).

A very important part of international soya bean trade is the futures market, which is a method of buying or selling by contract. Briefly, a soya bean futures contract binds the seller to deliver a fixed amount of beans at a specified time, or binds a buyer to accept delivery. In practice, dealers in futures generally do not intend to make or take physical delivery, but are speculating on a rise or fall in prices. Both buyer and seller can avoid making or taking delivery by executing an equal and opposite transaction for the same futures month. This balances the contracts and relieves the contractor of all responsibilities under the contract. A soya bean futures contract is for 5000 US bushels (US soya bushel = 60 lb or 27.216 kg). Two important uses of soya bean futures are fixing the price of a crop before sowing, and earning a return for storing beans; both are forms of hedging. The US futures market sets the world price for soya.

World soya bean production has increased at a faster rate than any other oilseed except rape. Between 1960 and 1980 production quadrupled from around 25 million tonnes (Mt) to almost 100 Mt, due to an increase in yield per hectare by established producers, and the introduction or major expansion of soya growing in other countries.

The USA is traditionally, and remains, the biggest single country producer of soya beans, and although its domestic production has expanded, it has been overtaken on a world basis by the combined production of South American countries. USA production has steadily increased from 42 Mt in 1988/89 to 80 Mt in 1998/99. If the average yield could be raised to the 1994 level of 2.8 t/ha, production could easily reach 85 Mt with no increase in the area planted. Major producer states are Illinois, Iowa, Missouri, Arkansas and Minnesota usually in that order, which together account for about half the US total. Exports averaged 16 Mt between 1988 and 1993, then rose to 22 Mt in 1997 and 1998.

Canadian soya area rose slowly but steadily from 1 million hectares (Mha) in 1985/86 to 2 Mha in 1996/97 and will probably remain at this level. Production has risen over the 5 years to 1997 from 1.5 Mt to 2.2 Mt (Soya Bluebook, 1998). Mexico, which produced nearly 1.0 Mt in the mid-1980s has steadily reduced the area under soya since joining the North American Free Trade Agreement (NAFTA).

The area planted to soya in Brazil increased dramatically between 1960 and 1980, again between 1980 and 1990 but levelled off in the next decade. Area sown was 10.8 Mha in 1995/96 and 11.6 Mha in 1996/97. Bean production rose from 0.72 Mt in 1966, to 12.5 Mt in 1976, to 17.3 Mt in 1986 and record 27 Mt in 1996, equivalent to 2%, 21%, 18% and 20%, respectively, of total world production. Brazil's total has also benefited from an increase in the national average yield, up from 1.7 t/ha in 1988 to 2.3 t/ha in 1997, the reverse of Argentina. Major Brazilian soya growing states with their average production in 1994–96 in Mt are; Parana 6.4, Rio Grande do Sol 5.4, Matto Grosso 5.2, Matto Grosso do Sol 2.3, Goias 2.3, Sao Paulo 1.3, Minas Gerais 1.2, Bahia 1.0, with the first three normally accounting for 70% of the total crop.

From 1993 to 1996, Brazil's bean exports were (in Mt), 0.3, 2.6, 3.3, and 4.0, accounting for 1%, 23%, 33% and 11% of world exports; the decline was due to the temporary imposition of the Differential Export Tax favouring domestic crushing. Production in 1997 was 25 Mt with 7.7 Mt exported, and 31 Mt in 1998 and 1999 with 10 Mt exported. Exports in 1996 and 1997 created a domestic shortage, and Brazilian crushers imported almost 1 Mt of beans in 1996 and about 2 Mt in 1997, mainly from the USA. In 1997 the Brazilian government also approved the import of genetically modified soya bean for use in the crushing industry, and for inclusion in animal and human foods. Total crushing capacity in 1998 was 470 000 t/day, with Bunge y Born, a subsidiary of Bunge International of the USA, being the leading crusher (Knight, 1998).

The potential for further expansion of soya production is huge as major infrastructural developments are implemented, which include new railways, roads and port facilities. For example, in 1998 it cost $40/t less to ship beans down the Amazon from Matto Grosso to

Rotterdam than by comparable routes, and 5 Mt could be shipped from Amazonas in the future. This state produced only 300 t of beans in 1996, but there are optimistic plans to plant 1 Mha above Manaus on the Madeira River within 5 years.

Argentina is the world's third largest soya bean producer, second to Brazil in South America, with production rising from 3.3 Mt in 1980 to 18 Mt in 1999, following a drought in 1996 which reduced the crop by 3 Mt. The area under soya in 1996 and 1997 was about 6 Mha. The main producing area is centred on Santa Fe Province, which accounts for 55–65% of the country's total. The national average seed yield has steadily fallen from 2.42 t/ha in 1991 to 1.86 t/ha in 1996, and this decline could continue. With the substantial increase in domestic crushing capacity from 6 Mt in 1990 to 11.2 Mt in 1997, seed exports have fallen steadily and were only 2 Mt in 1996 and 0.5 Mt in 1997, but rebounded to 1.5 Mt in 1998. Meal and oil exports are now more important, and local crushers can handle about 15 Mt of beans annually. An industry survey in 1998 suggested it could be more profitable to concentrate on crushing and import low-cost beans from neighbouring countries, rather than attempting to increase the generally low domestic yields.

Mainland China's soya crop has risen from 10 Mt in 1992 and 1993 to average around 14 Mt in the last 5 years, but has been unable to keep pace with expanding domestic demand for oil and meal. Imports rose from 0.2 Mt in 1992 to 1.5 Mt in 1996 and 1997 with the USA and Brazil major suppliers, while exports have varied from 1.0 Mt to virtually zero over the same period, as both are controlled by political and not market decisions.

Soya was a relatively new crop to Indian farmers, but with official support and breeding varieties with higher heat tolerance, the area sown and seed production increased annually; from 10 kt in 1970, to 1 Mt in 1985/86, 2.6 Mt in 1990/91, averaged 4.5 Mt in 1995/96 to 1997/98, reached 5.8 Mt in 1998/99 and forecast at 6.5 Mt by 2000. Annual production varies substantially according to the monsoon rains. India does not import soya beans, but imports or exports substantial amounts of oil and meal depending on whether the national harvest exceeds local requirements.

Indonesian production could increase substantially from the current 1.4 Mt and require an additional 0.5 Mha plus a small increase in average yield to become self-sufficient, not unachievable since soya is a common crop and soya products popular foods. Current demand is around 2.5 Mt, and the shortfall is satisfied by imports, which ranged from 0.8 Mt to 0.5 Mt between 1994 and 1997, when trade was disrupted by the national economic collapse. Indonesia has concentrated on greatly increasing the area under oil palm, and palm oil is now the major domestic edible oil, with substantial amounts exported. Soya production in other ASEAN countries, with the exception of Malaysia which is the world's leading producer and exporter of palm oil, depends on the degree of official support (subsidies) given to local farmers, as imported beans or meal are generally cheaper.

Soya growing in African countries with the exception of Egypt and Nigeria, has for the last decade depended mainly on European-run large scale farms. The three main producers are South Africa, with total production in the years 1990 to 1997 rising from 50 kt to 100 kt; Zimbabwe with 40 kt to 95 kt; and Zambia steady at around 30 kt over the same period, but future prospects in all three are unpredictable. Egyptian production has steadily fallen over that period from 120 kt to 35 kt, while Nigeria's has remained static at around 80 kt.

The European Union (EU) production has ranged between 0.8 and 1.4 Mt of beans annually from 1992/93 to 1997/98 and is far below domestic requirements, which are satisfied by imports. Russian production has fallen steadily from 0.65 Mt in 1992 to half that in 1997, and this trend is expected to continue; imports over the same period decreased from 0.5 Mt to virtually zero.

The EU is a major soya bean importer, the total rising from 10 Mt in 1980 to 19 Mt in 1989 and remaining around that level for the next decade. With the expansion of its membership that total is expected to rise substantially. Other significant importers are Japan, which averaged 4–6 Mt annually in the last 20 years, and China which imported 0.2 Mt annually between 1980 and 1990, rising sharply in 1996 to 2.2 Mt and generally continuing at this level since.

World soya meal production rose steadily from 30 Mt in 1970 to 70 Mt in 1990, and the upward trend continues, with a forecast of 110 Mt by the year 2000. This increase is directly related to a similar expansion in demand from the intensive livestock industry, as a great advantage of soya meal is that it can be freely fed to all livestock. Total exports of meal in the years 1992 to 1997 were, in million tonnes, 27, 30, 31, 34, 34, 35, and the three major exporters in recent years with their average share of world trade, are Brazil 33%, Argentina 25% and the USA 20%, although the ranking has varied. In 1997/98/99 the USA and Brazil each exported about 9.0 Mt. India consumes domestically about 0.5 Mt of meal and exported 2 Mt in 1990 and 3 Mt in 1998.

The EU is a major meal importer taking about 12–14 Mt annually and locally producing another 10–12 Mt, and as previously noted, both are expected to rise. Brazil is the main supplier accounting for about half of the total. Russian imports have fallen from about 2 Mt in 1990 to a negligible amount. South East Asian countries, with vastly expanded intensive livestock industries including aquaculture, have become major importers of soya meal, particularly Thailand, Indonesia and the Philippines. Meal consumption in China has also risen for the same reason from about 1 Mt in 1990 to 10 Mt in 1998, aided by an expansion and modernisation of the crushing sector, and thus exports have virtually ceased, except for political reasons in a few years to earn foreign exchange.

Soya oil is basically a by-product of meal production, and thus its price is usually less than other edible vegetable oils. Main exporters of oil in recent years, with their share of world trade, are Argentina 33%, Brazil 25% and the USA 17% although this ranking has varied due to local adverse weather conditions reducing local bean supplies. Brazil's oil exports in the period 1988 to 1997 varied between 0.7 Mt and 2.0 Mt, while Argentina's expanded from a few thousand tonnes in 1980 to 2.7 Mt in 1999.

Oil exports from the USA reached 1.3 Mt in 1997/98 and the USDA predicted a similar level in 1998/99 and 1999/2000. A breakdown of annual soya oil consumption in the USA showed a rise from 3 Mt in 1960 to 10 Mt in 1990, increasing to 12 Mt since then. Food usage in the last 5 years included; shortening 4 Mt, margarine 1.5 Mt, and salad dressings 5.5 Mt. Non-food usage included paint and varnish 29 kt, resins and plastics 73 kt, with a substantial but undisclosed use in fatty acid production (Soya Bluebook, 1999).

BOTANY

The taxonomy of species within the genus *Glycine* is still somewhat confused although revisions have removed many spurious taxa. As an example of the taxonomic problems, when Verdcourt (1966) was preparing a revised treatment of the genus *Glycine* for the *Flora of Tropical East Africa*, he found the type specimen for the entire genus was a Pueraria!

Currently *G. max.* (L.) Merrill (syn. *G. hispida*, *Soya max*, *Dolichos soja* etc.) is placed in the subgenus *Soya* (Moench) F.J. Herm. together with *G. soja* Sieb & Zucc. (syn. *G. ussuriensis*). Both have $2n = 40$ together with most of the subgenus *Glycine* L. Verdc. The subgenus *Bracteata* which included *G. wightii* and its various subspecies, has been removed from the genus *Glycine*. The chromosome number and geographic distribution of species in the genus *Glycine* (Willd) are shown in Table 6.1.

It is postulated that soya is an ancient polyploid that has now stabilised, and it is considered a diploid for all practical purposes, but doubt as to the ancestry of *G. max.* remains. Interspecific hybridisation within subgenus *Glycine*, or between the subgenus *Soja* is usually unsuccessful. Cytogenetic evidence suggests that *G. max* and its putative ancestor *G. soja* (*G. ussuriensis*) are the same species, but as a practical aid to plant breeders they have been maintained as separate by taxonomists, since the first is a wholly domesticated erect annual, the second basically a wild annual twining vine. A study of seedcoat charac-

Table 6.1 Chromosome number and geographic distributions of species in the genus *Glycine* (Willd).

Species	2*n*	Distribution
Subgenus Glycine		
1. *G. clandestina* Wendl.	40	Australia, S. Pacific Islands
var. *sericea* Benth.	—	Australia
2. *G. falcata* Benth.	40	Australia
3. *G. latifolia* (Benth) New & Hym	40	Australia
4. *G. latrobeana* (Meissn.) Benth.	40	Australia
5. *G. canescens* F.J. Herm.	40	Australia
6. *G. tabacina* (Labill.) Benth.	40	Australia, S. China, Taiwan
	80	S. Pacific Is., Ryukyu Is.
7. *G. tomentella* Hayata	38, 40	Australia, S. China
	78, 80	Taiwan, Philippines, P.N.Guinea.
Subgenus Soja (Moench) F.J. Herm.		
8. *G. soja* Sieb & Zucc.	40	China, Taiwan, Japan, Korea, former USSR
9. *G. max* (L) Merr.	40	Cultigen

teristic provided evidence that cultivated soya probably evolved from a wild ancestor via intermediate types (Li, 1995), and wild soya has a thicker seed-coat more resistant to disease and deterioration in storage (Shou *et al.*, 1998).

In common with other crops long domesticated and used intensively by man, modern soya has lost much of its original genetic diversity through continuous selection for high yield and uniform maturity. Thus in spite of superficial differences between modern cultivars, they are genetically similar. This loss of genetic diversity is particularly apparent in North America where 80 ancestors contributed all foundation germplasm to some 250 commercial cultivars released between 1947 and 1988, with 50% contributed by only six (Burton, 1997).

The USA possesses major collections of soya cultivars; some 2000 grain-type strains are maintained at Urbana, and nearly 1000 forage types at Stoneville. China also has substantial collections, some 1500 at the Sichuan Crop Institute and lesser collections elsewhere, while less extensive collections are maintained in Russia, Taiwan, and several other countries. Indonesia has established a germplasm collection holding about a thousand local soya cultivars, partly to identify disease resistant strains (Mejaya, 1994).

The soya plant grown for seed production is a leguminous annual, normally bushy, erect, usually less than 75 cm, much branched, with well-developed roots, and producing numbers of small pods containing round, usually yellow or black seeds. The various growth stages of a soya plant have been well defined, and an international standard published, as shown in Fig. 6.2. (Courtesy FAO, Rome). These defined stages are of value to research workers who can directly compare their observations with those published elsewhere, but also have an equally important use in crop assessment for hail damage and similar insurance claims.

The root system is extensive and consists of a tap root to 1.5 m, with many lateral branches in the 0–30 cm horizon. Considerable cultivar variation exists in respect of rate of growth, total amount, spread and degree of penetration of roots, factors to be considered when choosing cultivars for specific conditions. Generally at physiological maturity, 50% of dry root weight of rain-grown soya is in the 0–15 cm horizon and 30% in the 15–90 cm horizon; compared with 70% and 20% when irrigated. Thus a cultivar with a quick-growing, deeply penetrating tap root should be selected for rain-grown crops or where subsoil moisture is available.

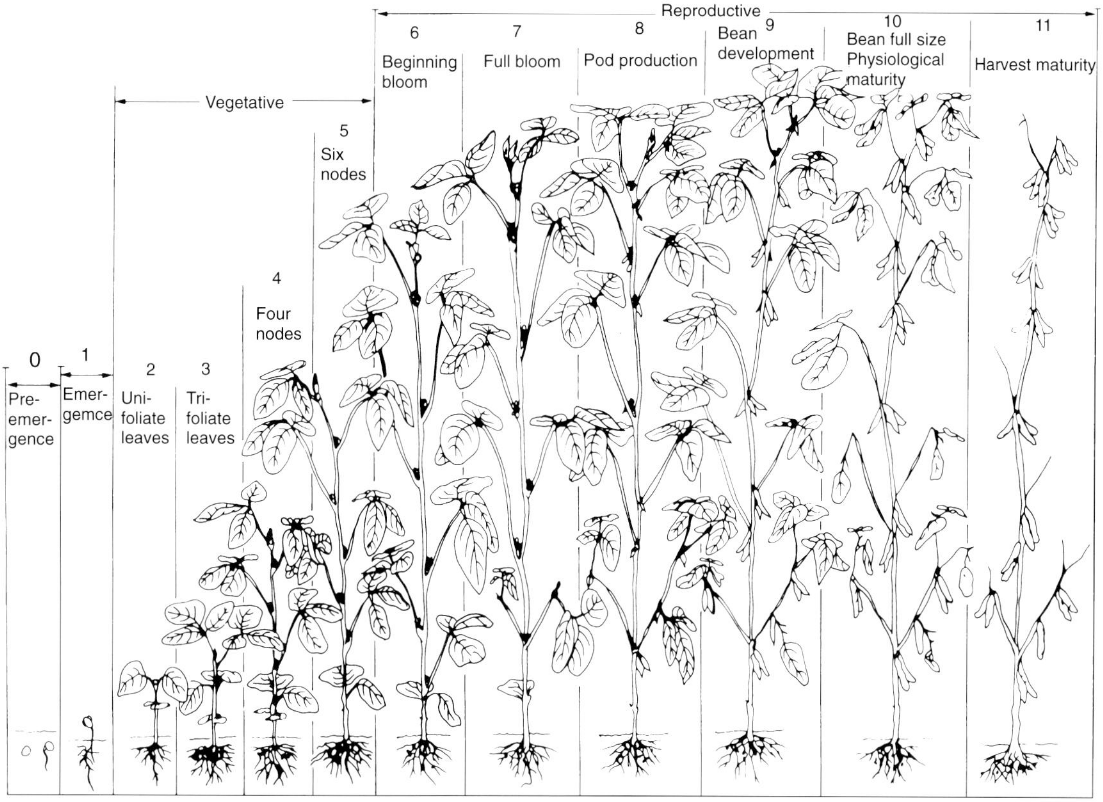

Fig. 6.2 Growth stages of soya.

Root development in the field has three separate phases corresponding to vegetative development. Extension of the tap root and shallow horizontal laterals accompany vegetative top-growth; root development to 75 cm is related to flowering and pod formation; extensive and deeper penetration of lateral roots occurs during seed maturation. Roots initially elongate faster than above-ground growth, and generally the roots of rain-grown plants will be twice as long as above-ground plant height at the six-node stage. Under irrigation this relationship extends to the pod-development stage. The effect of irrigation is also discussed in that section. Soil temperature affects the rate of root growth, and 27–32°C is the optimum for quick initial growth (Stoyanova, 1997).

Neutron radiography allows *in situ* studies of roots without visible damage. This technique has shown that 48 hours after sowing, the tap root was 14 mm long, by 72 hours it was 18 mm, by 120 hours it was 44 mm with several laterals, and at 170 hours it had reached 88 mm with laterals half this length. Thus small seedlings in the field can have extensive and spreading roots, which are easily damaged by incorrectly set or operated mechanical weeders (Fig. 6.3).

Roots carry nodules containing a species-specific strain of *Rhizobium japonicum* bacteria, which normally make the plant independent of applied nitrogen. Nodules are small, approximately spherical but occasionally lobed, smooth-surfaced, and readily produced when the specific strain of bacteria is present. When bacteria are not present in the soil, seed must be inoculated with the specific strain, which is often supplied with the seed by merchants. Quick initial growth of the tap root is important, as its nodules which

Fig. 6.3 Stages in soya seed; solid line indicates soil level.

appear 7–10 days after planting, are the main contributors to early plant development, and can finally number several hundred. Lateral root nodules become more important at flowering and seed formation. Light intensity apparently affects nodules, and can inhibit or increase nitrogenase activity (Vidal *et al.*, 1996). The effect of herbicides and fertilisers on nodulation is discussed in the relevant sections.

Current cultivars provide about 20 kg residual N/ha but the discovery of a gene that regulates nodulation should allow the development of hypernodulating plants that produce twice this ammout of N/ha. This would supply about half the current nitrogen requirement of many maize cultivars, and is also discussed in the fertiliser section. This gene is apparently controlled by a chemical signal from young leaves in the first 4 weeks of growth.

The stem is normally round, often hairy, with colour a cultivar characteristic, to 75 cm, the lower internodes become woody with age. Plants that achieve maximum height at flowering with subsequent minimum growth, produce higher seed yield than plants that significantly increase their height after flowering. A relationship may exist between stem thickness and final yield (He, 1996). Three stem types occur in cultivated soya; determinate, semi-determinate, and indeterminate, which is apparently controlled by two major gene loci. Cultivars commonly sown in the northern hemisphere are indeterminate, those in the south mainly determinate.

Plants are naturally much branched although modern cultivars usually have fewer than six branches. The spatial arrangement of branches affects growth and seed yield (Asanome & Ikeda, 1998), and fuzzy logic can be applied to evaluate plant shape to assist in selecting cultivars suited to specific environments or production systems (Ambuel *et al.*, 1997). Nodes which develop simultaneously on the main stem and branches tend to act similarly; they flower at the same time and have a similar number of flowers and pods. The effect of radiation on growth and development has been reported (Yu *et al.*, 1996). Fasciated types have shorter internodes, more and smaller leaves per plant, and long terminal inflorescences with numerous flowers.

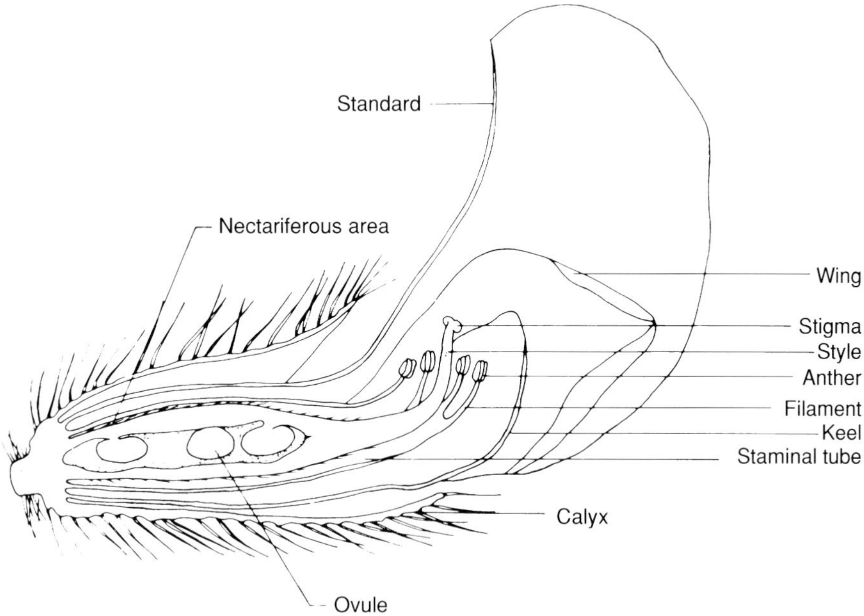

Fig. 6.4 Longitudinal section of soya flower.

Plants of medium height with the first pod well clear of the ground are most suitable for mechanical harvesting, but taller cultivars with more seed pods tend to give higher yields. Unfortunately they are also highly susceptible to lodging. Plants with stronger stems generally had lower yield, but this has been partially overcome (Lee *et al.*, 1997). Plant and pod height are also discussed in the section on harvesting. Tall cultivars to 200 cm and recumbent types, are generally grown for fodder or as interplanted smallholder crops.

Leaves are alternate, normally trifoliate, the three ovate or lanceolate leaflets borne on a long petiole, and stipulate; hairy in some varieties. Sinuate leaf types also occur. Leaves are commonly dark green, but can appear tinted with shades of brown, red or blue when growing in the field. Maximum leaf area index values of 5–8 are usually attained at main flowering, and reduce to 4–6 at physiological maturity. Leaves are normally shed as the seed-pods ripen, a valuable characteristic for mechanised harvesting. The effect of defoliation by insects or disease is noted in the appropriate section.

Leaves have characteristic Calvin cycle photosyntheses, and there is initially little variation between leaf photosynthetic rates until main flowering and pod filling, when new, upper leaves have much higher rates. The extent of this difference and the general rate of photosynthesis is also directly related to cultivar and environment. Pods and stems also contribute to carbon dioxide uptake especially during seed development.

Flowers are borne on short racemes in leaf axils, each inflorescence bearing up to 20 small, typically leguminous, purple or white flowers (Fig. 6.4) (McGregor, 1976). Cool periods, 15–20°C, adversely affect flower development, but have little effect on yield once seed has been formed. Plants can flower for 6 weeks, usually 3–4 weeks. Self-pollination is normal, pollen being shed just before or when flowers open, and normal outcrossing is 0.5–1.0%. Although insects are regarded as a minor factor in pollination, honey bees have been used in soya hybrid production since field studies showed they can increase seed yield by 20%. Also important in hybrid development is pollen production and shedding, with considerable variation from 200 to 800 pollen grains per flower.

Flowering in indeterminate lines begins on the main stem at the fourth to eighth node and flowering progresses upward, the branches flowering some days later than the main stem. Determinate lines flower at the eighth to ninth node and flowering proceed rapidly, so that within a few days every node flowers, including the branches and terminal raceme. The ratio between flowers produced and pods set is very low, often less than 25%, and the reason still imperfectly understood. It may be that, as with groundnuts, it is an evolved survival mechanism.

Growth regulators, in particular TIBA, which modify development and flowering have produced useful results, but such compounds are not cleared in all countries for use on crops grown for human consumption. Plants sprayed with TIBA are usually more compact and the number of pods per plant or full-sized seeds per pod are increased. There are however, cultivar differences in response to growth regulators, and these should be determined before any large-scale use.

Soya, as noted, is basically self-fertile, the amount of outcrossing under natural conditions is usually less than 1%, although in certain genotypes it can be 10%, probably due to the greater attractiveness to insects of flowers of some cultivars. Breeding methods are dominated by this self-fertilising property, but development of male-sterile lines has assisted in promoting natural hybridisation.

The fruit is normally a short, hairy pod, 2–10 cm × 2–4 cm, usually brown or black, but can have a green, red or purple tint. Pods normally shatter when ripe, the rate and degree being a cultivar characteristic, and early or high shattering cultivars are unsuitable for mechanical harvesting. A force of 1.7 kg was necessary to open pods of a non-shattering cultivar compared with 0.7 kg for shattering cultivars. Pod number per plant can vary from a few dozen to several hundred, and although the number is basically a cultivar characteristic, it is heavily dependent on climatic conditions during growth and flowering.

Fig. 6.5 Soya pods and seeds.

The effect of pod loss on final yield has been investigated, and the extent to which plants can compensate depends mainly on when the major loss occurs (Rocha *et al.*, 1996).

Pods usually contain three, occasionally more, small, hard, round or ovoid seeds, 5–10 mm in diameter; testa is smooth and shiny with a small distinct hilum (Fig. 6.5). A detailed comparison of seed coat structure between cultivated and wild soya showed the latter tended to be rough, the former smooth (Li, 1995). Seed colour can be yellow, green, red, brown, black, slightly mottled or occasionally bi-coloured according to cultivar. Pale yellow is the most commercially acceptable colour for beans intended for human consumption and oil production (Fig. 6.6). One hundred seed weight varies from 5 to 40 g depending on cultivar. Important to breeders is the requirement that beans destined for the food industry use must have yellow seed-coats and cotyledons. The most recent development aimed directly at

Fig. 6.6 Small, yellow soya beans, the most commercially acceptable type.

food processors, is the release of clear hilum varieties bred at Iowa State University.

Seed-oil content, usually in the range of 15–22%, is a cultivar characteristic and influenced by environment and climate. Oil normally contains approximately 10% linolenic, 55% oleic, and 30% linoleic acid, with up to 50% variation in a specific component. The relatively high linolenic acid content, compared with maize oil at 1%, is the main cause of poor flavour or flavour instability in soya oil. Protein content, of greater importance, is usually 40–50% and contains almost the entire complex of amino-acids including those considered essential in human diets. Protein content is inversely related to oil content, and there are indications high protein content can be associated with lower seed yield. The rate of increase in, and total seed-oil content is temperature sensitive, with high temperature generally favouring high oil content; conditions which normally have the opposite effect on protein content. The more rapid rate of oil accumulation probably reduces the assimilates available for protein conversion.

Seeds are rich in digestible nutrients, have a high calcium, iron and vitamin content, and are of great value in human diet. The chemical composition of soya seed compared with other tropical pulses is shown in Table 6.2. The low incidence of heart diseases in Asian countries with a high soya content in diets resulted in research to establish a link. It was determined that the isoflavones in soya proteins, phytoestrogen, could substantially lower cholesterol levels.

Radiation and other mutagenic seed treatments have increased the variability of yield components, resistance to moisture stress or disease. However, the seed of some new cultivars may contain higher levels of toxic and antinutritional compounds than is allowable in human diets, as in Brazil (Vasconcelos *et al.*, 1997).

Non-destructive analysis of seed is important when stocks are limited, as in breeding programmes, and nuclear magnetic resonance (NMR) is a common method. Where seed is not limited, ground seed may be analysed by conventional methods, but near infrared reflectance spectroscopy (NIRS) used for amino and fatty acid analysis could be an asset in the gross screening of entries in breeding programmes (Pazdernik *et al.*, 1997). Methods of reducing the growth cycle can be of importance to plant breeders, and one technique germinates immature seed (Roumet & Morin, 1997).

Soya breeding and selection programmes were generally directed to large-scale temperate country producers, but current targets include production of cultivars for tropical and subtropical regions, while cultivars are being developed in China for intercropping, an objective of value elsewhere. Extensive breeding programmes have been carried out in the USA, Russia, India, Taiwan, China, and increasingly by private companies since the introduction of Plant Breeders' Rights.

Plant breeders and geneticists are becoming increasingly concerned at the rapidity with which long-established local varieties are being lost. This is particularly so in Asia, where many of the original 'land races' have been replaced, quite naturally, by high-yielding cultivars. Hymowitz (1976) noted that the last major soya seed-collecting expedition from the USA was nearly 50 years earlier, and that less than one-third of

Table 6.2 Chemical composition of tropical pulses (%).

Legume	Moisture	Protein	Fat	Carbohydrate	Fibre	Ash
Black gram	9.7	23.4	1.0	57.3	3.8	4.8
Chickpea	9.8	17.1	5.3	61.2	3.9	2.7
Cowpea	11.0	23.4	1.3	56.8	3.9	3.6
Dolichos bean	9.6	24.9	0.8	60.1	1.4	3.2
Groundnut	5.4	30.4	47.7	11.7	2.5	2.3
Lima bean	12.6	20.7	1.3	57.3	4.3	3.8
Pigeon pea	10.1	19.2	1.5	57.3	8.1	3.8
Soya bean	5.0–9.4	29.6–50.3	13.5–24.2	14.0–33.9	2.6–6.3	3.3–6.4
Tepary bean	9.5	22.2	1.4	59.3	3.4	4.2

the original 4000 samples existed. As is the case with other oilseed crops, the establishment of collections to preserve gene pools of wild and cultivated soya species is of great importance, but is currently somewhat neglected.

ECOLOGY

Soya is basically a warm-temperate, short-day plant, but the proliferation of breeding programmes has so extended the range of soya that it is now virtually worldwide, with cultivars selected for day lengths between 12 and 16 hours and for growing from sea level to 3000 m. Data from the International Soya Variety Experimental Trials of 1973–82, show that protein and oil content was highest between latitudes 0 and 20°; protein content was highest between 500 and 1000 m; oil content between 0 and 500 m, and there were seasonal variations (He *et al.*, 1996), confirmed by later trials (Wan *et al.*, 1998).

The low yield often recorded from soya crops in the tropics is, from personal observation, mainly due to bad management rather than any inherent plant deficiency. The yield potential of soya in tropical areas is similar to that in temperate regions, and once the environmental and commercial parameters within which a crop must be produced have been established, it is the author's experience that a suitable cultivar can usually be selected.

Soya is naturally highly photoperiodic, and a variation of 15 minutes in day length sufficient to inhibit flower development in a specific cultivar. In general, short-season cultivars require longer days to flower and mature than long-season types. This extreme photoperiodic sensitivity was a major restriction on wider commercial cultivation of soya, and the subject of extensive and continuing research outside the scope of this book, but published in the literature.

To assist growers and breeders in the late 1970s, 12 maturity groups were recognised in the USA to identify regions of adaptation of soya cultivars. Group 1 was the northernmost to include Canada, and Group 12 the most southern to include the semi-tropical countries of Central America. Since then cultivars have been bred for almost any suitable region, sometimes using the 'long juvenile' trait that delays flowering in short-day environments, or using a combination of short season cultivars and planting dates (Kane *et al.*, 1997; Tomkins & Shipe, 1997). Similar information, or area suitability maps for soya growers, have been produced in many countries including Japan (Kohyama, 1996), Brazil (Seguy & Bouzinac, 1996), and Italy (Danuso, 1997).

A 50-year cycle in the USA determined that two other important factors influenced domestic soya yield; the level of maximum air temperature and average rainfall during the 3 months of the main growing season. The climatic parameters for extending soya cultivation in Russia showed an average air temperature of 13°C for 10 days after sowing, 22°C from emergence to flowering, and 15°C from flowering to maturation was desirable (Stepanova & Korsakov, 1977).

Table 6.3 Effect of daylight and night temperature on time to first flowering (days from sowing) in soyabean and cowpea – Nigeria.

Cultivar	Day length/night temperature treatment combination				Sensitivity to:		
	1 h 40 min		13 h 20 min		Day length	Temperature	Both
	24°C	19°C	24°C	19°C			
Soyabean							
Grant	25	27	25	27	–	–	
Hshi Hshi	28	33	28	34	–	+	
TK5*	29	35	35	44			+
Belgium Congo*	31	37	37	44			+
Improved Pelican*	32	37	39	59			+
Cowpea							
TVu 1503	39	41	39	40	–	–	
TVu 167	44	54	44	54	–	+	
TVu 6	47	48	100	100	+	–	
TVu 859*	49	53	53	62			+
TVu 1020*	40	46	45	53			+
TVu 2744*	49	56	53	69			+

*Note the consequence of equal and opposing night temperature and day length effects on onset of flowering.

The optimum mean daily temperature for most cultivars is a 28–32°C, and although plants appear to be generally unaffected by a fairly wide variation in day temperatures before flowering, maximum development usually occurs around 30°C. The optimum temperature varies with the life of the plant, and at certain periods night temperature has a significant effect; around 24°C will decrease time to flowering, and the converse applies (Piper *et al.*, 1996). The interaction between day length and night temperature on the time to first flowering on soya and cowpea is shown in Table 6.3. (Summerfield *et al.*, 1979). A daytime temperature of 25°C or less also delays flowering, and prolonged periods of cloudy weather tend to extend the vegetative phase at the expense of seed yield. Low night temperatures during germination or just after emergence can slow later growth, and may accelerate the reproductive cycle.

Periods of high temperature in the last 45 days of seed maturation adversely affect viability and vigour in some cultivars, resulting in lower emergence, poor growth and reduced yield (Spears *et al.*, 1997). Such periods can also affect seed oil composition, especially in the low linolenic acid cultivars, or oleic acid levels in early maturing cultivars (Gibson & Mullen, 1996a; Kane *et al.*, 1997). In the tropics, viability of seed kept in uncontrolled storage falls rapidly, especially during periods of high humidity, and germination can be very low after 2 years.

Little data is available on the effect of altitude on growth and yield of soya, except that of low temperature. In China's Yaan and Fulin provinces, cultivars from mountainous regions generally had a higher average seed protein content, 50%, than lowland types (Wang, 1995).

Soya is slightly more drought resistant than maize, but much less so than sesame. A rainfall of 500–750 mm is necessary to produce high seed yield, and although soya can tolerate dry soil conditions before flowering, adequate moisture is essential once the buds have formed and until the pods have filled. However, it appears that cultivars with dense, coloured pubescence are more drought resistant than cultivars with little pubescence (Morrison *et al.*, 1997).

High humidity during seed maturation lowers viability and reduces storage life. Waterlogging is detrimental to growth and yield, but especially so at germination and emergence. However, mature soya is considered to be more tolerant of wet conditions than maize or groundnuts in Queensland, Australia, and more so than groundnuts, cowpeas or velvet bean in South Africa.

Frost normally kills soya plants at any stage of growth before maturity, and a frost-free period of 120 days is desirable for optimum yields. However, short-season cold-tolerant cultivars have been bred for use in Russia and Canada, which could be of value in the high-altitude tropics. Cultivars vary in their resistance to low temperatures, and most cultivars are substantially unaffected once pods fill. Hail severely damages seedlings; on maturing plants a substantial number of buds or flowers can be destroyed by hail, but plants are capable of compensating for such loss if other conditions are favourable. Strong winds increase the danger of shattering when pods are ripening. Lodging caused by termite damage to drying roots and stem can be severe in arid regions, and in East Africa the author lost the major portion of a seed-multiplication planting for this reason.

The growth and development of soya plants can be significantly influenced by plant growth regulators, irrespective of climate, and has been widely reported in the literature. The effect of these chemicals and their interaction with the environment is important to tropical growers, as they may be able so to influence growth as to allow a crop to mature in marginal situations. This could be of special value on breeding and trial farms, when temporary adaptation to local conditions is more important for research purposes than high yield.

SOILS

Soya is tolerant of a wide range of soil conditions, with the highest yield obtained on well-drained fertile loams. In general, soya will produce good yields in areas where maize also produces a good yield, although very sandy or heavy clay soils require skilled management when used for commercial soya production. A soil reaction of pH 5.8–7.0 is acceptable, between pH 6.0–6.5 is desirable, since acid soils reduce nodule bacteria activity, and also magnesium and calcium availability. Soils with a boron content of only 1 ppm will adversely affect yield, as this element is toxic to soya. Country or regional soil maps indicating the most suitable areas for soya are now available, assisted in some countries by computer models, as in India (Jagdish & Bradley, 1997).

Soya has low salinity tolerance, less than half that of cotton but slightly more than maize. A quick method the author has used for initially screening cultivars for salinity resistance is to germinate seeds in various salt solutions, including samples from the local soil or irrigation water to be used, and sowing in the field those cultivars that showed the highest survival rate. This *ad hoc* approach allows an agronomic programme to begin while awaiting the results of more comprehensive salinity screening. The variation in response to minor elements and salinity is of considerable assistance in the selection of cultivars tolerant to levels that would normally be toxic. Wild perennial species native to Australian coastal areas for example, may provide genes valuable in increasing salt tolerance.

Tropical red earths grow good soya crops, but in general black-cotton soils have proved extremely difficult to manage. The surface can dry and cake with great rapidity and the crust formed prevents emergence, or results in the cotyledons being held in the soil while the hypocotyl continues to grow until it fractures, resulting in a fatal condition known as 'bald head'. Within a particular area soil type can considerably influence seed yield, and also affects profitability of fertiliser usage. In Kenya unfertilised soya grown on a sandy loam derived from granite produced a higher yield than fertilised soya grown on a red clay derived from phonolite (Table 6.4) (Weiss, 1967b). Both crops followed well-fertilised hybrid maize.

Table 6.4 Effect of soil type on yield and fertiliser – Kenya.

P_2O_5 (kg/ha)	Phonolitic red clay		Granitic sandy loam		
	Single super.	Triple super.	Single super.	Triple super.	Diammonium phosphate
0		496		1076	
22	560	568	1305	1248	1235
44	560	549	1345	1260	1265
66	588	556	—	—	—

Values in table are seed yield in kg/ha.

FERTILISERS

Soya has a moderately high nutritional requirement and in the USA removed more phosphorus, potassium, magnesium and calcium than a comparable maize crop. Total uptake of the three major nutrients at two yield levels is shown in Table 6.5 (University of Illinois data). Of the total amount of nutrients utilised, 70–80% of nitrogen and phosphate and 60% of potassium are usually in the seed at harvest. However, the rate of mineral nutrient accumulation and the rate at which individual nutrients accumulate in the maturing seed, are not constant, although 80% of final nutrient content of seeds is usually present at the beginning of leaf senescence. Analyses of a crop yielding 3400 kg/ha is shown in Table 6.6, which also shows the main distribution within the plant (Sale & Campbell, 1980). A more detailed analysis is shown in Table 6.7, which gives percentage uptake of all nutrients at 120 days (Mascarenhas, 1973). The rates of uptake and distribution of major nutrients within the soya plant are shown in Fig. 6.7, which is compiled from Brazilian data, but similar data is available from other countries (Bataglia *et al.*, 1977; Dong and Zhang, 1997). Soya readily makes use of fertile soils to produce excellent vegetative growth, and although the yield initially increases with total plant dry weight, there is an optimum level beyond which seed yield decreases.

Table 6.5 Total uptake of NPK at two yield levels – USA.

Nutrient	Seed yield (kg/ha)	
	3500	6500
Nitrogen (kg)	340	620
Phosphorus (kg)	30	56
Potassium (kg)	170	225

When soya is grown in rotation with other highly fertilised crops and inoculated seed is used, little additional nutrient is necessary. Phosphate is most frequently required, with wide dispersal throughout the root zone. Application of lime to acid soils normally increases yield and may be all that is necessary for profitable seed production. Current cultivars differ not only in their response to varying levels of nutrition, but also to individual nutrients. Fertiliser should be placed approximately 10–15 cm deep, below or to the side of, and not in contact with the seed. Rock or similar phosphate and potassium fertilisers should be broadcast and ploughed in.

Soya is often grown on bunds around paddy fields on which were burnt crop residues, weeds, etc. The ash provides fertiliser and acts as a soil ameliorator for the heavy clay of these bunds, and high individual plant yields are obtained. Sewage or similar sludge is often beneficial, improving seed yield and soil conditions, but it can also increase the heavy metal content of seeds, leading to toxic contamination of resulting food products.

Table 6.6 Analysis of soyabean plants to show nutrient distribution – Brazil

Component	N	P_2O_5	K_2O	Mg	S (all kg/ha)
Grain	210	50	70		
Straw	70	17	45		
Stubble/roots	33	11	22		
Total	313	78	137	14	11
Per 100 kg seed	9.2	2.3	4.0	0.4	0.3

Note: Seed yield was 3400 kg/ha.

Table 6.7 Total nutrient uptake and distribution in soyabean (%) – Brazil.

	Plant part		
	Stems	Leaves	Pods and seeds
N	13.8	20.4	65.8
P	20.6	14.4	65.0
K	34.6	12.2	53.2
Ca	28.0	39.8	32.2
Mg	44.0	20.2	35.8
S	24.4	10.0	65.6
B	17.5	20.9	61.6
Cl	5.0	9.3	85.7
Cu	26.0	19.2	54.8
Fe	40.4	48.2	11.4
Mn	19.3	44.4	36.3
Mo	23.2	7.2	69.6
Zn	34.2	32.9	32.9
Al	38.3	55.1	6.6

Age of plants = 120 days.

Nutrient deficiency symptoms have been demonstrated in laboratory trials and there can be cultivar differences, which, as noted in the literature, may also be masked by herbicide/pesticide reactions.

Nitrogen

The nitrogen (N) uptake by an average soya crop reaches a daily peak of 5 kg N/ha at the end of 3 months, then declines rapidly (Fig. 6.8), and is substantially supplied by *Rhizobium* bacteria in root nodules (AVRDC, 1978). Commercial growers in the USA consider well-nodulated soya plants can supply nitrogen sufficient for 3500 kg/ha seed yield; the development of hypernodulating plants could double this. Adequate nodulation can also substitute for nitrogen applications in less developed countries. In Bangladesh for example, the local inoculant strains produced dependable and adequate nodulation, and were an effective and cheap substitute for urea fertilisers (Hoque, 1993). In India biogas slurry is an effective substitute for diammonium phosphate (Kharat *et al.*, 1997).

Part of the N fixed by soya plants remains in foliar residues and roots, and is often available to a following crop. This is discussed in the section on intercropping and rotations. Young plants can exhibit severe nitrogen deficiency symptoms especially when over-irrigated or when heavy rain falls at emergence. Nitrogenous top-dressing of seedlings may thus be necessary before nodule supply is adequate. However, different types of N fertilizer can affect the rate and level of nodulation in seedlings, as noted in China (Liu *et al.*, 1998).

Nitrogen at planting is generally unnecessary, can suppress nodulation, and although seedlings may appear to benefit, the yield increase is invariably unprofitable. Some cultivars respond to nitrogen top-dressing at flowering and the yield increase is due to an increase in seed number. In arid areas, ploughing in crop residue or green manure can produce a soil moisture and nitrogen deficit, and both will require correction. Planting after good rains, seed inoculation, and a small top-dressing with nitrogen when seedlings are some 10 cm high is usually sufficient.

Nitrogen application can increase seed protein at the expense of seed-oil content, and although

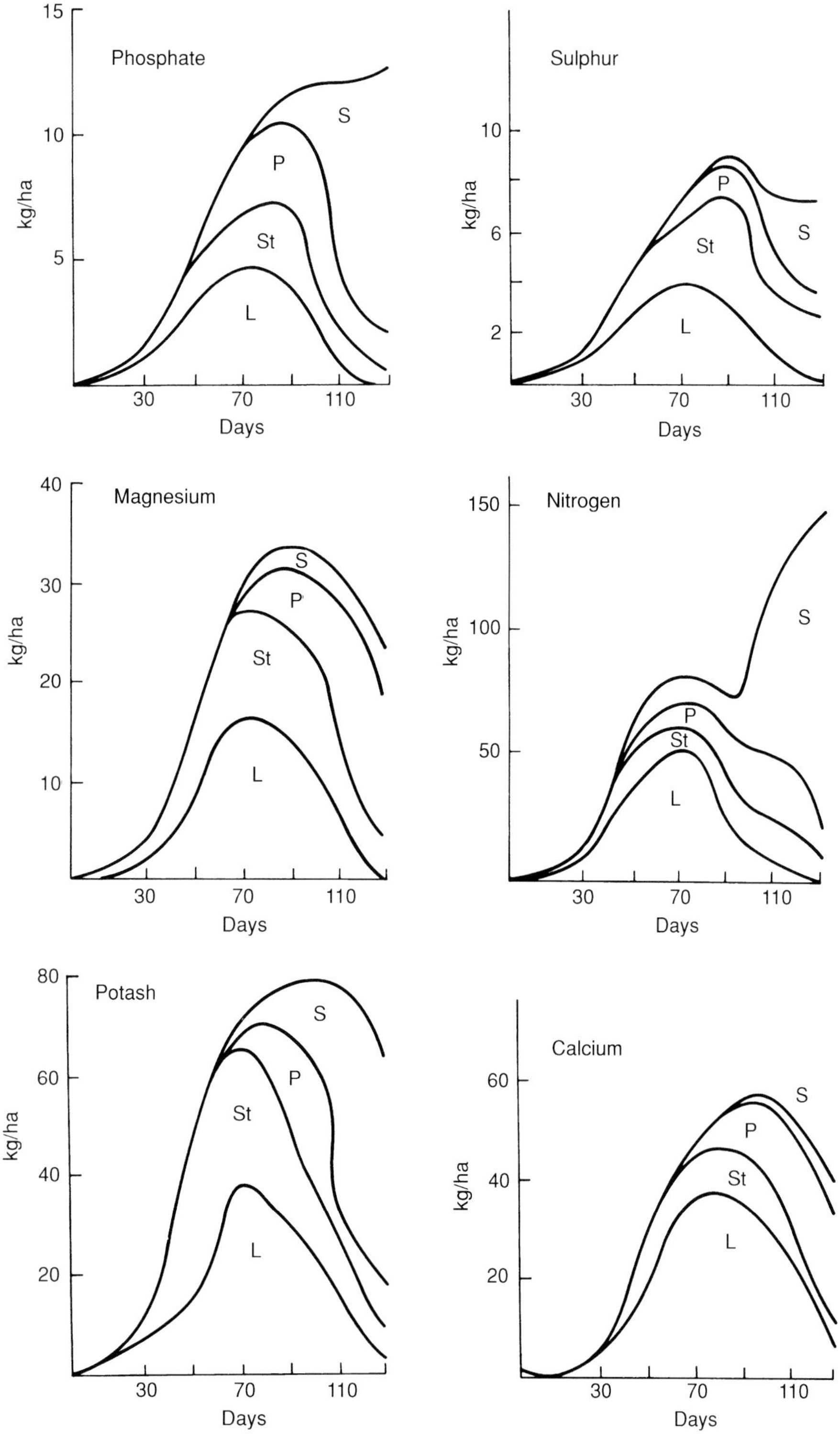

Fig. 6.7 Accumulation of nutrients in soya plants (determinations made at plant maturity, 130 days after sowing). S = seeds, St = stems, P = pods, L = leaves.

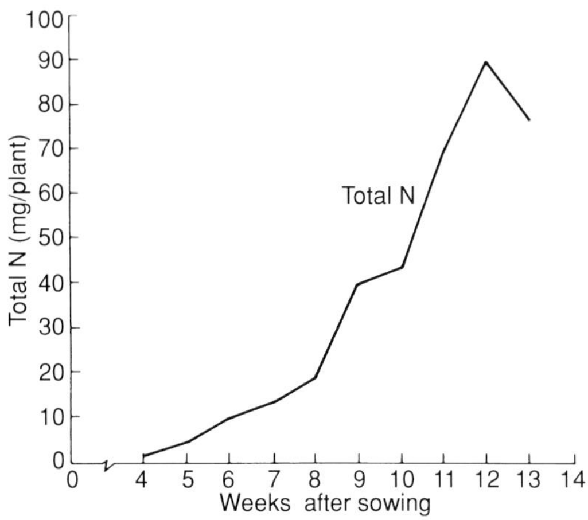

Fig. 6.8 Nitrogen accumulation in soya.

the type of nitrogen is generally of little importance, urea has adversely affected yield and protein content. Plants with a high nitrogen status can be affected by low temperature before flowering which induces sterility and reduces yield. Root temperature can also affect N fixation (Stoyanova, 1997). Plants generally fix less atmospheric nitrogen when soil nitrogen is readily available, than in its absence. A good rule to follow in respect of nitrogen application to soya is, 'When in doubt, don't!'.

The type of nitrogenous fertiliser used is generally unimportant, with urea and ammonium sulphate the most popular, or calcium ammonium nitrate and diammonium phosphate in the seedbed. Occasionally a special compound is to be preferred; in China, urea containing borate enhanced the uptake of nitrogen and phosphorus and increased the nitrogen utilisation rate by 4–7% compared with urea without borate (Wang *et al*., 1992). However, urea may adversly affect germination and seedling growth, and react with fungicides and herbicides (Bremner, 1995).

Phosphate

Soya requires little additional phosphate (P) or other nutrients when sown after other well-fertilised crops; in Queensland, Australia for example, a yield response is unlikely unless soil tests show less than 30 kg available P/ha. In the USA, a crop yielding 3500 kg/ha contained in its seed and vegetative parts nearly 30 kg P, compared with 12 kg P in a similar wheat crop, and 45 kg P in a 10 000 kg/ha maize yield. Higher-yielding cultivars tend to give greatest response to fertiliser applications; which is of economic importance since a given seed yield may be obtained at a lower level of application. The effect of phosphate on nodulation is erratic as both a deficit and a surplus can adversely affect nodule numbers; the level of potassium is apparently an important factor. Conversly, innoculating seed with *Pseudomonas striata* increased the uptake of P in India, and also increased seed yield (Dubey, 1996).

Excess phosphate may depress seed protein and oil content, or induce zinc and iron deficiencies. Until soil analyses or formal trial results are available, half the annual application rate of P applied to maize is usually sufficient; in non-maize growing areas the application rate is 20 kg P/ha.

Phosphate is best applied in the seedbed, since seedling requirements are high, and where potassium levels are adequate, single superphosphate with its slightly greater sulphur levels have produced higher yields in many regions of Africa and India (Weiss, 1969) (see Table 6.4). Adequate supplies of phosphate are necessary to ensure optimum seed-protein content and amino acids in the protein, which is particularly important where soya is a major protein source. Worldwide crop response has shown little difference in the effectiveness of various phosphatic fertilisers at the same level of availability.

Potassium

Soya requires a large amount of potassium (K) during vegetative growth; the highest concentration occurs in the stem and lower leaves. A similar level occurs later in the pods, and potassium content of the seeds is initially higher than at maturity. Potassium must therefore be easily obtainable by plants during rapid growth, with a slowdown in uptake as seeds begin to form. The nutrient is readily translocated within the plant, and a large deficiency is necessary before seed requirements are affected. A crop with a seed yield of 3500 kg/ha removed 170 kg K/ha compared with the 40 kg K/ha removed in the grain by a 10 000 kg/ha crop of hybrid maize. Soya can take up luxury amounts of potassium, and analyses do not necessarily indicate the replacement level required, or initial availability.

A potassium deficiency retards nodulation, but why is still to be accurately determined. Conversely, adding potassium produces more healthy plants with more nodules, and it may be that stronger plants can sustain more nodules. It is difficult to make field recommendations for potassium usage; in Africa and India for instance, crop response has been most erratic and soil analyses and local field trials essential to determine K availability. Generally the amount necessary equals the phosphate, but since residual soil potassium is an important source for crops grown in rotations, the annual application is usually less. Broadcasting and ploughing-in produces the highest return per unit of nutrient applied, since wide dispersal in the root zone encourages uptake. There is little data comparing the various types of K fertilizers, but similarly to phosphate, potassium sulphate would probably give a greater yield response in areas where soil sulphur is deficient. There is some evidence that adequate potassium reduces damage caused by fungal diseases, especially *Diaporthe soyae*.

Sulphur

Sulphur (S) is deficient in many tropical soils; in Australia it restricted soya seed yield at all levels of nitrogen and phosphate; in India the sulphur requirement is high in comparison with other legumes, and up to 50 kg S/ha applied as gypsum. An interaction between sulphur uptake and other minor elements occurs, which may affect yield; in Brazil an interaction between iron, zinc and sulphur limited plant growth (Fontes & Cox, 1998). Where the sulphur deficiency is small, single superphosphate will normally provide that necessary. Soil analyses to determine sulphur availability can be unreliable, and where other crops have responded to this element, trials should be conducted with soya. The uptake and distribution of sulphur within soya plants has been investigated (Sunarpi & Anderson, 1996).

Calcium

Soya is tolerant of relatively high soil acidity, to pH 5.0, but yields are progressively depressed as the acid level increases. Calcium application greatly improves plant growth and yield on acid soils, since it not only alleviates a calcium deficiency but assists in maintaining a balance between calcium and magnesium–potassium availability. Calcium is also necessary for growth and normal development of nodule bacteria and effective nodulation of seedlings. A method of screening cultivars for tolerance to low calcium has been developed (Spehar & Galwey, 1997) which can be of great value. Large amounts of lime are usually not available or cannot be easily transported in many tropical areas with acid soils, and in these circumstances such cultivars can be sown, or moderate yields be accepted.

Manganese

Soya is very sensitive to manganese (Mn) deficiency which frequently occurs on soils with a high pH, and plants remain stunted with yellowish leaves. Cultivars react differently to similar Mn levels, in the accumulation and distribution of Mn within the plant (Abreu *et al.*, 1996: Oliveria *et al.*, 1997). It appears that plants which have a higher distribution of Mn to leaves are more sensitive to a Mn deficiency. A deficiency is easily corrected by adding the required amount of manganese to other fertilisers, or by spraying with manganese sulphate or oxide. The manganese : iron ratio is important, and a deficiency of one can be indicated by toxicity symptoms of the other. A desirable ratio is manganese 2.5 : 1.5 iron. Symptoms of iron chlorosis can also occur when plants are relatively well-supplied with this element. Manganese toxicity usually appears on soils below pH 5.0, which may also contain iron–manganese concretions. Liming corrects this toxicity in the majority of cases.

Minor nutrients

Similarly to many other leguminous crops, soya can benefit from application of minor plant nutrients, but it is essential to establish which are limiting plant growth. The line between too little and too much is easily crossed; either can adversely affect yield. Soil and foliar analyses can determine availability and uptake, but many current soya cultivars vary considerably in their response to, and ability to extract soil nutrients. Deficiencies can be alleviated by seed dressing, fertiliser additives or foliar spray.

A magnesium deficiency reduces plant nodule bacteria activity, and is most common on acid soils, but can also occur following a heavy potash application. The extent to which a magnesium deficiency reduces seed yield has not yet been fully determined. To alleviate a magnesium deficiency, dolomitic limestone is recommended, or serpentine as in China (Wu, 1998).

Boron is important in soya nutrition as both a deficiency and toxicity affect plant growth, but the specific reactions have yet to be fully determined. Boron appears to be more mobile in plants than was originally believed, and occurs in foliage and seeds at harvest (Bukhoriev, 1997; Reinbott *et al.*, 1997). Boron content is related to germination, and a low content in seed can reduce viability, seedling vigour and survival (Rerkasem *et al.*, 1997; Rodrigues *et al.*, 1997). In China urea containing up to 4% borate increased yields, and response to phosphorus and nitrogen (Wang *et al.*, 1992).

Seed treatment with cobalt, copper, or molybdenum can increase yields where these trace elements are deficient, and may also reduce Septoria infection. In India, a zinc deficiency was alleviated by up to 10 kg Zn/ha as zinc ammonia chloride, which was more effective than zinc sulphate (Raghuwanshi *et al.*, 1997). Cultivars also differ in their iron utilisation ability, and on alkaline soils inefficient lines show marked iron-deficiency symptoms. An excess of aluminium can reduce nodulation and general plant growth, but is seldom encountered in the field. The effect of beryllium has been demonstrated (Sajwan *et al.*, 1996). Selenium increased spider mite resistance in China (Liu *et al.*, 1998) and selenium was also detected in plants from Japanese trials to determine the absorption of atmospheric radionuclides (Shinonaga & Ambe, 1998).

CULTIVATION

To obtain the potential yield from soya, especially the newer cultivars, they must be planted at the optimum period, correctly fertilised, adequately weeded and properly harvested. Obvious maybe, but frequently not practised! As has been noted on other oilseeds, cultural and environmental factors have a far greater influence on field yield than genetic factors. An important early evaluation directly affecting profitability and which is often overlooked, is a realistic assessment of the probable yield in a particular area, with inputs planned accordingly. A number of techniques can assist, including a method of predicting field emergence levels from seedbed conditions (Egli & TeKrony, 1996).

The pre-planting cultivation necessary for cotton, maize or beans is suitable for soya, with a seedbed that allows even planting depth and spacing. In drier areas subsoiling can materially increase yields by allowing greater moisture penetration into the subsoil. In China the moisture storage capacity at a depth of 100 cm increased from 15 mm to 52 mm following this operation, and soya yield rose by one-third (Wang *et al.*, 1996). Direct sowing of soya into stubbles is described in the weed control section.

Innoculation of seed with a cultivar-specific strain of bacteria is necessary where soya has not previously been grown, and inoculants and application techniques have been reviewed (Smith, 1995). On high-yielding commercial crops in the USA annual inoculation is practised, since the cost of so doing is far less than potential yield loss through lack of adequate nodulation, and seed merchants normally supply seed ready inoculated for immediate planting. However, where soya has previously been grown or where an indigenous *Rhizobium* strain is present, inoculation with a different strain may be ineffective; antagonism between a specific soya cultivar and *Rhizobium* bacteria also exists. Pelleting seed with various compounds including a *Rhizobium* culture can be beneficial, particularly when soil conditions are not ideal for germination; in India activated charcoal or peat impregnated with *Rhizobium* increased plant health and subsequent seed yield. Soaking seed in hot water or a sulphuric acid solution enhanced germination percentage in Nigeria (Agboola, 1996), but such pre-treatment is not usually necessary.

Treating seed with a fungicide or insecticide may reduce the effectiveness of the inoculant, but does not interfere with the action of bacteria already in the soil. Such dressing should be considered only in those areas where high seedbed losses are to be expected; a higher seed rate may be a better method of offsetting the latter. Seed treatment should be routine when introducing soya into new areas as this can assist in controlling the spread of disease, as a number of pathogenic fungi are seed-borne. Treating seed with growth regulators, including GA and NAA, generally has adverse effects on emergence and shoot growth, and is of value only in the laboratory.

For commercial crops only seed less than one year old should be sown, since viability of stored soya beans falls rapidly, as does that of damaged seed (Table 6.8) (University of Illinois data). Viability is an important factor in obtaining good stands, but equally important is seed vigour, agronomic methods of increasing it, and techniques for evaluating it, which have been developed (Dias & Marcos, 1997). Where the growing season is short or seedbed disease/pests abundant, cultivars with quick, even and vigorous germination are a major asset. Where large areas are to be sown, the use of several cultivars with differing maturity rates will extend the harvesting period even though their potential yields are lower. This reduction will be offset by more economical use of equipment and reduced shattering loss.

Good-quality seed is uniform in size and weight, and flows easily through a drill or planter. Seed should be sown 3–5 cm deep, and generally not more than 7.5 cm, although in sandy soils or where the top few centimetres are dry, up to 10 cm is possible if soil at this depth is moist. In these circumstances, the seed furrow should not be completely filled. Adequate soil moisture is essential as seed requires a moisture content of about 50% for germination compared with 30% for maize. Large seed cultivars can be planted deeper than those with small seed. Within a cultivar there is little difference in percentage emergence between large and small seeds. Germination is epigeal and under optimum conditions seeds emerge in 3–5 days, the primary leaves unfolding by the 5–7th day. The first trifoliate leaf unfolds in 9–11 days, and is fully expanded by the 15–17th day.

A seedbed temperature of 25–33°C is the optimum, below 15°C and above 37°C reduces emergence, although some cultivars germinate below 10°C but with the period to emergence prolonged. Depth of planting can modify temperature effects; deep planting and low temperature can substantially reduce the emergence percentage, while high temperature and shallow

Table 6.8 Effect on emergence of sowing mechanically damaged seed – USA.

Seed condition	Emergence (%)	Vigorous seedlings (%)
Whole sound seeds	80	72
10% cracked seed-coats	70	65
100% slightly cracked seed-coats	57	41
100% severely cracked seed-coats	43	23

Table 6.9 Effect of sowing date and spacing on seed yield – Kenya.

Date	Spacing (cm)				Mean	Mean yield*
	60 × 5	60 × 10	90 × 5	90 × 10		
25 May	1524	1520	1637	1704	1596	1437
15 June	1502	1300	1350	1569	1430	1294
6 July	1323	1121	1009	1256	1177	1001
27 July	695	964	852	493	751	665
Mean	1261	1226	1212	1256		

*Mean yield of all trials in all years on approximately similar dates; all yields in kg/ha.

Table 6.10 Recommended planting times in selected countries.

USA		Egypt	April
North	May		
Central	May	Turkey	July
South	June		
		India	
Brazil		Summer (wet)	June
North	May	winter (dry)	
South	November		
		Japan	May
Argentina	November		
		China	
Former USSR		North East	May
West	May	South East	March or July
East	July		
Australia	November	New Zealand	November

planting will dry out germinating seed. Such situations frequently occur in the tropics, and the time of planting in relation to onset of the rainy season must be accurately determined (Table 6.9) (Weiss, 1967b). Similar results are obtained in most countries where soya is grown.

Yield falls rapidly outside the optimum planting period, and increasing seed rate or fertiliser levels cannot normally compensate for the loss. This general observation can be modified, where it is possible to do so, by using newer cultivars that are very short-season or resistant to a local disease favoured by cool, moist conditions, and can thus be planted at times unsuitable for cultivars normally grown. The factors influencing the use of early-maturing cultivars in the USA have been reviewed (Kane *et al.*, 1997).

The recommended time of sowing in west Kenya is from mid-May to mid-June, and in years of high yield the financial gain from early planting, even within the recommended period, can be substantial. A yield increase of only 100 kg/ha over several hundred hectares can have a major effect on the crop's profitability. The recommended time of planting in selected countries are shown in Table 6.10.

The optimum plant population is a major factor influencing yield and thus profitability, and should be determined locally. Row-widths are often limited by the machinery available, are not as critical as plant population, and can vary from 25 to 100 cm (Table 6.9). As a guide, seed rates of 50–65 kg/ha viable seed should suffice, to give a population of 250 000–400 000 plants/ha. High plant population is to be preferred to low for mechanised crops, since at greater densities plants tend to grow taller, producing their lowest pods higher above ground, reducing loss when direct combining (see Table 6.12). Close spacing, 20–50 cm, is better than 60–90 cm if soil conditions and equipment are suitable. The yield in narrow rows is generally higher at the same plant population with better weed suppression. Cultivars specifically adapted to narrow rows should be selected; i.e. plants with a more open canopy, earlier flowering and resistance to lodging.

Twin rows 60 cm apart with 120 cm between pairs or similar combinations may give higher yields than regularly spaced rows on rain grown crops. However, weed growth in wide rows is generally greater and more costly to control. Wide rows may also restrict growth, for there is some evidence that soya leaves grow faster when partially shaded than when exposed with long periods of full sunlight. For rain grown crops, the earliest plantings can be at closer row-widths than later plantings, as can short-season cultivars compared with long-season. Plants in narrow rows also cover the ground more quickly, thus increasing sunlight interception and suppressing weed growth (Fig. 6.9).

It is not always possible to plant soya mechanically, and many smallholders prefer to sow by hand. In pure stands planting three to six seeds together can produce yields of 80–90% of drilled crops at the same row-width (Fig. 6.10).

Planting machinery must be accurately calibrated and frequently checked, especially when soil and climatic conditions vary during sowing. Allowance must also be made for cultivar differences in seed weight, size etc. Electronically controlled precision seeders ensure complete accuracy of both planting depth and sowing rate, and are recommended for large-scale commercial plantings. Grain drills, maize, cotton and groundnut planters can sow soya provided they are correctly adjusted, but should not place fertiliser in direct contact with the seed. Replacement of grain drill shoes with double disc openers is recommended, and the feed mechanism should be modified to ensure an even flow without damage to seed. Grain drills are quite adequate as demonstrated in Ohio, USA, where some 500 000 ha were sown by these drills in one season. If a grain drill must be used, sowing seed or fertiliser down alternate spouts is effective; the author used this technique successfully to sow large areas of soya in Tanzania, although the locals operating the drill had never previously seen one!

Fig. 6.9 Close-spaced, weed-free commercial soya bean crop, South Africa.

Methods of increasing soya yield or profitability using existing machinery or cultivars should be continually under review. It was stated in 1979 that average yields could easily be increased if then current technical know-how was applied by the majority of soya bean producers. This remains applicable, especially in less technologically developed areas.

Fig. 6.10 Hill planting of soya in rice paddy (no-till technique), Republic of China.

WEED CONTROL

Cultivating soya immediately before and just after emergence is usually unwise, and every effort should be made to reduce the weed population before planting. Soya seedlings are unable to compete with many fast-growing weeds, whose control at this period is most important. Mechanical cultivation is less damaging once plants reach 5–10 cm, although it is necessary to ensure that large amounts of soil are not thrown against plants stems. After 6–8 weeks soya can usually compete with all but the most persistent weeds, which it may be necessary to spot-spray or remove by hand.

Plant population affects the ability of soya to compete with weeds, but spatial arrangement of plants can modify its effect. Within a given population, row-width is the most important factor reducing the depressing effect of weeds. In Australia yield depression in 25 cm rows was slight, rose to 20% in 50 cm rows, 26% in 75 cm rows and nearly 40% in 100 cm rows. Associated with row-width is the rate of canopy growth, and the period necessary to cover the inter-row, as this affects not only weed growth but also the timing and type of weed control operations. Increasing row width from 25 cm to 50, 75, then 100 cm in the USA increased the time for the canopy to cover the inter-row from 36 days to 47, 58, then 67 days (University of Nebraska data).

Rotary hoes or finger weeders give excellent results in row crops with little damage to plants, and two or three cultivations are usually sufficient. Soya plants depend more on their vertical than lateral roots, but overdeep weeding operations can be damaging.

Chemical control of weeds in soya crops has been extensively studied and well covered in the literature. Commercial compounds used pre-emergence include acifluorfen, alachlor, bentazone, butralin, chloramben, clomazone, fluchloralin, glyphosate, linuron, metolachlor, metribuzin, nitralide, oxyfluorfen, penoxalin, prometryn, terbutryn and trifluralin. Combinations of one or more of these can be effective against different weed complexes, and frequently allow lower levels of their active ingredient to be used. Some compounds require to be soil-incorporated pre-sowing, with others this technique allows them to be used at slightly higher active ingredient levels which improve weed control. Sensitivity to herbicides exists, with some cultivars extremely susceptible to fluchloralin, metolachlor and metribuzin. Herbicides can also affect the extent and degree of nodulation. Herbicide residues in soya beans have been reported, including metolachlor in China (Zhang *et al.*, 1998).

Some compounds adversely affect soya's resistance to fungal disease, for instance chloramben enhanced the severity of *Thielaviopsis* root rot, and trifluralin had a similar effect in increasing root rot damage. Conversely, aciflourfen, bentazone, methabenzthiazuron and oxadiazon can increase resistance to *Macrophomina phaseolina* (Mushtaq, 1996).

Soya can also be injured by herbicide residues from previous crops, and when these were applied in bands, damage to soya will form a pattern of strips across a field. Atrazine has produced this effect, and can also interact with other agrochemicals to produce adverse reactions in plants. Severe damage to soya occurred when sown after a failed cotton planting to which fluridone and cyanazine had been applied. Toxic interaction between herbicides/insecticides used in the previous season and those applied to the current crop can occur; picloram precludes growing of soya for at least 3 years. Such reactions may also occur when insecticides and herbicides are applied together at planting, and synergism has also been reported.

Soya will be damaged by drifting spray from 2,4-D, 2,4, 5-T, dicamba and similar chemicals. There are few recommended post-emergence herbicides, but directed sprays of acifluorfen, bentazone, diclofop, mefluidide and oxyfluorfen have proved effective. Simple methods of application including the dribble-bar, wax-bar, wet canvas or rope-wick are useful for spot weeding or in smallholder crops.

Sowing soya into lands that have not been mechanically cultivated (chemical or minimum tillage) is most successful in temperate regions especially following small-grain cereals, and frequently allows double cropping when fast-maturing soya cultivars are grown (Fig. 6.11) (Triplett, 1978). The herbicides required are a significant cost factor in profitable soya production.

Fig. 6.11 No-till sowing of soya into cereal stubble, USA.

However, wheat straw contains phytotoxic substances which can adversely affect growth of soya plants, and cultivar trials are necessary to select tolerant types (Keisling *et al.*, 1996: Vyn *et al.*, 1998).

Minimum tillage techniques are less successful in many tropical and subtropical regions, where the major problem is the subsequent greatly increased weed growth. The author's experience with this technique proved soya was unable to compete when pernicious grasses such as *Digitaria scalarum*, *Sorghum halepense* (Johnson grass) and *Imperata cylindricum*, or perennial broad-leaved weeds were common, and the amount of post-emergence weed control necessary was unprofitable.

Weed control during emergence and in seedlings is the main problem, and in the Philippines the longer weeds were left to compete with the crop, the greater the yield depression. Sowing soya into harvested rice paddies is more complicated than following other cereals, but can be equally successful (Fig. 6.10). A major cause of poor stands is not weed growth but excess water, with rainfall immediately following planting the most damaging.

IRRIGATION

The majority of soya crops are rain-grown, and the main factor affecting the expansion of irrigated crops is the cost of water, its supply and distribution. In dry regions of many less developed countries, yield increases obtained under irrigation are frequently offset by the greatly increased cost of production, including bulk transport and storage facilities for fertiliser and seed produced.

A rainfall equivalent of 600–1000 mm is required to produce a seed yield of around 2500 kg/ha, and it is doubtful if a yield below this is economic. As a field guide to the amount of water/rainfall necessary, soya had a evapotranspiration rate 13% higher than sorghum for the same leaf area, and in Australia, sorghum outyielded soya when irrigation water was restricted. Thus if rain-grown sorghum yield is only moderate, it is unlikely that rain-grown soya will thrive. Water used by soya in relation to other crops is shown in Table 6.11 (University of Illinois data).

Table 6.11 Water use efficiency of soyabean and other crops – USA.

Crop	kg of water/kg dry matter produced
Sorghum	140
Corn	160
Wheat	250
Oats	260
Soyabean	300
Alfalfa	390

A substantial pre-planting irrigation wetting the soil to 60–100 cm is recommended, and moderately large amounts at longer intervals are preferred to frequent small applications. However, the irrigation technique should be related to the cultivar grown or vice versa, for as noted previously, there can be considerable difference in growth-rate and type or roots, and varying reaction to seedbed temperature. The rate of root growth of irrigated soya determines the wet soil depth necessary to ensure continued root growth, and a ratio between plant height and root penetration is a valuable field guide. From the six-node stage until pod development began, root penetration on a clay loam can be twice that of plant height. Such a ratio is particularly important when irrigation is limited by time, the Sudan Gezira for instance, where water is strictly allocated for specific periods and a tenant reduces the area covered by using unnecessarily high water rates. On irrigation projects in East Africa under the author's direction where soya was grown on ridges, watering every second, then third furrow once the canopy had almost covered the inter-rows, gave highest yield per unit of water applied.

The system selected must ensure no moisture stress occurs in plants from flowering until the pods have filled, as stress at this period has the greatest adverse effect on seed yield. A major reason for this reduction is that a moisture stress when pods are filling reduces seed size by accelerating leaf senescence and thus the total period for seed to reach its maximum size (Choi *et al.*, 1996: De Souza *et al.*, 1997). Where water is the limiting factor and must be rigidly conserved, tensiometers are of great value to regulate and time applications. Soya plants may wilt on hot, windy days when soil moisture is adequate, probably due to the high resistance to water movement through soya plants.

Flood irrigation is seldom successful, except bunding across wadis to impound and spread irregular rainfall in arid areas, with soya grown on subsoil moisture and roots following the moisture horizon down. A system described as 'saturated soil culture' (SSC) where water is maintained in furrows between beds from the early vegetative stage to maturity has proved of value in semi-arid Northern Australia, but not elsewhere.

Waterlogging for any reason is generally detrimental to soya, and when sown into paddy fields in Brunei, no plants survived on fields flooded to a depth of 20 cm for short periods compared with unflooded plantings which yielded 1.5 t/ha. Rainfall following planting, which left standing water was a similar problem in Taiwan, while adequate drainage was the most important factor in a soya–rice rotation in Brazil. In India between 4 and 8 days of waterlogged conditions in early growth reduced yields by up to one third depending on cultivar. Fungal infections, nodule damage and herbicide toxicity can also become serious in such conditions, and may result in greater loss than that caused by the physical effect of water. A high water-table directly affected growth and seed yield of soya in Egypt, where a watertable at 40 cm in a sandy loam was more detrimental than at the same depth in a more clayey loam. However some cultivars are highly tolerant of temporary waterlogging.

Soya has little salt tolerance and saline water will seriously reduce yield, especially where drainage is restricted. Moderately saline water can be used to supplement rainfall, especially where storms are heavy and plants are grown on contour ridges, since leaching will then materially reduce surface concentrations. Straw mulching reduced the effects of soil salinity in China, but is only possible on small plots (Liu, 1996). Fluorine is a common constituent of soil and irrigation water; its effect on plants varies and often increases when the water is also saline. Soya is more susceptible to high fluorine content than associated salinity, and root growth is reduced more than top-growth.

Irrigated soya is usually grown on ridges or raised beds, which if small can be constructed during the planting operation, but if substantial as is common on many irrigation projects, are best constructed before planting. Planting is thus more complicated than on the flat, but in general involves cutting off or flattening the ridge top to expose moist soil from pre-planting irrigation. Seeds are sown in either 1 or 2 rows/ridge, or multiple rows on beds, into this seedbed. Ideally, irrigated lands should be level but this is seldom possible, and the author has found that ridging on the contour is more practical and involves less topsoil movement. Variation in water volume is less dangerous, and retention of rainfall and erosion control improved. Sprinkler irrigation of soya has usually proved unprofitable, unless the installation already exists primarily for more profitable crops. It also tends to increase the incidence and severity of leaf diseases.

INTERCROPPING AND ROTATIONS

Soya is not usually grown by African smallholders either as a cash or an intercrop, since they prefer other beans. In Asian and Pacific Rim countries it is very popular, and can be found growing on bunds around rice paddies, strip-cropped, and intercropped in a range of combinations. Climbing types are often sown into maturing maize or sunflowers, but grain soya yields are suppressed when underplanted in either. There are strong regional preferences for particular cultivars, especially in China and Japan, and unripe pods are commonly used in South-East Asia, generally as a vegetable. This use for soya pods is also being introduced into Argentina (Kantolic & Gimenez, 1996). Intercropping sugarcane and soya was not successful in Egypt, as the optimum planting dates did not coincide, and selecting one depressed yield of the other (Eweida *et al.*, 1996).

The incidence of pests and diseases affecting soya is often less in intercrops, and in India there was a significant reduction in the number and species of insects present and degree of plant damage (Gangwar *et al.*, 1994), although this varied with the crops involved (Shukla, 1995). A review of intercropping soya in India has been published (Mehtre *et al.*, 1997)

Underplanting soya can reduce the amount of light reaching plants, and although there is a difference in the effect on yield between short and long maturity types (Egli, 1997), there is also a cultivar interaction and results are unpredictable. Underplanting soya in coconut plantations near Madras, India, was not successful due to increased seed loss from lodging (Jayaraman & Subramananian, 1994). However, when coffee growing in East African countries became unprofitable due to civil disorders or political manipulation of marketing boards, coffee was successfully underplanted with a variety of low growing food and cash crops including soya.

Cereals grown in a rotation with soya and other leguminous crops, generally benefit from residual nitrogen especially if plant residues are ploughed in (Vance & Graham, 1995). This effect can substantially reduce the amount of nitrogen required for a following crop, as in South Africa (Bloem & Barnard, 1995), and the USA (Kessavalou & Walters, 1997). The author visited a farm in Victoria, Australia, in 1998, which practised a rotation of 3 years of rice followed by a fallow year, then wheat, irrigated soya, wheat, oilseed rape, irrigated soya, then back to rice; which significantly reduced the nitrogen requirement of the cereal crops, and was a major factor in overall farm profitability.

A comparison of single, relay and double cropping soya in the USA showed that when possible, double cropping was the most profitable (Jacques *et al.*, 1997). However, the damage caused by diseases usually increases after two successive soya crops, especially when another crop in the rotation is an alternative host. Soya root material remaining in the soil after harvest may also adversly affect growth of a following soya crop, and in China reduced yield by up to 25% (Han & Xu, 1998).

Soya appears to be immune to plant parasites, but like sunflower has the ability to germinate seeds of witch-weed, *Striga* spp. It can thus be a valuable cleaning crop in cereal rotations, especially sub-Saharan Africa, where *Striga* causes a major reduction in yield.

HARVESTING

Mechanised harvesting became possible after the introduction of compact, erect, shatter-resistant cultivars, but lower seed pods were often close to the ground. Increased height of the lowest pods was initially linked to delayed flowering and low yield, but this has now been overcome with recent high yielding cultivars producing the first pod well above cutter-bar height.

Soya is ready for harvest when the majority of leaves have fallen, the lowest pods are yellowish and dry, and the seed (in the main commercial varieties) wholly yellow. Seed should have a maximum moisture content of 14–15%; at higher levels beans will not store without drying, and below 12% mechanical damage increases. Seed moisture will usually be 50–60% at physiological maturity, and 12–24 days are necessary for it to fall to 14%.

Plants can be harvested when the majority are mature, but it may be an advantage to harvest slightly earlier. Frost or alternating periods of rain and sun can cause extensive shattering, and to await full maturity can increase seed loss. Defoliation or desiccation is an alternative, since the yield reduction can be substantially less than loss from shattering (Saraswathi *et al.*, 1995). Magnesium chlorate, ammonium nitrate and dipyridyl phosphate have been so used. The operation is usually unnecessary except in special circumstances, since final maturation is so rapid that defoliation or desiccation normally reduces seed yield or quality.

The author has harvested soya by a variety of methods from wholly manual to fully mechanised, and all were equally successful, provided the timing of the operation was accurate. The basic criteria are cost and effectiveness. In regions where labour is plentiful, plants are pulled by hand, thrown into heaps and threshed with sticks or a portable thresher (Fig. 6.12); or windrowed after pulling and threshed by a combine fitted with a pick-up reel. In general windrowing is not to be recommended, except where field conditions normally preclude natural drying, since losses from shattering are increased.

A factor that often limits the use of combine harvesters is the height of pods above ground, as noted. In East Africa, especially in very dry seasons which retarded growth, some 20% of pods were less than 10 cm from the soil surface, below the cutter-bar minimum setting. Pulling plants and throwing them into the machine auger after immobilising the cutter knife, reduced loss to the minimum. Gleaned beans were collected and paid for at an agreed price.

Combine harvesting is the simplest method, and provided the cutter-bar is set low enough and guides or air jets fitted, shattering losses will be low; a forward speed below 5 km/hour is recommended initially (Fig. 6.13). Because of its importance as a source of seed loss, considerable research has been carried out into reel and cutter-bar design, use and positioning. The effect of incorrect cutter-bar setting is shown in Table 6.12 (Iowa State University data). These data not only indicate the importance of correct setting, but also the selection for mechanised production of a cultivar whose lowest pods are formed at least 10 cm above ground level.

Modern combine designs include floating heads, double-acting cutter-bars, lifters and air jets, all of which materially assist in reducing seed loss. Drum speeds and concave clearances must be adjusted to prevent cracking. As with other crops, a slightly dirty sample of whole beans is to be preferred to a clean sample of damaged beans, or increased losses over the straw walkers. Although special soya harvesters are available to producers who grow considerable areas, any properly adjusted grain combine is adequate (Kowalczuk, 1995).

The importance of close supervision of harvesting operations cannot be over-emphasised. Seed losses can reach 20% of total yield; from shattering in the field, in the combine, or during hand threshing, cleaning and bagging. As a field guide, 10 seeds/m behind the combine represents about 200 kg/ha, and the reason for this loss must be established. It may occur at the cutter-bar, and reversing the machine after a few metres of cut will indicate how serious this is by the number of seeds lying on the ground. Common sources of seed loss when combining are shown diagrammatically in Fig. 6.14, and in Table 6.13 (Ohio State University data).

When grown for seed production, plants must be harvested when moisture content falls between 12 and 14%, combine forward speed must be slower than normal and the drum setting

Fig. 6.12 Hand harvesting and threshing soya, Republic of China.

Table 6.12 Effect of combine harvester cutter-bar height on yield – USA.

Height of cut (cm)	Crop yield (kg/ha)	Loss (kg/ha)	Loss (%)
Hand-harvested	2260	0	0
9	2135	121	6
12.5	2051	215	10
16.5	1983	229	12

Fig. 6.13 Combine harvesting soya, USA.

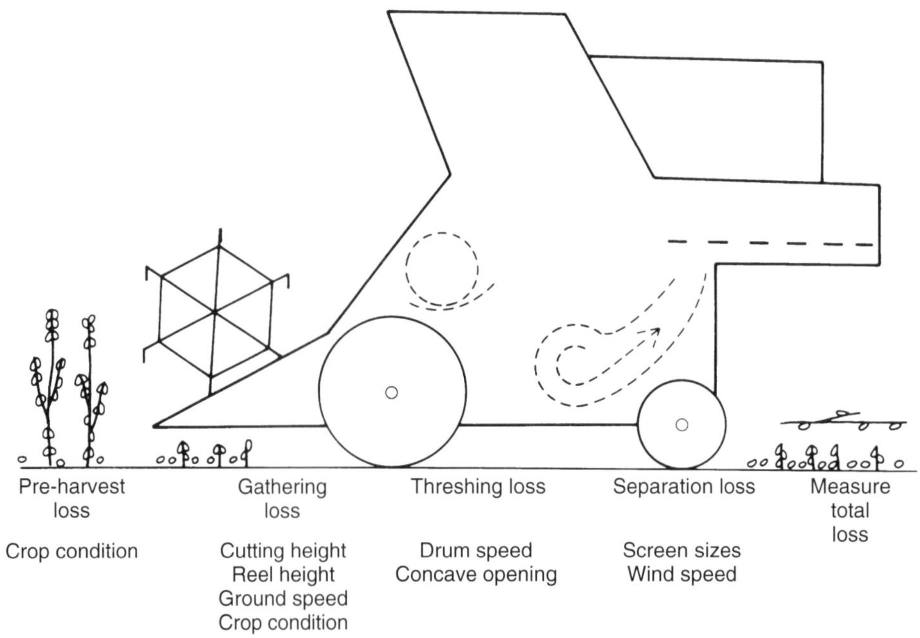

Fig. 6.14 Sources of loss when combining.

Table 6.13 Estimated harvesting loss in kg/ha on 3500 kg/ha seed crop.

Source of loss	Average harvesting operation	Excellent harvesting operation
Shattering	121	47
Stubble	40	20
Lodged	67	27
Cylinder	13	0
Cleaning	13	7
Total loss	254	101

accurate. A direct correlation exists between poor germination, damaged seed, and high drum speeds. The author found that hand-harvested and threshed seed invariably gave a higher germination percentage than mechanically harvested seed from the same field, and also remained viable longer in store. Where seed from communal crops or tenant farming projects must be kept for sowing the following season, seed collected just before the main harvesting period is usually superior in terms of soundness, storage potential and viability.

The effect of field weathering on seed quality and vigour is especially important in commercial seed production; seed quality is highest at physiological maturity, but since this corresponds with high seed moisture, harvesting is impracticable. Thus seed remains on plants up to 3 weeks to attain harvest maturity, and the longer the period between the two maturities the greater is the loss in seed quality. Seed vigour declines more rapidly than viability, and environmental conditions and time of harvesting may well determine the suitability of a crop for use as seed. Also adversely affecting seed quality is weed growth present when seed is ripening; tall weeds caused high relative humidity around maturing pods and reduced seed quality.

Seed yield obtained in selected countries is shown in Table 6.14. The relatively low average should be judged against yields of nearly 6 t/ha achieved by commercial growers in the USA. This also indicates the huge potential that exists for increased soya production, and the general necessity for better cultural procedures.

Table 6.14 Soya bean yields in selected countries (t/ha).

	Average	High
Africa		
West	0.5	2.0
Central	1.0	2.0
East	0.5	1.5
South	1.0	2.0
Australia	2.5*	4.5*
	1.5	2.5
Canada	1.5	2.5
USA	2.5	4.0
	3.0*	5.5*
Brazil	1.5	3.0
Central America	1.5	2.5
China		
North	2.0	3.5
South	1.5	2.5
Taiwan	2.0	3.5
Japan	2.5	3.5
Turkey	1.5	2.5
Former USSR	1.5	3.5
Eastern Europe	1.5	3.0

* Irrigated; other yields from rain-grown crops.

The substantial yield increase achieved since large-scale commercial cultivation began is unlikely to be maintained, as current yields are almost treble those of the 1930s. Interactions between environment and plant growth are now as important as genetic factors, including cultivars with increased ability to assimilate, or more efficiently use applied nutrients and irrigation water. A strong negative correlation exists between the two important factors of oil and protein content. Reported heritability of yield would appear to be low, averaging some 50%, while maturity, lodging, height and seed size is between 70 and 90%, with protein and oil content around 75%, implying that selection for yield is more difficult than for other attributes.

STORAGE

Correct harvesting procedures that produce whole sound beans will materially reduce subsequent storage problems. Dirty samples of sound beans are easily cleaned and require less protection than clean samples with a high proportion of damaged beans. Oil from the latter will also generally have a higher level of free fatty acids. Clean dry beans, with a moisture of not more than 10% are easily moved in bulk and store well. Beans harvested above 14% moisture must be dried before bagging or bulk storage. In hot dry regions, standing the bags open-necked in the shade is sufficient, otherwise artificial drying is necessary. Cold air should preferably be used, particularly for seed crops.

Hot-air dryers require close supervision to prevent seed damage, with 40°C generally the highest; 60°C reduces germination and affects seed quality. With properly supervised artificial drying there is little difference in seed viability, quality, oil content or characteristics between hot and cold air blasts, although hot air dried seed may require more careful storage. If the seed is harvested at high moisture content, above 20% and dried to 14%, the higher the initial moisture content the lower is viability.

For on-farm storage of seed a moisture content of 10–12% must be maintained, and as moisture can migrate to the top of storage bins, these must be checked regularly and the top layer turned. Fungicidal seed treatment can significantly assist the maintenance of viability if no special storage facilities are available, but for best results it is essential to treat the seed as soon after harvesting as possible. In general, beans with hard testa retain their viability longer when so treated than seed with soft testa.

Storage of small quantities of seed on isolated trial or experimental farms with limited facilities is usually a problem, especially in more humid areas, since high relative humidity generally reduces seed quality and viability quicker than high temperature. A most suitable packet for long-term storage of soya is a laminated bag made of cellophane, aluminium and polyethylene. In these packets soya retained almost 100% viability after 30 months at room temperature. Paper or polyethylene bags are unsuitable. Controlled storage at 5°C is also effective. See the next section for storage pests and their control.

PESTS

Many insects that attack the various legumes commonly grown for food or forage worldwide, also damage soya, and when the crop is introduced into a rotation it is to be expected that local legume pests will also attack soya plants. For instance, *Etiella zinekenella*, the lima bean pod-borer, is widespread and a major cause of loss in many countries. The larvae are about 15 mm in length, pale bluish green with light-reddish spots on the upper body. They bore legume pods and seeds, including soya, and attacks can occur throughout the season (Fig. 6.15).

Thus there is an indication of the pests likely to occur; for example, 540 species were collected over a 3-year period from soya fields in the southern USA; in Japan 220 species were collected but only 30 caused economic damage. On a world basis estimated total crop loss in the 1970s in million tonnes (Mt) from pests was 2 Mt, diseases 5 Mt and weeds 6 Mt, which reduced total potential production from 45 Mt to 32 Mt. Greatest proportional loss occurred in the then USSR and China, followed by North and South America. The total calculated value of that loss

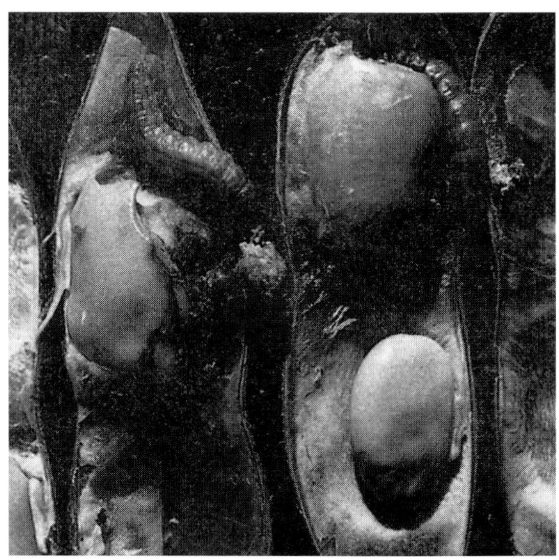

Fig. 6.15 Larvae of *Etiella zinekenella* in soya pods.

using market or on-farm prices was a massive US$1380m. It is considered the current loss in volume is little different, but the value has greatly increased. The USDA estimated the annual yield loss in the domestic soya crop due to insects at approximately 14%. Intercropping smallholder crops can significantly reduce insect numbers and degree of crop damage (Gangwar *et al.*, 1994).

Growers generally allow too high an initial infestation or follow a fixed spraying programme, but the increasing cost of agrochemicals and their application, combined with a growing resistance by insects (and consumers to their use), makes this a debatable and arbitrary procedure. The level of insect damage that can be tolerated in soya, or the 'threshold value' of crop destroyed compared with the cost control must be determined. Biological control of many leaf-eaters by viruses and pathogenic bacteria (entomopathogens), have the added advantage of being non-toxic to humans and bees, and are becoming increasingly important with growing pesticide resistance. Recent cultivars may have a degree of resistance to a particular pest, but incorporating insect resistant genes into current high-yielding cultivars can adversly affect seed yield, as reported in the literature. One important aspect of chemical pest control in soya is that residues from pesticides can accumulate in seeds and are extracted with the oil.

In Japan the threshold value for smallholders was calculated at 15–20% crop damage, with two or three applications controlling 50–70% of insects attacking pods and seeds. In India the time of main attack was found to be more appropriate; greatest yield depression resulted from defoliation at first flowering, and control should be planned to reduce pest incidence before flowering. As a field guide in the USA, one pentatomid bug per metre of row is considered sufficient to justify control measures. Although damage to plants is slight, seed is more severely affected; above 2% damaged seed in a commercial consignment results in downgrading and lower price. In China 4 to 10 *Melanagromyza sojae*/100 plants, or 8–11 *Megacopta cribraria* nymphs, justified control (Pan, 1996; Wang *et al.*, 1996). Insects not only directly reduce yield by damaging plant parts, they can also do so indirectly; bean leaf beetle larvae substantially reduced the extent and development of nodulation, and thus nitrogen fixation.

Probably the most injurious to seedlings and young plants are the flea beetles, which also attack more mature plants. *Epilachna* spp. in the Americas, especially *E. varivestis* in North America (Fig. 5.12); *Colaspis flavida* in South America; *Pagria signata* in Asia and the Far East; *Aphthona* spp. in East Africa. All are small shiny beetles, often brightly coloured, the adults eating the foliage and their larvae eat the roots. Infestations can build up with great rapidity, and unless controlled can destroy large numbers of young plants. Cutworms, especially the ubiquitous *Agrotis* spp., true and false wireworms, grasshoppers, particularly *Chrotogonus* spp. in Africa and Asia, and mole crickets *Gryllotalpa* spp., generally cause damage wherever soya and other edible legumes are grown (Fig. 7.17). Most insects attacking the seed and seedling can be controlled by systemic insecticides placed in the

Fig. 6.16 *Nezara antennata* attacking soya pods.

seedbed, band spraying, or by baiting in smallholder crops.

Stem-borers, especially *Melanagromyza* spp. and particularly *M. sojae*, can be very damaging in China, India, the Middle East and Africa, but less so in the New World. Also commonly reported are *Agromyza phaseoli*, *Elasmopalpus lignosellus*, and in Asia, especially India, the girdler beetle, *Obereopsis (Oberea) brevis* can reduce yield by a third. Resistance to stem-borers is apparently related to lignin content of stems.

Butterfly and moth larvae cause varying degrees of defoliation, the particular species differing in importance both annually and regionally, with noctuids and pyraustuids among the most widespread. The level of defoliation and at what growth stage it occurs affects the degree of subsequent leaf replacement, and thus the final yield. This varies between cultivars, but in general defoliation at or approaching flowering is the most damaging. When it occurs in plague proportions the armyworm, *Spodoptera litura*, can virtually destroy whole fields of soya and groundnuts, but is easily controlled by spraying with a bacillus (noted later). Several methods are used to monitor moth occurrence, including sex pheromone traps (Singh & Sachan, 1993).

The bollworms, *Helicoverpa (Heliothis)* spp., larvae (see Fig. 4.13) often attack the pods, but can also cause severe defoliation if unchecked; *Plathypena scabra*, the green cloverworm, and *Vanessa cardui*, the painted lady, become important locally in North and South America. *Anticarsia gemmatalis*, the velvet bean caterpillar is common in the USA, and other *Anticarsia* spp. are minor but common pests in many other countries. Planting to avoid the major period of infestation is recommended in the USA, together with selection of cultivars tolerant to defoliation (McPherson *et al.*, 1996). *Plusia oo* and *P. nu* are most damaging in South America, although the species occur widely on various legumes. *Maruca testulalis*, the mung moth, is widespread on a range of legumes throughout the tropics, but is particularly damaging to developing soya pods in many parts of Africa, Asia and the Far East. Also in Africa *Cydia ptychora* can be locally important.

Sylepta derogata, the cotton leaf roller, is a major pest in Africa and Asia through to Australia, and *S. ruralis* in the Pacific region, with degree of damage caused to soya often directly related to the presence of more favoured food plants. *Pyrrhia umbra*, the striped tobacco caterpillar, is a common pest in Japan, China, and occasionally in neighbouring countries, although *Matsumuraeses phaseoli*, the Azuki podworm, is generally a far more serious pest. The larvae eat leaves and also bore stems, doing severe injury and often causing death of plants. *Hedylepta indicata*, leaf roller, *Lampides boeticus*, bean butterfly, and *Laspeyresia glycinivorella*, soya moth, can all become important locally.

Leaf-eating beetles or their larvae are major local pests, with *Plagiodera inclusa* one of the most widespread. In the Americas the Japanese beetle *Popillia japonica*, the bean leaf beetle, *Cerotoma trifurcata*, and *Epilachna varivestis*, the Mexican bean beetle cause varying degrees of damage, which can be seasonally extremely severe. First signs of an infestation are holes, or ragged edges to leaves, which spread rapidly as pest numbers increase. Young leaves and shoots are often the first to show damage, and on more mature plants the flowers and young, succulent pods are eaten. The leaf miner, *Aproaerema modicella*, is a major pest in India and Pakistan, and as it also attacks groundnuts, both crops can suffer increased damage if planted in close proximity. In China the pod-borer, *Encorma glycinivorella*, and larvae of *Megacopta cribraria*, are major pests and can cause substantial damage unless controlled. Some control of herbivorous insects can be achieved by partial defoliation at the middle of the pod-filling stage, which reduces leaf area and allows greater coverage with less pesticide (Board *et al.*, 1997).

Pentatomid and other plant and sucking bugs are common pests of soya. *Nezara* spp. adults and nymphs suck sap from developing seeds and pods, which often shrink and die; those remaining usually contain underdeveloped and damaged seed (Fig. 6.17). *Nezara viridula*, the

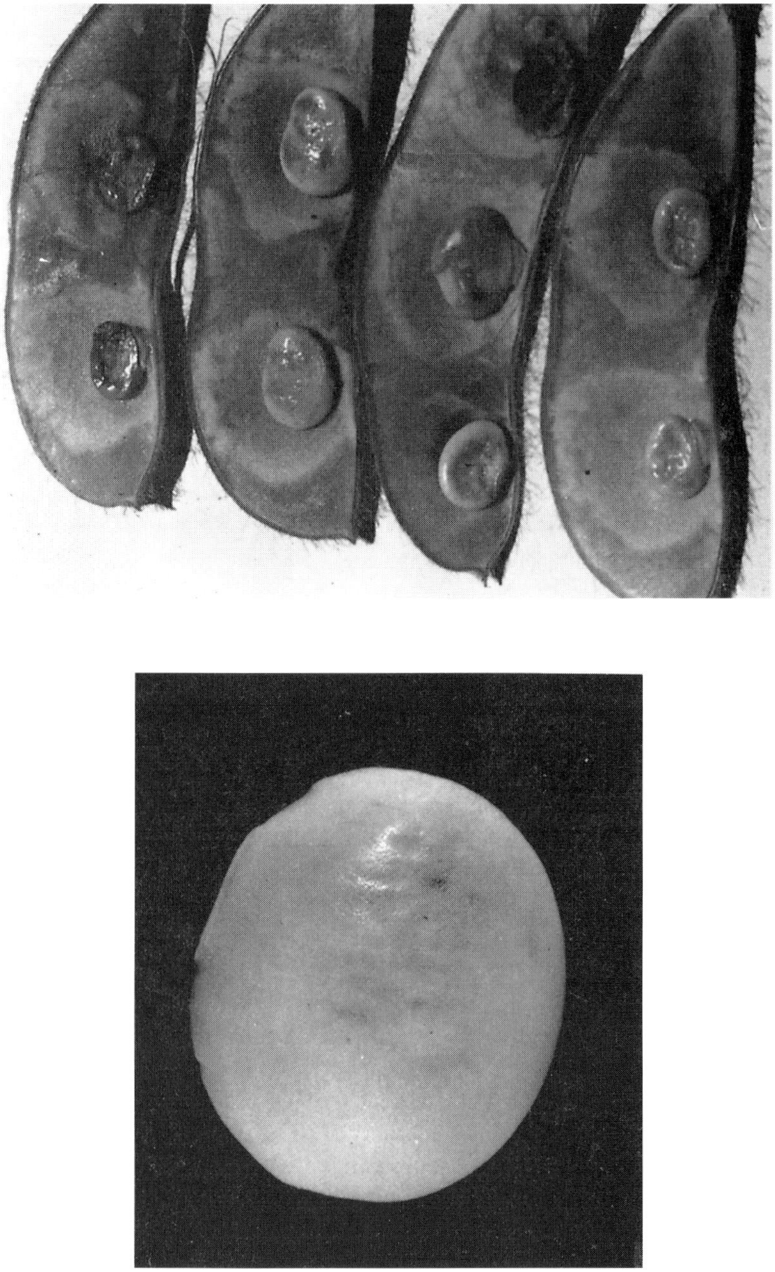

Fig. 6.17 Damage caused to soya pods and seed by *Nezara viridula*.

green shield bug, is most common and widespread (Fig. 5.11), and *N. antennata* in the Pacific region (Fig. 6.16). *Acrosternum hilare*, the green stink bug, and *Piezodorus guildinii*, pod borer, are similarly damaging in the Americas, but *P. hybneri* occurs from Africa eastward to Australia, China and Japan. Crops badly infested with sucking bugs when pods are filling should not be used for seed, since viability following such attack can be greatly reduced.

Thrips can be damaging in specific crops in Africa and Asia, but are generally of minor importance elsewhere, although they can be more numerous on soya in the USA than any other insects. *Frankliniella* spp. are those most often recorded. Leafhoppers, jassids, whiteflies and aphis often occur in large numbers on soya, but it is often difficult to assess the amount of damage that would warrant control measures. Young plants generally suffer more than mature ones. Some species can transmit virus disease, the aphis *Myzus persicae* in particular. In Japan, *Nesophrosyne orientalis*, is a vector of witches' broom disease of local soya, and in Indonesia the common soya aphis *Aphis glycines*, transmits yellow mosaic virus and soya dwarf virus. Mealy bugs, especially *Pseudococcus* spp. and *Dysmicoccus* spp. are common and widespread, but like mites and scales appear to cause no great damage. In Australia the brown leaf hopper, *Orosius argentatus*, common on many local plants, is a vector of little-leaf, *Mycoplasma*.

Nematodes are generally of little importance, but soya should not precede or follow highly susceptible crops in a rotation. Nematode infestations are often more severe on light sandy soils in the tropics; damaged plants are usually stunted, chlorotic and under moisture stress tend to wilt earlier than healthy plants. Above-ground symptoms are not diagnostic, and plants suspected of being attacked must be carefully removed for specialist examination. However, the presence of nematodes *per se* may not always be undesirable, since a relationship between a mycorrhizal fungus and root-knot nematode may be beneficial. It is thus important to identify the major nematode species present, as cultivar differences exist in degree of nematode resistance (Wang, 1995; Young, 1996; Wheeler *et al.*, 1997). Mites are frequently reported on soya but apparently seldom require control. Application of selenium increased resistance to spider mite (*Tetranychus* spp.) in China (Liu *et al.*, 1998)

Termites can cause major crop loss, especially in seasons of below-average rainfall when plants mature early and roots become dry and fibrous; not only are plants weakened, their stems fracture, and they are then too low for the combine pick-up reel. Pods in contact with the ground are also eaten. Loss from small animals in pre-harvested crops has been little investigated, although it can be significant in Africa and Asia, and may be more serious elsewhere than currently appreciated (Peles *et al.*, 1997).

The majority of storage pests which infest soya beans are non-specific, attack most stored legumes and grains, and are noted in other chapters. Rodents and similar small animals cause serious in-store losses in less developed regions, India for example (Ishwar & Ghosh, 1992). In these regions vegetable oils, including soya oil, are used as grain protectants, and in some instances to protect stored oil seeds (Obeng-Ofori, 1995). Soya beans are one of the few crops handled in the same bulk quantities as grain and can therefore be treated similarly. Gas fumigation of seed stocks should be avoided since some can seriously affect viability. For additional information on control of storage pests and pest control in general see this section in Chapter 3 on groundnuts and Fig. 9.10.

DISEASES

The expansion of soya production worldwide increased the incidence of, and damage caused by disease. The extent of yield loss is not always apparent until formal trials are carried out; in Taiwan for example, application of dithane increased average yield of four cultivars from 1.2 to 1.9 t/ha, with a 50% increase from the most susceptible variety; in India a 22% loss of yield was due to two rusts, *Cercospora sojina* and *Colletotrichum destructivum* (Mittal, 1996). Loss of yield due to disease in the southern USA has been quantified (Pratt & Wrather, 1998).

Diseases recorded have also increased, some formerly of little importance have become more significant, and some 120 pathogens attack soya, with about 40 of economic importance. In India 35 species of fungi were recorded from soya (Murthy & Raveesha, 1996), plus three more on stored beans; similar and sometimes longer lists have been compiled in many countries.

Some diseases are seed-borne and cause severe seedling losses, particularly when soil temperature during emergence is 30°C or above, as in Nigeria, where all plants were killed in field trials when infected seed was planted with a soil temperature of 33–37°C. Dressing seed with a fungicide substantially reduces the damage caused by these diseases, and should preferably be carried out in advance of sowing to improve its effectiveness; however too long an interval before planting dressed seed may reduce germination, as previously noted. A number of techniques to detect pathogens on seed are described in the literature.

A particular rotation can also increase disease damage, including wildfire disease of tobacco, *Pseudomonas tabaci*. Many legume diseases can be transmitted to soya, and comments in previous chapters apply in respect of identification and control. Interactions between diseases and fertilisers, which either increase or decrease the incidence or damage caused by a specific pathogen, have been recorded from a number of countries, including India (Sharma *et al.*, 1996) and the USA (Pacumbaba *et al.*, 1997).

The most damaging diseases are those that attack the root or seedling, and defoliating leaf spots. Chemical control is often neither profitable nor successful. Cultural methods aimed at restricting the incidence and severity of attacks, and sowing resistant cultivars should also be used to provide an integrated protection programme. Where chemical control is effective, preventative sprays will reduce the incidence of disease. Many pathogens inhibit formation of anti-pathogenic activity in the host, and thus a reduction in initial infection results in a generally lower incidence of disease. In some regions, or soil types, fertilisers can assist; application of potassium has been shown to reduce the damage from *Diaporthe phaseolorum* and *Cercospora kikuchii* in North and South America. In Russia cobalt chloride and ammonium molybdate reduced the incidence of leaf spots, but also slightly decreased yield. Seed treatment can reduce seed-bed losses, but its efficiency is impaired if seed is mechanically damaged.

Genetic resistance to many diseases reportedly exists in germplasm collections, for example the rust resistance of Australian glycines. Resistance to a regional disease is often high in local cultivars, but this resistance is difficult to transfer to other cultivars, or to maintain if the local types are grown elsewhere. Increased disease resistance has generally proved difficult to maintain, but proved successful against *Phytophthora* root rot.

Probably the most serious disease of soya worldwide is bacterial blight, *Pseudomonas syringae* (*P. glycinea*) (Fig. 6.18). The bacterium causes small, angular, coalescing lesions with dark, reddish-brown to black centres surrounded by a water-soaked margin with a yellowish-brown halo. Centres of older lesions drop out and tear away, giving the leaves a ragged appearance. Pods, petioles and stems may be infected. Bacterial pustule, *Xanthomonas ampelina* (*X. phaseoli*) causes very small to large irregular brown spots with small raised centres on either side of the leaf which develop into light-coloured pustules with small creases. The lesions are not water-soaked in the early stages. Symptoms of bacterial pustule in later stages of development may be confused with those of bacterial blight. However, blight is favoured by cool, moist conditions, while pustule is most damaging in warm, moist conditions. Both can be found on the same plant, and may result in serious defoliation; resistance exists in some cultivars.

Rust due to *Phakopsora pachyrhizi* is the most serious disease of soya in Asia and the Far East, and has a wide host range. The rust can be present in a country for a long period before

Fig. 6.18 *Pseudomonas syringae* (bacterial leaf spot) on soya leaf.

becoming economically important; it was recorded in Japan in 1903, but only reached epidemic proportions in 1940; in Australia it was first recorded in 1934, and became serious in 1970. The disease was first observed in Puerto Rico in 1976 and is now widespread. Symptoms, which initially appear on leaves and petioles, are small, angular, grey-brown spots on both sides of the leaves, on which pustules develop. Mature pustules are smooth, and break open to give a 'volcano-like' appearance with the dispersal of spores. Symptoms generally appear first on the lower leaves, progressing up the stem. A relationship may exist between susceptibility to rust infection and the level of and activity of peroxidase isoenzymes (Fei *et al.*, 1997).

A number of leaf spots occur on soya but only three are of major importance. *Septoria glycines*, brown leaf spot, causes severe defoliation, with reddish-brown angular spots first appearing on the lower leaves, spreading upward. The colour of the spots distinguishes the disease from the watery, reddish-black lesions with a yellowish halo of bacterial blight, which causes similar defoliation. Pods, petioles and stems may also be affected. Frog-eye leaf spot, caused by *Cercospora sojina* has a characteristic light-coloured centre with a narrow reddish-brown border, mainly confined to leaves. It attacks plants earlier in growth than *Alternaria* leaf spot, whose brown spots with concentric rings often coalesce to form large dead areas. Purple seed stain is caused by *C. kikuchii*, and although seed discolouration has given the disease its name, the fungus also attacks the leaves, stems and pods. It is most obvious later in the growing season when angular red-brown spots appear on the leaves, which later coalesce to form large areas of dead tissue. Other leaf spots which can occasionally be damaging include *Corynespora cassiicola*, *Colletotrichum truncatum* and *C. destructivum*, *Fusarium oxysporum*, *Pleosphaerulina sojicola* and *Pyrenochaeta glycines*.

Various stem and seedling rots, often known colloquially as damping-off, occur in most soya crops. Soil temperature can have a considerable effect on the severity of infection, as noted previously. Charcoal rot caused by *Macrophomina phaseolina*, (*Rhizoctonia bataticola*), which has a very wide range, is often most destructive to seedlings although it can affect plants at all stages of growth. The hypocotyl develops a reddish-brown discoloration at emergence, or at the stem base which subsequently fractures. Stems and roots on older plants show black streaks when split. The fungus is seed- and soil-borne in the form of sclerotia, and often appears in periods of hot, dry weather or when growth is checked by unfavourable conditions. Resistance to infection may be increased by some systemic herbicides (Mushtaq, 1996).

Phytophthora rot is one of the most destructive diseases, and first noticeable field symptoms are a yellowing and wilting of leaves. Uprooted plants show a dark-brown main root and a mass of dead and drying secondary roots. *Pythium* rot more often attacks seedlings at or before emergence, and generally only scattered individual plants or small groups are affected. Severe attacks will cause characteristic gappy stands. The most conspicuous symptom is 'bald head', which is usually associated with wet soil.

Another stem rot, which can become very destructive, is due to *Sclerotinia* spp. which cause large black sclerotia on stems and may also invade pods. The sclerotia are usually much larger than those caused by charcoal rot. *Sclerotium rolfsii* occasionally attacks soya; initial symptoms are white fungal growth on lower stem, which rots. *Rhizoctonia* root and stem rots caused by *Thanatephorus cucumeris* occur in all soya growing areas, and usually attack seedlings and young plants. On older plants, infected leaves have a water-soaked appearance initially and become greenish to reddish-brown, later turning very dark. Brown lesions form on petioles stems and pods. Another very damaging and widespread root rot is caused by *Thielaviopsis basicola*, which has a large host range.

Pod and stem blight, and stem canker are caused by *Diaporthe phaseolorum* var. *sojae* (Fig. 6.19) and *D. phaseolorum* var. *caulivora*, respectively; both may occur on the same plant. Symptoms initially appear on lower leaf petioles or damaged lower branches during periods of warm, wet weather at mid-pod stage or later,

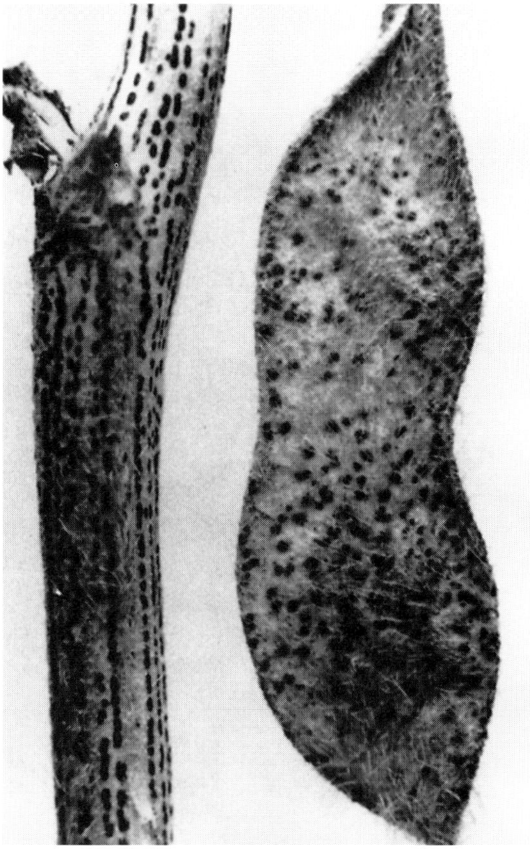

Fig. 6.19 *Diaporthe phaseolorum* var. *Sojae* stem and pod blight of soya.

even though the fungus can be isolated from month-old seedlings. Infected plants may be stunted and develop blackish spots or blotches on the stems and pods. Infected seeds are shrivelled, cracked and often covered with a white mycelium. Heavily infected seeds generally do not germinate.

Powdery mildew, *Microsphaera diffusa*, can be a limiting factor in soya production, particularly in the tropics during certain seasons. Early infection on the leaves appears as ash-grey spots with a dull-green halo, which enlarge and are soon covered with powdery masses of fungal growth on both sides of the leaves. Plants can be severely stunted or killed. *Peronospora manshurica* causes downy mildew of soya. but is seldom a threat to yield. Seeds from infected pods can be smaller than normal and often encrusted with a whitish mass of mycelium. The disease appears on the upper surface of young leaves as pale-green to yellow spots that enlarge into pale to bright-yellow spots, which darken with age. Severely infected leaves may drop prematurely.

The fungi most often associated with stored seed, and which may also reduce viability, are *Alternaria*, *Aspergillus*, *Colletotrichum*, *Penicillium* and *Rhizopus* spp. Seed can also introduce into stores other infections originally contracted in the field, and fumigation may be wise.

A large number of viruses have been reported from soya but few are of economic importance. Insects involved in virus transmission have been discussed in the previous section. Viruses not only affect vegetative parts, they can also reduce oil percentage and protein content of seed. Some, including soya mosaic and tobacco ring spot viruses are seed-borne, but apparently not yellow mosaic virus; the three have a wide host range and are sap-transmissible. Tobacco ring spot is occasionally transmitted by nematodes and thrip nymphs, the other two are transmitted primarily by aphids. *Aphis glycines* transmits both soya dwarf virus and soya yellow mosaic virus in Indonesia. Cultivars exist which are resistant or immune to regional strains of soya bean mosaic virus (SMV), which increased their profitability in double-cropping systems in the USA (Ren *et al.*, 1997).

7
Sunflower

The sunflower, *Helianthus annuus* L., is a member of the Compositae, a large and successful family of flowering plants occurring throughout the world. The genus *Helianthus* is named from the Greek *Helios*, sun, and *anthos* flower. *Helianthus annuus* was so named by Linnaeus to whom the only sunflower known lived but a single season, although there are many perennials among the 70 recognised species in the genus. Another well-known member is *H. tuberosus*, Jerusalem artichoke, native to North America. In this text sunflower refers to *Helianthus annuus* cultivars grown for their seed, oil and derivatives, unless otherwise specified.

Sunflower probably originated in the south west United States–Mexico area (Heiser, 1976), and its seed was early used for food by the local inhabitants (Fig. 7.1). That it was deliberately sown is doubtful, since it grows naturally in profusion, but sunflower was under cultivation when Europeans arrived in North America. The seeds were roasted or made into a flour, and a yellow and purple dye was extracted from florets or achenes. Sunflower as a cultivated plant was reintroduced to North America from Europe in the late nineteenth century, thus explaining the many differences between eastern and western continental races. Improved Russian varieties were obtained in 1893 via the American Consul-General at St Petersburg, and within 3 years yields of 1500–2500 kg/ha were obtained in Oregon (Hurt, 1948).

Sunflower was introduced to Europe in the sixteenth century, and the first published description as *Chrysanthemum peruvianum*, was in 1568 by the Belgian, Rembert Dodoens. The Peruvian origin implied is incorrect, since it almost certainly reached Europe from Mexico via Spain, and became very popular as an ornamental. In his *Herbal* of 1597, Gerrard makes the astonishing claim that 'There has been seen in Spain and other hot regions a plant sowne and nourished up from seed, to attain to the height of 24 feet in one yeare.' However he also stated that most plants averaged about 12 feet, and were much branched 'from their roots to their very tops,' with an illustration of one such plant. In 1716 an Englishman, Arthur Bunyan, was granted Patent Number 408 on a method of expressing oil from sunflower, the first European record.

Imported from Holland into eastern Europe and subsequently to Russia in the eighteenth century, the first commercial oil production was in 1830–40, and the crop steadily grew in importance until in the mid-1970s there were approximately 6 million hectares (Mha) under sunflower. The reason for its rapid acceptance and popularity in Russia is interesting. Apparently the Church had severely restricted use of oily food on many Holy Days, but sunflower was a later introduction and not included on the banned list! In Germany the leaves were used as a tobacco substitute, and the inner fibres of the stalk to produce a fine writing paper.

In the 1960s the development of Russian varieties suitable for mechanised harvesting, including Peredovik, Mennonite and Sunrise, had an immense impact on the development of sunflower as a commercial crop. The first commercial hybrid in 1972 increased yield by 25%,

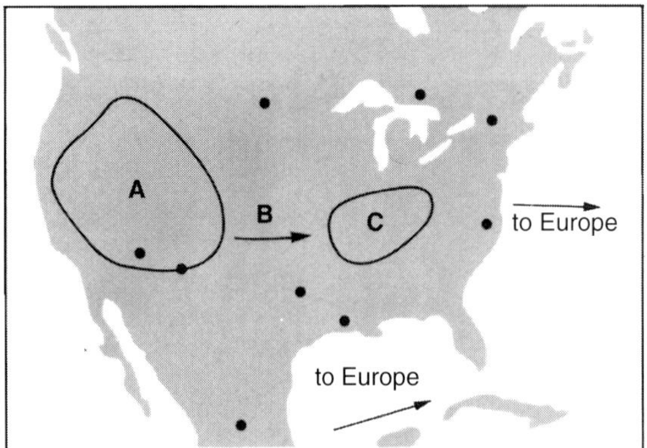

Fig. 7.1 Probable centre of domestication of sunflower.

and sunflower subsequently became one of the world's important oilseeds; in 1979–80 there was a record global oil production of nearly 15 million tonnes (Mt). Sunflowers are also cultivated as ornamentals, and often deliberately sown on estates or in natural gardens to attract birds.

Sunflower became a major international oilseed due primarily to the introduction of short-stemmed, high-yielding hybrid cultivars as noted. These cultivars not only increased the oil yield/ha but also allowed more efficient mechanised cropping. Emphasis on the advantages of polyunsatated fats in human diet widened the use of sunflower oil (sunoil), in edible fats and processed foods. However, as with sesame seed, total world production of sunflower seed (sunseed), cannot be accurately quantified, since a significant proportion of developing countries production is for personal consumption. Anyone who has travelled extensively in rural areas of these countries will have noticed how frequently sunflower occurs in peasant cultivation. Only that portion marketed is assessable.

World sunflower production was originally dominated by the USSR which accounted for some 50% of commercial crops, and harvested some 6 Mt annually for many years reaching 7.4 Mt in the 1973–74 season. Since the mid-1990s however, Argentina has become the world's largest exporter of sunseed.

Sunflower growing in the former USSR was concentrated in the Ukraine, Moldavia, Kazakhstan and Grusia, with highest yield/hectare in Moldavia. Since the break-up of the USSR some of these regions have become independent countries and continue to grow the crop. The area planted in the USSR in 1980 was 2.4 Mha yielding 2 Mt, in 1990 the figures were 2.7 Mha yielding 3.4 Mt. In its successor, the Russian Federation total seed production averaged 3.0 Mt to 1994, rose in 1995 to 4.2 Mt, fell to 2.8 Mt in 1996, but rose to 3.0 Mt in 1997. The Ukraine formerly produced about 40% of total USSR oilseed production, with sunflower accounting for 90%; 2.5 Mt in the years 1989 to 1993, but production has since been erratic. Exports were about 1 Mt in 1997 to 1999.

The area sown to sunflower in Argentina has expanded continuously since 1960 when about 1 Mha was planted, but the introduction of hybrids in 1975 boosted the crop's profitability and the area rose steadily to 2.2 Mha in 1990 and reached 3 Mha in 1998, although 4 Mha was targeted. Total seed production in 1960 was 0.8 Mt, in 1990 it was 4 Mt and in 1998 it reached 5.2 Mt, down from a forecast 5.6 Mt due to heavy El Nino rains. Most seed is processed locally.

The European Union (EU) is considered one region, and sunseed production developed from 1 Mt in 1980 to about 4.0 Mt in 1998 due to the EU's support policies. Production is expected to again increase as more Iberian growers switch from maize, while Eastern European growers are also planning a similar expansion. In 1998 the Romanian crop was nearly 1 Mt, and Slovakia at 0.4 Mt doubled the 1997 total. The EU is the world's largest importer of sunseed.

The main sunflower growing regions of the USA are the Red River valley region of Minnesota, North and South Dakota, and Texas. The area sown rose from 0.8 Mha in 1988/89 to 1.4 Mha in 1994/5 and reached 3 Mha in 1997/8; seed production rose from 0.9 Mt to 2 Mt over the period. Average yield/ha is very variable due to seasonal conditions, and not directly related to the area planted.

Sunflower production in India has been affected by political decisions regarding the relative status of oilseed crops and the level of oil imports. With increasing official support production rose from 0.3 Mt in 1985/86 to 1.4 Mt in 1995/96, and nearly 3 Mt in 1997/98 and 1998/99. Since the area available for expansion is limited, more efficient extraction methods would probably increase oil output more quickly. Main producing regions are in the states of Andra Pradesh, Haryana, Karnataka, Maharashtra and Uttar Pradesh. The area under sunflower in Pakistan, mainly in Sind Province, has varied widely from 30 000 to 60 000 ha for no apparent reason, and the official target of 250 000 ha by 2005 is unlikely to be achieved. Total seed production in 1996 was officially stated as 60 000 t (60 kt) from 45 000 ha.

Australia became a significant producer with the area planted fairly constant between 1970 and 1980 at 200 000 ha, but fell substantially to 1990 despite the introduction of short hybrids, due mainly to bird damage and disease reducing the yield/ha. By 1995 the area was down to 83 000 ha but rose to 140 000 ha in 1998 and forecast to remain at that level in the short term. Seed production totalled 70 kt in 1995/96, 142 kt in 1997/98, but fell to 110 kt in 1998/99.

The substantial increase in intensive livestock production in South-East Asia, particularly pigs and poultry, has increased demand for oilseed meals, and sunflower production is being officially supported in many countries. Thailand for example has a domestic demand for 60 kt of sunseed, but produced only 13 kt in 1995/96 and 20 kt in 1996/97, the shortfall satisfied by imports. Total sunseed production in China is officially stated to have risen from 80 kt in the 1975/76 crop year to 1.0 Mt in 1985/86 and 2 Mt in 1995/6, wholly consumed domestically.

The considerable local African production is now virtually consumed domestically, with specific countries exporting a small surplus in particular years. No reliable country statistics are available with the exception of South Africa, where production for the 1995/96 season was a record 730 kt but fell substantially in 1996/97 to 425 kt. General comments on Africa in Chapter 3 regarding groundnut are relevant.

The proportion of sunseed that enters international trade has varied from 10% to 20% of world production, since local expression and sale of oil and meal is more profitable; thus major South American producers have established a large and efficient crushing industry. A major market for South American sunseed is the EU. Argentina exported about 200 kt of sunseed annually in the late 1970s and early 1980s and this rose steadily to 884 kt in 1995, fell substantially to around 100 kt in the following seasons, but rose to 700 kt in 1998.

The EU sunseed availability was nearly 4.0 Mt in 1996, down 200 kt on the previous season. However, growing internal demand for oil and meal plus increased production will require European crushers to substantially expand their capacity, as little is used as whole seed. Russian sunseed exports reached a maximum in the late 1960s of some 400 kt, dropped sharply in the 1970s, ceased in 1975, but resumed in the 1990s; seed availability was around 650 kt in 1996 and rose to 1 Mt in 1997 mainly to the EU.

The largest market for sunflower oil (sunoil) is Europe which absorbs approximately two-thirds of oil traded. Domestic demand in the EU in 1997 and 1998 was 1.8 Mt and rising. Argentina is a major producer and exporter of sunoil; in 1979–80 exports were 250 kt and rose steadily as the planted area and crushing capacity increased, to 1.3 Mt in 1990 and fluctuated around this level to 1998. USSR exports of sunoil expanded sharply in the 1960s, declined from 225 kt in 1976–77 to 80 kt in 1979–80 and have since virtually ceased, although domestic production in the CIS remained around 1.5–2.0 Mt USA sunoil production in 1998/99 was 1.7 and 2.4 Mt.

World production of sunflower meal (sunmeal) has risen as production of sunoil increased and domestic crushing capacity in producing countries expanded. The main markets for sunmeal are the EU and Eastern Europe. Sunmeal consumption in the EU more than doubled in the decade to 1990 from 1.5

to 3.5 Mt and is currently around 5.0 Mt. Argentina exported about 75 kt in 1979, rose steadily to 1.4 Mt in 1990s as the domestic crush increased; Argentina remains the EU's major supplier.

BOTANY

Three main subspecies (subsp.) are considered valid; *H. annuus* subsp. *annuus* Dougl., *H. annuus* subsp. *lenticularis* Dougl., and *H. annuus* var. *macrocarpus* Dougl., with the second considered a close wild relative of the oilseed sunflower. Russian geneticists recognise *H. lenticularis* and *H. annuus*, but include *H. petiolaris*, and further divide *H. annuus* subsp. *annuus* into subsp. annuus, pustovojtii, armeniaca and australis.

The majority of *Helianthus* spp. have $2n = 34$, 68, 102, exceptionally $2n = 14$ and 28 or $2n = 32$, with *Helianthus annuus* $2n = 34$. Although these are all diploids, autotetraploid sunflower has been produced in breeding programmes. *Helianthus annuus* will cross with related species to produce interspecific and intergeneric hybrids, and one of the most notable was with *H. tuberosus* (Jerusalem artichoke) in the former USSR, which introduced into sunflower some important disease resistance genes. *In vitro* culture techniques not only allow very quick establishment of breeding lines, they are also used to produce the more difficult interspecific hybrids. Techniques used on sunflowers and other oilseeds have been reviewed (Palmer & Keller, 1994).

About 20 species can be considered as cultivated sunflowers, mainly ornamental, and identification is difficult. Commercial varieties grown for seed are considered to be *H. annuus* var. *macrocarpus*. A wild species closely related to *H. annuus* is *H. argophyllus*, native to and common in Texas. Most ornamentals should apparently be referred to *H. annuus* subsp. *annuus*, the common sunflower of western North America.

The cultivated sunflower is a tall, erect, unbranched, coarse annual, with a distinctive large, golden head, whose seeds are eaten or crushed for oil. Since it is often necessary to define the development stage of a crop in the field, easily recognisable changes in plant growth are shown diagrammatically in Fig. 7.2 (Courtesy FAO, Rome), field stages are shown in Fig. 7.3 (Anderson, 1978), and their relationship to various plant characteristics in Fig. 7.4.

Roots are substantial, often shallow, and although the tap root can be up to 3 m in length it rapidly reduces in diameter with depth and offers little support to the mature plant. In the dry tropics it is often partially destroyed by termites before harvest, so increasing lodging. Roots of mature giant cultivars can penetrate to 2.5 m, although the bulk generally remain in the first 50 cm (Fig. 7.5) (Weaver, 1926). In suitable soils the tap root grows at approximately the same rate as the stem for the first 5–6 weeks, then more slowly. Root growth is generally prolific and since the majority remain near the surface, too deep inter-row cultivation can be most damaging. Lateral roots of plants one-month old and about 30 cm tall will usually have covered the inter-row at 60 cm row spacing, and many roots will be severed by ill-adjusted weeders.

Root distribution and penetration is most important with rain-grown crops, for late vegetative growth, flowering and seed filling are generally dependent on soil moisture from below the 1-m zone. The rate of root growth is most important, and this aspect of sunflower development has been somewhat neglected; probably because it is harder work to dig holes in fields than measure height, head diameter and leaf area! Root development in sunflower is directly affected by soil type, for the author examined plants growing in a sandy loam whose roots were in the first 12 cm only (Fig. 7.6) while others in a nearby clay soil had tap-roots 40 cm long.

The stem is robust, circular in section, usually 3–6 cm in diameter but occasionally reaching 10 cm, bearing rough hairs, and may have slight longitudinal ridges. Colour is usually green or greenish-yellow, occasionally mottled, or with a bluish tinge. The stem elongates rapidly to 1–2 m for dwarf types, up to 5 m for giant varieties. The woody exterior is filled with a stiff white pith, and frequently becomes hollow with age. The main stem constituents are cellulose 50–55%, lignin 17–20%, pentosan 15–20%, crude protein to 5%

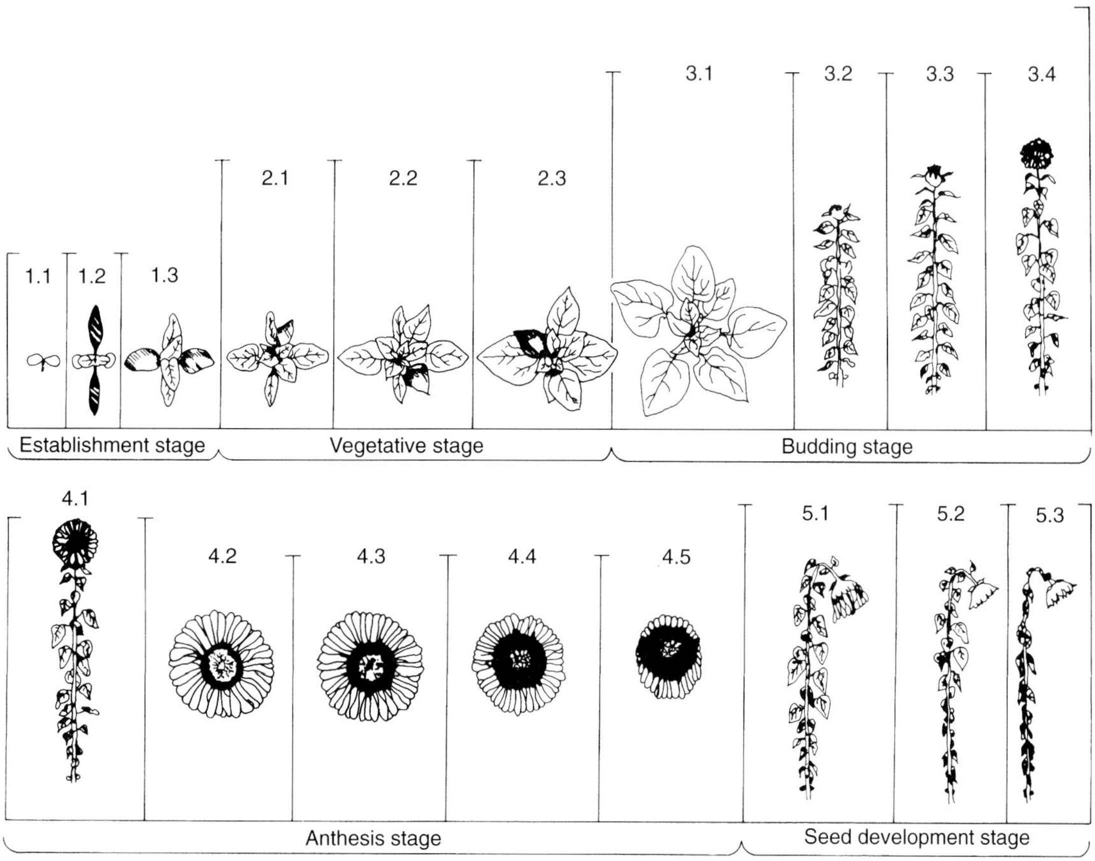

Fig. 7.2 Growth stages of sunflower.

and ash 8–10%. Saline soil or irrigation water can reduce both stem diameter and height. New hybrids are remarkably even in growth and final height, and in some breeding fields 90–95% of plants are within a 5% deviation. Branching is common in the species, but has been removed from commercial cultivars.

Growth regulators generally reduced mature height by affecting internode length, and increased leaf area, although there may be differing cultivar reactions to the chemical used or dosage. Growth regulators may also affect seed composition, as in Poland (Musnicki *et al.*, 1996), or seed numbers and onset of flowering as in Argentina (Hernandez, 1997). Irradiation can stimulate or increase growth-rate of seedlings and immature plants, and subsequent seed yield.

Leaves are usually alternate, occasionally opposite on the lower stem and alternate above, large, ovate, cordate, frequently heavy and carried on long petioles; usually a dark green, sometimes with a bluish or reddish tinge. About 20–40 leaves are produced per plant, their rate of development and number being a cultivar characteristic. A comparison between two hybrids in Australia showed one produced a leaf every 28 hours until heading, the other every 40 hours. The total leaf area, as measured by a leaf area index, can be correlated with various aspects of general growth, transpiration, light interception, and also with the incidence of disease (Debaeke & Raffaillac, 1996).

Within a cultivar the final leaf number per plant is highly influenced by population, soil

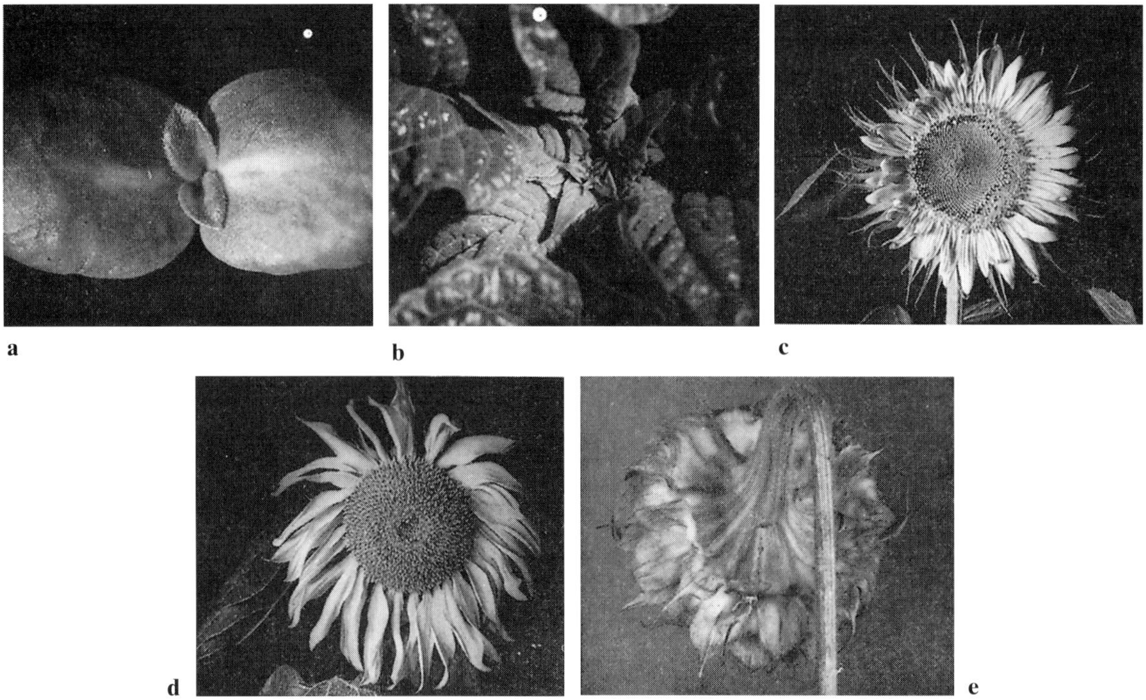

Fig. 7.3 Growth stages of sunflower: (a) first true leaves; (b) bud emergence; (c) first anthesis; (d) last anthesis; (e) physiological maturity.

moisture or temperature. Plants generally produce two to three times the leaf area when irrigated than when rain-grown. Leaves are highly heliotropic before anthesis, but this diminishes with plant maturity. Leaf production and stem elongation continue until the inflorescence opens and flowering begins, when the number of active leaves decline. Photosynthesis by leaves is related to age and stem position, and is greatest in the highest leaves.

Artificial defoliation to simulate severe insect or hail damage, showed about 50% of leaf area must be lost to cause any significant depression in seed yield or seed-oil content. However, loss of half the top leaves has a more severe effect on seed-oil content than loss of half the lower. The most critical period for defoliation is between bud formation and flowering (Table 7.1). (Schelotto, 1978). The five lowest leaves on the stem contribute little, and the ten lowest about 10%, to final seed yield.

The disc-shaped head, capitulum, is borne terminally on the main stem, varies in size between cultivars, seasons, soil type, etc., and is commonly 10–30 cm in diameter, occasionally much larger; the greatest diameter recorded was almost 80 cm (Heiser, 1976). Within a cultivar, head diameter has the most significant effect on seed yield, but with an optimum diameter for maximum seed

Table 7.1 Effect of defoliation on growth and yield.

Characteristic	Var. – INRA 4701 Defoliated*		Var. – HS 18 Defoliated*	
	Early	Late	Early	Late
Plant height	91[‡]	96	104[‡]	—
Head diameter	67	72	75	78
Seed yield	53	72	52	59
Oil yield	58	73	51	58
Grain weight	72	74	73	90

* Early = in mid-vegetative period; late = just before flowing.
[†] Treatment means as a percentage of untreated controls.

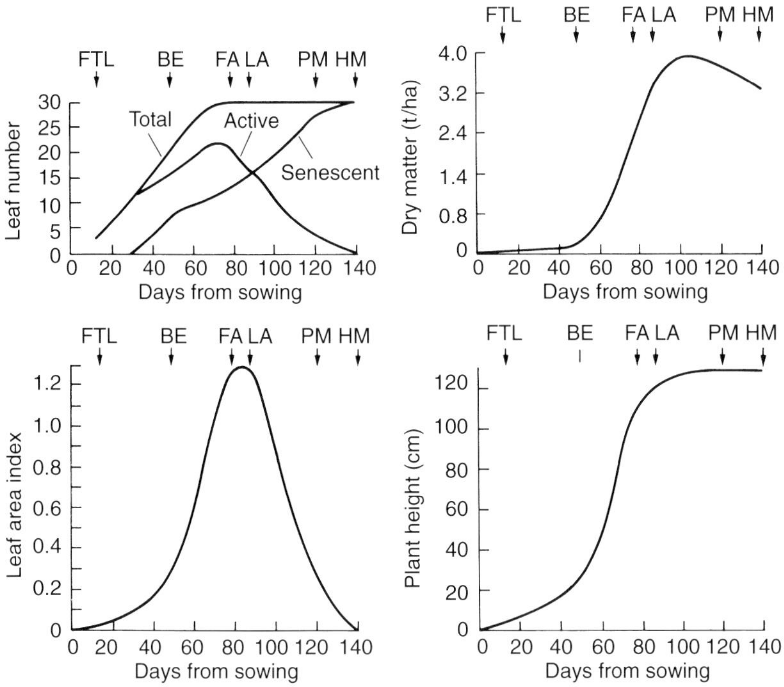

Fig. 7.4 Relationship between phenological stages and growth. FTL = first true leaves; BE = bud emergence; FA = first anthesis; LA = last anthesis; PM = physiological maturity; HM = harvest maturity.

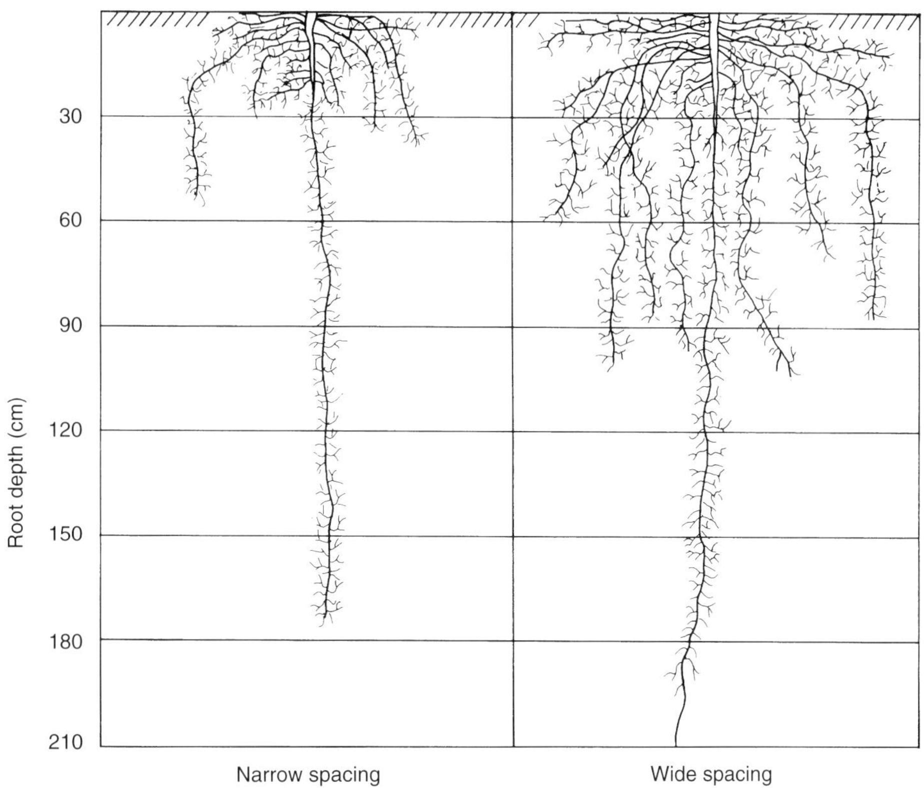

Fig. 7.5 Effect of spacing on root development of sunflower.

Fig. 7.6 Restricted root development of sunflower in field, Australia.

production in the field. Probably the most important head/seed factors affecting yield are the number of sound seeds per head, and sound seeds per unit area of the head. Within these parameters, selection for a long, thin-hulled seed would appear theoretically desirable, but in practice such seed is difficult to sow in commercial plate and air seeders. Composition of mature heads has been determined, and includes glucose and various pectins (Lou, 1996).

The head is heliotropic until the majority of flowers are fertilised, after which it generally remains fixed facing to the east. This movement is often quoted as the reason for the plant's common name, and is related to auxin level and distribution in actively growing plant parts. The easterly aspect of flowering heads probably reduces the maximum daily temperature to which florets are exposed and contributes to increased seed set.

Flowers of the capitulum are of two types: an outer row of brightly coloured, sterile, ligulate flowers, usually yellow but varying from deep yellow to red, and the brown or purplish disc flowers (Fig. 7.7) (McGregor, 1976). A head can contain 1000–4000 individual florets arranged in spiral whorls originating at the disc centre, maturing progressively from the outer to the centre. The whorls follow a numerical Fibonacci Sequence which allows theoretical construction of a sunflower head of any required size (Mathai & Davis, 1974) (Fig. 7.8).

Opening of all florets on a single head is usually completed in 5–10 days, but individual florets can remain receptive for up to 14 days. Flowering within a crop of hybrid sunflower is remarkably uniform, with 80–90% of heads opening within 3–4 days of the first. Introduction of the reproductive phase is related to day length, and although the relationship appears complex, short days favour floral initiation. As a field guide, floral initiation occurs at about the eight leaf stage of development.

Growth regulators not only affect plant growth as noted, they can also affect seed characteristics. TIBA sprayed on flower heads as ray florets open may increase seed yield and 1000 seed weight, and reduce the number of 'pops', while gibberellic acid or ethephon applied when heads are forming may produce almost complete sterility, an effect of value to plant breeders.

Flowers are protandrous and normally cross-fertilised, insects being the main agents and wind pollination rare (Table 7.2) (Low *et al.*, 1975). Honey bees are sometimes the only pollinators, and a major cause of poor seed set in commercial plantings may be inadequate pollination. Centre florets produce less pollen than ray florets and are thus less likely to be visited by insects, reducing the number of sound seeds. Also determining the extent of pollination is that cultivars differ in the amount of nectar produced, and thus their attractiveness to insects, particu-

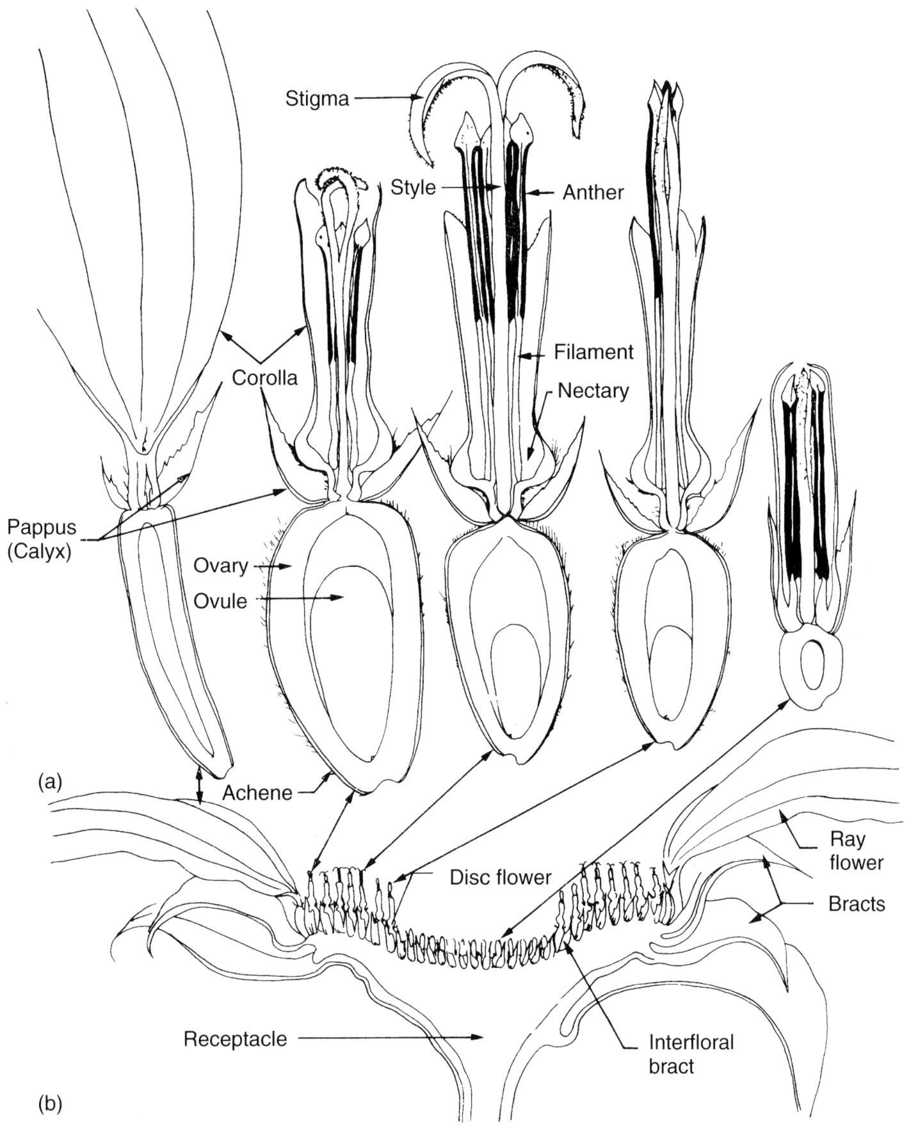

Fig. 7.7 Longitudinal section of sunflower head with individual florets: (a) ray floret; (b) disc florets in different stages of development.

larly bees. Nectar production is also affected by date of sowing which influences time of flowering, and in India maximum nectar sugar content occurred around 10 a.m. (Sharma & Sharma, 1997). Older sunflower cultivars are basically insect-pollinated, and vary greatly in the degree of cross-fertilisation, or self-pollination, with quoted figures ranging from 5 to 95%; most commercial hybrids are virtually self-fertile.

The fruit is an achene, commonly known as sunflower seed, and the colour of commercially grown cultivars ranges from black through to white, but can also be brown, striped or mottled (Fig. 7.9). A colour rating system widely used in breeding or selection programmes is black; black

Fig. 7.8 Fully developed sunflower head.

with grey stripe; black with white/light-grey stripe; dark grey with white/light-grey stripe; grey with white stripe; grey with grey stripe; white. Many growers still equate dark seed with the oil market and light or striped seed with birdseed or confectionery sectors. Although there is often a relationship between a dark hull and high average seed-oil content, an increasing number of hybrids have light-coloured seed with high oil content.

Seed from individual heads generally decreases in size, weight, viability and oil content from the rim to the centre, the degree of difference being a cultivar characteristic. Variation in seed-oil content and fatty acid composition of the oil, can also occur in seed from different sized heads; in India seed from heads less than 20cm in diameter frequently had up to 40% more oil than seed from 30cm diameter heads, and differed in seed characteristics from the rim to the centre. At physiological maturity, the seed of hybrid sunflower comprises about half of the head dry weight. The kernel:husk ratio is of prime importance for profitable oil production and utilisation of the meal, and larger seeds generally have the lowest kernel percentage. Black and striped seeds usually have a lower

Table 7.2 Effect of pollination method and plant density on head size and seed yield of three cultivars – Australia.

Characteristics	60000pph			90000pph			125000pph		
	OP	WP	SP	OP	WP	SP	OP	WP	SP
Chernianka									
Head size (cm)	11	11	11	8	9	10	8	9	7
Seed wt (g/head)	14	14	7	7	6	3	7	7	3
Seed yield (t/ha)	0.9	0.9	0.5	0.6	0.6	0.2	0.8	0.8	0.4
Sunfol									
Head size (cm)	13	13	11	12	12	12	12	12	12
Seed wt (g/head)	55	19	9	50	22	10	23	21	7
Seed yield (t/ha)	3.3	1.1	0.5	4.3	1.9	0.9	2.9	2.7	0.8
Hysun 20									
Head size (cm)	16	16	16	14	14	12	13	12	12
Seed wt (g/head)	90	87	38	75	76	34	46	30	22
Seed yield (t/ha)	5.4	5.2	2.3	6.5	6.6	2.9	5.7	3.8	2.8

pph, plants/ha; OP, open-pollinated; WP, wind-pollinated; SP, self-pollinated.

Fig. 7.9 Striped sunflower seeds.

kernel:husk ratio and total oil content than white, and a negative relationship between oil content and hull thickness exists.

Seed varies greatly in size and weight, but is generally a compressed, flattened oblong, the top truncated and base pointed, roughly diamond shaped in section, and usually within the range 10–25 mm long, 7.5–15 mm wide and 3–7.5 mm thick; 1000 seed weight varies from 50 g to many times this. The rate of development of seed characteristics, especially of oil formation is important as it affects many agricultural operations, including irrigation, defoliation or desiccation, harvesting date (Table 7.3) (Harris *et al.*, 1978). Oil biosynthesis in developing seeds has been determined (Mazar *et al.*, 1998). There is little difference between low and high oil content varieties, but total seed-oil content is related to rate of cotyledon cell formation. This is faster in seed with a high oil content than in seed with a lower oil content, and may continue longer; thus emphasising the importance of adequate nutrition and soil moisture at this stage of development.

Table 7.3 Composition of seed oil during development – Australia.

Days from pollination	Total oil (%)			
	Palmitic acid 16:0	Stearic acid 18:0	Oleic acid 18:1	Linoleic acid 18:2
3	19.6	1.8	10.4	49.4
9	19.4	3.5	18.1	46.3
15	13.3	7.7	32.8	42.4
21	8.2	7.0	28.4	54.4
27	6.7	6.2	23.7	61.9
33	6.2	5.3	17.9	69.2
39	5.9	5.2	17.0	70.1
45	6.0	4.9	16.2	71.6
51	6.2	5.4	14.7	72.4

The oil is currently the most commercially important derivative of sunflower seed, with the meal a major contributor to mill profitability. Of increasing interest are other derivatives including protein isolates, and these are further discussed in Chapter 9 on oilseed processing.

Seed constituents are normally a cultivar characteristic; oil content ranges between 25 and 48% but can reach 65%, with a basic difference between seed produced under hot or temperate conditions. High temperature during seed development can reduce total oil content; for example oil content fell to 24% from the usual 40–42% following a heat-wave in Australia. Average oil composition of temperate crops is usually 55–60% linoleic and 25–30% oleic acid, but seed produced under high temperature may contain 65% oleic and 20% linoleic acid; protein content is usually 15–20%. The considerable effect of environment is shown in Tables 7.4 and 7.5. The first demonstrates the effect of location on oil composition, indicating variation in oil obtained from the variety Peredovik when grown in Canada and to different regions of the USA (Robertson & Russell, 1972). Similar variations due to location are reported elsewhere (Wang *et al.*, 1997). The interaction between variety and soil is shown in Table 7.5, and is from data obtained in Kenya (Weiss, 1966a). It also illustrates the variation in seed components; Comet, Black and White being grown for oil, Grey Striped for the birdseed market. Similar effects of environment/location on seed charac-

Table 7.4 Effect of location on seed-oil composition – USA.

Location	Fatty acid composition (%)					
	Palmitic	Stearic	Oleic	Linoleic	Linolenic	Other
Canada	6.1	3.7	16.4	73.7	—	—
Minnesota	5.6	6.5	19.1	67.0	—	—
California	7.0	4.5	25.0	61.9	<0.1	1.5

Table 7.5 Interaction between variety and soil, seed and oil characteristics – Kenya

Component	Comet		Black		White		Grey striped	
	G	P	G	P	G	P	G	P
100 seed wt (g)	7.27	6.87	8.18	8.90	9.07	9.21	10.90	11.04
Husk (%)	41.2	42.7	45.3	44.8	40.9	42.1	48.1	49.8
Kernel (%)	58.8	57.3	54.7	55.2	59.1	57.9	51.9	50.2
Moisture (%)	6.47	6.24	6.25	6.17	5.85	5.95	6.98	6.65
Protein (%)	19.4	17.0	16.8	14.5	15.5	15.8	16.1	15.0
Crude fibre (%)	23.8	24.3	25.2	24.8	25.7	27.2	29.9	31.4
Oil (%)	27.1	29.5	26.1	28.3	31.0	28.5	24.9	26.1
Moisture-free basis								
Protein (%)	20.7	17.9	17.9	15.5	16.4	16.8	17.3	16.1
Crude fibre (%)	25.4	25.9	26.9	26.4	27.3	28.9	32.2	33.6
Oil (%)	29.0	31.5	27.8	30.2	32.9	30.3	26.8	28.0
Oil								
Oleic acid	1.9	1.3	4.9	4.4	6.0	9.0	5.5	4.0
Iodine value	112.3	116.6	108.5	110.1	98.5	78.8	111.0	107.3
Colour	VPY	VPY	WW	WW	VPY	VPY	VPY	VPY

P, Phonolitic clays; G, granite sands; VPY, very pale yellow; WW, water white.

teristics have been reported from other sunflower-growing areas.

Fertilisers may also affect seed characteristics; phosphate can increase seed-oil content and nitrogen reduce it, and there is an optimum level for both to achieve highest oil content. Viability, and subsequent seedling vigour can also be affected by the nutrient status under which the parent plant was grown. Saline soils or irrigation water decrease both seed yield and oil content. Dormancy in sunflower seed exists, and may vary with harvest date, storage temperature and between cultivars.

Initially the most important aim of plant breeders was to increase seed-oil content, reduce height and introduce pest and disease resistance, and the first two proved more easy to achieve. A strong, positive correlation exists between seed yield, and plant height, head diameter and seed weight, while oil content is negatively related to hull thickness. Considerable variation for many commercially desirable characteristics exists, for analyses of cultivars and wild species in Canada, Russia, and the USA showed ranges for seed oil of 12–20%, hull content 40–44%, protein 29–35%, and seed weight 3–19 g. It would also appear that simple recurrent selection procedures are effective in increasing both oil content and desirable constituents. However, many plant breeders believe that with present known resources, a seed-oil content of approximately 55% is the upper limit and hull content of around 20% per cent the lower limit for this characteristic in commercial cultivars.

Oil constituents also vary, with linoleic acid to 89% or oleic acid to 85%. To supply consumer demand for spreads and cooking oils containing mono-unsaturated fats, cultivars were introduced whose seed oil was high in oleic, above 80%, and low in linoleic acid, less than 10%, which became commonly known as mono-hybrids. Breeders aim to widen this ratio.

Rapid, and non-destructive, analysis of seed may be important and a number of methods have been developed. Screening for specific components is more difficult; for non-protein and protein amino acids a rapid gas chromatographic method has been developed (Oh *et al.*, 1995), and dietary fibre content is assessed by enzymatic/gravimetric analysis (Cardozo & Li, 1994).

The well documented finding of cytoplasmic male sterility (CMS) in crosses between *H. petiolaris* and *H. annuus* by Leclerc in 1968, and fertility restorer genes by Kinman in 1970, allowed breeders to produce hybrid sunflowers and cultivars with short stems suitable for mechanical production. Male and female sterility can also be chemically induced. Branching is undesirable in commercial sunflower crops, however, in breeding programmes the use of a branched male with a prolonged flowering period has definite advantages. A short, sturdy, single-stemmed plant with leaves mainly at the top of the stem is a most desirable type for mechanised production, especially in drier areas.

The expanding use of these hybrids and their rapid replacement of traditional varieties, poses the same problem for sunflower breeders as has occurred with safflower. The dilution or disappearance of local cultivars and indeed some wild *Helianthus* spp. is almost certain, and their genes are lost. Sunflower gene pools are essential to ensure availability of the widest possible range of material; however, the maintenance of such germplasm collections is more complicated than for other crops, since sunflower species are mainly cross-pollinated with high self-incompatibility. Sib-pollination maintains germplasm purity, and works well with cultivars; the method is successful even with wild *Helianthus* spp. in which self-incompatibility is usually more complete. Additionally, seeds of these species often have long, strong, dormancy periods that require special treatment to break, although the natural cycle of freezing and thawing is sufficient.

A method developed at UCLA, California, has proved adequate to maintain entries; dry seeds are soaked in a 10% aqueous solution of ethylene at 50°C for 1 hour, then blotted dry and stored for 10 days at 5°C before planting. Irrespective of the difficulties and despite the expense involved, governments must establish or at least cooperate in the establishment and maintenance of such gene pools to enable future plant breeders to apply techniques that will inevitably be discovered.

ECOLOGY

The main commercial production of sunflower is in the warm-temperate regions, but breeding and selection produced cultivars adapted to a wide range of environments. The crop is cultivated from 40°S to 55°N, but greatest production is between 20 and 50°N and 20 and 40°S. It will grow from sea-level to 2500 m, but generally gives highest yield of oil/ha below 1500 m. Sunflower like sesame, is considered highly adaptable and there is no inherent reason why the crop's range should not be considerably expanded. Vernalisation to extend the crop's northward boundary was not very successful; more successful are the early maturing dwarf cultivars adapted to higher latitudes, including Sunola, commercially planted in Canada in 1995. A method of determining the suitability of sunflower cultivars for subtropical environments has been developed in Australia (Bange *et al.*, 1997).

An unusual use for sunflower at the northern limit of its range, is the planting of tall, strong-growing cultivars as snow-breaks to protect spring wheat in Russia. Not only do the plants give protection from drying winds, accumulated snow significantly increased soil moisture available to the subsequent wheat crop. Sunflower has been similarly used in Canada to trap snow, and in the USA as windbreaks for soya.

Russian sunflower cultivars are divided into groups differing significantly in photoperiodic reaction; short-day, long-day and insensitive types. Under controlled environmental conditions, development is more rapid at 12 hours daylight equivalent than at 16 hours or longer, but in Australia day length influences time to flowering only in the period from emergence to budding. In Italy yields tended to be higher in the long-day north than further south (Monotti *et al.*, 1997). For all practical purposes sunflower can be considered day-neutral, and the effect of photoperiod on sunflower is negligible compared with the influence of other environmental factors; as noted, high temperature can reduce the time to maturity. However, the type of light has a significant effect, with red light most destructive (Ushakova *et al.*, 1997), while high levels of ultraviolet radiation adversely affected general growth (Rajendiran, 1996; Mark & Tevini, 1997).

Mature plants are more resistant to low temperature than soya bean or maize, while young plants in the four- to six-leaf stage may withstand –5 to –6°C for short periods. However, a hard frost after floral initiation, about the eight-leaf stage, will usually reduce yield by affecting head development, although visual damage may appear slight. Frost will also damage immature seed and substantially reduce viability, but mature seed is less affected. This may be of little concern to commercial oilseed producers, but can influence a site for seed production. A frost-free period of about 120 days is usually necessary for commercial crops. Recommended sowing dates for selected countries are shown in Table 7.6.

Sunflower grows well at temperatures of 20–30°C, with 27–28°C the optimum. A range of 8–34°C is tolerated without significant yield reduction, indicating adaptation to regions with warm days and cold nights. A major visual effect of temperature is on the rate of development, and prolonged high temperature can halve the time to maturity. In Japan the sum of effective temperature (defined as daily mean temperature –5°C) required from sowing to maturity for a tall Russian cultivar was 1300°C, and an Australian of 1367°C (Anderson, 1978).

Temperature is known to affect seed weight, oil content and characteristics, but its effects on plants in the field are often masked or modified by other environmental factors. In general, a temperature which remains above 25°C at flowering reduces seed yield and seed-oil content, and although there is lack of trial data, it is certainly so in the eastern states of Australia. Conversely, temperature below 16°C at this period can reduce seed set, and oil content, except in some low-temperature resistant cultivars.

Table 7.6 Recommended sowing dates in selected countries.

Country	Month	Country	Month
Asia		Europe	
Bangladesh	August and January	Spain	March
India		Russia	
North	July and October	East	March
South	April and November	West	April
Iran	July		
Iraq	April–May	Australia	
Pakistan	February	North-East	January
		South-East	September
		West	May
Africa			
Central	June		
North	March		
South	December–January		
Americas			
Argentina	October		
Brazil	August		
Canada	March and November		
USA	April–May		

A strong relationship exists between linoleic acid content of seed and daily mean minimum temperatures from flowering to harvest, with high night temperature especially important; linoleic acid increases as temperature during development decreases, while the converse produces an increase in oleic acid. The marked reduction in linoleic acid is due to temperature acting on the desaturase enzymes, which convert oleic to linoleic acid. It may thus be necessary to relate time of sowing to temperature at flowering when a particular type of oil is required. Temperature may not only affect seed-oil constituents, the basically different oil types produced can directly influence local manufacturers and their product range. If customer demand is for 3:1 polyunsaturated margarine for instance, sunflower growers in high-temperature areas may not be able to comply, as some current cultivars are not sufficiently heat tolerant.

Sunflower is considered to be drought resistant and while this may be so, oil yield is substantially reduced if plants are allowed to become stressed during the main growth period and at flowering. A major symptom of stress in the vegetative phase is a reduction in the number and size of leaves. Should it continue, lower leaves are shed and plant height will be substantially reduced. Plants resist moisture stress by wilting, as in limp leaves water loss was reduced to a greater extent than photosynthesis. Solute accumulation in the leaves assists in maintaining leaf activity, but incomplete closure of leaf stomata prevents optimum water conservation.

Plants can be largely independent of rainfall provided subsoil moisture is adequate, and good yields are obtained on only 500 mm of irrigation water. Sunflower will produce a moderate yield on 300 mm rainfall, but at this level of production is unlikely to be commercially viable. The relationship between low rainfall and seed yield is often almost linear from 200 to 500 mm, with 1 t/ha usually achieved around 300–350 mm. The steadily increasing average yield of sunflower in Argentina to 1.5 t/ha in the decade to 1995, is attributed to the increase in summer rainfall in the last two decades (Pascale & Damario, 1997).

The author found the yield that can reasonably be expected from rain-grown sunflower in more arid areas is approximately half that of a good local sorghum crop. However, in arid conditions seed-oil content is adversely affected. Between 500 and 750 mm of rain evenly spread over the growing period and ceasing just before main flowering and seed filling, will normally produce excellent crops. Where regional rainfall is insufficient for an annual crop, use of a cultivated fallow to store moisture allows two sunflower crops in 3 or 4 years. As a field guide to growers there must be one metre of subsoil moisture at sowing. Losses from disease and lodging can be severe when rainfall is above 1000 mm unless the soil is free-draining, especially if there are heavy falls when plants are fully grown.

Sunflower plants are susceptible to excess water and the degree of damage is directly related to duration of waterlogging. Very short periods, 2–3 days, affect young plants more severely, but several such periods are less damaging than a single period of 7–10 days. Data indicates that flooding reduces photosynthesis and stomatal conductance of leaves.

In South Africa a formula relating rainfall and soil type to potential yield has been developed, which may be more widely applicable. It has proved sufficiently accurate over a number of years to be used by local sunflower growers, particularly when evaluating alternative cropping or land use. The basic formula is: yield (kg/ha) = rainfall (mm) × soil depth (mm) × F, a constant related to the cultivar selected. In Brazil and Argentina where there is substantial smallholder production as a cash crop, suitable areas are clearly defined on official maps.

Plants are susceptible to wind damage from when half-grown, more so when irrigated as moist soil gives less support. The considerable variation in ability to resist wind damage is related to root development, and this should influence selection when choosing cultivars to be grown in windy locations. The effect of wind, in terms of lodged plants, is shown in Table 7.7. (Weiss, 1966a). Hail can be extremely damaging to young seedlings, which seldom recover if the terminal shoots are destroyed. Mature plants are little affected even though leaves may appear shredded or badly holed.

Table 7.7 Comparison of lodged plants per hectare – Kenya.

Variety	Number of lodged plants		Yield (kg/ha)[‡]
	46 cm*	92 cm*	
Black-seeded	1410	830	1380
White 655	1520	920	1345
Grey-striped	2420	1030	1230
Mean	1450	930	

*Between plant spacing in 92 cm rows; [†]final yield shown is that from standing plants.

SOILS

Soils that grow good maize crops are generally suitable for sunflower, and in more temperate northern areas the southern edge of the wheat belt is also suitable. In undeveloped regions, however, there are large areas basically suitable for sunflower on which the crop is not sown, and even though local growers may have produced poor crops, it is frequently possible to select a cultivar more suited to these soils. In general, soils with an appreciable sand content produce higher yields than more clayey soils in the same area under similar standards of cultivation. Whatever soil type is available, good drainage is more important than basic fertility, for it is usually easier and cheaper to supply nutrients than improve drainage. Sunflower grows well on neutral to moderately alkaline soils, pH 6.5–8.0, but dislikes acid conditions.

Salinity affects plant growth, development and some seed characteristics, mainly oil content, and also influences nutrient uptake. Where salt concentrations are low to moderate, the first visual symptom is often a thin stem and stunted growth. Salinity can also reduce disease resistance to disease, particularly *Macrophominia*, and charcoal rots. Seedlings are sensitive to salt concentration especially under irrigation when surface accumulation can occur, they become more tolerant when well grown, but less so than cotton. The activity of sunflower roots is considered important for increasing this tolerance, as they act as accumulators of sodium rather than a barrier to its assimilation. Plants grown on Russian soils containing 0.2% salts had yield reductions of nearly 40%, but at 1.0% salts the reduction was only 47%, while exchangeable sodium above 16% delays germination and development of flower heads, as recorded in many countries.

The depressing effect of high salinity levels on emergence is also related to temperature, but there are cultivar differences; some are adversely affected by high temperatures some by low. The degree of salinity also affects the rate of emergence, and total emergence at different temperatures. It would thus appear that there is considerable scope for increasing salt tolerance in sunflower. Selection of sunflower cultivars for more saline soils would be of great value and salt tolerance exists in some local selections and wild species. Salinity tolerant cultivars currently produce lower yields, but are very valuable as the proportion of salt-affected soils is tending to increase rather than diminish.

Plants grown in soil contaminated with heavy metals may accumulate levels in foliage and seed, which can be toxic but not necessarily fatal to stock and humans, or impart an unpleasant taste to seed and oil; cf safflower.

FERTILISERS

Sunflower is considered an inefficient converter of plant nutrients as measured by its low harvest index; weight of seed produced often less than one-third of total plant weight. Thus the price received for sunflower seed or oil must be substantially higher than for other more efficient crops which could be grown, and accurate determination of optimum fertiliser application is important. Adequate nutrients are essential for high seed yield; in the USSR 100 kg seed contained 6 kg nitrogen, 1 kg phosphate, and 18 kg potassium, and three-quarters of the total phosphate uptake by the plant was contained in the seeds.

An examination of world data indicates that local fertilizer trials to determine the optimum rates to be applied are essential. The author has supervised trials in a number of African, Asian and Central American countries, and results were conflicting, often owing more to climatic influence than true crop response. Interpreting data from intercountry comparisons is also made more difficult by misnaming the cultivars under test and the lack of information on the original soil nutrient status, soil moisture, etc. Other essential requirements often omitted are: location (latitude and longitude), altitude and rainfall distribution.

A major factor initially affecting the level of fertilisers applied to sunflower was the relatively low profitability of the crop, but with higher-yielding cultivars, expanding oilseed demand and price increases, fertiliser use has become more general. In the absence of local trial results, an initial guide for sunflower would be to use the NPK recommendation for cotton or hybrid maize, with the nitrogen content reduced by at least half. On acid soils, liming will increase sunflower seed yield, but in general this is not profitable.

Individual nutrients generally have little effect on seed-oil percentage and less on seed-oil constituents compared with the gross effects of cultural methods and environment. Seed size, weight or kernel-husk ratio can be affected, but again seasonal effects frequently modify such results. An interaction also occurs between fertiliser response and seasons, reflected in varying growth-rates, period to maturity and seed yield. Since new hybrids differ, often radically, from older cultivars in their response to cultural methods and fertilisers, it is unwise to rely on recommendations derived from trials with these older types when growing the newer, as the author found in East Africa after an interval of 20 years!

Nitrogen

Nitrogen (N) must be related to the availability of phosphate and potash, for if these are deficient, nitrogen will usually depress yield and seed-oil content. In general, nitrogenous top-dressing should be applied only when local trials have determined its profitability, and above 50–80 kg N/ha is seldom necessary. When nitrogen produces a yield increase in the field, it is usually by increasing the size of flower heads, less often by an increase in 1000 seed weight. Although nitrogen may decrease seed-oil content as later noted, the increase in seed yield increases total oil yield per hectare, and nitrogen application is usually profitable.

One application when plants are some 30 cm in height or at the eight-leaf stage is usually sufficient. When applied mechanically, the machine's clearance will normally determine the timing, which can be combined with a weeding. The type of fertiliser and method of application is of little importance providing the coverage is even and plants are not mechanically damaged. There is little data on field comparisons between the various nitrogenous fertilizers: in Russia, ammonium nitrate reduced infection by downy mildew, and ammonium chloride reduced growth; in Kenya there was no yield difference between ammonium sulphate and calcium ammonium nitrate (Weiss, 1966a).

On soils of average fertility, i.e. those which produce profitable crops of maize or wheat, small amounts of seedbed nitrogen can be useful to assist early seedling growth and thus increase uptake of other nutrients. This is most easily supplied as an NP combination i.e. diammonium phosphate. Anhydrous ammonia drilled before planting is widely used by growers of irrigated sunflower in North America.

Field application of nitrogen generally has little effect on seed weight, kernel : husk ratio, or oil constituents, although there may be a slight reduction in seed-oil content. However, some of the new high oleic acid content cultivars show a direct correlation with nitrogen levels, the oleic acid percentage in seeds increasing with increasing nitrogen, while the linoleic acid percentage remains steady or falls. Thus it is possible to influence the oil type produced.

Protein accumulation in seeds appears to be directly related to soil nitrogen availability, and this affects viability and seedling vigour; poor germination is frequently the cause of low sunflower stands and therefore low yields. Emergence of seed grown from plants supplied with varying amounts of N and P is shown in Table 7.8 (Srivastava, 1978). Inadequate soil nitrogen is probably also responsible for a proportion of unfilled seeds in the centre of sunflower heads, with a soil moisture deficit a lesser factor.

Table 7.8 Emergence of seed produced from plants grown under various nitrogen and phosphate levels – India.

Nitrogen	Phosphate (kg/ha)				Mean
(kg/ha)	Nil	30	60	90	
Nil	45	48	45	40	45
25	60	65	72	56	63
50	75	81	85	71	78
Mean	60	65	67	56	

Values in body of table are percentage emergence.

Phosphate

Once the basic requirement has been supplied, additional phosphate (P) gives no significant increase in yield, while an excess adversely affects seed composition. Data indicates that an increase in phosphate tends to increase 1000 seed weight, but that kernel percentage frequently falls; a deficiency will adversely affect total seed-oil content. Where sunflower is grown in rotation with crops receiving an annual phosphate application, relatively small amounts at sowing are normally adequate. A basic seedbed application of 20–30 kg P/ha is common, except when using rock phosphate in which the nutrient is less readily available to plants in the season of application. The residual effect of this type of phosphate, however, is frequently similar to that of other phosphatic fertilisers.

The greatest increase in seed yield obtained from phosphate has invariably occurred in the presence of adequate potassium and this interaction is important in commercial sunflower production. Single superphosphate, which often has a higher sulphur content, can give highest yields per unit of phosphate on soils where sulphur is deficient, and in Australia superphosphate with a zinc content of 800 ppm is recommended where there is a zinc deficiency.

Potassium

Potassium (K) is essential for healthy growth and high yield, and in its absence response to other nutrients is reduced. As a general guide, the amount of P and K should be approximately equal, unless sunflower is grown in a rotation with crops requiring a large amount of potassium, when it will be necessary to determine the optimum application rate, since the residual effect of previous applications can be substantial. In some circumstances potash alone will produce good yields, for instance where there are high soil phosphate residues. In general, absence of adequate nitrogen and phosphate will reduce the effectiveness of applied potash.

Little information is available regarding the response to a particular potassic fertiliser, but as sunflower shows a response to calcium it is probable that commercial potassium chloride or potassium sulphate fertilisers with an average calcium content of 18% and 14%, respectively, would be superior. Mature sunflower plants contain significant quantities of potash, which should preferably be returned to the soil. In smallholder crops, shelled heads and stalks should be burnt and the ash used as fertiliser. In pot trials, K increased seed oil content and number of whole seeds per head, but there are no field data in support.

Sulphur

Sulphur (S) is deficient in many tropical soils, but the amount present in some superphosphates is often sufficient to alleviate all but a major deficiency. In India, 0, 20, 40 or 60 kg S/ha as gypsum, ammonium sulphate or single superphosphate increased seed yield with increasing S, and ammonium sulphate produced the highest yield of 1.32 t/ha. A substantial residual effect, greatest with gypsum, was noted on following crops of green gram, safflower and sorghum (Sreemannarayana & Raju, 1993; Intodia & Tomar, 1997). Sulphur may increase seed protein content, but at above average levels can also reduce oil content.

Zinc

Lands newly levelled for irrigation may require small amounts of zinc to correct an imbalance caused by topsoil dispersal. In Australia this deficiency is frequently alleviated by the zinc content of local superphosphate; elsewhere zinc can be mixed with superphosphate at the rate of 1–2 kg/ha, or a foliar application of zinc sulphate heptahydrate and urea at about 1 kg/ha each in 100 litres of water is usually effective.

Boron

Sunflower is highly susceptible to a boron deficiency; symptoms in young plants are a thickening of the stem apex where brown lesions appear, and upper leaves may die. On more mature plants the stem is deformed below the head before and during flowering and be so weakened it fractures. There can be cultivar differences in the visual symptoms of a boron deficiency, although the rate of uptake, and total amount may be similar. A deficiency can be alleviated by spraying or adding borax to fertilisers, and up to 12 kg/ha may be necessary. In areas of known boron deficiency, as in Yugoslavia (Plesnicar *et al.*, 1997), fertilisers are supplied with the amount locally required.

Molybdenum

On acid soils, pH 4.0–5.0, a molybdenum deficiency can occur. Symptoms are pale plants with stunted growth; leaves develop interveinal yellowing, margins die back, and leaves assume a typical cupped shape.

A deficiency may be alleviated by liming, or sodium molybdate as a spray, mixed with fertiliser, or as a seed dressing. Rates of 100–400 g/ha sodium molybdate are normally adequate, and on very acid soils must be applied to each crop. A large application of ammonium nitrate can induced a molybdenum deficiency.

Other elements

Sunflower generally does not respond to the application of minor or trace elements, unless there is a gross deficiency, and foliar analysis is necessary to determine the status of individual nutrients. Conversely, residual copper resulting from spraying tobacco with fungicides for instance, can cause local toxicity problems; aluminium toxicity has also been reported. Applying lime to supply calcium may occasionally be profitable; in southern Australia gypsum up to 4 t/ha is commonly applied to acid soils to increase the rate and regularity of emergence.

Sunflower frequently accumulates manganese in foliage to levels which are toxic to other crops, and why this is so has not been fully determined. Toxicity symptoms are similar to those caused by molybdenum.

The minor effects of magnesium, iron and other elements on plant growth and yield have been described. Since these data were obtained from small experimental or pot trials, their relevance to field conditions should be confirmed locally. Heavy metal contamination of soils is becoming more common, and in Hungary cadmium accumulation and distribution in sunflower was determined, together with its effect on plant growth (Simon, 1998).

CULTIVATION

The cultivation necessary for maize or wheat is adequate for sunflower, and provided drainage is not impeded there are no special requirements; deep or chisel ploughing is necessary only to break up impacted layers and allow unrestricted root development. In regions where wind erosion can be a hazard, cultivations should aim to preserve a somewhat rough soil surface, as in the Ukraine (Nikitchin *et al.*, 1995).

Since sunflower seeds vary considerably in size, those used for mechanical planting should preferably be graded to produce an even sample, to allow free flow in seeders and even stands. Germination will also be more even, allowing more efficient weed and pest control. The optimum plant population is essential for profitable yield, since sunflower with its single head is less able to compensate for gross variation in plant number than, for instance, soya or groundnut (Table 7.9).

Combine drills, maize or cotton planters with suitable sunflower seed plates can be used, but precision seeders will ensure even depth of planting and in-row spacing. Uniform spacing ensures uniform growth, and thus ease of working for cultivators and harvesters. When using electronically controlled seeder units, depth of planting and seed rate are very even, and grading of seed may be unnecessary.

Although seed grading is not essential, since large or small seeds of equal quality produce similar plants at maturity, clean, bold seeds should be selected since they germinate evenly and produce vigorous seedlings, allowing earlier and more accurate inter-row cultivations. Very large seed however, may have low embryo vigour (Saranga *et al.*, 1998). Whenever possible fresh seed should be sown, since sunflower seed rapidly loses viability when improperly stored. As a field guide to final stand, most modern cultivars have 80–85% emergence 2–5 days after first seedlings appear. Seed should be dressed with a fungicide and preferably an insecticide; the local groundnut or maize seed dressing is suitable. Dressing seed some months before planting helps to retain viability and reduce fungal infection. However, some insecticides may reduce germination if the seed is retained in storage after treatment.

Seed should normally be planted 3–8 cm deep, depending mainly on cultivar, and soil in the vicinity of the seed must remain moist throughout germination. A combination of dry topsoil and moist subsoil is most common in many dry-land areas at the optimum time for sowing sunflower, and deep sowing, 10–20 cm, is possible provided the furrow opened is not completely filled. On light soils a press wheel should be set to give a firm seedbed, since this will ensure that the soil does not dry out too quickly and the small furrow made will stop soil blowing off the seed. This technique should also be used in areas of low rainfall to assist moisture penetration and absorption. However, on more clayey soils too heavy a press wheel loading will cause emergence problems, and in Australia it was found necessary to reduce axle loading by half for this reason.

Table 7.9 Sunflower sowing guide.

Row-width (cm)	Desired population (plants/ha)				
	28 000	32 000	40 000	50 000	75 000
35	75	66	53	43	28
45	60	51	41	33	21
53	50	44	35	28	19
60	44	39	31	25	16
68	40	35	28	21	15
70	37	33	26	21	14
75	36	31	25	20	13
83	33	29	23	18	13
88	30	21	16	11	11
90	29	25	20	16	11
100	27	24	19	15	10

Note: Values in the body of the table are distances between plants in centimetres.

Water injection at planting is becoming increasingly popular with dry-land farmers, and special machines are available which place a narrow band of water under or around the seed at planting, to ensure quick, even germination and rapid initial growth. Speed of sowing is reduced by some 25% since 500–2500 litres water/ha are required.

A seedbed temperature of 25–30°C is preferable, although higher temperature is tolerated in the presence of adequate moisture. Below 15°C generally retards germination and below 10°C inhibits many cultivars. High oil-content seed is more susceptible to low temperature damage than seed with low oil-content, especially when germination has commenced. Cultivars have been developed that will germinate at very low temperatures, to 5°C, but the period from sowing to emergence is extended.

When using standard planters with 75 cm row-width, seed rates will vary from 3 to 8 kg/ha depending on seed size, to achieve a final spacing of approximately 30 cm between plants. Seed rates can be calculated from the following formulae, or reference to Table 7.9.

(a) Seed number / ha

$$= \frac{\text{desired population / ha} \times 100}{\text{expected germination percentage}}$$

(b) Seed required, kg

$$= \frac{\text{seed number / ha} \times \text{area to be sown (ha)}}{\text{seed number / kg}}$$

(c) Number of seeds required / 10 m of row

$$= \frac{\text{seed number / ha} \times \text{row-width (mm)}}{1\,000\,000}$$

Where spacing is critical, tests should be carried out on a firm surface to allow machines to be calibrated. On large-scale commercial plantings, or under irrigation, optimum seed rates must be determined by formal trials, since both short and tall cultivars may be grown. These trials should preferably be confounded with the date of planting, since there can be a significant interaction; however, a higher seeding rate cannot compensate for sowing outside the optimum planting time. The practice of sowing sunflower on lands that remain unplanted to maize or wheat is unsound, since the yield tends to fall significantly and quickly outside the optimum period (Table 7.10) (Weiss, 1966b; Barkla & Pritchard, 1980). Similar results have since been obtained in many other countries.

Correct time of planting not only affects yield but also the kernel : husk ratio. Seed weight tends to decrease with delay in sowing, and the

Table 7.10 Effect of time of sowing on sunflower seed yields.

Australia		Kenya	
Date	Yield (kg/ha)*	Date	Yield (kg/ha)†
27 Aug.	2522	4 June	986
27 Sept.	3640	26 June	885
25 Oct.	3345	17 July	549
22 Nov.	3427	7 Aug.	235
20 Dec.	3192		
18 Jan.	2512		
15 Feb.	790		

* Mean of four hybrids; † mean of four varieties.

greatest loss is in the kernel. However, sowing outside the optimum period may be necessary in some areas to avoid flowering at periods of high temperature, to obtain a particular grade of oil.

Within the optimum sowing period, plant population can vary to a small extent as plants produce larger heads at wider spacing (Table 7.11) (Barkla & Pritchard, 1980). However, large heads also tend to have more pops, and yield of oil per hectare may be significantly reduced. Seed size may also be affected, and thus the kernel:husk ratio. The effect of spacing on this ratio is shown in Table 7.12, and the rise and decline in kernel percentage directly affects the oil yield (Gaev & Gaponenko, 1996). Thus the optimum plant population must be determined in terms of the product required, oil or birdseed, and not solely by final seed yield. An added advantage is that the amount of seed required per hectare may be reduced, since hybrid seed is relatively costly.

A population of 30000–40000 plants/ha at harvest is acceptable for the tall varieties and 35000–60000 for hybrids in areas where annual rainfall of 500 mm or above is normal. Population can be 25–50% higher under irrigation in the same area, but the possibility of increased loss from lodging must be considered. Under dryland farming the plant population is usually between 12000 and 15000 plants/ha, mainly dependent on the rainfall expected during crop growth.

Where empirical estimates of optimum plant population are necessary, the size of heads at maturity is a good guide. If the average head measured across the seeds in an even crop is less than 12 cm, population is probably too high; above 25 cm it is too low; large heads may look good, but the centre seeds are often mainly pops. Populations in the range 20000–30000 plants/ha are common when sunflower is grown to provide seed for the confectionery trade or birdseed market. The large white or striped seed commands a price premium but is usually lower in oil content. Hybrids have been specially bred for these markets, but cultural practices are identical with crops grown for oil.

The introduction of hybrid sunflower in the late 1970s radically improved profitability, and long-term prospects for the crop. Hybrid sunflower has important advantages over open-pollinated varieties. It produces very high seed yield, has increasing seed-oil content, flowers over a short period and has a short, sturdy stem with a single head. All these characteristics increase the amount of mechanisation possible and add to its efficiency. Many hybrids are also drought resistant, or resistant to some insect pests or diseases as has been mentioned in the relevant section. Also important, hybrids can be self-compatible and not therefore wholly dependent on insects, especially bees, for pollination. Thus chemical control of insect pests and disease is facilitated, with less risk of low yield from killing insects while plants are flowering.

Hybrids of inbred lines have given the biggest increase in yield and oil content and are in general commercial use. Named hybrids of known performance are now available in many countries from reputable seed companies. Synthetic varieties, a combination of inbred lines, are also available and have produced high seed and oil yields.

The production of hybrid sunflower seed, although tedious, is not difficult, and is basically similar to the production of hybrid maize seed (Figure 7.10) is a light-hearted illustration of this procedure. Foundation seed is produced using a male sterile line, A, crossed with a male fertile line, B, which is again crossed with a male fertile

Table 7.11 Effect of plant population on sunflower seed and yield – Australia.

Plants/ha	Seed yield kg/ha at 9% moisture	Harvest index	100 seed wt (g)	Plant height (cm)	Head diam. (cm)	Lodging score	Oil (%)
20500	2310	0.40	8.98	123	21	1.0	42.6
41300	2960	0.45	7.02	122	16	1.0	47.3
62200	3290	0.44	5.83	133	14	1.3	48.3
81800	3110	0.43	5.17	126	12	1.5	48.2
100600	3020	0.42	4.70	134	12	1.7	48.2
148600	2930	0.39	4.12	129	10	2.3	47.9
194900	2640	0.36	3.69	132	9	2.8	44.3

Values in table are mean of three varieties. Lodging score 1 = erect, 5 = horizontal.

Table 7.12 Effect of plant population on sunflower seed characters – Kenya.

Spacing in 85 cm rows (cm)	Kernel (%)	100 seed wt (g)
30	52.6	7.6
35	54.5	7.7
45	55.4	7.9
55	54.3	8.1
60	53.8	8.3

Note: Values in table are mean of four varieties.

line, R. Since these may have different maturity periods, planting dates must ensure both flower together. An isolation distance of 3 km is a normal requirement to ensure purity. Roguing is usually necessary, and checking blocks for any increased resistance to pests or disease recommended, since some major advances in these areas have been developed from odd plants noticed in breeding fields.

Many sunflower cultivars are open-pollinated with bees usually the main agents, and to ensure a high seed set there must be a high insect population. In Australia for example, only 11% of florets on heads from which bees had been excluded produced seed, compared with virtually 100% on non-bagged heads. A small number of empty seeds, pops, can usually be found in the centre of most sunflower heads, but a high proportion of such unfilled seed will drastically reduce oil yield per hectare. Where large areas of

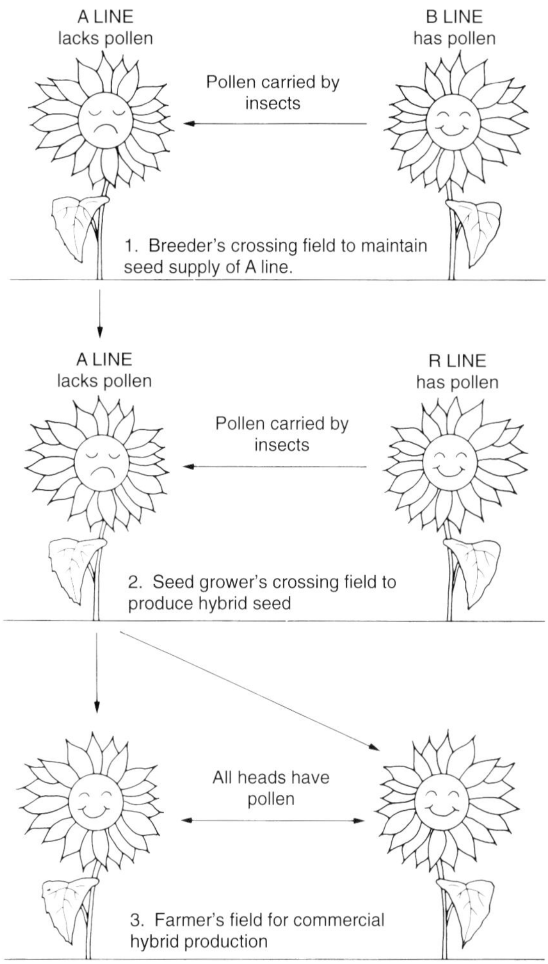

Fig. 7.10 Cytoplasmic male sterile restorer method of hybrid sunflower seed production.

open-pollinated sunflowers are grown in pure stands, the fertilisation percentage as measured by the number of sound seeds per head decreases rapidly from the edge to the centre of fields, and is directly related to the number of bees active in the crop.

Since local wild bee populations are seldom adequate to pollinate extensive sunflower fields and beekeepers state that bees seldom fly more than a mile or so from their hives, it is strongly recommended that hives be placed at strategic positions. In many tropical areas bees may be restricted in their working hours by high temperature; at 35°C or above, pollen and nectar gathering tends to be progressively reduced as workers are diverted to water collection and hive cooling. In India three bee species were most active around 10.00 and 13.00 hours (Sharma & Bichoo, 1996). A minimum of two to four strong colonies per hectare of crop is recommended in Russia and France, and in Australia one bee to four sunflower heads is considered the minimum to ensure complete pollination; generally the number of bees must equal the plant population. In developed agricultural areas contracts can usually be arranged with professional beekeepers, but in less-developed regions it may be necessary to introduce beekeeping as part of normal agricultural operations, as in Zimbabwe (Coleman, 1997). There is usually little difficulty in so doing, for the honey produced is always readily saleable, or consumed domestically.

Insects as pollinators are less important on many current hybrids, but data indicates the presence of bees in commercial crops significantly increases seed yield. The decision to hire or maintain hives must be related to the cost of so doing compared with the increased return, since sunflower is a relatively poor nectar producer and the pollen low in protein. By servicing a sunflower crop the beekeeper may thus have to forgo a more profitable honey producer, which in Australia will be the red gum, *Eucalyptus camaldulensis*.

WEED CONTROL

The yield loss in sunflower crops due to weed competition can be substantial but is seldom quantified. In Italy, 38 weed species reduced seed yield by 17% to 40% depending on the local weed complex (Covarelli, 1996), while in East and Central Africa the author frequently encountered inadequately weeded smallholder crops virtually overwhelmed by rampant grasses.

Sunflower seedlings are poor competitors to weeds, and young plants are easily damaged by mechanical cultivation. For this reason sunflower should not be used as a breaking crop on new land with its high weed potential, and large amount of roots and other debris. The minimum tillage techniques very successful with soyabean, have been less so with sunflower, due mainly to the increased weed control necessary in young plants, with consequent damage to developing roots.

Weed control in sunflower should ideally begin with the previous crop. Where a persistent weed occurs which is difficult to eradicate in sunflower, the previous crop should be one in which this weed can be controlled. Conversely, sunflower can be used as a cleaning break, and in Australia the Bathhurst burr (*Xanthium spinosum*) is a major cereal weed easily controlled in sunflower. Pre-planting operations should aim at a weed-free seedbed, and post-emergence tillage should be as shallow as possible, with implements set to throw soil towards the rows as soon as seedlings are strong enough to withstand this treatment. Sunflower's rapid growth quickly produces a heavy shade which reduces the numbers of annual weeds, although perennial grasses tend to be more tolerant.

Mechanical weed control is normal in commercial crops, with 20–45 days after emergence the most critical time, but the period of effective machinery use is limited by sunflower's rapid growth or unsuitable soil conditions. In Argentina where sunflower is an important oil crop, adequate mechanical weeding is made difficult by seasonal rains, with consequent yield reduction. Over-the-top harrowing with fixed or spring-tined weeders is possible in colder areas, but is not to be recommended in warmer regions where plants tend to be more succulent. Spring-tined or rotary weeders are most effective when used between rows of young seedlings and are least damaging to developing roots.

Pre-emergence herbicides are effective but should be used with care, as results are unpredictable and there are also cultivar differences. Sunflowers are extremely susceptible to hormone herbicides, and these compounds should not be used in or near sunflowers. The following herbicides have been used pre-emergence in commercial sunflower crops either singly or in combination; chlorpropham, dinitramine, EPTC, glyphosate, linuron, monolinuron, pendimethalin, prometryne, simazine and trifluralin, some of which must be soil-incorporated before sowing.

Rain-grown and irrigated crops differ in their reactions to herbicides, and also the amount of active ingredient necessary to control weeds. In general, irrigated sunflower appears to be more susceptible to herbicide damage at the same active ingredient level than rain-grown, and this is probably due to leeching of the chemical into the main root zone. The author found hand weeding, where possible, invariably gave the highest yield when compared with chemical methods of weed control; cf. castor and sesame.

A major parasite of sunflower in Russia and to a lesser extent in Eastern Europe and Turkey, is broomrape, *Orobanche cumana*, but it is of little importance elsewhere, though it attacks safflower in Afghanistan. At least 3–4 years between crops is recommended in Russia, since yield is reduced by 20% when there are four shoots of the parasite per sunflower plant. The leaf miner *Phytomyza* spp. has been used to control *Orobanche*, while partially resistant strains of sunflower have been developed in Russia and Bulgaria. Resistance was linked to plant mineral levels in Turkey (Anac *et al.*, 1996). Another parasite, *Cuscuta indecona* (large-seed dodder), which occasionally occurs in sunflower fields can be controlled by herbicides.

Witchweed (*Striga* spp.), a major parasite in many tropical countries, does not attack sunflower, although its root exudates cause witchweed seeds to germinate. Sunflower can thus be valuable in a rotation where *Striga* is a problem, and materially assist in its reduction. In West Africa for example, the *S. hermontheca* seedbank (18 seeds/100 g soil) was reduced about 30% by sunflowers compared with an average of 12% by a fallow. Yields of maize and sorghum were significantly increased by preceding crops of sunflowers compared with yields after cereal monoculture (Sprich, 1994).

IRRIGATION

Extremely high yields can be obtained from irrigated sunflowers, and with suitable cultivars and watering techniques the crop can be very profitable. Probably the most important factor is to obtain optimum yield per unit of water use by control of watering schedules, and some system of accurately determining water need or soil moisture level is essential. Sunflowers do not wilt under moisture stress as quickly, or as obviously as many other irrigated crops, and visual assessment of water need is unreliable. This may not be obvious at harvest, and only become apparent after detailed examination of seed characteristics, i.e. percentage hull, oil content, etc. Another important factor is time of planting, which can differ significantly between rain-grown and irrigated crops. In Zimbabwe for example, rain-grown sunflower is sown in late December but irrigated crops in May.

A major disadvantage of irrigated sunflower compared with other crops is height of plants in relation to a relatively restricted root system. When grown under irrigation in areas where high winds are common (and this combination occurs in many countries suitable for sunflower), plants are easily uprooted because wet soil gives little support. Since the last watering is usually applied as the heads are maturing to ensure the seeds are filled, plants are exposed to maximum leverage on the roots when most vulnerable (Fig. 7.6). Lodging losses can be very heavy even in smallholder crops, since the heads fall on to wet soil and may rot before collection. Short cultivars should be used whenever possible, even if seed-oil content may be lower than taller local cultivars. Seed-oil and protein content of some hybrids may be higher when grown under irrigation than when rain-grown.

Total water use and water-extractive ability of cultivars can be similar but seed yields may differ, apparently due to a difference in internal

Fig. 7.11 Irrigated giant sunflower, Tanzania.

water-use efficiency. This variation in plant response to irrigation emphasises the importance of local cultivar/water use trials before a final selection of either is made. Short cultivars have the additional advantage of allowing the soil to dry out more quickly before harvest since there is less cover, and harvesting machinery can be used earlier than would be possible with taller types, a considerable asset where bird damage is appreciable.

As a general guide, water use by sunflower is about twice that of sorghum in most localities where the two are grown. In Chile evapotranspiration of sunflower was 706mm, with a maximum of 8.5mm/day, and in Romania 512 mm. When calculating the water need of well-grown plants, isolated rainfalls of 10mm or less should be ignored, since this will normally suffice for only 1–2 days and there is little to be gained by disrupting irrigation schedules.

When following a heavily watered crop, or where there is naturally high subsoil moisture, sunflower will produce good yields with only 300–500mm of added water. In normal conditions, water use by sunflower is mainly confined to the upper 150cm of the soil, but roots frequently penetrate to at least twice this depth. When surface-irrigated, 600–750mm is considered the minimum including rainfall, and generally 75–85% water utilisation is from the 0–60cm layer. A watering before sowing and as heads are forming is essential, with at least two other applications at appropriate intervals between.

Data indicates that water stress at anthesis results in a greater reduction in seed yield than stress in later stages of development. Additionally, oil content and kernel percentage can be reduced. The most obvious field symptom is an increase in the number of unfilled or aborted seeds in the centre of heads, although inadequate pollination or soil nitrogen may also be partially responsible. Field data indicates a final irrigation 14–21 days after 50% of heads have flowered gives maximum yield; later waterings tend to keep the surface wet, and cause problems for harvesting equipment.

It is very important to water at stated intervals rather than when plants show stress, as is common with other crops. On irrigation schemes in Africa under the author's supervision, watering alternate furrows reduced lodging and water loss, and when this procedure became necessary due to a water shortage, gave higher yields per unit of water used than watering all furrows at longer intervals.

Although many tall types of sunflower are considered drought resistant, the potential yield of newer cultivars is quickly reduced by a water stress. Experimental data indicates a reduction in photosynthesis and wilting at relatively low soil moisture deficits, but since the majority of these trials were in greenhouses or pots, their relevance to field conditions should not be taken for granted. However, work on the diffusive resistance of sunflower leaves has shown up to 50% differences between cultivars, and could be an important characteristic when selecting types for arid areas. A stress tolerance index to evaluate cultivars for drought tolerance developed in Iran, is partly based on leaf characteristics (Mozaffari *et al.*, 1996).

Chemicals can also reduce the effects of moisture stress; chlormequat applied to plants in the 10-leaf stage significantly reduced water use in the following 21 days. Although this may be of little commercial value, it could be of assistance on special seed or breeding crops.

Excess moisture is detrimental to sunflower at any growth stage as noted. Soil type can also directly influence the degree of damage, and sunflower crops grown by the author on more sandy soils were seldom affected to any degree, even in those areas where high-intensity tropical storms frequently flooded ridged fields. Sprinkler irrigation is now common, especially when short hybrids are grown; a disadvantage is an increase in fungal infection by building up in-crop humidity.

Saline water is generally unsuitable for irrigating sunflower, although there are cultivar differences in response to salinity levels (Ashraf & O'Leary, 1997). Where it must be used, moderately saline water can reduce seed-oil content by up to 50%, and also tends to increase the severity and incidence of root rots (Table 7.13) (Raju and Ranganayakulu, 1978). The planting and watering techniques used for irrigating safflower in saline environments are equally effective for sunflower. Another common constituent of soils and irrigation water is fluorine, which can accumulate under continuous irrigation, and depress plant growth. Contamination of water supplies by factory effluent or sewage wastes can be toxic to sunflower, or taint seed and oil with heavy metal residues.

Table 7.13 Oil yield of sunflower as influenced by saline irrigation water – India.

Treatment	Soil salinity EC × 10^3 (1:2) 0–15 cm layer	Oil (%)	Oil yield (kg/ha)
Least saline water available EC × 10^3:			
1.1	0.5	41.1	5590
Salinity level of:			
2.0	0.9	41.3	4980
4.0	1.2	39.5	4180
6.0	1.6	39.0	3760
8.0	2.1	38.2	2960
12.0	2.1	37.1	2790
16.0	3.2	35.8	1920

EC in mmhos/cm (SI units 1 mmhos/cm = 1 × 10^5 siemens/m).
Note: Sandy coastal soil, pH 8.2, initial salinity level 0.3.

INTERCROPPING AND ROTATIONS

Sunflower is invariably intercropped by smallholders, with the proportion it occupies reflecting the needs of the grower. It is frequently sown in clumps in fields of groundnuts, beans, or as part of a mixed crop grown to supply food, when it acts a a living support for beans and gourds.

When grown as a cash crop it is often sown in strips between groundnuts, pulses, and millets, but only in those districts where the seed price is higher than the crop it displaces (Jadhao *et al.*, 1994; Shahbaz & Ibrar, 1996). Raingrown cultivars are usually giant types, but in India dwarf hybrids are preferred under irrigation. Giant types may be used as windbreaks around field margins, and the smaller cultivars can frequently be seen growing on banks around irrigated fields.

Sunflower is frequently grown in rotation with cereals, soyabean, safflower, sorghum and maize, but should not be sown for more than two consecutive crops. It may also be included to enable cereal weeds to be controlled, to take advantage of its ability to reduce the numbers of parasitic weeds, or to avoid the build-up of pests and diseases, as noted in the relevent sections. In Russia and North America rows of giant sunflower may be used as snow barriers.

Intercropping sunflower can also significantly increase the number of pollinating insects; interplanting with niger in India nearly doubled sunflower yield due to the greatly increased insect population attracted to niger. A similar effect was noted in Indonesia with intercropped maize and sunflower, provided both were of a similar height. Intercropping with forage legumes in the USA reduced soil erosion, with no adverse effect on seed yield or oil content (Kandel *et al.*, 1997).

HARVESTING

Tall cultivars are usually ready for harvesting in 120–160 days and short types at 90–120 days, the variation being mainly due to seasonal effects, especially temperature and soil moisture availability. Evenness of growth, ripening and moderate height are essential requirements for successful mechanical harvesting. Thus it is essential to obtain guaranteed seed of a known cultivar.

Leaving mature plants standing in the field because many others are unripe can increase bird damage, which is a major cause of seed loss in many countries. Seed is also lost from semi-ripe heads by shattering, caused by a combination of high temperature and dry wind. Delayed harvesting also allows unwelcome changes in seed-oil composition, for once seeds are physiologically mature, free fatty acids steadily increase.

Time of harvest may also be dictated by buyer specifications, which can require seed to have a certain moisture content, a minimum of admixtures, a minimum oil content, and oil with a stated minimum of certain constituents, for example linoleic acid. All these requirements are within the control of growers, who must monitor their crop to ensure compliance. Substantial penalties usually apply to seed that does not meet the specifications.

Plants are ready to harvest when the back of heads turn yellow or brown, and seed moisture content is 10–12%. Selective manual harvesting is usually possible only on smallholder crops unless labour is cheap and plentiful (Figs 7.12, 7.13 and 7.14). Where labour is available, plants

Fig. 7.12 Harvesting smallholder sunflower, India.

Fig. 7.13 Threshing and winnowing sunflower, India.

Fig. 7.14 Drying harvested sunflower heads, Central Africa.

can be cut and stooked in the field until dry, then threshed by a harvester moving from stook to stook. The heads are cut from the plants and thrown into the tray of the machine, from which the cutter-bar and reel have been removed; this may increase harvesting costs, but substitutes for an even stand. Alternatively, the heads are cut from the plants, the stem is lopped about 1 m from the ground and the heads spiked on the stems, seeds down, to dry. Seed heads can also be threshed with modified hand or mechanically powered maize shellers provided these are accurately adjusted.

Chemical defoliation or desiccation using magnesium chlorate or dipyridyl phosphate can accelerate drying of standing crops and materially assist combining. However, this operation requires aircraft, for ground application is usually extremely difficult. Seed-oil content of defoliated plants is similar to untreated, and germination is unaffected when this operation is correctly timed, i.e. when seed moisture content is approximately 35–40%; around 30% on irrigated crops. Too early application results in loss of yield, oil content and oil quality.

Grain combines can be fairly easily adapted to harvest or thresh sunflowers (Fig. 7.15); a front-end modification known as the 'head snatcher' considerably reduces green material reaching the drum, and allows a lower setting of the cutter-bar (Fig. 7.16). Accurate fitting and operating is essential, since losses at the cutter-bar can reach 10% of the crop. Seed trays are also necessary to reduce cutter bar losses. Modifications may also be necessary to screens and straw walkers to avoid blockages and the risk of fire.

Provided the recommended speed, drum setting, and attachments are used, direct combining of sunflower requires no special skill. It is essential accurately to relate the forward speed

Fig. 7.15 Combine harvesting sunflower.

of the machine to each crop, and settings will usually vary during the day; fast speeds overload elevators and the threshing drum. Crops grown for seed must be handled with much greater care than commercial crops, since even minor seed-coat damage will reduce germination by 25%, and seed crop production left to specialists.

Although tanker harvesters are in general use they are not recommended in less-developed areas, since small portions of the head, which cannot easily be separated from the seed without excessive loss, cause heating in bulk. Bagging allows the contents to dry more quickly, and makes transport easier where bulk movement of crops is also uncommon.

Threshed seed should be cleaned as soon as possible and dried to less than 10% moisture or oil quality will suffer. In hot countries, standing sacks with the necks open preferably under shade, is usually sufficient; elsewhere artificial drying may be necessary and cold air is generally effective. A review of harvesting and post-harvest techniques for principal foodstuffs, including oilseeds, in developing countries has been reviewed (Lucia & Assennato, 1994).

Threshed sunflower heads are usually burnt or ploughed in after harvest, but are occasionally ground and fed to stock, although they have limited food value. Analysis of Egyptian sunflower heads gave the following percentage composition: ash 12, lignin 12, crude protein 12, glycosides 0.8, total lipids 2, unsaponifiable lipids 14, total sterols 1, pectic substances 12, galacturonic acid 8, galactose 3.4, glucose 1.7, arabinose 0.6, xylose 0.7, rhamnose 20 (Edrees *et al.*, 1976).

It is generally unprofitable to grow commercial sunflower crops that yield less than 3 t/ha unless there is a specific market available, for

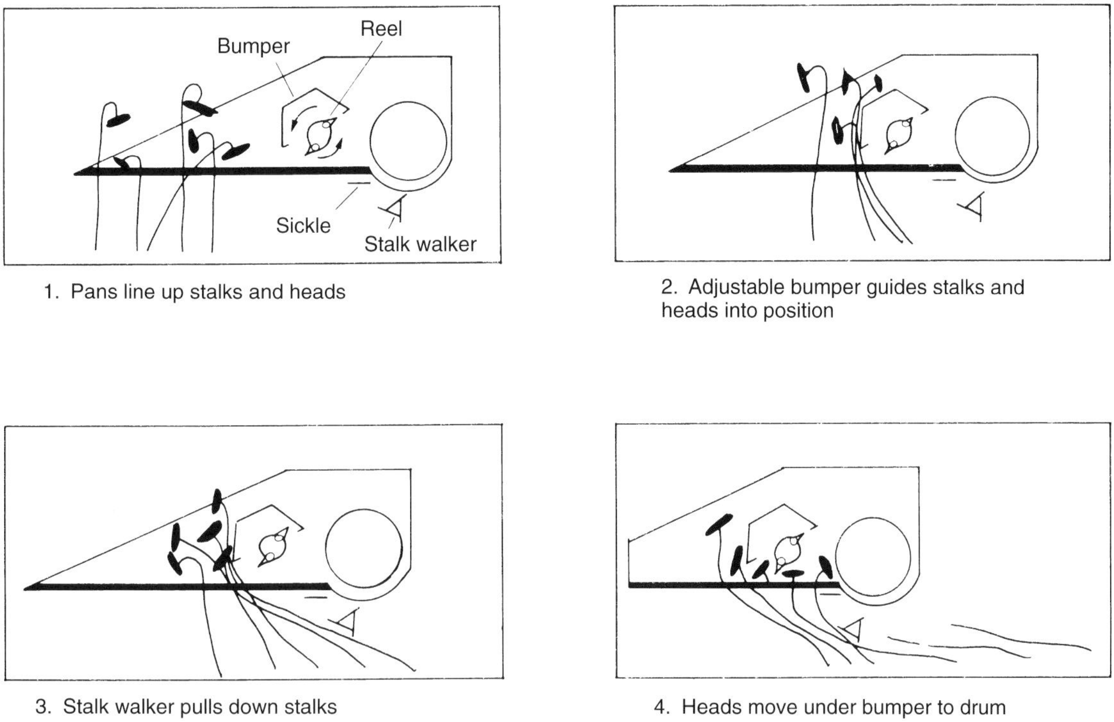

Fig. 7.16 Head snatcher attachment for sunflower harvesting.

instance as birdseed or for roasting. For crushing, high seed yield of high oil-content seed is essential. The author believes that an average yield of 4 t/ha is possible using current hybrids, and any commercial grower averaging less than 3 t/ha should review his methods, varieties or location. It may be necessary to change some or all of them!

Yields of over 5 t/ha have been reported from Russia, but 1.5–3 t/ha is more usual; 1.5–4 t/ha is commonly achieved in the USA and 1.0–3.0 t/ha in Australia. In India around 1.0 t/ha is considered very good, with less than 0.5 t/ha more common. Higher average figures are quoted in Chile, Brazil and Argentina, where some large growers normally harvest 2–3 t/ha. In southern and eastern Europe, 2–3 t/ha is a good crop, with half this more common. In Africa yields are generally low, except in plantation crops when 1.5 t/ha is normal.

A general increase in average field yield can be obtained with improved agricultural methods, since environment has a greater effect on seed and oil content than genetic factors. However, problems of hulling sunflower seed remain, since it is the difficulty of separating hulls and kernels that limits production of flours and protein isolates. Breeding cultivars whose seed has a space between the hull and kernel without loss of oil content, would materially assist seed processors, and increase the commercial attractiveness of sunflower seed in the crushing industry.

STORAGE

Sunflower seed dries easily, either in-sack or artificially, and once dry to below 9% moisture content, stores well but is susceptible to insects, vermin and other pests. Storage time for seed with high moisture content is critical, at 25% it is negligible, at 16–18% should not be more than 50–60 hours. On-farm storage must take these factors into account, or losses can be extremely high. The author has seen bins of sunflower seed which had not been properly cleaned before filling and not protected by insecticides, literally heaving from the activities of insects and mice. Seed with an increased phytomelan layer is less susceptible to insect damage in store, as is low oil-content seed. Since the seed is bulky, extended on-farm storage is unprofitable, and seed should be moved as soon as is practicable. Storage pests are discussed in the section on pests.

Seed required for special purposes or breeding, should be stored in metal containers rather than polythene or paper bags, and placed in cold storage. This is especially necessary in periods of high temperature combined with high humidity, for in these circumstances viability rapidly diminishes. Large seed apparently maintains its viability longer than small seed under all conditions. When insecticidal treatment of bulk seed is not possible, or undesirable, forced aeration with cold air can substantially inhibit insect infestation. The factors affecting seed storage in Romania are probably applicable to other eastern countries (Beratlief & Iliescu, 1997).

A range of fixed or mobile dryers is now available to producers, including fixed batch, recirculating batch, and continuous flow dryers, using hot or cold air. When hot-air drying is necessary, a temperature above 50°C tends to reduce viability, but has little effect on seed constituents; too low a temperature increases cost of the operation. Higher temperature to 80°C can be used for seed destined for crushing, with no increase in free fatty acid content. However, data are conflicting on the effect of drying temperature on viability, and seed from individual driers should be tested to establish the upper temperature limit. Germination tests should always be carried out prior to planting, since some temperature effects may be long term.

PESTS

Assessments of crop loss caused by insects or disease in sunflower are limited, but indicate the size of the problem. When *Helicoverpa (Heliothis) armigera* and other insects were controlled in India, yield rose from 909 kg to 1378 kg/ha, and the loss caused by *Helicoverpa* spp. alone was calculated at 120 kg/ha seed (Panchabhavi & Krishnamoorthy, 1978). On a world basis the calculated loss due to pests, disease and weeds in 1967 was nearly 2 million tonnes with an estimated market value of $240 million, as shown in Table 7.14. Since these figures were from older cultivars whose potential was considerably lower than current high-yielding hybrids, the loss from pests is now correspondingly greater. Irrespective of the relative accuracy of these assessments, the yield loss caused by pests and disease has reached billions of dollars and their control not only more desirable but more profitable.

Complicating the effective control of insect pests is the lack of detailed studies on the degree of infestation requiring control measures. In Australia for example, some growers spray when there are 6–10 Rutherglen bugs (*Nysius vinitor*) per plant, others when there are around 50. Similar instances can be quoted for other pests, or other countries. Adding to the problem is the considerable variation in damage caused by an insect in different countries, or at different stages of plant growth. This information is important, for extensive data indicates a narrow margin between the insect population a plant can toler-

Table 7.14 Assessment of crop loss due to pests and disease (thousand tonnes).

	Actual production	Potential production	Losses due to:			Total loss
			Insects	Disease	Weeds	
Africa	120	156	16	4	16	36
Asia	170	221	22	7	22	51
Former USSR and China	6056	7865	787	236	786	1809

ate, and that which is economically damaging. The effect of defoliation has been noted in the botany section.

Few insects are specific pests of sunflower, and an insect which may cause economic loss in one region is often of minor importance elsewhere. A comprehensive bibliography of sunflower pests, which remains relevant, is that of Rajamohan (1976), which also quotes references to many individual countries. With the increased interest in growing sunflower, Departments of Agriculture in many countries have produced lists of locally important pests.

Sunflower is generally insect-pollinated, principally by bees, and chemicals toxic to these insects should not be used once flowering has begun. This applies equally to newer cultivars and hybrids, which still require the presence of insects to realise their full yield potential. Spraying in the late afternoon will cause fewest casualties among bees, which are usually most active between 8 and 11 a.m. Once seed has set, control of seed-eating insects may be necessary, but by this time bees are active elsewhere.

Trap cropping is of great value where the main infestation period of a particular pest is known. Strips or small plots on the boundary of the cropped area are sown in advance of the main crop, or sown with a crop more attractive to the pest. The trap-crop is then sprayed or destroyed thus reducing pressure on the main crop. This technique is particularly useful for smallholders or co-operatives, where the main date of sowing can be mutually agreed and supervised and where pesticides are little used. Similarly to castor, varying the date of planting can materially assist in limiting the damage from a major local pest, as in India (Ramesh *et al.*, 1998).

A major cause of poor emergence and gappy stands is the larvae of various cutworms, *Agrotis* spp., and of the tiger moth, *Amsacta lactinea*, in Asia; scarabs especially *Schizonycha* spp. in Africa and *Euphoria boliviensis* in South America; true and false wireworm, *Gonocephalum* spp. Mole and other crickets especially the widespread *Gryllotalpa* spp. Fig. 7.17, grasshoppers particularly *Zonocerus* spp. in Africa and *Chrotogonus* spp. in Asia; flea beetles especially *Diabrotica* spp. in South America. The sunflower beetle *Zygogramma exclamationis* is a common pest in the Americas, and the coumarin content of sunflower leaves is considered to reduce their attractiveness to the beetle (Roseland & Grosz, 1997). If soil insecticides are not used, increased seed rates or delayed thinning of smallholder crops can partially compensate for such loss.

Fig. 7.17 Mole cricket, *Gryllotalpa* spp.

Stem-borers are a widespread pest, and are important in specific crops in particular years. They appear to cause most damage in those areas where alternative hosts are scarce, or where sunflower is newly introduced into rotations with more favoured crops. For instance, *Agapanthia dahlii* larvae attack sunflower and safflower, and rotations including these crops could increase overall damage. The most frequently reported stem-borer in eastern Africa is *Heteronychus* spp.; in Russia and Asia *Melanagromyza* spp., *Pyrausta nubilalis* and *A. dahlii*; while *Rhodobaenus tredecimpunctatus, Strauzia longepennis* and *Suleima helianthana* are common in the Americas.

Foliage-eating insects are of importance mainly on immature plants, and heavy infestations require control. However, although defoliation can be visually severe, the effect on yield is often small, as noted in the botany section. Bollworm, especially *Helicoverpa* (*Heliothis*) spp. (Fig. 4.13), the various grasshoppers previously mentioned and specific locust species, for instance *Austracis guttulosa* in Australia, commonly cause some degree of damage.

Aphis and whiteflies are often found in large numbers on mature plants with little apparent effect, but *Aphis gossypii* and *Anuraphi helichrysi* can become locally important. The first is a widespread pest of many crops, the latter more common in Russia and eastern Europe. The white fly, *Trialeurodes vapovariorium*, can cause severe damage in Australia when present in very large numbers. Where sunflower is grown adjacent to areas of cotton, jassids may require control, particularly *Austroasca* spp. in Australia, *Empoasca kerri* in India and *Amrasca bigutulla* in Asia. Resistance to aphid and other insects has been recorded, and aphid resistance is most pronounced in perennial species and is probably related to their greater pubescence.

The most damaging pests are those that attack the bud, flower head or developing seeds, and heavy infestations are difficult and costly to control. Feeding punctures can also allow secondary infection by pathogenic fungi, in particular *Rhizopus* head rot. Probably the most widespread and important single pest is the sunflower moth *Homoeosoma* spp.; *H. nubulella* in Europe, Asia and China, *H. electellum* in North America and *H. heinrichi* in South America. Resistance to *Homoeosoma* exists in American wild *Helianthus* spp. apparently related to their higher levels of diterpenoid acids, which apparently restrict growth of the moth larvae. This resistance has been incorporated into oilseed cultivars. Also locally important in many areas is the cosmopolitan bollworm, *Helicoperva* spp. Larvae of these and similar insects cause most damage when feeding on developing buds and seed heads, and once concealed within the bracts or the fleshy portion of the head are very difficult to destroy; also some species have developed resistance to insecticides. Resistance to the sunflower moth exists in some cultivars, and to *Helicoperva* spp. in others.

Lygids and pentatomids cause damage in the majority of countries where sunflower is grown (Fig. 7.18) and can occasionally reach incredible numbers. For instance in central Tanzania in 1949, the author harvested one particular sunflower crop on which blue bugs, *Calidea dregii*, made up 25% of the volume passing through the combine drums! Infestations of this magnitude normally preclude growing any host crop, since control would be unprofitable. The cotton stainer, *Dysdercus* spp. also feeds on sunflower, and since sunflower is increasingly grown under irrigation in rotation with cotton, this bug could become more important. The andat bug, *Agonoscelis pubescens*, causes severe

Fig. 7.18 Typical pentatomid bugs: top, harlequin bug; bottom, *Calidea* spp.

damage in the Sudan in 'outbreak years'; the green shield bug, *Nezara viridula* (Fig. 5.11), causes some degree of damage generally in Africa and elsewhere, but *Dolycoris penicillatus* is more important in West Asia and Russia. In Australia the Rutherglen bug, *Nysius vinitor* (Fig. 4.14), and the related *N. clevelandensis*, are most damaging when the heads are forming, and in some years can require six insecticidal applications to control.

The development of seeds with a very hard testa by increasing the phytomelan layer can reduce the damage caused by insects to maturing heads in the field. It is possible that other *Helianthus* spp. can be a source of insect resistance that could be transferred to commercial varieties.

The leafworm *Spodoptera littoralis*, is generally damaging in Africa, but *S. litura* in Asia and Australasia, and *Rachiplusia ou* in South America; in Russia and eastern Europe the meadow moth, *Loxostege sticticalis*, and *Mamestra brassicae*; the hairy caterpillar, *Diacrisia obliqua*, in India and neighbouring countries; the seed maggot, *Neotephristis finalis*, in North America; while the hornworm, *Agrius convolvuli*, and the silver moth, *Autographa gamma*, are widespread pests.

Damage caused by many insects is usually more severe on moisture-stressed plants, while the punctures made by bugs in sunflower heads and seeds often allow invasion by pathogenic fungi, and this secondary infection frequently results in greater loss. It is important to prevent the build-up of these and similar pests, since control of fungal infection on susceptible cultivars is generally impractical. The free fatty acid content of seed damaged by sucking bugs can also increase during storage, reducing oil value and exacerbating financial loss.

A major cause of seed loss in maturing sunflower is various bird species; not only by eating seeds, but their weight and movements frequently cause stems to break behind the heads, which fall to the ground and rot. In Tanzania, doves and love-birds devastated irrigated dry-season crops grown by the author, and in Yugoslavia 80% heads can be damaged. An official figure of 100000 t/year destroyed by parakeets and doves was quoted to the author in Uruguay. In some districts of Australia bird damage is so severe it virtually precludes growing sunflower. Some local cultivars in Iraq with thick-hulled seed and incurving peripheral head bracts are partially bird resistant, but also have a low yield and seed-oil content.

Bird control is more difficult than insect control, since many bird pests are beautiful; for example, love-birds in Tanzania, sulphur-crested cockatoos and galahs in Australia. Wholesale destruction as is practised on *Quelea* spp. in cereal-producing areas of Africa is both unpopular and undesirable. Chemical repellents, sonic generators, scare guns, desiccation of maturing crops to reduce the susceptible period, decoy crops and a uniform district planting date to spread the damage, all assist in reducing loss to individual growers.

In Central Africa large animals were once a problem, for ripening sunflower is very attractive to wildlife. A bull elephant with the yellow heads of sunflower hanging from the sides of his mouth may look comical, but damage to the author's seed-multiplication plots was catastrophic!

Sunflower seed is an attractive food for most cosmopolitan storage pests noted in previous chapters, few are specific to sunflower. Cultivars with large, hard seeds are less liable to damage than current types with thin hulls or high oil content. Since damaged seed is more liable to infection, the greater care exercised during harvesting the lower the storage loss. For seed production and breeding stocks, use of insecticides and sealed storage is recommended; for commercial crops handled in bulk, stores and godowns should be protected against vermin, with fumigation of bulk stores or shipments. Smallholder crops usually suffer most heavily from storage pests, and from personal observation 25–50% loss of harvested seed is quite common. Similarly to sesame, it is often necessary to save 50% more seed than necessary to ensure an adequate stock for the following year's crop. On a national basis, in India and Pakistan for instance, this results in a food loss worth millions of rupees. For additional information on the control of storage pests and pest control in general, see this section in the Chapter 3 on groundnut and Fig. 9.10.

DISEASES

Sunflower diseases were previously considered of little importance compared with the depressing effect of adverse environmental and agricultural techniques. However, with the greatly increased area sown to pure stands, disease can now be a limiting factor in commercial sunflower production since there is no practical or economical method of controlling most of them. (See also remarks at the beginning of the section on pests.)

In Canada yields were reduced from an average of 1.4 to 0.8 t/ha by moderate to heavy infection of leaf diseases, which in Queensland, Australia, could also reduce yields by 50%. Symptoms of many common sunflower diseases are visually very similar, and thus field identification is difficult. Where a disease becomes important it is essential that it be accurately identified, for diseases with similar symptoms may require quite different control methods. A further problem has been the reorganisation and renaming of many groups or individual species of fungi, and references in the literature can be misleading or difficult to correlate.

The incidence of disease can be substantially reduced by cultural methods; disposal of crop residues by burning as soon after harvest as possible; ensuring that sunflowers are not grown successively on the same or adjoining land; destruction of volunteer or wild plants which can act as a reservoir of infection; seed dressing; and most important of all – selecting resistant cultivars when available. When introducing sunflower into an area where a specific disease is troublesome, it is often possible to reduce its adverse effect for example, by planting outside the optimum sowing period. Although this will in itself reduce potential yield, it may give a higher field yield. Since many diseases are carried on seed and plant material, imported pollen will reduce the risk of introducing disease, and can be safely transported below 30°C with little loss of viability.

The major genes controlling resistance to mildew, certain rusts and wilts have been identified and used to produce resistant cultivars, including resistance to rust *Puccinia helianthi*, and *Plasmopara* spp. Since there are local races of the pathogens, 19 of *P. helianthi* have been identified in Australia (Kong, personal communication 1998), such cultivars may have only local resistance. Exacerbating the problem is that new races of the rust quickly develop, several may co-exist, but resistance limited only to one or two.

Sclerotinia is probably the most serious disease of sunflower on a worldwide basis, and limits production in many areas. In Europe, particularly France, it is recommended that sunflower be grown only once every 5 years in a rotation, and every 8–10 years in the former USSR. This last restriction is probably very effective, but could only be maintained where there is central control of agriculture and no land shortage. One crop in 4 years is more common, although in Australia sunflower is grown for several years in succession.

Sclerotinia spp. are responsible for many sunflower root and stem rots, have a wide host range and where alternative hosts are grown in rotation with, or in close proximity to sunflower, damage caused will be greatly increased. *Sclerotinia minor* is often present in crops but is apparently not of economic importance. *Sclerotinia sclerotiorum* is very common; the first sign of infection is often a sudden wilting of leaves, and inspection of the stem base shows light-coloured or bleached spots where the fibres have disintegrated (Fig. 7.19). On the heads, the fibres break down into a light-brownish stringy mass (Fig. 7.20). *Sclerotinia* may be seed-borne or sclerotia (small, hard, black dispersal bodies) may be mixed with seed. It is strongly recommended that whenever possible only seed from disease-free stock should be sown, especially where sunflower has not previously been grown. This will materially assist in reducing damage from the disease. Some control is possible with the hyperparasite *Coniothyrium minitans*, which attacks the scle-

Fig. 7.19 Damage to sunflower due to *Sclerotinia* (stalk rot).

Fig. 7.20 Damage to sunflower due to *Sclerotinia* (head rot).

rotia of *S. sclerotiorum* in the soil. Another widespread and damaging stem rot is caused by *Phoma* spp., the black stem rot.

A common and serious disease is red rust caused by *Puccinia helianthi*, which reduces both seed yield and seed-oil content, and can be a major factor limiting sunflower production, as in Australia. It occurs worldwide and the author has seen it on the majority of crops visited in the Old and New Worlds (Fig. 7.21). The small, circular, reddish-black pustules, mainly on the undersurface of the lower leaves are unmistakable. The degree of damage is related to growing conditions and cultivar, since some are extremely susceptible to the pathogen and its effects can be devastating. Some control is possible using fungicides but several applications will probably be necessary. Sowing to avoid periods of high temperature and humidity, using resistant cultivars, and removing all volunteer and out-of-season plants, are useful methods of reducing the severity of an infestation.

A related disease, *P. xanthii* known as Noogoora burr rust in Australia, is of importance because of its ability to infect hybrids resistant to *Plasmopara halstedii*. Another rust, *Albugo tragopogonis*, grey white blister, can be locally very severe in specific crops, especially in South Africa, where resistance to this disease exists in breeding material.

A very harmful and widespread leaf spot is caused by *Alternaria helianthi*. First described from Uganda in 1943, it rapidly spread to most of the world's sunflower growing areas, and its symptoms have often been attributed to other causal organisms, especially *Septoria helianthi* and *Helminthosporium* spp. The brownish-black,

Fig. 7.21 Rust caused by *Puccinia helianthi.*

usually interveinal spots occur on the leaves and stem, developing faster on the former, and coalescing to form large black, necrotic areas. Plants can be attacked at any growth stage, but in the field major infestations usually occur in more mature plants. In very severe attacks the leaves wither and die, stems fracture and seed heads rot. Since the fungus has a fairly wide host range, including castor, control is difficult. *A. zinniae* is of less importance.

Other leaf spots of wide occurrence are *Septoria helianthi* and *Helminthosporium helianthi*, very similar in appearance and distinguished from *Puccinia* by producing brown angular spots, which may coalesce, forming large irregular dead patches. Both occur on lower leaves initially and frequently cause defoliation, but since these leaves are senescent, their loss is of little importance.

A wilt caused by *Verticillium alboatrum*, which attacks many tropical plants, can be serious locally, and susceptible crops should not be grown in succession. The disease can be seed-borne, and seed from disease-free stocks will materially reduce its occurrence. *Rhizopus* spp. especially *R. oryzae*, cause a common head rot, which is most severe on heads that have been previously damaged by insects, birds or hail. The most common symptom is formation of brown spongy tissue around the wounds or the rear edge of heads, which later spreads into large patches and may cover the whole of the head. When opened, affected areas show a grey furry mould in which are small, black fruiting bodies. A less obvious but more important effect of *Rhizopus* head rot is that oil from infected seed has a very much higher free fatty acid content.

Soil-borne fungi can be very damaging wherever they occur, in particular *Macrophomina phaseolina*, charcoal rot, not only attacks the stem base can also cause a neck rot leading to failure of the stem immediately behind the head. The stems of affected plants become brownish, very dark towards the base, with stunted and grey roots. Both become very dry and break easily. Susceptibility of sunflower to *M. phaseolina* can be increased by salinity, either soil or irrigation water. *Sclerotium rolfsii*, stalk and collar rot, commonly occurs in odd patches in sunflower fields. It attacks the roots, and its effects are exacerbated when these are damaged by mechanical cultivation. *Aspergillus* spp. and *Botrytis* spp. attack the developing head, and can also affect seed-oil content and its constituents, either the free fatty acid or iodine value.

Mildews cause major damage when they infect young plants or the seed head, and when these diseases are prevalent seed dressing gives some control, as does avoiding badly infested fields. Powdery mildew due to *Erysiphe cichoracearum* has a wide host range and is cosmopolitan. Resistance to *Plasmopara halstedii*, downy mildew, which is a major disease of sunflower in Spain, exists in many recent hybrids and some control has been obtained by dressing seed with systemic fungicides. Local sunflower cultivars can be more or less resistant to a local race of mildew, but may have little or no immunity against other races. An unusual feature of *P. halstedii* is that although the seed may be infected, plants grown from it will show few, if any, symptoms of the disease. These occur in the *following* sunflower crop.

Sunflower is also susceptible to the aster yellow virus, AYV, and tobacco leaf curl virus which can be transmitted by whiteflies, specifically *Bemisia tabaci* in India. In Africa sunflower is frequently infected by *Verticillium rugose*, sunflower mosaic. A symptom that is frequently attributed to disease is loss of heads by plants with no obvious injury to stems. It is most often due to a boron deficiency, can be partially genetic, or occasionally caused by stem girdling insects.

8
Crambe, Niger and Jojoba

CRAMBE

Crambe, *Crambe abyssinica* of the genus *Crambe,* is a member of the Cruciferae, tribe Brassicae, is closely related to the rapes and mustard. The number of species in the genus is in dispute, between 20 and 30. Most are perennial herbs, although a few are shrubs or annuals, which are distributed mainly in the Mediterranean area, Euro-Siberia and the Turko-Iranian region. The annual species occur mainly in the Mediterranean region, and Ethiopia to Tanzania, while shrubs are apparently confined to the Canary Islands–Madeira area. One species *C. filiformis* occurs in Chile (Fig. 8.1) (Leppik & White, 1975).

Considerable misidentification occurred within the genus *Crambe,* particularly between *C. abyssinica* and *C. hispanica,* and many references in the literature and distribution maps may be misleading. For this reason doubtful references have been omitted, unless of a general nature or interest. The only cultivated member of the genus is *C. abyssinica,* also known as Abyssinian kale, and reference to crambe, its products and derivatives in the text is restricted to this species.

Evaluation of the literature indicates the most probable centre of origin of the genus as the Turko-Iranian area, despite the species designation *C. abyssinica,* with migration to the Mediterranean area during ancient periods of geological and climatological change. Further distribution can be associated with early human migration and more recently by trade. The small population of *C. filiformis* in Patagonia and Chile is almost certainly due to Spanish or Portuguese exploration and settlement. Similarly the distribution of *C. maritima* is most extensive on the Black Sea, northern Mediterranean and Baltic shores, an environment in which it thrives, and its distribution was probably widened by ancient mariners.

North and East African species of crambe are closely related to *C. hispanica,* a Mediterranean species which is believed to have migrated to North Africa and become naturalised in Ethiopia. A logical movement, since there was ancient communication via the Nile Valley and the Red Sea. Wild crambe in North and East Africa usually forms a low bushy plant found as a weed in fields, along edges of paths and cultivated areas, or scattered through local bush and scrub. It normally occurs as isolated plants or small clumps, and seldom forms a large or continuous growth.

Crambe has no natural varieties, but a range of recent cultivars have distinct differences in growth, yield, seed size and constituents (Lessman, 1975). A limiting factor would appear to be the narrow range of germplasm available. The high erucic acid content gives the seed oil many valuable industrial applications, and where the oil can be so utilised the crop is profitable. Crambe production is thus limited to areas in close proximity to end users.

There are apparently no records of crambe having been domesticated or grown for its oil on

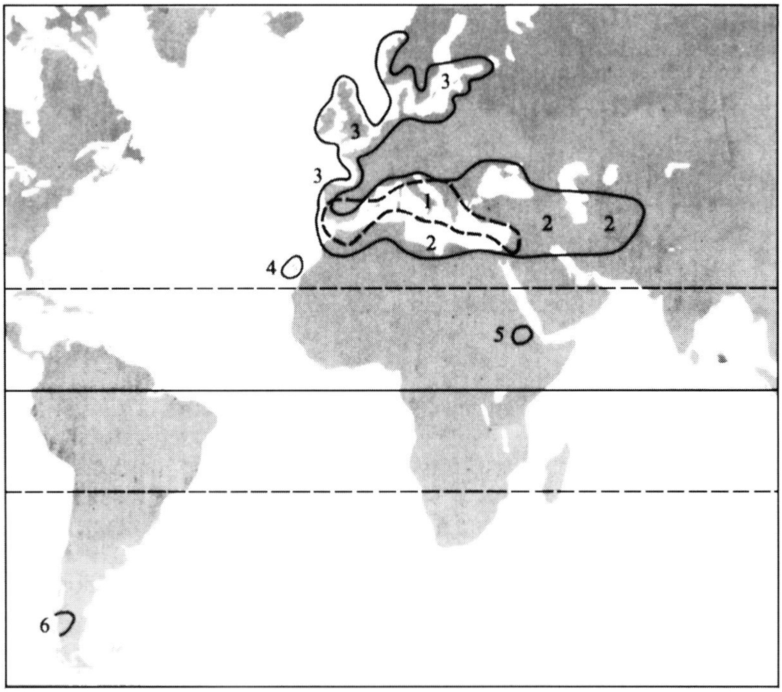

Fig. 8.1 World distribution of crambe: (—) general distribution of genus; (- - -) distribution of *C. hispanica*. 1. *C. hispanica*; 2. Section *Sacrocrambe*; 3. *C. maritima* Atlantic area; 4. Section *Dendrocrambe*; 5. *C. abyssinica*, *C. sinuatolenta*, *C. kilimandscharica*; 6. *C. filiformis* outpost in Chile.

any scale in the former USSR before 1932, but it reached some 25000 ha in Poland after the Second World War II. Trials indicate that crambe is a suitable crop for northern Italy (Lazzeri *et al.*, 1997).

Crambe seed was first officially introduced into the USA in the early 1940s, improved Russian and Swedish strains were introduced in the late 1950s, and the area planted subsequently reached 12000 ha. Crambe is reportedly the only oil-bearing crucifer to produce mature seed in the northern, tropical areas of South America. Recent trials indicate that crambe could be a promising oil crop for China (Wang *et al.*, 1995).

BOTANY

The genus *Crambe* is currently divided into three sections: *Sacrocrambe* DC. perennials occurring mainly in Eurasia; *Leptocrambe* DC. annuals and biennials mainly occurring in the Mediterranean, North and East Africa; *Dendrocrambe* DC. shrubs and semi-shrubs of Madeira and the Canary Islands

Those *Crambe* species of interest as oilseed producers are placed in the *Leptocrambe* and include:

C. abyssinica Hochst. ex R.E. Fries	$2n = 90$
C. filiformis Jacq.	$2n = 30$
C. hispanica L.	$2n = 60$
C. kilimandscharica O.E. Schulz	$2n = 60$
C. kraliki Cosson	$2n = 60$
C. sinuato-dentata Hochst & Schimp.	$2n = ?$

It would appear that the basic chromosome number for the genus *Crambe* is $x = 15$, and was probably derived by aneuploid loss from a tetraploid based on a basic set $x = 8$, according to a survey of nine different species of *Crambe*, and approximately 250 species of 80 genera of Cruciferae (Meier & Lessman, 1973).

Several varieties of *C. hispanica* have been described; *C.h.* var. *hispanica*, *C.h.* var. *major*, *C.h.* var. *edentula*, and *C.h.* var. *glabrata* recorded as having $2n = 30$. However this classification has been questioned and a more detailed study of the *Leptocrambe* is necessary. Confusion between *C. abyssinica* and *C. hispanica* is common, as has been noted, and seed accessions should be positively identified before their use in breeding or evaluation programmes. Genetic variability in *C. abyssinica* appears to be restricted, and it may be necessary to go outside the section *Leptocrambe* to find additional germplasm to increase its potential and profitability. *Crambe hispanica* has been described as an excellent source of such improvement, and it crosses readily with *C. abyssinica* to produce fertile seeds.

The relatively small amount of crambe material presently available, together with restricted genetic variability in agronomic characteristics as noted, are major constraints on improvement programmes. Hybridisation, however, has resulted in cultivars with larger and heavier seeds or disease tolerance, indicating that more agronomically valuable cultivars can be produced. For this reason it is important that wild crambe germplasm resources be identified and protected for long-term use.

Of importance to growers, especially in the high-altitude tropics, would be an increase in frost or low-temperature resistance, or strains which matured in 75–90 days. Plants with fewer primary branches and a more upright growth would be of value in mechanised production, since this would allow sowing in narrow rows. Alternatively, increased resistance to lodging in existing branching types would be an asset. However, since shattering is not a major problem, this objective could be less important in the short term than breeding for more pods per raceme irrespective of plant type. Increased disease resistance is a perennial objective, but it is most likely that good cultural practices can maintain the relative freedom of crambe from major infection. Pest resistance could possibly be increased via *C. hispanica*. Reduction or removal of the toxic glucosinolates in seedmeal would also be a major advantage.

Crambe is an herbaceous annual, usually less than 1 m in height, naturally much branched, bearing long racemes of yellow flowers, and producing a large number of small, brownish seeds with an oil content around 40% (Fig. 8.2).

Fig. 8.2 Typical crambe plant at flowering.

The plant normally produces a substantial root system typical of the Cruciferae, with a tap root which may penetrate to some depth. This extensive root system enables the plant to draw moisture from a considerable volume of soil, and it is thus drought resistant in the wild.

The stem is stiff, roughly circular in section, the lower portion covered with fine, short hairs, the upper smooth and hairless. Stem colour is normally a typical cabbage-green, becoming greenish-yellow to yellow when mature. Height varies with soil and rainfall, and is generally around 1 m when cultivated, occasionally to 2 m. The main stem branches close to the ground to form secondary stems which may number 30 or more, and these again branch to form tertiary branches. The extent of branching is affected by cultivation methods, plant population, etc., and several hundred branches can be produced on individual plants under suitable conditions.

Extensive branching is a disadvantage for mechanical harvesting, since each terminates in a raceme. Thus, although branching can increase individual plant yield, by extending the flowering period it may increase loss from frost and pests, or number of immature seeds harvested. However, where frost does not limit growth, highly branched plants could be advantageous, as windrowed plants would remain fairly high off the ground and so be more accessible to pick-up combines. An indication of the range of plant types possible is shown in Table 8.1 which details plant characteristics as affected by spacing (Lessman & Meier, 1972).

Leaves are glabrous and basically ovate lyrate, although varying in shape and size up the stem, with margins shallowly dentate. The lower leaves are carried on stout, long, 20 cm, U-shaped, somewhat hairy petioles. The leaf blades are usually 10–12 cm long, 6–8 cm wide, becoming progressively smaller up the stem and branches. The shape of the basal leaves is a species characteristic, and is probably the simplest method of distinguishing between the two; in *C. abyssinica* it is ovate, in *C. hispanica* cordate (Fig. 8.3).

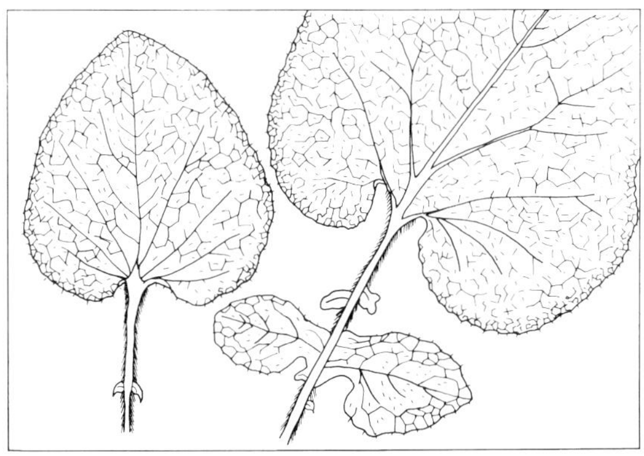

Fig. 8.3 Lower leaves of *C. abyssinica* (top) and *C. hispanica* (bottom), showing different shape of basal leaf and base of lower leaves. *Crambe abyssinica* is truncate, while *C. hispanica* is cordate.

Table 8.1 Effect of spacing on crambe characteristics – USA.

Row spacing (cm)	Plant height (cm)	Primary branches	Secondary branches	Yield (kg/ha)	Pods + seed wt (g)	Seed wt (g)	Oil (%)
50	102	23	108	1170	56	40	38
100	101	24	126	1016	55	39	39
Range	68–128	13–36	39–264	256–2269	45–70	22–58	26–41

Results are means of trials, cultivars and years.

Identification based on the length of the lower fruit segment is not considered accurate, nor that based on leaf hairyness, as the two species cross readily and this characteristic is easily transferred. Maximum leaf area index and maximum daily increase in biomass occurred at the onset of flowering in the USA (Kcem *et al.*, 1998).

Flower buds develop from the 10 to 13 leaf stage, and are mainly dependent on rainfall and temperature following emergence. Temperature also influences the time from emergence to flowering; in Poland it was around 50 days, but in the southern USA may be only 30–40 days. The inflorescence is borne terminally on branches, and with progressive growth of the cyme becomes an elongated raceme of normally white, occasionally yellow-tinted flowers, with four petals about 3 mm long, and six stamens, producing one seed. Flowers on recames vary in number from 20 to 50, and bloom progressively from the lower to the highest over a period of 7–10 days. Flowering can continue over a long period, since inflorescences on subsidiary branches flower sequentially after the main stem.

The total number of flowers is directly related to the number of branches, and can vary from a few hundred to several thousand on a plant that has developed naturally under suitable conditions. Flowers are generally self-pollinated, but outcrossing in the range 7–10% has been recorded from plants 1 m apart, and up to 30% in the field, which indicates a high degree of natural cross-pollination. High temperature at flowering, 30–35°C, is believed to affect pollen fertility, since there can be a considerable reduction in seed set. Low temperature, 12–13°C, for any period will affect bud/flower development, and also reduce subsequent seed production.

The fruit is a capsule (siliqua), initially pale green but becoming yellow with maturity, when

Fig. 8.4 Crambe flowers.

it also develops a well-defined network of small ridges (Fig. 8.5). The reticulation varies within the genus and also between cultivars, and is not a reliable means of identification. It is, however, more reliable as a rough guide to maturity, becoming more pronounced as seeds ripen. The fruit is in two parts, a terminal spherical capsule and a much smaller cylindrical lower segment by which it is attached to a relatively long pedicle. This segment was originally used to differentiate species, but is too highly variable to be accurate.

Fig. 8.5 Crambe fruits.

Each capsule contains a single spherical, greenish-brown or brown seed, and although capsules usually have two embryos initially, only one develops into a mature seed. Seed size varies from 0.8 to 2.6 mm in diameter, being mainly affected by the number of seeds per plant, soil fertility and rainfall. For this reason 1000 seed weight is most variable and can range from 25 to 50 g. The seed is composed of approximately 40% pericarp, 4% testa, 6% hypocotyl and 50% cotyledons. These proportions may vary slightly with locality in a particular region, but are affected by climate to a much greater degree. Seeds harvested in a very dry year for example, have a much higher percentage of hull, with lower kernel and oil content. Detailed analyses of seed are shown in Table 8.2 (Earle *et al.*, 1966). Analysis of Indian seed gave the following (in %); moisture 6.5, oil 54, protein 18.5, allyl-isothiocynate 0.22, ash 4; with analysis of cake as N 6.4%, P 0.8%, K 1.4%, Ca 0.65%, Mg 0.62%. Characteristics of the oil were refractive index 1.4711, iodine value Wij 88.0, saponification value 173.4, ester value 172.7, oleic acid 0.37 (Pathak *et al.*, 1973).

The seeds of *C. abyssinica* and *C. hispanica* are very similar in appearance, and most seed obtained by Vasilev (1950) from sources in Europe and the former USSR and described

Table 8.2 Analyses of crambe seed – USA.

Constituent	Pericarp	Seed	Seed-coat	Hypocotyl	Cotyledon
Proportion of sample as harvested (%)	40	60	5	6	49
Proportion of seed (%)	—	—	8	10	82
Component analyses, % dry basis:					
Oil	0.4	47	17	38	55
Crude protein	4	25	23	34	23
Ash	8	—	4	5	4
Oil-free meal analyses, % dry basis:					
Crude protein	4	51	29	56	52
Phosphorus	0.2	—	0.3	1.8	2.1
Sulphur	0.0	2.4	0.7	3.1	2.6
Thioglucoside, as *epi*-progoitrin	0.1	8.8	2.1	13.0	10.9
Calcium	2.2	—	1.0	0.7	1.0
Magnesium	0.2	—	0.1	0.6	0.7

as C. *hispanica* was later identified as C. *abyssinica*.

Crambe seeds also contain glucosinolates, and untreated meal remaining after oil expression is of limited value as stock feed; however they can removed by suitable processing. Crambe seed has a high fibre content, and a reduction in hull thickness would add to the meal's palatability. The proportion of husk to kernel also directly affects profitability, and this proportion rises significantly if there is moisture stress in the plant when the capsules are formed and filling. The residual meal after pressing averages some 50–60% of total seed weight, and contains 30–60% crude protein, the higher level being obtained from dehulled seed, and up to 11% thioglucosides.

The oil content of hulled seed ranges from 35 to 60% with 40–50% most common, protein content is 20–40% with 25–30% most common; erucic acid content of the oil is in the range 35–65% with 50–60% most common. A comparison of the physiochemical properties of oils from nine *Crambe* species showed that C. *abyssinica* had an erucic acid content at 57% and *C. hispanica* at 52% (Dolya *et al.*, 1977). The oil can also contain up to 10% linoleic and 20% oleic acid.

The erucic acid content of oil in developing seed increases from approximately 25–30% about 14 days after fertilization to 40–45% one week later, and reaches 55–60% at maturity. Iodine number (about Wij's 106 at maturity) and unsaturated fatty acid content of the oil remain fairly constant during ripening, but an initial high content of palmitic acid drops rapidly from 30% to 10% within two weeks of fertilisation. Cultivars have been developed which have a high content of a particular component, i.e. very high erucic acid content (Wang *et al.*, 1996, 1997).

The most important commercial chemical derivatives of erucic acid for industrial use are erucamide and oleamide, world demand for which is growing. Increasingly important oil constituents are the oleochemicals that have 20 or 22 carbon atoms in the fatty alkyl chain. Classified by environmentalists as 'green chemicals' compared with those obtained from fossil fuels, they are considered to have the potential to greatly increase crambe oil demand. A method of density estimation for crambe and other seed oils based on fatty acid composition has been developed (Halvorsen *et al.*, 1993).

ECOLOGY

Crambe is a plant of moderate-rainfall, warm-temperate regions, which has become adapted to colder or drier areas. Its natural range is restricted to a Mediterranean-type climate, but it has been grown commercially and experimentally over a much wider climatic range, and thus has considerable adaptability. It can be found growing from sea level to 2000 m, and the author established pilot scale plantings in the Kenya Highlands at 2500 m. From the limited data available it appears that within a region, location has little effect on the oil content of seed compared with its effect on seed yield. Since crambe is grown only for its oil, concentration on those cultural procedures resulting in the highest local seed yield would currently be the most profitable.

A temperature range of 15–25°C is required over the main vegetative period, although higher temperatures are well tolerated except at flowering. Seedlings have reportedly survived temperatures of −4 to −6°C for short periods, but one degree of frost will normally severely damage or kill plants at any stage. A frost-free period of 60–90 days from emergence is desirable, and is a major main factor currently limiting the crop's range. In North Dakota, USA, the heat unit accumulation from sowing to physiological maturity was approximately 1350 degree-days above 2.5°C (Kmec *et al.*, 1998).

Crambe is independent of rainfall to a considerable extent, and provided sufficient falls during germination and until plants are well established, can thereafter make good use of residual soil moisture. A rainfall of 800–1500 mm or its equivalent is necessary for commercial crops, dependent mainly on soil type or residual moisture, and there must be no water stress at flowering and pod formation. Moisture availability has a major effect on plant growth, particularly branching, and the number of racemes and subsequent fruit are determined by the extent this occurs. Thus crambe's stated drought-resistant properties are in fact an ability to draw on subsoil moisture, since in its absence or with lack of rain, the plants remain stunted or die. As a commercial crop crambe is not suited to arid areas of low water availability.

Hail causes severe damage to young plants; on more mature plants the loss of terminal shoots causes branching and delays flowering. This may result in plants being killed by frost before full maturity, or a large proportion of immature seed at harvest.

SOILS

Crambe will grow on a variety of soils, with highest yield obtained from sandy loams. Heavy clays are unsuitable, as are soils with an impacted layer which restricts root growth as occurs on many otherwise suitable lateritic and alluvial soils in the tropics. The impacted layer is very difficult and expensive to fracture without major disturbance to the soil structure.

Crambe can withstand short periods of waterlogging, or grow in areas of high rainfall if the soil is permeable, but should not be grown on soils with inadequate drainage. Crambe grows best on neutral to slightly alkaline soils, but below pH 6.0 growth and yield are usually adversely affected; however, locally acceptable yields have been obtained in Ethiopia on acid soils of pH 5.0. Introduction of salt tolerance would increase crambe's attractiveness and this might be possible via *C. maritima*. The natural habitat of this species is the salt-impregnated sandy sea-shore, and it has many varieties that adapt readily to northern and southern hemispheres.

FERTILISERS

Reports of fertiliser trials on crambe are few, and application levels, types and timing must be locally determined. Crambe can be deep-rooted, and thus draw on soil nutrients at depth. Additionally, if subsoil moisture is the main source of water, fertiliser placed on the soil surface or incorporated in the 10–20 cm horizon will be largely wasted. Fertilizers recommended for use on local small grain crops in the USA, Russia and Poland are used on crambe, and where there are large areas of grains and a small area of crambe this is a simple and economical solution. In Kenya the author used the standard fertilizer recommendation for wheat, plus a top-dressing of nitrogen.

Nitrogen

Nitrogen (N) is the most important nutrient in terms of increasing seed yield, and is best supplied in the seedbed. In the presence of adequate phosphate, 20–40 kg N/ha is usually profitable, and there are no data concerning the merits of different types of nitrogenous fertilisers. Top-dressing, as with other oilseed crops, requires field trials in a specific region to determine its profitability. In Europe there have been conflicting results, both yield increase and depression followed one to two top-dressings of ammonium sulphate at 20–120 kg/ha. Application was at various stages of growth from budding to first flowering, and again there appears to be no correlation in respect of timing and stage of growth.

High levels of nitrogen generally depress seed yield and sometimes also reduced seed-oil content, which indicate that soil and climate probably influenced crambe's reaction to nitrogen application. This was certainly so in East Africa, where the author's trial plantings showed a pronounced water availability/nitrogen interaction. In Ethiopia top-dressing with approximately 15 kg N/ha increased growth, improved general plant health and tripled yield over plants which received no nitrogen. However, these were observation plots and not statistically valid. In the Czech Republic 40 kg N/ha is considered sufficient in most areas (Strasil & Skala, 1997).

Phosphate

Phosphate (P) applications of around 75 kg P/ha have been recommended for commercial crambe crops in Europe and Canada, or similar amounts as applied to other crops in the rotation.

Other nutrients

Potassium (K) is not generally required except where it limits growth of all local crops, and crambe grown in a rotation with other crops receiving potassium usually requires no additional amount. There are no reports of minor element deficiencies in the field, although greenhouse trials have indicated a response to copper at the rosette stage.

CULTIVATION

Cultivation necessary for small grains is suitable for crambe, but it may be necessary to break up a plough pan or similar layer, especially in areas of low rainfall, where residual soil moisture is an important factor in plant growth. Crambe's ability to exploit soil moisture at depth is a valuable characteristic, and should be fully utilised.

Seed should be sown 1.5–2.5 cm deep, but up to 4.0 cm is acceptable in loose soil, or to ensure the seed is sown into moist soil. Crambe seedlings are vigorous, and some unevenness in planting depth is unimportant at normal sowing rates, provided there is no danger of soil crusting. Adequate soil moisture is far more important to achieve even stands, for crambe seed is usually sown unhulled and the slight space between the seed and hull acts as a barrier to moisture penetration. Thus there can be a delay before the seed absorbs sufficient moisture to germinate, and this delay is a major reason why reliable rainfall, adequate soil moisture or irrigation water should be available at sowing. A seedbed temperature of 10–12°C is preferable at sowing, but up to 15–20°C tolerated; below 8°C emergence is retarded, at 5°C normally inhibited. Seedlings normally emerge some 10–14 days after sowing at the optimum temperature, provided there is adequate soil moisture. In the high-altitude tropics crambe can thus be planted early in the rainy season, when soil temperatures are usually too low for other oilseed crops.

Crambe can be sown by modern combine grain drills fitted with attachments recommended by the manufacturers. Precision seeders give accurate seed rates and planting depth, but unless they are already available, are not essential. Because of its free-branching habit crambe can compensate for a wide variation in plant population, the main disadvantage for mechanical harvesting being in the extended maturation period that results.

Seed rates range from 10 to 40 kg/ha, but in general 20–25 kg/ha is recommended for unhulled seed. At very low seed rates the yield per plant increases, but total yield per hectare falls. Crambe seed is normally sown unhulled, and if obtained from a commercial crop usually contains a significant proportion of damaged seed. Poor conditions at harvest, and unsuitable harvesting techniques or storage quickly reduce viability. Seed for sowing should be cleaned and graded to remove broken seed and pops, for large, sound seed generally emerges faster and more evenly than small, sound seed. The germination percentage must be determined, since this will directly affect the sowing rate. Visual assessment of seed quality is unreliable, as seed may be damaged by insects or fungi without any apparent injury to the hull.

Row-width is most profitably related to the machinery available and method of harvesting, as most cultivars available are branching types, and when planted in narrow rows produce fewer branches. Row-width trials indicate the degree of primary branching is relatively stable, the effect of reducing row-width being to reduce secondary branching. If narrow rows become desirable then breeders may have to produce tall plants with few, upright, primary branches. Row-widths of 20–30 cm are recommended for commercial mechanised production, have consistently given highest yield, and tend to reduce flowering and maturation periods.

Solid drilling in 15 cm rows, can be successful when there is adequate soil moisture and a basically weed-free seedbed, or where weeds are readily controlled by a pre-emergence herbicide. Wide rows of 40–60 cm allow hand and mechanical weeding where this is necessary or herbicides are not used. Ribbon or band sowing has been recommended in the former USSR and eastern Europe, but there is no obvious advantage in this method. Row-widths greater than 60 cm are acceptable for non-mechanised crops, but the larger plants produced tend to lodge and make direct harvesting difficult. Of greater importance is the widely varying range of seed maturity from such plants at harvest.

Seed dressing is generally unnecessary, and where soil insects are a problem, increased seed rates are usually effective in compensating for loss. Soaking seed for 30 minutes in water at 55°C has given good control of some fungi infesting the seed-coat without impairing viability. Time of planting is important, since sowing outside the optimum period quickly reduces yield and can also affect seed-oil content (Table 8.3) (Dembinski *et al.*, 1962; Weiss, unpublished). It is thus essential to determine the optimum date by local trials, for crambe's profitability is directly related to maximum seed yield. A delay of 10–14 days can be sufficient to reduce yield by 25%, therefore pre-planting and seeding operations must be well planned and quickly completed.

Two main factors are important when considering the time of sowing, and both have been previously mentioned; soil moisture and frost. In drier regions where soil moisture may not be high, rainfall before sowing will determine when this operation can take place to ensure a good stand. In regions where frosts limit the growing period, sowing must ensure that plants have sufficient time to mature. Sowing may thus have to be delayed in those regions where there are late spring frosts, or advanced in others where there are autumn frosts. A period of 100 frost-free days should be available from planting to harvest. Some recommended planting dates in selected countries are shown in Table 8.4.

WEED CONTROL

Crambe grows vigorously and once past the seedling stage can compete with any but the most rapidly growing weeds (Fig. 8.6). It quickly branches to cover the inter-row and seldom requires more than one cultivation. This operation should be as shallow as possible and finger weeders, rotary hoes or similar implements are most effective. Provided weeds are not suppressing seedling growth, inter-row cultivation should be delayed as long as possible so that crambe plants will cover the inter-row before weeds again become numerous. Ideally, crambe should follow a crop in which the major local weeds are easily controlled.

The following herbicides used pre-emergence gave good control of weeds and little effect on yield: desmetryne, glyphoste, monolinuron and prometryn/simazine mixtures. Directed sprays of benazolin have proved successful post-emergence. Herbicides proved locally safe on

Table 8.3 Effect of sowing date on yield.

Date	Yield (kg/ha)
Poland	
Mid-Apr.	2933
End-Apr.	2734
Early-May	2646
Mid-May	2426
Kenya	
Mid-June	1896
Mid-July	1255

Table 8.4 Some recommended planting dates.

Canada	Early May	USA Iowa	Early April
Czechoslovakia	Early April	Texas	Early Feb.
Germany East	Early April		
Kenya (Highlands)	Mid-June	Russia St. Petersburg area	Late May
Poland	Early April	Harjov area	Late April

Fig. 8.6 Crambe seedling prior to commencement of stem elongation.

rape should also be suitable. Crambe is very susceptible to the chlorophenoxy herbicides, 2,4-D and related compounds, and may also be adversely affected by herbicide residue remaining in the soil from the previous season, atrazine for example.

IRRIGATION

Little information is available on irrigation requirements, nor assessments of total water consumption by crambe. However, application of approximately 1200 mm would appear to be a reasonable minimum for commercial cropping, and a few substantial applications preferred to many light waterings. A pre-planting irrigation to wet soil to depth is recommended, with a similar watering at first flowering. Depending on rainfall, an intermediate irrigation may be desirable to maintain growth. From the author's experience irrigation should not be prolonged as this encourages regrowth, and the watering at flowering should be the last. Irrigation of crambe is usually uneconomic in most areas where it would be desirable.

INTERCROPPING AND ROTATIONS

Crambe is seldom intercropped by smallholders and usually grown as a cash crop in a pure stand, or by commercial growers as a field crop. Little data is available on specific rotations, although it can be an oilseed break in rotations, as it is a specialised commercial crop and grown only by individual producers.

HARVESTING

Plants are mature when the majority of leaves have fallen, the upper part of the stem is yellow, and at least 75% of capsules are yellow, usually 90–100 days after planting. Plants can be cut, windrowed and threshed with a combine using a pick-up reel (Fig. 8.7), direct combine harvested (Fig. 8.8), or cut with a binder and stooked to be threshed when convenient. In general, the operations and machinery used for harvesting oilseed rape are equally successful for crambe. When grown by smallholders it is usually cut, piled in the field and threshed by hand

Fig. 8.7 Modified pick-up reel on combine, suitable for rape or crambe, Australia.

Fig. 8.8 Combining crambe, USA.

with sticks or flails. Since shattering is not a major problem crambe plants should be left as long as possible before harvesting, as fully mature seed has a lower percentage of hull to seed.

Crambe may be windrowed with a seed moisture content of 25–35%, but should not be threshed until the moisture content has dropped to 12–15%; below this hulls can be damaged and oil quality in stored seed is affected. Rain falling immediately before cutting or on newly windrowed plants in warmer regions can cause an increase in fungal disease. It may be necessary to turn windrows and hasten drying to prevent seed becoming damaged, as fungal infection reduces viability and can affect oil quality. Time of cutting can be critical for another reason; Russian reports indicate that cutting too early can seriously reduce seed-oil content, which continues to accumulate to full maturity. The effect of frost on mature seed has not been reported, but experience with other crucifers indicates that viability will be affected. However, oil content of sound mature seed will be less affected, although oil constituents may be changed.

Standard grain harvesters are suitable for combining or threshing crambe, provided that the drum, concave and fan are correctly adjusted. High drum speeds will cause excessive seed damage with consequent oil loss, and high fan speeds increase seed loss at the back of the combine. No attempt should be made to separate the hulls from the seed during harvesting. Yields of 1500–2000 kg/ha are commonly achieved on a field scale in Europe, Russia and Canada, slightly less in the USA. Trials in many countries have consistently yielded 2000–3000 kg/ha, which would indicate that 2500 kg/ha can reasonably be expected by specialist growers and should be the target.

STORAGE

Crambe seed must be cleaned and dried to 10% moisture content for storage before processing, and any commercial drier used for oilseed rape is suitable. Seeds in stooked plants, built into ricks or left unthreshed in barns quickly lose their viability, although the oil content is basically unaffected. Storage tests in Poland showed that seed with low moisture content withstands low temperatures without major loss of viability. Seed lost viability when stored for 2 years at 21°C and 71% relative humidity, but in controlled storage at 10°C and 50% relative humidity, or −1°C and 40% relative humidity, seed remained viable for 8 years.

PESTS

Crambe grown in pure stands is attacked by many of the insect pests damaging other cruciferous crops, *Meligethes* and *Ceutorynchus* spp. for example, but in the wild is relatively unaffected by endemic pests. When grown in the vicinity of large rape fields, the level of infestation is higher and the incidence of pests is also higher when the main weeds of a cultivated crop are cruciferous. The apparent resistance of crambe to certain pests is worthy of further study, and may be increased by crossing with other crambe species. *Crambe hispanica* for instance is considered more resistant to *Phyllotreta nernorum*, and this species crosses easily with crambe (Nielsen, 1977). Crambe can also withstand substantial leaf loss after anthesis without significant loss of yield (Kmec *et al*., 1998).

Where no trials have been carried out on chemical methods of pest control, a good guide is to use compounds recommended for rape as insects attacking crambe will generally be similar; recommendation for active ingredient level and method of application will also apply. It may be necessary to modify these locally to suit a specific pest complex, and determine the time of application to achieve maximum control. In naturally occurring crambe populations many endemic pests are highly parasitised, and this may be one reason for the generally low level of insect damage noted. It is also most probable that evolutionary pressure has eliminated strains very susceptible to specific insect species.

Flea beetles are probably the most damaging pest to crambe seedlings, and in Kenya occurred in such large numbers over so prolonged a period as to render commercial crambe growing unprofitable. Although control with commonly available pesticides was effective, there was continuous reinfestation from surrounding uncultivated lands. *Phyllotreta mashonana* and *P. cheiranthi* are commonly recorded in eastern Africa, but *P. cruciferae* is more important in South-West Asia and *P. nemorum* in Europe. Aphis, especially *Brevicoryne brassicae* and *Rhopalosiphum pseudobrassicae*, are widespread and very damaging to young plants. Thrips appear to be most serious on buds and flowers, with *Thrips tabaci* and *Franklinliella* spp. most often recorded. Spring tails, *Entornobrya purpurascens*, have been recorded as infesting crambe in Ethiopia, but with no note of the damage caused. A sawfly, *Athalia colibri*, is a widespread pest in southern Europe, North Africa to Asia. It may cause severe defoliation in specific years.

Butterfly larvae, especially those of the whites and sulphur butterflies, are serious throughout Europe, the Middle East, South-West Asia and Africa. *Pieris brassicae*, *P. daplidice* and *P. rapae* are the most important, and widespread (Fig. 8.9). Two pyraustid moths, *Crocidolomia binotalis* the cabbage moth, and *Hellula indalis* the oriental cabbage webworm, are usually found on crucifers in the semi-tropics, have been recorded on rape and crambe, and could become more important if crambe planting expands. *Plutella* spp. (diamond-back moths) are common pests of a wide range of crucifers; *P. xylostella* is the most often recorded and can cause severe damage to crambe plants, especially pods and seeds.

Plant and stink bugs can become serious pests as the plants mature, most damage being caused to developing inflorescences and fruits. *Eurydema* spp., *Bagrada* spp., *Nysius* spp. or *Lygus*

Fig. 8.9 Small white butterfly.

Table 8.5 Common pests of crambe.

Pest	Common name	Distribution
Athalia colibri	Sawfly	Europe–Asia
Ceutorynchus assimilis	Pod and seed weevil	Europe–SW Asia
C. quadridens	Stem weevil	Europe–SW Asia
Brevicoryne brassicae	Cabbage aphid	Cosmopolitan
Dolycoris baccarum	Stink bug	Europe–Asia
Eurydema oleraceum	Cabbage bug	Europe–Asia
Eurydema ornata	Cabbage bug	East Africa–SW Asia
Frankliniella spp.	Thrips	East Africa
Lixus spp.	Weevils	Europe–Asia–Africa
Lygus rugulipennis	Plant bug	Europe–Asia
Meligethes aeneus	Pollen beetle	Widespread
Myzus persicae	Green peach aphis	Widespread
Nysius cyrnoides	Lygus bug	Europe–SW Asia
N. erica	False cinch bug	Americas
Phyllotreta mashonana	Flea beetle	East Africa
P. nernorum	Flea beetle	Europe
Pieris brassicae	Large white butterfly	Widespread
P. daplidice	Cabbage white butterfly	Europe–SW Asia
P. rapae	Small white butterfly	Widespread
Plutella xylostella	Diamond-back moth	Widespread
Rhopalosiphum pseudobrassicae	Turnip aphid	Widespread
Tephroclystia oblongata		Europe–former USSR

spp. all become important locally, the particular species varying regionally. Weevils, either *Baris* spp. or *Lixus,* commonly occur on crucifers, usually as a minor pest. *Baris coerulescens* and *L. anguinus* have been recorded in Africa and South-West Asia and *L. latro* in Ethiopia.

DISEASES

In its natural habitat crambe appears to be little affected by disease, and although this absence of infection persists under cultivation, it is probable that new cultivars will be attacked by some of the more common diseases of cultivated brassicas. A range of pathogens has been reported on crambe species in their native habitat. These pathogens appear to be generally in equilibrium, since susceptible plants will have been gradually eliminated by selection pressure over time. However, some pathogens can become more serious when crambe is sown in different environments; for example, higher soil moisture will allow increased infection by *Alternaria* leaf spots or *Fusariurn* wilts.

As with insect pests, many diseases of rape are capable of attacking crambe and the same remarks regarding identification and control are applicable. Chemical control measures, apart from seed dressing, are currently neither practical nor profitable, as once plants have covered the inter-row only aerial application is feasible and seldom available, unless used on other more profitable crops.

The most serious diseases reported are those caused by *Alternaria brassicicola* leaf spot, and *Plasmodiophora brassicae* clubroot, both being common and widespread on brassicas. Mildew, *Erysiphe cichoracearum* in particular, can occasionally cause some loss in specific crops, as can bacterial and root rots when climatic conditions favour their incidence. Crambe is reportedly susceptible to tobacco and turnip mosaic viruses. Information on disease in commercial crambe crops is sparse; most records or reports come from experimental or breeding work, and thus of limited value to growers.

NIGER

Niger, *Guizotia abyssinica*, is another oil-producing member of the Compositae, which originated in the northern Ethiopian highlands, where it grows wild and is cultivated. Niger is more widely grown in India than any other country, but is found only under cultivation or as a weed of cultivated crops. In the text, niger refers to *G. abyssinica*, its seed and derivatives.

The ancient Agau people of Ethiopia had a well developed society and agriculture, and traded with contemporary Egypt. Local crops from the two countries, including niger, were probably interchanged along the Nile valley, and subsequently more widely distributed along traditional trade routes. Another possible method was via Arabian or Persian Gulf traders who in very early times had direct contact with both the East African and Indian coasts.

Niger was once widely but locally grown in East and West Africa and Republic of Congo (Zaire), but as an oilseed crop it is now almost wholly confined to Ethiopia where it is known as *neuk*, *noog* and *hug*. One of that country's earliest cultivated crops, it still provides over half total edible oil production. Ground niger seed together with chillies, are main components of the traditional fiery stew known as wat. Niger has been grown for green manure and silage in Central and South Africa, but as an oilseed it was not considered profitable enough to be included in rotations. Plantings in Kenya also proved unprofitable.

It is very widely grown in India and Pakistan where it is an important local oilseed (Fig. 8.10), but when it was originally introduced is unknown. Niger's wide dispersal in India is indicated by the great variety of local names including *kalatil*, *karala*, *ramtil*, *sarguia*, *tilangi*, many indicating a resemblance to sesame. Niger averages about 3.5% of the total oilseeds area, production currently averaging 135 000 t annually, with the States of Bihar, Madhya Pradesh, Maharashtra and Orissa accounting for nearly 90% of the total. Plants are threshed in the field, the seed winnowed by the wind, and groups of people so employed are a characteristic sight in the Indian landscape at harvest. The women do this work most gracefully, and their actions form the basis of the *bhangra*, a famous Punjabi country dance.

Fig. 8.10 Smallholder niger crop, Pakistan.

In the nineteenth and early twentieth centuries niger was grown in Germany, Switzerland, France and Liechtenstein as the ramtil plant, and more recently the Czech Republic and Slovakia. Introduced into Russia from Ethiopia it was tried on a field scale in 1926, but never achieved any success as a commercial crop. Niger has been grown on an experimental or minor scale in many other countries, including the West Indies. It was tested in Canada, for instance, in a general evaluation of potential oilseed crops, but was considered unsuitable for commercial production.

However, niger should be considered primarily as a valuable oilseed crop for smallholder production, especially in those areas where other oilseed crops are few. It has great potential as a source of edible oil, and could support small-scale extraction plants. Niger seed oil is suitable for edible and industrial purposes, the press cake as animal feed and manure, and the whole plant for fodder or green manure.

Production of niger seed on any scale is presently limited to India and Ethiopia, which together produce an estimated 350 000–375 000 t (375 kt) annually. Indian production has been officially estimated as varying between 80 to 100 kt but it is believed this figure excludes a significant proportion consumed directly by growers. Official statistics are based on niger seed that is commercially crushed or marketed. Major producing states are Madhya Pradesh, Maharashtra, Andhra Pradesh and Orissa. Madhya Pradesh, usually produces around half of India's niger seed and the four states together produce about 80% of the total. Exports of Indian niger seed reached 20 kt in the middle 1950s, but since then have been affected by various government regulations which banned the export of oilseeds. Niger oil production is approximately 40–50 kt annually. Average yield per hectare of Indian niger seed is very low, and there is considerable potential for greatly increased seed yield with no increase in the area sown, a very valuable asset in a country where population pressure on available agricultural land is continually increasing.

Niger is Ethiopia's most important oilseed, with an estimated annual production of 200–250 kt which represents nearly two-thirds of all oilseeds produced. Niger is grown throughout the country where conditions permit, but the main producing areas are Begemdir, Gojam and Welaga. The Ethiopian government actively supported development of the oilseed industry and production increased annually until the civil war in 1974. Since then the unsettled political situation has substantially reduced all oilseed production. Exports of niger seed varied between nil and 20 kt annually, and the main markets were Europe, especially Italy, and Japan. With the present fall in production, almost all Ethiopia's niger seed harvested is consumed domestically. Similarly to India, there is great potential for increased yield per hectare.

BOTANY

Guizotia (Cassini), named for the French historian Guizot, of the family Asteraceae, is a small genus containing six species with *G. abyssinica* (L.f.) Cass. (syn. *G. oleifera* DC.) the important oilseed, which together with other members of the genus has the somatic number $2n = 30$. Crosses between *G. abyssinica* and *G. scabra* subsp. *schimperi*, *G. scabra* subsp. *scabra*, and *G. villosa* produced well developed seeds, but not crosses involving *G. zavattarii*. Data obtained indicated that *G. abyssinica* and *G. scabra* are more closely related to *G. villosa* than to *G. scabra* subsp. *schimperi*, and that *G. zavattarii* has relatively low affinity to the rest of the taxa (Dagne, 1994; Murthy *et al.*, 1995). *Guizotia scabra* subsp. *schimperi* is a common weed of niger and other crops in Ethiopia.

Analyses of the seed oil of Ethiopian *G. abyssinica* and wild *Guizotia* spp. showed the differences were too small to be used in classification (Dagne & Jonsson, 1997). Three different types of niger are recognised in Ethiopia and are adapted to different seasons and soil conditions; *abat noug* is grown in the rainy season (June to December) on free draining soils, *mesno noug* later (September to January) in high rainfall areas where waterlogging occurs, and *benenge noug* in more dry lowland areas between July and October.

Significant differences exist between Ethiopian and Indian cultivars, and the former are much more variable than the latter, since there has been little selection in Ethiopia (Alemaw & Wold, 1995; Dutta *et al.*, 1994). Considerable local pressure has occurred naturally in India through district preferences (cf. sesame), and there are also many improved cultivars including Ootacamund and Alsi-1. Indian trials comparing Indian and Ethiopian types showed differing reactions to photoperiods, temperature and plant populations. Field trials in Ethiopia confirmed these basic differences, and under local conditions Ethiopian cultivars outyielded Indian. The reverse occurred in India, indicating a long-acquired continental adaptation.

Plant variability can be easily seen in any monocrop of niger in Ethiopia; plants show considerable variation in height, branching, number and location of flower heads, and period to maturity. This is also true of local types in India, but improved cultivars are more homogeneous. Semi-mechanised agriculture requires shorter cultivars with uniform flowering, less branching, increased seeds/head, higher oil content, and non-shattering heads. Not all these need to be incorporated in one cultivar; non-branching plus uniform flowering combined with a high seed number/head would reduce loss from shattering, since plants could be harvested at the optimum period. Alternatively, non-shattering heads alone would significantly increase yield from existing cultivars.

The considerable progress achieved shows the great scope for further improvement. The range of phenotypes available indicates that selection for desired plant characteristics should be both relatively fast and rewarding, and the existence of male sterility will assist breeders (Singh & Trivedi, 1993; Sujatha, 1997). However, selecting for a single characteristic may not be as efficient as selection based on related traits; in India these were seed/capitula, capitula/plant, and days to harvest (Mathur & Gupta, 1992; Goyal & Sudhirkumar, 1993; Adda *et al.*, 1994; Mishra, 1995; Alemaw & Wold, 1995). Such improvements are essential if niger's considerable potential as an oilseed is to be realised. A collection of 1200 local cultivars is maintained in India, about 900 in Ethiopia, and is available to plant breeders.

Fig. 8.11 Typical niger growth in pure stand, India.

Niger is a stout, erect, moderately branched annual herb, usually less than 1 m, with attractive yellow flowers, and small round black glossy seeds. Although, as noted, there are no distinct varieties of niger, there is considerable variation in local cultivars in colour of stem and leaves, height and degree of branching, time to maturity, seed size colour and constituents. The following description of niger is generally applicable and reference should be made to the literature for details of a specific local cultivar; for example the current Indian *Handbook of Agriculture.*

The root system is usually well developed, with a central tap root and branching laterals, but since the plant is often grown on poor or stony soils, roots seldom develop to their maximum. There are few studies on the rate or total extent of niger roots, but gibberellic acid applied to young seedlings doubled the rate of root elongation. Under experimental conditions, niger roots showed exceptional resistance to waterlogging, attributed to the existence of aerenchyma, which appeared to vary with the degree of soil saturation. Research indicates that niger roots secrete a compound that inhibits the germination of monocots.

The stem is usually round, approximately 2 cm in diameter when mature, up to 2 m in height but usually less, bearing small, soft hairs. Colour is green to reddish-green, depending on cultural conditions, and becomes more or less yellow with age. Local cultivars may have a distinctive stem colour, the green being replaced to some degree by red or reddish-yellow. The anatomy of the stem is described in Chavan (1961). Stems are usually moderately branched, and the degree of branching is mainly dependent on environment, as is the height at which branching commences; frequently at the six to eight leaf stage. Dwarf plants have occured in breeding programmes and could be used to reduce stem height in cultivars for mechanised harvesting (Dagne, 1994).

The extent of branching directly affects seed yield, since the capitula are borne terminally on stem and branches. Exision of growing shoots of seedlings in India increased branching, but depressed seed yield, while the number of branches and their capitula accounted for 50% of variation in seed yield. Thus the spatial arrangement of plants in the field has a major effect on potential yield. In general, stem and leaf biomass increase until the first capitulum appears, then steadily decrease. Shoot biomass was greatest at 120 days after emergence on selected Indian cultivars (Pandey, 1997).

Leaves are opposite, sometimes alternate on the upper stem, sessile, lanceolate, 10–20 cm long and 3–5 cm wide, softly hairy on both surfaces. The margin is dentate, and the blade tapers to a point. The usual colour is dark green, but lower leaves frequently show a distinct yellow tinge. The leaves of a local cultivar may be more or less yellow, occasionally with a reddish tinge. The capitula are surrounded by leafy involucral bracts arranged in numerous rows, the outer bracts being lanceolate, and up to 2 cm long. The inner rows of bracts become progressively smaller and finally merge into the flattened paleae of the receptacle, usually 1.5–3.0 cm in diameter.

The period from emergence to flowering is approximately 90 days, but as niger is grown in a wide range of environments and this period is directly affected by climate, it is more often shorter than longer. It can also be shortened by a reduction in day length. A plant may produce 30–40 flower heads under suitable conditions, exceptionally twice this number. The flowers develop from leaf axils in clusters of two to five. The author grew one plant from Indian seed in a container for demonstration purposes, and this produced 82 heads over a 26-day period, and grew to 1.25 m with numerous branches. The heads flower over an extended period, 15–30 days, and since this results in uneven ripening, losses from shattering can be high.

Flowers on the capitula are of two types, disc and ray florets. The latter number eight to fifteen, are a bright yellow becoming more golden with age, and sterile. Disc florets are usually fewer than 45, but occasionally more, and arranged in three whorls, those at the edge open first, followed progressively by the next in line to the centre of the head, as with sunflower. The process is usually completed in 5–8 days. A floret opens and liberates pollen early in the morning, the style emerges about midday and the stigma lobes separate and curl backward in the evening. The stigma lobes rarely curl sufficiently to touch the style, and the plant is thus basically self-sterile, although self-pollination has been recorded.

Insects are more important than wind as pollinators, and in India plants that flowered in cages to which bees were introduced had three times as many seeds per head than plants from which bees had been excluded. Similar results with bagged plants have been obtained elsewhere, indicating an adequate insect population is essential for high seed yield. Niger flowers are extremely attractive to insects, and for this reason it is often interplanted with sunflower, which doubled its seed yield in India from such an association.

Although niger is basically cross-pollinated, self-fertilised seed is viable, and dusting the capitulum with pollen 3–4 days after flowers open without emasculation, is a most efficient and simple technique. Isolation of breeding plots is important, and at least 250 m is recommended, and shielding with tall crops such as castor, sunflower or sorghum gives an added protection. Since a large number of seeds is produced per plant, selection for required characters can be rapid. Plants can also be regenerated in vitro from callus, and when transferred to a growing medium, 98% of seedlings so produced survived (Nikam and Shitole, 1997).

Temperature affects the duration at flowering and also the fertility of pollen, and on Indian cultivars a temperature of 30°C or above adversely affects both. At 25–27°C the flowering period for individual heads is 5–16 days, and the total flowering period for a single plant can range from 15 to 30 days.

The fruit is an achene, typical of the Compositae, small, some 3–5 mm long and 1.5 mm wide, almost lanceolate in shape, without pappus. The testa is hard, glossy and usually black, but mottled seed occasionally occurs. Normally 1000 seed weight is 3–5 g. There are usually 15 to 30 mature seeds/head, occasionally more, and a varying number of immature seeds or pops at the centre as occurs with sunflower. Unlike safflower, mature niger seeds are easily dislodged and shattering in the field before harvest is a major problem. Although not a practical solution, soaking seeds in diethyl sulphate produced plants with heads less liable to shatter. A reduction in hull content would make the seed more attractive to stockfeed manufacturers.

Seed oil content varies widely from 25 to 45% for unimproved types and 50 to 60% in selected cultivars; protein content can vary from 12 to 25%. The expressed oil is pale yellow, virtually odourless or with a slightly sweet scent, a nutty taste, and contains up to 75% linoleic acid, the reminder is mainly oleic acid, although selected strains have up to 85% linoleic acid (Dashagiriah & Nagaraj, 1993). A small amount of lignoceric acid, to 2%, may be present in some cultivars. Analyses of the seed of wild *Guizotia* spp. indicated that their oils were probably safe for human consumption (Dagne & Jonsson, 1997). Niger oil is frequently used in India as a substitute for, or extender, of sesame oil which it closely resembles. Analyses of niger seed and seed oil are shown in Table 8.6 and further analyses can be found in the chapter on processing.

ECOLOGY

Niger is basically a short-day, temperate-region plant that has adapted to a semi-tropical environment. From visual impressions gained in the field, the author believes that niger strains could quite easily be selected for a much wider climatic range.

Grown between 200 and 2500 m above sea-level, it can be found at much higher altitudes, in Nepal for instance. However, the highest seed yields are produced between 500 m and 1000 m under semi-tropical conditions with adequate moisture. In Ethiopia the main growing areas are between 1600 m and 2500 m bisected by the 10°N latitude, although it is sown in small patches throughout that country. In India it grows well on the plains and up to 2000 m in more hilly districts. There is thus a basic difference in environment between Ethiopian and

Table 8.6 Component acids (%) of niger seed oil.

Origin	Saturated				Unsaturated		
	C_{14}	C_{16}	C_{18}	C_{20}	Oleic	Linoleic	Linolenic
India	3.3	8.2	4.8	0.5	30.3	52.9	—
India	1.7	5.0	2.3	—	39.4	51.6	—
India	—	12.4	11.6	2.8	4.5	67.7	1.0
India		10 (total)			16.0	72.0	2.0
West Africa	1.1	7.0	3.2	0.6	17.0	68.1	3.0
Zimbabwe	—	7.9	10.6	1.0	7.0	72.6	0.9

Average composition of niger seed (%)

Indian		Kenyan	
Moisture	5.53–6.03	Oil	40.06
Oil	38.53–39.95	Crude protein	23.24
Albuminoids	17.94–19.02	Crude fibre	10.27
Soluble carbohydrates	17.99–19.32	Carbohydrate	22.40
Woody fibre	11.94–12.54	Ash	4.03
Soluble mineral matter	4.64–5.29	Calcium	0.37
Total nitrogen	3.02–3.26	Phosphate	0.36
Albuminoid nitrogen	2.87–3.04		

Indian types, which are sufficiently differentiated to vary in their reaction to specific climatic conditions, day length, temperature, rate of growth etc., confirmed by greenhouse trials (Abebe *et al.*, 1978) (Fig. 8.12).

Niger requires moderate temperature during growth, 18–23°C, and above 30°C the rate of growth and flowering are adversely affected and maturity accelerated. Ethiopian types are apparently more severely affected by high temperature than are those from India, and in California flowering of Ethiopian types occurred only in the range 15–21°C (Abebe *et al.*, 1978). For both regional types an increase in temperature of 5–10°C for other than a very short period reduces the number of blooms per plant and the fertility of pollen. Locally high temperature can also shorten the period from flowering to seed maturity, which is usually some 40–45 days.

In the Ethiopian highlands, night temperature to 9°C had no apparent effect on growth, but frost will kill young seedlings. Its effect on older plants is not well documented, but growers in Ethiopia sow to avoid frost. Ripe and dry seed on mature plants is little affected by short exposure to frost, in terms of viability or seed constituents, but pods often shatter on warming with substantial seed loss. Recommended sowing dates for selected countries are shown in Table 8.7, and are directly related to commencement of the rains or monsoon.

Niger requires adequate rainfall over the main growing period to produce a commercially acceptable yield. It is not primarily a dryland crop although it is often grown in low rainfall

Table 8.7 Recommended sowing dates.

India	
Maharashtra	Mid-June and mid-August
Madhya Pradesh	Mid-July or mid-December
Orissa	Mid-September
Karnataka	Mid-July
Bangladesh	Early-November
Ethiopia	Mid-May to mid-July
Tanzania West	Mid-November
Kenya West	Mid-June
Europe	March–April

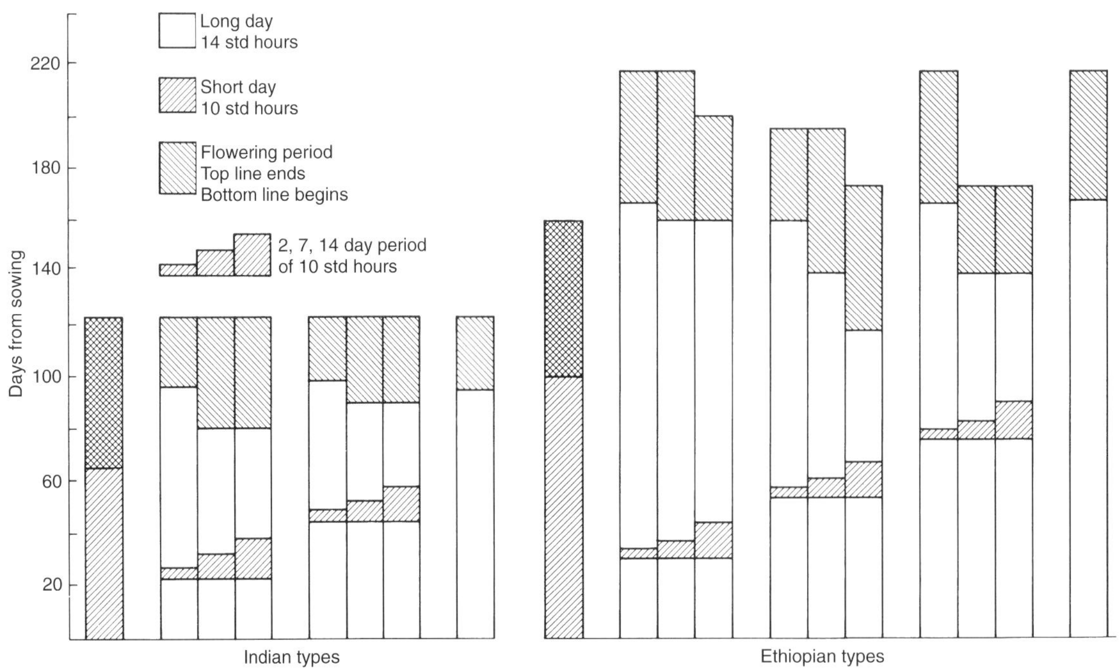

Fig. 8.12 Effect of varying photoperiod on flowering of niger.

areas, but this is usually due to the necessity to use more fertile land for food crops. A rainfall of 1000–1300 mm is the optimum, 800 mm will produce a reasonable yield if spread over the growing period, above 2000 mm growth may be depressed. However, plants can withstand periods of very high soil moisture during the vegetative phase as previously noted. For this reason it is considered a most suitable crop for hill regions of high rainfall and humidity in India. In Ethiopia cultivars that will withstand waterlogging are planted in high rainfall areas, or in soils with impeded drainage. High wind or hail when the seed is mature will cause severe shattering, but at other periods is of little importance.

SOILS

Niger can be found growing on a wide range of soil types, but appears to thrive best on clayey loams or sandy clays. Niger does not prefer poor soils but the more fertile soils are needed for food crops – as in India, where it is frequently grown on stony or lateritic soils. Invariably when demonstration plots of niger were planted on more fertile soils in the same locality, or given adequate nutrients and water, the difference in growth and yield was remarkable, especially in Ethiopia.

Referring to niger in India, an official report quoted in Chavan (1961) stated:

> It is grown during the rains, generally on the flat and stony lands which lie along the summits of trap hills, and which at first sight would seem to be entirely unproductive. The surface of the land is covered with nodules of red laterite soil of the thinnest description. Yet with resting fallows after every three or four years, it can yield very fair crops of niger or kodon, (*Paspalum scrobiculatum*).

In Ethiopia it is grown in the highlands on the dark-brown clays of the south Begemdir regions, the red-brown clay loams of Gojam, and more loamy clays around Addis Ababa. The common heavy clays and black cotton soils are not the most suitable for niger, although they are often utilised. Pilot plantings grew well on the sandy clay halfway down the catena in western Tanzania. It also grew well on sandy loams in north-eastern Congo (Zaire), and on similar soils in the southern Sudan. In Zimbabwe and South Africa it was grown as a green manure in the wetter maize-growing areas, on sandy clays or tropical red earths.

Many of these soils are very acid, and samples taken from fields growing niger in Ethiopia and India gave a pH range of 5.5–6.5. The most suitable pH range for niger has not been determined, but since acid soils are widespread in the highland tropics and crops that grow well on them are relatively few, it would probably be profitable to select for this characteristic within local niger types. Some niger selections are also moderately salt tolerant, a valuable attribute in an oilseed, especially so on the Indian subcontinent where the area of salt-affected soil is increasing. In saline soils of 4 mmhos/cm emergence was reduced by less than 10%, at 8 mmhos/cm emergence was halved, but in India existing cultivars are generally not sown on saline areas. (In SI units 1 mmho/cm = 1 × 105 siemens/m.)

FERTILISERS

The fertiliser requirements of niger are ignored in most smallholder sowings, the crop benefiting from fertiliser applied to suit the major component of the mixed crop. When sown in a rotation as a monocrop, niger receives the minimum of added nutrients, plus residual fertility from previous crops. Small growers do not consider niger important enough to warrant the use of precious cash resources needed more urgently elsewhere, with the result that yields/hectare are often extremely low. This may not be the best criterion on which to judge niger's potential, as in many instances it is not grown for sale but to supply domestic needs.

Where it is grown as a cash crop, in India for example, it frequently follows grain or cotton which received whatever fertiliser was available, usually crop or animal wastes. The ash and plant residues of the previous crop are usually all that niger receives, sometimes with a small amount of phosphate broadcast at final ploughing. Data from fertiliser trials are few, the majority from India, where a review of 20 years of niger production in India to 1988 showed little change in average yield or in fertiliser usage. However, the Department of Agriculture has recommended an integrated breeding and fertiliser use programme, since current cultivars show much greater responses to added nutrients.

The inoculation of leguminous plants with species-specific *Rhizobium* bacteria to promote nitrogen production and uptake through root nodulation is well known, and it appears that niger can benefit from a similar root association. Mixing soil in which niger had previously been grown with soil used to grow plants in pots considerably increased yield, and indicated that mycorrhiza which grew on the roots of plants growing in the inoculated soil contributed by increasing uptake of phosphate. This beneficial association may also explain why succeeding crops of niger grow well on relatively infertile soils. Inoculating seed with *Azotobacter* in India followed by 30 kg N/ha top dressing increased seed yield (Halder *et al.*, 1997).

The most common fertiliser used on niger in India is animal manure of indifferent quality, but the profitability of so doing is questionable. Assessing available data it appears farmyard manure at any level is unprofitable and coincides with the author's experience, since relatively small amounts of nitrogen in the same areas produced substantial yield increases.

Nitrogen is the nutrient that most often increases niger yield, with the response to potassium and phosphate extremely erratic and generally very small. Thus it is essential to carry out local trials when considering introducing niger. In East Africa for instance, observation plots sown by the author on a sandy clay showed excellent responses to both nitrogen and phosphate. However, since this was the first year niger had been grown in the area, subsequent response to phosphate might not have been so high if the root association previously mentioned became active.

Nitrogen

Nitrogen (N) is the only major plant nutrient that consistently increased seed yield, irrespective of country. In India it is considered essential for high yields from pure stands of niger, and in Ethiopia a similar result was obtained from nitrogenous top-dressing as occurred with sesame. A major reason for the very small amounts of nitrogen actually applied to the crop, irrespective of local trial results and official recommendations, is the difficulty many small farmers have of obtaining and paying for it! The situation is similar in Ethiopia, where farmers have great difficulty moving any fertiliser over the rugged and mountainous terrain, where mules are often the only method of transport.

Nitrogen apparently improves yield by increasing the number of flower heads per plant, the number of seeds per head and the weight of seed per plant, but once the optimum level is reached, the rate of increase of these yield components becomes non-significant. At higher application rates, seed-oil content can be depressed, as occurs with other annual oilseeds.

The amount of nitrogen applied should be directly related to niger's local profitability, and the yield increase to be expected. Such calculations are rare, and results of a trial in India illustrate their importance (Table 8.8) (Patil, 1979b).

Table 8.8 Increase in yield and cash return from applied nitrogen – India.

Nitrogen (kg/ha)	Seed (kg/ha)	Oil (1) (kg/ha)	Oilcake (2) (kg/ha)	Total return (1) + (2) (Rupees/ha)
0	172	68	104	489
20	213	84	129	608
40	247	96	151	699

Mean yields in table in kg/ha.

There was a very slight decrease in the seed-oil content at the highest nitrogen level. It is preferable to apply nitrogen as a top-dressing, since there are no data on its use in the seedbed. Application should be made when the first buds are noted, or when plants are some 30 cm high, whichever is the earlier. Applications above 50 kg N/ha are generally unprofitable given the present low average yield. However, some newer Indian cultivars such as Ootakmund, gave a greater response (Singh *et al.*, 1991; Sharma & Kewat, 1994).

Phosphate

Soil phosphorus (P) levels are low in many areas in which niger is grown, but published data shows a lack of, or low response to P applications in terms of yield increase. In general, phosphate alone is of little value in the absence of nitrogen. Conversely, the response to nitrogen is often greater in the presence of added phosphate. However, it is possible that the impression of a lack of response to phosphate is due to a lack of published results rather than the reaction to be expected in a particular district.

In East Africa, the author's observation plots of niger at several locations responded well to triple superphosphate at planting, and unpublished reports of Peace Corps workers in Ethiopia confirmed that this also occurred north of Addis Ababa. One reason for the lack of response to phosphate may be the mycorrhiza–root relationship previously noted, since if this is a general effect it can partially explain the irregular results obtained in India. Niger in that country is usually grown on the same land for many years either in pure stands or intercropped, and soil will contain the fungi concerned.

Seedbed application of more than 40 kg P/ha is unusual, and where niger has not previously been grown, this level should be used as a guide. Lower levels, down to 10 kg P/ha, can be sufficient when niger follows well-fertilised crops in a rotation. There are no published results comparing different types of phosphate fertilisers, but where a sulphur deficiency is known to exist, a situation common in Africa and India, single superphosphate is preferred. The alleviation of a phosphate deficiency has slightly increased seed-oil content, similar to other oilseeds. The effect of varying levels of phosphate on several plant characteristics is shown in Table 8.9. The low overall yield is typical of results obtained in India.

Other nutrients

The only responses noted to application of potassium has been from areas where there was a known deficiency limiting general plant growth. Most trial results on niger give either no, or non-significant yield increases to potassium. Sulphur has increased seed yield and seed oil content, but only when there was no deficiency in a major nutrient, as in India (Haldar *et al.*, 1997). No other nutrients are reported as necessary, but this is probably due to lack of research. Data from pot trials showed responses to Mg, Ca, Mn, Fe, Si and Na, and the content of these nutrients in plant parts at various growth stages (Nalawade & Chavan, 1990; Guggari *et al.*, 1995).

CULTIVATION

Land preparation for small grains such as sorghum, wheat and millet is suitable for niger,

Table 8.9 Effect of varying levels of phosphate on yield components – India.

Phosphate (kg/ha)	Yield (kg/ha)	Oil (%)	Yield (kg/ha)	Heads/plant	Seeds/head	Seed wt plant (mg)
0	194	39.1	167	20	7.2	505
20	213	39.2	196	23	7.4	585
40	224	39.4	191	21	7.6	604

Note: Consolidated results of several trials.

but since it is generally grown as an intercrop, lands are cultivated to suit the most important component, with the result that niger frequently suffers from unsuitable land preparation. This is particularly so when intercropped with pulses in India, and a striking improvement in growth can be seen when it is interplanted with eleusine millet in the same district. In Mysore, India, fields are ploughed without weeding or manuring, and seed is sown in July or August after the first heavy rains. Three months after sowing the crop is cut, left to dry for a few days, then threshed in the field.

Niger seed is small, therefore a level seedbed is essential to ensure even depth of planting and subsequent emergence. The most usual method of sowing is by broadcasting, and seed is sometimes mixed with sand to assist even distribution. The land is then harrowed to cover the seed, since few niger crops are mechanised. When a drill is used it is frequently animal-drawn, and often a locally made wooden type similar to the common Indian four-row drill, the goru (Fig. 8.13).

When mechanical planters or drills are used the seed should be sown 1–3 cm deep, depending on soil type and amount of seedbed moisture. Up to 10 cm is possible in loose soils, in which case the drill furrow should not be completely filled. Grain drills fitted with grass-sowing attachments, grass-boxes on fertiliser distributors, or precision seeders, can be used to sow niger. Since the seed is small, a method of accurately adjusting seed flow is essential, or seed can be mixed with sand or ash to give the required bulk. The seed should not be mixed with fertiliser, nor placed in contact with it in the seedbed.

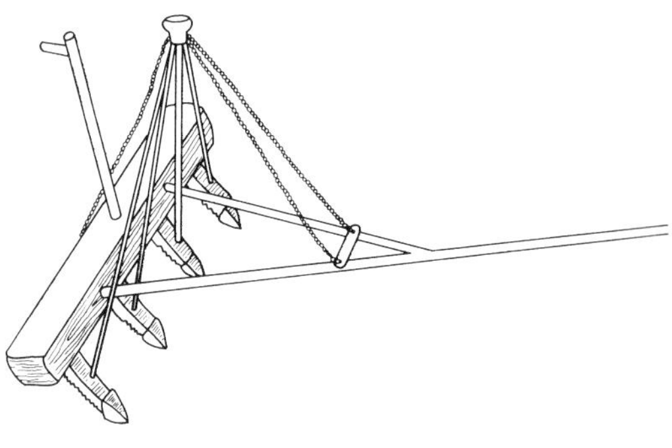

Fig. 8.13 Indian four-row wooden seed drill, goru.

Seed must be sown into moist soil, but the level of available soil moisture adequate to sustain germination and emergence varies with soil type. A higher level is necessary in sandy soil than clay, as radicle elongation after germination of niger seed is faster in clay than sandy soil. A seedbed temperature of 17–25°C is the optimum, above 35°C emergence is uneven, below 10°C impairs germination; thus there should be no danger of frost during germination or in the seedling stages. Niger germinates very quickly under optimum conditions, and can be above ground in 3–5 days in India, in Ethiopia 5–10 days, and comparison trials have shown this difference to be constant.

When sown in pure stands, row-widths vary from 30 cm to 50 cm and are mainly dependent upon soil moisture availability. In low-rainfall areas the wider row-width should be chosen. Seed rates of 6–12 kg/ha are usual, but seed rate is not critical as niger plants branch more profusely to fill gaps. In Assam for example, rates of 4–10 kg seed/ha had no significant effect on yield (Paul, 1995). However, in Orissa, populations of 333 000, 444 000 or 666 000 plants/ha showed the highest density and gave the highest yield at similar levels of N and P (Paikaray *et al.*, 1997).

Sowing rate for interplanted crops depends on the proportion of land a grower wishes niger to occupy in relation to the major crop; frequently 20–25%. When strip-cropped, it is often thinned after emergence to 15–20 cm between plants. Ethiopian cultivars give greater yield increases at higher plant populations than Indian, and this could become important in breeding programmes aimed at producing cultivars more suited to mechanised agriculture.

Most recommendations for niger growing are based on smallholder production, and mechanisation will greatly modify many of them. For instance, wide rows and in-row spacing produce tall, bushy plants that flower over an extended period, with seed heads at many levels. These plants are excellent for selective manual harvesting, but precisely the opposite type of plant to that required for mechanical harvesting. More appropriate was a fully mechanised Kenya crop, sown in 30 cm rows at a seed rate of 12 kg/ha, and harvested with a grain combine. Seed dressing as for sesame is appropriate, but where damage from soil insects is high, increasing the seed rate may be as effective.

WEED CONTROL

Weed control in niger is important to ensure high yield, and in India neglect which allows substantial weed growth can reduce niger yield by two-thirds, and is a major reason for the low average yields. Major weeds of niger in India include: *Solanum elaeagnifolium, Oxalis latifolia, Avena fatua, Imperata cylindrica, Parthenium hysterophorus, Alhagi comelorum, Pluchea lanceolata* and *Saccharum spontaneum* (Saraswat & Rabha, 1993).

In a clean seedbed niger seedlings grow rapidly once established, and their dense growth suppresses most annual weeds. Provided pre-planting operations have removed the majority of persistent weeds, two further weedings are generally sufficient – the first when seedlings are about 10 cm tall, the second not later than first budding or before the plants have spread across the rows, whichever is the earlier. Over-the-top harrowing is not recommended as seedlings are easily uprooted.

There is little information on chemical weed control, but those compounds that have proved harmless to sesame should be tried on a small scale to test their effectiveness. The parasite *Cuscuta chinensis* is damaging in India, especially in the hill country which is most suitable for niger. Not only is the yield depressed, but a dense growth of the parasite hinders cultivation and harvesting. The traditional method of control by hand weeding is seldom wholly successful, and should be combined with a herbicide which can be manually applied with a wand or similar applicator. Niger roots contain a water soluble compound which has an allelopathic effect on monocotyledons, and thus reduces weed numbers in succeeding crops. This weed-suppressing effect is known and utilised by Indian and Ethiopian farmers.

IRRIGATION

Niger is invariably grown in the wet or monsoon season, and little data available on the crop's irrigation requirements, nor consumptive water use. From personal observation of rain-grown crops, a soil moisture shortage in the seedling stage affects the growth of niger more severely than at other times. In contrast to other oilseeds, established niger plants can withstand high soil moisture levels and recover to produce a good seed yield. Slightly saline water may be tolerated by niger, and there are indications of increased tolerance to salinity in some cultivars, Alsi-1 for instance in India. Niger is considered more susceptible to calcium chloride than sodium chloride.

INTERCROPPING AND ROTATIONS

Niger is commonly strip-cropped as noted, and in India pulses and millet are favoured. In general a niger + low pulse mixture often results in a higher total seed yield than from either alone, as niger is attractive to insects and pollination is high. However, where the legume has a high local cash value, i.e. soyabean or groundnuts, the return from sowing them in pure stands may be greater. In Ethiopia niger may be interplanted with teff (*Eragrostis tef*), or sown in rows around field margins, but is usually grown in pure stands.

Millets sown following a niger crop in the Indian dry season can suffer moisture stress, since niger's deep roots reduce subsoil moisture. Conversely, niger seed yield was unaffected when sown after a variety of monsoon crops; a rotation of maize–niger gave the highest total cash return in Orissa. Niger is often grown for two or three years on the same plot, and under the local system of production maintains its yield, due probably to the mycorrhizal association previously mentioned. Niger is also sown into drying paddies following the rice harvest.

HARVESTING

Niger normally matures 120–180 days after emergence in Ethiopia, Indian cultivars mature in 75–120 days. The maturation period stated is often arbitrary, frequently being the mean of that period over which the heads become mature. This may well be 20 days, and since the seeds are not held firmly in the head as with safflower, those which ripen first begin to shed their seed before the last heads are mature. This is a major factor affecting mechanised harvesting. Shattering also substantially reduces profitability, as up to 25% of seeds can be lost.

Manual harvesting is normal, plants then being dried in the field for 2 or 3 days, bundled and taken to a threshing yard or corner of the field (India), laid in rows to dry and either threshed in the field or on the traditional threshing area (Ethiopia). When to cut the crop is also often an arbitrary decision, with other factors such as labour requirements elsewhere on the holding, or the necessity to harvest the dominant crop in a mixture being considered of greater importance than selecting the optimum period for niger.

The time of harvesting is more accurate when niger is strip-cropped (India) or grown in pure stands (Ethiopia), but even in these circumstances field losses are high. Thus the very low average yields often quoted are no real guide to niger's potential in a particular country. In Orissa, India, a crop planted at optimum spacing, fertilised and weeded yielded 1000 kg/ha compared with the local average of around 400 kg/ha. When such yields are demonstrated locally, as they have at many experimental stations in a number of countries, they can completely alter niger's profitability in district cropping programmes.

Plants are normally threshed by hand in India, but in Ethiopia where there are more often pure stands of niger, the plants may be threshed by oxen either treading out the seeds or pulling a small threshing sledge. All these methods result in soil contamination of seed, with the inevitable result that the crop's commercial value is drastically reduced. The use of special concrete threshing floors, or a heavy tarpaulin or plastic sheet is essential to obtain a clean sample. Hand threshing is not difficult, does little damage to the small, hard seeds, and in areas where labour is plentiful or has to be found employment, is probably the best method. The author demonstrated the effectiveness of small pedal-operated threshers, similar to the common rice thresher of South-East Asia; they are also portable and could be taken to the crop rather than *vice versa* with a substantial saving in labour.

Direct combining of niger presents no major problems, and most grain combines are suitable with the necessary adjustments. Niger seed is small, smooth and flows easily, thus all holes and cracks must be sealed. The author used a grain combine to thresh both standing and stooked crops, and accurate setting of the drum and fan is essential to prevent loss or cracking of seed. A well-threshed dirty sample requiring recleaning is far preferable to a clean sample with high losses from the back of the machine. It was necessary to wait for the crop to dry off before direct combining, and shattering loss was high, which a defoliant would probably have reduced. It was possible, although there was unfortunately no opportunity for tests, that a reaper-binder would have reduced this loss, together with closer in-row spacing to produce plants with fewer heads, which would have ripened more evenly. Reduction in potential yield would have been compensated for by the far more important increased in-bag yield.

When grown in pure stands as a cash crop, a seed yield of at least 1000 kg/ha should be the target. A yield of 500–600 kg/ha has been obtained from crops grown in pure stands on moderately fertile soils, and around 700 kg/ha at the Debre Zeit Experimental Station. Between 250 and 400 kg/ha is more common in India, and at this level niger cannot compete with other crops. Yields of 1000 kg/ha have been achieved by sowing newer cultivars without any major additional input, and there is no reason why this level should not be more widely obtained. Intercropped niger generally gave highest yield when the other component was sunflower, and least when inter-planted with sorghum. When the local niger cultivar continues to be grown, relatively

Table 8.10 Effect on niger yield of various cultural techniques – India.

	Seed yield (kg/ha)	Capitula/plant	Seeds/capitula	Seed wt/plant (mg)
Sowing date				
15 June	235	41	4.5	710
15 July	160	22	5.0	661
15 Aug.	238	21	7.1	899
Plant pop/ha				
500 000	198	16	7.8	476
250 000	183	25	8.8	815
150 000	153	29	10.7	1022
Weeding				
Nil	76	9	7.6	290
Hand	157	5	8.6	506
Hoe	107	10	9.0	366
Sown				
Broadcast	106	6.6	9.4	232
Behind plough	127	8.4	9.2	314
Row 20 × 10 cm	209	11.4	11.2	478
Row 30 × 10 cm	192	12.0	9.4	438

Note: The poor yields, which are the mean of 3 years, do not detract from the value and validity of the results, but are an indication of the low potential of unimproved local strains.

minor alterations to traditional practice can double the yield, i.e. planting at optimum period, drilling instead of broadcasting, or more efficient weed control. The effect of some of these techniques on yield is shown in Table 8.10 (Patil, 1979a & b).

A most important objective of breeding programmes was to increase seed yield, since this was considered a major factor limiting niger's profitability, although the author considers the improved agronomic methods already noted will dramatically increase yield from existing cultivars. Yield is significantly and positively correlated with the number of branches and capitula per plant, and both are also correlated with node number, while multiple correlations indicate these characters accounted for about 52% of variability for yield. Seed-oil content is negatively correlated with yield, but the degree of depression is small compared with the increase in potential yield, and thus oil yield/ha. Since there is also considerable variation in seed-oil content, selection could well balance this disadvantage.

STORAGE

Clean, whole niger seed stores easily, and being small it occupies relatively little space, but runs easily and must be stored in new sacks or holeproof containers. Niger seed is attractive to insect and other storage pests and must be protected or moved quickly to a bulk store. For smallholder and over-season storage, seed treated with an insecticide can be easily and cheaply stored in 200 litre drums with tightly fitting lids.

Seed required for special projects or experimental trials should be dried to 12% moisture content, and when kept in strong brown paper bags in a moisture free environment, retains its viability for at 2 to 3 years. For longer periods controlled storage is recommended.

PESTS

It is frequently stated in the literature that niger is attacked by few pests, but these opinions probably reflect the lack of published material rather than an accurate assessment of field conditions. That niger may be more resistant to many common pests attacking other crops is possible, since it often appears to suffer less damage than other crops when interplanted. Should resistance exist it would be a valuable asset, and a good reason to introduce niger into areas when pest damage is high and feasibility of chemical crop protection low. Resistance to the capsule fly, *Dioxyna sororcula*, may exist in some Indian cultivars (Ganguli *et al.*, 1992).

Leaf-eating caterpillars, aphids, mole crickets (Fig. 7.17), grasshoppers and cutworms (Fig. 2.17a), are most often recorded. Many of the important pests of sesame and safflower may also attack niger, and where these are grown in proximity or growing seasons overlap, damage is exacerbated. Insecticides are not commonly used on niger unless it is sown in pure stands, or the mixed crop or cultivated area is sprayed. Hand picking of large caterpillars, or baiting against soil insects is effective in smallholder crops.

The moth *Perigea capensis*, is common in India, Pakistan and South-East Asia, and can cause severe defoliation in particular crops. It can usually be found as odd caterpillars in the majority of niger crops, and is most damaging on young seedlings. The adult moth is dark brown, the caterpillar initially green but becoming more brownish with age, when it is about 2.5 cm in length. There can be several generations in a season. Since it also feeds on a number of crops specific control is impractical. Caterpillars in India are said to be highly parasitised by a number of flies and wasps. It also occurs in East Africa, but has not been recorded on niger in Ethiopia.

A hawk-moth, *Acherontia atropes*, causes minor damage in Asia, and *Diacrisia obliqua* larvae may move from more favoured host plants to niger when interplanted. Bollworms, *Helicoverpa (Heliothis)* spp. (Fig. 4.13) and the leafworms, *Spodoptera* spp., have been noted as damaging niger in common with other local crops in India, Ethiopia and East Africa. In Tanzania *S. littoralis* usually occurred only on niger growing close to cotton. In India *H. armigera* larvae bore and feed within the flower heads, but niger is not a preferred host.

Macrosiphum (Dactynotus) spp. are the commonest aphid recorded from niger, and can become a major pest. They are small black insects, often occurring in large numbers on the stem and young leaves. Plants become stunted, and the heads may be undeveloped. *Macrosiphum carthami* is most common in India, with *M. compositae* later in the season. The latter is also common in East Africa and the lower elevations of Ethiopia. Thrips, *Frankliniella schultzei*, have been noted as infesting flowers in Ethiopia, and damaging buds. The extent of damage has not been estimated, although thrips can frequently be found on niger in Ethiopia. Thrips were not noted in any numbers in East Africa, and the species concerned not identified. Leafhoppers of unknown species occured in Ethiopia, Zaire and East Africa, but in India the leafhopper *Orosius albicinctus* transmits phyllody disease.

Pentatomid bugs can be seriously damaging in eastern Africa, where they feed on the young shoots and developing flower heads. *Sphaerocoris annulus* is most common in Ethiopia, and *Calidea dregii* and *Nezara viridula* in East Africa (Figs 5.11 and 7.18). A lace bug, *Galeatus scrophicus*, reported feeding on niger leaves in India, may have migrated from neighbouring sunflower; it is a pest of the Compositae generally in East Africa. A stem-borer considered to be *Melanagromyza* spp., was found in odd plants in Tanzania. In Ethiopia the fly *Dioxyna sororcula* is an important local pest whose maggots bore pods and are very damaging in specific crops: it also occurs widely in East Africa. In India it may be responsible for more damage than is currently recorded.

Grasshoppers, mainly *Chrotogonus* spp., can become damaging to interplanted niger in India, less so to pure stands. Grasshoppers are not generally pests in East Africa, with the exception of *Cyrtacanthacris tatarica* which is also a minor pest of other oilseeds. Emerging seedlings were eaten off at soil level by mole crickets, *Gryllotalpa* spp., (Fig. 7.17) and local damage in a specific crop could be severe. Developing heads were found to be infested with small maggots, probably larvae of *Larinus* spp., in East Africa, but have not been reported from elsewhere. Termites can attack roots of maturing plants, and in India *Odontotermes wallonensis* caused significant damage to niger and other oilseed crops.

Birds can become a pest when pods mature, often attracted by fallen seed. Damage is exacerbated by birds attempting to perch on plants, which causes additional shattering.

DISEASES

Reliable reports on niger diseases are lacking, and it may well be that the plant is resistant to some of the more widespread pathogenic fungi infecting other oilseeds. It certainly appeared to be so in India, where it was often visually less affected than neighbouring crop plants. Most records of niger diseases occur as passing references in literature dealing with other aspects of the crop, or general reviews of agriculture in a particular country of district. The information quoted includes personal observations or conversations with oilseed growers and research workers. There is a great need for more work on the subject.

Leaf spots are commonly seen in niger plants, but are seldom accurately identified. *Cercospora* spp. and *Alternaria* spp. are the most frequently recorded from India and Africa, but with no comments on the degree of damage. A leaf spot caused by *Macrophomina phaseolina* has damaged young seedlings in India. Bacterial blight due to *Pseudomonas* spp. occurs in Africa and India, and a root rot is caused by *Phytophthora* spp. on young plants in India. Powdery mildew, *Sphaerotheca* spp., has been reported attacking niger in India, but not from East Africa or elsewhere. It is apparently more damaging in pure stands than intercrops; in the USA *Erysiphe cichoracearum* attacked plants in Missouri.

JOJOBA

Jojoba, *Simmondsia chinensis,* is the only species of the genus *Simmondisa,* one of six genera in the Buxaceae, family Euphorbiacae. *Simmondsia chinensis* is so named because it was mistakenly believed to have come from China, and although *S. californica* is more apposite having regard to jojoba's distribution, *S. chinensis* prevails on seniority. Jojoba, pronounced hohoba, is believed to be a corruption of the local Papago name, *howhowi.*

The origin of jojoba is obscure, although its present wild distribution is limited to desert areas of the south-western United States, Baja California and north-west Mexico (Fig. 8.14) (A1 Aini *et al.*, 1972). One theory suggests this distribution is due to the browsing pressure of herbivorous Pleistocene mammals (Gentry, 1972), and as jojoba has certain phylogenetically primitive characteristics it could well be a remnant of a prehistoric ecosystem. The earliest reference to jojoba is contained in the report of an expedition of Jesuit priests to New Mexico, Padre Luis Verlarde's *Description of Pimeria Alta.* Verlarde visited the area in 1716 and wrote of the medicinal use of jojoba fruits by local Indians (Burrus, 1971). These observations predate those of Francisco Clavijero in his *Storia della California* by nearly 75 years.

Jojoba was extensively utilised by the indigenous Papago people, with many folk and traditional medicinal uses for the seed. The plant has many names in its native habitat, including nutbush, bucknut, deernut, goatnut, pignut and lemon leaf. Most indicate the method of utilisation by man or animals. The first modern extraction of seed oil for analytical purposes was made at the Boyce Thompson Southwestern Arboretum in Arizona in 1929, from bushes planted there in 1925. Seed from the Arizona area was sown in Australia at Broken Hill in the early 1930s as part of a programme to revegetate areas denuded by mining operations, and two female plants remain. In 1933 it was discovered the seed oil was in fact a liquid wax, and in 1936 its unique

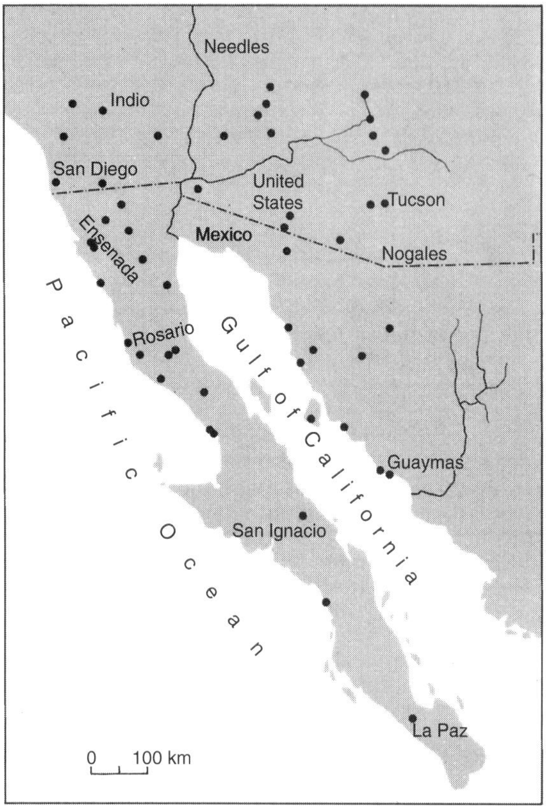

Fig. 8.14 Distribution of *Simmondsia chinensis.*

properties were described almost simultaneously in the USA and England.

The agronomy and possibility of commercial jojoba cultivation was seriously investigated from the 1950s, and in 1978 some 70t of seed was hand-harvested from wild plants in the USA. However, it is reliably estimated that total annual seed production in the Sonoran desert is between 15 and 50kt depending on the season. Most seed is virtually impossible to harvest due to the harsh climate and scattered distribution of plants. Subsequently jojoba plantations were established in a number of countries, with varying degrees of success. By 1981 about 6000–7000ha of jojoba plantations had been planted in the USA, 7500 in Mexico, and there were pilot plantings in Australia, Argentina, and South Africa, Israel, and India (Punia *et al.*, 1991).

However the original optimism for jojoba rapidly evaporated as the difficulties of commercial production became apparent. By the mid-1980s only serious growers were still in production, with world oil demand and supply roughly in balance from 1990; by 1998/9 world production was estimated at 1550t annually and steadily increasing. Jojoba's potential as a specialised oil crop remains to be established, as the total market will be basically determined by how much oil becomes regularly available and at what price. Important agronomic factors that affect this objective are early maturing plants, even flowering, high and stable yield, cold and salt tolerance, and acceptance of pruning.

BOTANY

Simmondsia chinensis (Link) C. Schneider, (syn. *S. californica Nutt.* and *S. pabulosa Kellogg.*) Originally described as *Buxus chinensis,* it was later renamed *S. californica,* and placed in the monotypic genus *Simmondsia.* The systematic position of the *Simmondsia* has also been questioned; it is generally placed in the Buxaceae, although a monotypic family Simmondsinceae is proposed. The chromosome number of *S. chinensis* is $2n = 52$, and it has been described as a relictual polyploid, whose chromosome number indicates a polyploid origin but which is nevertheless monotypic. Plant breeders in several countries have produced named cultivars which have been registered under Plant Breeders Rights (PBR) Acts; Baringji, Wadi Wadi and Waradgery in Australia for example.

Seed used initially to produce plants for research was obtained from cross-pollinated, heterogeneous, wild plants, with very wide variability in physical characteristics. Thus in Israel no superior seed producers were detected after a decade of selection and hybridisation. The monoecious strain of jojoba developed by Yermanos at the University of California was of great significance in the production of superior hybrids. However, it may be possible to select for gross physical difference in the short term, since plants from a specific locality are apparently adapted to the local environment; for example plants from very dry areas are considered more drought resistant.

Jojoba is a woody shrub or small tree, multi-stemmed, to 6 m but usually around 2–2.5 m, occasionally almost prostrate, evergreen, drought resistant, dioecious, the nut-like seeds produce a liquid wax possessing unique qualities for a vegetable oil. Wood from mature plants is hard and heavy, lemon yellow, and without distinctive scent or taste. The simple tree-ring dating technique cannot be used on jojoba, since it has a unique stem anatomy.

Most observations on the growth and development of jojoba were originally on wild plants, or plants grown from seed collected in the wild. Such seed, as noted, will produce a heterogeneous collection of plants highly variable in their characteristics; however current cultivars, some of which are clones, behave more uniformly. The following remarks should be accepted as generally accurate, and caution exercised in their specific interpretation.

Following germination the seedling quickly produces a penetrating tap-root, and this is well developed before the apical meristem becomes active. The rate of tap root growth under controlled conditions has been recorded at 1.5–2.5 cm/day with a length of 50–60 cm before shoot emergence, but after emergence the rate of root growth is reduced. When seedlings are raised in a nursery for planting out, each should be sown in an individual biodegradable pot or sleeve. Fatalities following transplanting are invariably high, and due primarily to root damage.

Under suitable conditions of temperature and adequate soil moisture, the root system rapidly proliferates, but with relatively little lateral spread. The tap root of mature plants can be 15–25 m below the soil surface, with substantial parallel laterals and secondary roots, giving an ability to draw moisture from a considerable volume of soil. Together with the sclerophyllous leaves this enables the plant to grow in arid regions.

Plants are usually multi-stemmed in the wild (Fig. 8.15) but the extent and degree to which this occurs is most often due to environmental pres-

Fig. 8.15 Multi-stemmed growth typical of wild jojoba.

sures, i.e. prevailing wind, or heavy browsing by animals as the plants are most palatable. There is probably no type which can be accurately described as typical, except in relation to a specific region. Local strains within the wild population differ in the type of stem and the degree and type of branching, from almost prostrate to upright types with only primary branches. Current cultivars are more uniform.

Jojoba seedlings grown in open nurseries are about 30 cm tall after 1 year, but similar plants pruned to a single stem and grown in a heated greenhouse can reach 200 cm, and growth regulators have increased the rate of stem growth in seedlings. Under favourable conditions in California, jojoba grew 15–30 cm annually for the first 10 years, but it is considered a higher mean minimum temperature would increase growth-rate. Since flowers are produced only on new growth, plants that rapidly increased in size could also be expected to have proportionately increased seed yield.

Leaves are faintly hairy, broadly ovate, opposite, vary considerably in size both between and within plants, in the range 14–40 mm long and 6–20 mm wide, but can be longer and wider. Leaf colour is usually a dull, dark green, lighter when young, but the leaves often have a bluish tinge and appear grey at a distance. A wholly yellow (albino) type has occurred. Leaves are sclerophyllous (rather hard and stiff), and remain on the plant year round, which is a somewhat unusual characteristic for a plant in a desert habitat. The ability quickly to reduce the rate of transpiration and maintain the lower level for long periods, is related to leaf structure and position of stomata, and considered an important survival mechanism.

Specific aspects of leaf physiology and composition have been investigated (Loreto *et al.*, 1992; Rais *et al.*, 1993; Mallet *et al.*, 1994). Jojoba is dimorphic in secondary sexual characters in some desert populations, where females on average have larger leaves and more open canopies than males. The advantages of this sexual dimorphism was examined in a Californian population (Kohorn, 1994).

Plants are normally dioecious, producing male and female apetalous flowers on separate plants, although hermaphroditic plants exist, and no practical method exists of determining the sex of fruits. Female flowers are usually single, silvery-green, similar to new leaves, while male flowers are yellow and occur in small clusters. Flowers are normally produced at alternate nodes, but far more female flowers are produced than seed set, although why this should be so is unclear. Individual plants may produce from 2 to 10 flowers/node or flower at every node. This last characteristic could be valuable in terms of fruit production per bush if it does not affect total weight of seed produced or seed-oil content.

Plants normally flower in 20–24 months, and cultivated male plants flower earlier than female; thus plants which flower in their second year are generally male. However, male plants flowered in 5 months and females in 9 months when grown in an Australian cyclotron. The ratio of male: female plants in the wild varies from 1:1 to 1:5, but is affected by age and environment.

Temperature and moisture are the main factors affecting time of flowering, which normally occurs during late winter/early spring. Although a temperature differential is necessary for bud development, low temperature or frost kills flowers or substantially reduces pollen viability. Individual plants vary in the number of flowers produced, and the period over which this takes place. Some have a short, intense flowering pattern, others produce a series of flowers over a period. Plants tend to flower earlier, more profusely and continue to flower for a longer period when irrigated than when rain-grown.

Flowers are wind-pollinated and being dioecious also cross-pollinated, thus a wide range of genetic variability must exist in any long-established wild population. Wind pollination is effective, since jojoba pollen readily travels at least 30 m on a relatively mild breeze. Stored pollen in India remained viable for up to 60 days (Chabra *et al.*, 1994). Insects in general are probably important pollinating agents and bees are frequent visitors to male flowers, thus beehives in plantations could be an asset.

The fruit is a green, elongated oval capsule, enclosing one to three seeds, although usually only one seed develops (Fig. 8.16). A four-carpelled mutant exists which is considered valuable as genetic marker (Estilai & Hashemi, 1993). Fruit size and shape is variable, on individual plants, between regional strains, and in single or multi-seeded capsules. When ripe the capsule splits to expose the seed, which can fall quickly or remain attached for several months. It is believed that this may be a strain-specific characteristic, although influenced by seasonal conditions.

If a dry period follows fertilisation, capsules do not complete their development until conditions again become favourable, and a prolonged drought will cause most plants to abort their fruit. Fruits mature some 3–6 months after fertilisation, and indications are that a longer period to maturity increases seed weight. The age at which a jojoba plant produces seed depends almost entirely on the environment, about 10–12 years in the wild. Irrigated plants produce seed within 3 years, but 6–8 years is generally considered necessary for trees to reach a sustained high yield.

Jojoba seed resembles a pecan nut or acorn, with a small, pointed apex, and flattened, slightly indented base. The seed is usually some 12–18 mm long and 6–12 mm in cross-section, which is often roughly triangular (Fig. 8.17). Seed varies greatly in size and weight, from 500 to around ten times this per kilogram (Fig. 8.18) (Yermanos, 1975). However seeds are generally constant from wild plants in a specific location. A commercial standard of 1600 seeds/kg has been widely accepted, and is also used as a standard in experimental work.

Jojoba seed contains little or no endosperm, consisting of an embryo and cotyledons enclosed in a hard testa. The testa when ripe is dark, sometimes reddish-brown, tough, leathery, more or less covered with very small, fine hairs (trichomes), and up to 0.2 mm thick. The cotyledons are large, their parenchymatic cells filled with liquid wax, which flows freely when the seed is

Fig. 8.16 Jojoba fruit capsules in situ, multi-flowered type.

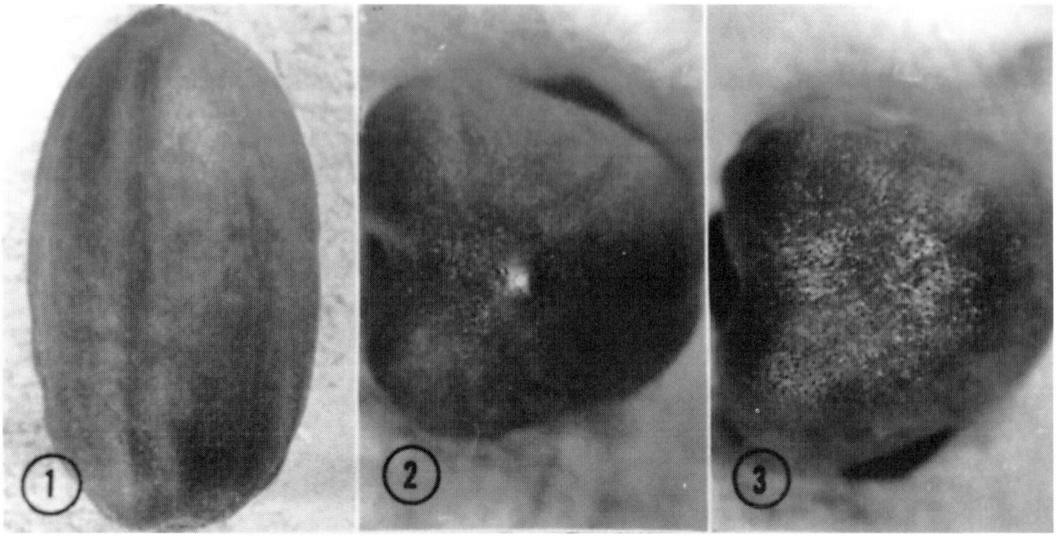

Fig. 8.17 Jojoba seed. (1) side view, (2) micropylar end, (3) chalazal end showing trichomes.

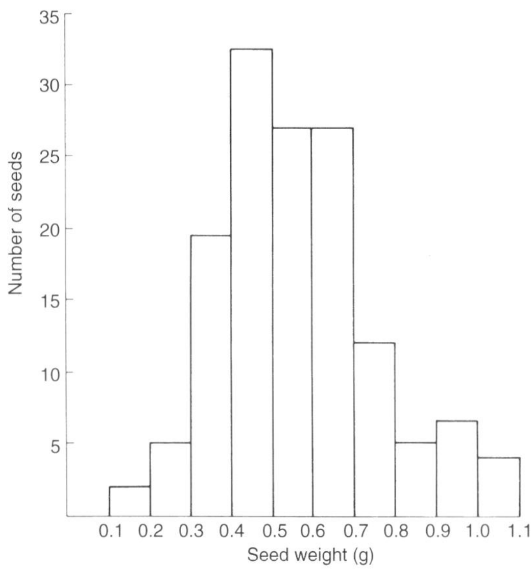

Fig. 8.18 Jojoba: mean seed weight.

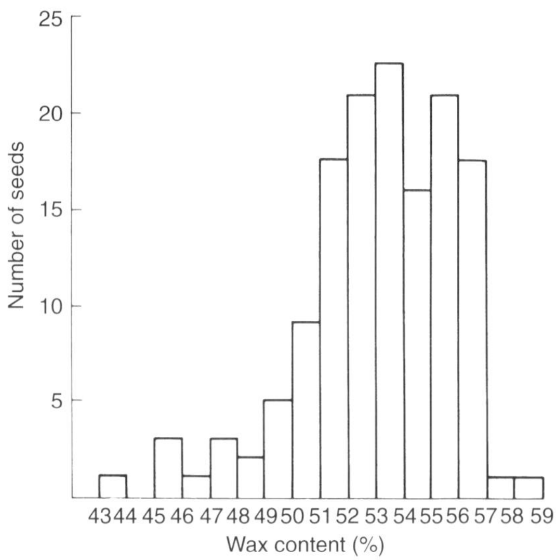

Fig. 8.19 Jojoba seed: per cent wax content.

Table 8.11 Composition of California jojoba seed (wt %).

	Moisture	Nitrogen	Fibre	Fat	Ash
Whole seed	4.7	2.3	6.6	43.6	2.2
Wax	0.1	0.04	—	100.0	0.0
Hull	8.4	1.4	14.0	3.5	4.2
Hull-free seed	4.2	2.5	5.8	51.4	1.7
Pressed meal	4.3	3.6	9.8	15.3	3.4

sliced. Wax content of the seed increases rapidly during the first 4 weeks of development from 13% to 40% and then more slowly to 50% or 60% at maturity. There is a significant positive correlation between seed weight and wax content. Wax composition appears to be very uniform from all localities (Fig. 8.19) (Yermanos, 1975).

Protein content is 26% to 34% and negatively correlated with wax content. Seed composition and constituents have been studied in detail (Wisniak, 1994; Shockey *et al.*, 1995). The composition and constituents of seed collected from Californian plants is shown in Table 8.11 but seed from other sources may vary.

The liquid wax is a clear golden colour, known commercially as jojoba oil, and consists almost exclusively of long-chain esters, predominantly C_{40-42} fatty acids and alcohols, but contains no glyceride esters, the major constituents of other vegetable oils. Analyses showed that jojoba oil molecules contain two double bonds separated by an ester bond, and thus are the source of a large number of intermediate or final products, including industrially important waxes. The oil has a wide range of uses including specialised lubricants, cosmetics and pharmaceuticals, and the solid wax where its specific characteristics are required. The only other natural liquid wax used in commercial quantities was sperm whale oil, and to a lesser extent carnuba wax from *Copernicia cerifera*. More recently the oil from certain fish species including orange roughy (*Hoplosthetus atlanticus*) found mainly in Australasian waters, is considered close to jojoba oil. Characteristics of jojoba oil are shown in Table 9.15. A number of synthetic substitutes for the oil and wax have been developed and described in the literature, but there is now sufficient natural oil available to render these unnecessary.

ECOLOGY

The natural range of jojoba is restricted to areas of south-western USA/north-western Mexico, between 23 and 35°N latitude, in basically arid, desert habitats (Fig. 8.14). However, it is believed this distribution has resulted from grazing pressure, historic and prehistoric, since jojoba is a favourite browse plant for wild and domestic animals. Jojoba is thus virtually eliminated from the habitat where cattle are allowed freely to roam, and grows only where protected. Outside its natural range jojoba plants have grown well in Israel, the Sudan, Saudi Arabia and Australia, and are being grown experimentally in many other countries.

Jojoba is usually found on dry slopes at elevations of 300–1500 m, is frost-sensitive, and the lower elevation is locally imposed by temperature; it can be found growing at sea level in frost-free Baja California for instance. Temperature affects germination, growth and flowering, and a lower limit of 12–14°C is probably critical for germination and growth, with root development restricted at 21°C (Table 8.12) (Reyes *et al.*, 1977). Night temperature around freezing will injure seedlings to some extent, and –4°C is usually fatal. Vegetative growth and flowering may be suppressed or destroyed by prolonged near freezing temperature.

Plants are able to compensate for this loss by producing more flowers later in the season, but if damage is extensive at the normal main flowering period, the yield will be substantially reduced. Optimum growth occurs in the range

Table 8.12 Effect of soil temperature on growth of jojoba seedlings – USA.

Dry weight (g) of:	Soil temperature (°C)		
	21	27	33
Leaves	20.3	21.2	25.0
Stems	25.8	27.6	32.2
Roots	8.1	8.9	9.3
Total	54.2	57.7	66.5

27–36°C, but wild plants can withstand very high temperature. A daily range of zero to 50°C has been recorded in the Mexican desert habitat, but above 50°C is believed to suppress growth, although it is not lethal. The specific effects of low temperature and freezing on jojoba has been investigated (Benzioni *et al.*, 1992; Ceccardi *et al.*, 1995).

In the wild a positive link exists between winter rain and early spring temperature, and onset and degree of flowering. A temperature drop to around 15°C is considered necessary to promote flowering, and this requirement may limit jojoba's introduction into some tropical and subtropical regions. However, where this criterion is met, as in the Sudan, plants grow vigorously and flower profusely.

Rainfall in the plant's natural habitat is 75–450 mm annually, which indicates high drought tolerance, and as rain may not fall every year the plant is truly drought resistant. Plants of up to 3 m were observed in the 250–400 mm rainfall area, and this is probably the minimum necessary for irrigated plants. Regular rain may be unnecessary to established plants since the root system of jojoba allows it to make extensive use of subsoil water. Under irrigation plants are able to utilise the higher moisture levels to produce greater seed yield, since moisture stress is a major factor limiting flower development and seed production.

The effect of photoperiods on mature jojoba plants remains to be fully determined. However, seedlings grown in controlled conditions reacted positively to an increasing light period up to 24 hours duration. This strong reaction is limited to the seedling stage, but can be utilised to produce large seedlings for transplanting to the field. However, jojoba plantations are probably best established in regions with similar conditions to their wild habitat.

SOILS

Jojoba will grow in a variety of soils, but a basic requirement is it must be free draining. In its natural habitat jojoba is usually found on sandy gravels, sandy clays and alluvial soils, and seldom on the more heavy clay in valleys. However, this may again be a temperature factor, since these low-lying areas are significantly colder at night than more sandy hillsides. There should be no hard pan, so often a feature of tropical laterites or river-deposited alluvia, to restrict growth of the deeply penetrating root system. Thus soil type per se is not a limiting factor, and soil structure and extent of soil moisture retention is more important. Soil reaction does not appear to be critical, since tests in wild stands of jojoba gave pH 5 to pH 8.

Jojoba is considered salt tolerant, a valuable characteristic in any crop, especially an oil producer. Healthy wild plants have been found growing within a few metres of the sea, and in one instance saline water with an electrical conductivity of 24 mmhos/cm occurred in the subsoil 2 m below plants. Plants grown in pots irrigated with a saline solution of 13.6 mmhos/cm for 2 years grew well and flowered, with no obvious symptom of major injury. (In SI units 1 mmho/cm = 1×10^5 siemens/m.) Specific aspects of salt tolerance of roots and seedlings have been investigated (Rasoolzadeghan, 1991; Burgos *et al.*, 1993).

FERTILISERS

Topsoil fertility is of little importance to wild jojoba, whose deeply penetrating roots explore a large volume of soil for both nutrients and moisture. Application of nitrogen and phosphate to mature cultivated plants has thus produced little response, although seedlings grown in a nursery usually do so. Response may also be related to root type, since seedlings produced from cuttings instead of seeds have a more fibrous, shallower root system and would therefore probably benefit from fertiliser application. Soil temperature also influences nutrient uptake by seedlings. Research into the specific or local nutritional requirements of jojoba and methods of determination continue (Moll, 1990).

CULTIVATION

Jojoba is basically an orchard or plantation crop, and land preparation necessary for establishment depends on location and environment. When grown in arid areas with minimum rainfall, around 200 mm, land preparation will probably be most concerned with conserving rainfall and soil moisture. In higher-rainfall regions soil conservation will be important since jojoba gives little soil protection. Jojoba can be a most useful cash crop in semi-arid regions presently underutilised or subjected to inefficient and wasteful grazing methods, and well-known dryland farming techniques can be applied. Plants can be grown in trenches, holes, or graded terraces in more underdeveloped areas, while methods of channelling rainfall, run-off or dew by means of plastic or soil-fixing artificial catchments can be used where practicable (Fig. 8.20).

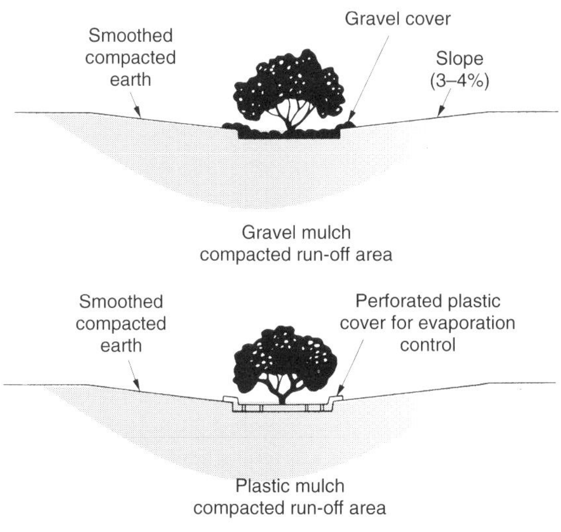

Fig. 8.20 Types of artificial catchment.

There appears to be little difference in viability between large and small sound seed, and unless a standard size seed is necessary for use in planters, grading for size is unnecessary. However, many growers consider large seed produces faster initial root growth and thus more vigorous seedlings.

Jojoba can be sown directly where it is to grow, or seedlings produced in a nursery for later transplanting. Direct seeding is normal, but warmth and moisture are essential, and a seedbed temperature in the range 25–35°C appears optimum for high germination; below 20°C and above 40°C normally reduce emergence. In suitable conditions emergence takes 5–7 days, but as the root initially grows much quicker than the shoot, emergence may be 20–25 days after sowing. The natural temptation to dig down to see how plants are progressing should be resisted, since tap roots are initially fragile and their loss or breakage severely retards or kills developing seeds. Where soil-borne pathogens are known to cause seedling damage, seed for direct field sowing should be treated with a fungicide. Seed should preferably be sown at least 5 cm deep, but depth depends primarily on soil moisture availability.

Where seasonal rain occurs sowing should be during this period, provided soil temperature is high enough. Direct hill-seeding requires a minimum of 4 seeds/site to ensure an 80% probability of having one plant of each sex. At flowering these can be thinned to leave one of the desired sex. Commercially available planters can sow jojoba, and provided the selector mechanisms are suitable for the relatively large, oily seed, this operation is no more complicated than sowing any large-seeded field crop. Adding a small amount of talc or similar substance to the seed hoppers, as for castor, will absorb any surplus oil and reduce the danger of clogging selector pawls and seed-plates. Using two seeders mounted on a tool-bar, some 25 ha can be planted in 1 day.

In more arid areas, or where the soil may be compacted, special units have been designed that will break up this pan, rotovate, water-inject and sow seed in one pass. In a relatively hard, somewhat infertile soil in the USA a tractor-mounted post-hole digger was used to drill down to about 1 m. Loose soil was removed, mixed with a small amount of compound fertiliser and replaced. Seedlings still in their biodegradable pots were then watered in.

Sowing seed in nurseries and then transplanting seedlings to prepared field sites speeds plantation establishment and saves seed. Seed may be pre-germinated in a suitable moisture-retaining medium at a temperature of 30°C, and immediately it proves viable should be transplanted into pots, which can be of any suitable biodegradable material. The potting mixture is apparently unimportant, provided it retains moisture without becoming saturated. Germination is hypogeal, with emergence after 10–20 days, and seedlings can be planted out at 8–10 weeks.

Since there is no efficient method of sexing seedlings and male plants are necessary for fertilization, all had to be planted out or retained in the nursery until 2 years of age. Any plants which flowered during this period were normally males. This delay was commercially unprofitable, and overplanting and subsequent thinning were far less expensive. However, several commercial methods of vegetative propagation have been developed, and cloned material is offered for sale by private companies, while tissue culture can quickly propagate progeny from selected parents (Kackar *et al.*, 1993; Palmer & Keller, 1994).

If orchards are to be irrigated, plants with a large spreading root system are to be preferred to the normal deep-rooted type. A method used in California produced seedlings with this type of root system. Seed containers were placed in boxes with an air space underneath. Seedlings were air-pruned, and produced many fibrous roots rather than a few long tap roots, and seedlings produced from cuttings also tend to a fibrous root system. Plants can be propagated from stem-tip cuttings of new growth; leaves should not be removed except from that portion of the stem in the rooting medium, preferably sand or vermiculite, which must be kept wet (Palzkill & Feldman, 1993). Rooting normally occurs from 30 days after potting, and a major proportion of cuttings should strike.

Seedlings may be transplanted into the field by hand or mechanically, and both methods have been used successfully to establish commercial plantations. Where seedlings have been raised in a glasshouse, they must be hardened off outside for 10–14 days before being planted out. Care is necessary when transplanting not to injure the brittle main roots, and it is recommended that seedlings be irrigated or watered at or as soon as possible after planting.

The optimum distance between plants in a plantation must obviously depend on the local environment, especially in arid regions where plants are expected to grow on rainfall alone. In Arizona 5 m between the rows is considered the minimum to allow efficient mechanical inter-row cultivations. Seedlings are planted at an in-row spacing of 45 cm or 15–30 cm if direct-seeded, but hill seeding with later thinning may be more efficient. With a row spacing of 3.5–5.0 m, the optimum distance between mature plants would appear to be 1.2–1.6 m. Since direct-seeded plants usually produce long tap roots rather than spreading roots, fairly close spacing in long single rows may be possible where there is subsoil moisture, i.e. along wadis and river beds, or where water-spreading techniques are practised. Non-mechanised harvesting and weed control will allow closer row-spacing, around 3 m.

A diamond planting pattern is to be preferred in orchards rather than on-the-square. Continuous rows of hedge-type plants will probably be 4–5 m apart, with trees spaced approximately 1 m apart, one in five being male. Another factor modifying planting pattern, or spacing, is the ratio of male to female trees necessary to ensure adequate pollination, and spacing may be closer if insects and bees are locally abundant or supplied. A ratio of 1 male to 5–7 females has proved satisfactory in the USA, but may vary elsewhere and should be determined empirically. Surplus male and underproductive female trees can removed when necessary, and this is less expensive than attempting to introduce additional male trees into an established plantation (Fig. 8.21).

Mechanised plantations usually require tree spacing to suit the equipment, and hedge-type

Fig. 8.21 Three year old jojoba plantation in Australia.

development of plants is popular. Multi-stemmed and hedge-type plants, once produced, can be mechanically maintained in the desired shape. Pruning of individual trees may reduce the time to flowering and increase the number of flowers, but there is wide variation in individual plant response to the operation. Growth regulators can affect type and extent of branching, and also flower production (Ravetta & Palzkill, 1992).

WEED CONTROL

Seedlings in dry regions should be protected from competition until they reach 0.5 m; subsequently annual slashing of weeds and removal of shrub and tree seedlings that threaten long-term competition is sufficient. This reduces the necessity for regular attendance at the planted site, a valuable attribute which should allow jojoba to be integrated into a semi-nomadic lifestyle where lack of rain precludes permanent settlement.

In orchards or plantations the extent of mechanical weed control will be determined by cultural techniques; for instance it may be very restricted where artificial catchments are used. In general, any implement normally used in orchards is suitable for use in jojoba with equivalent care in its use. For chemical control compounds applied locally to irrigated orchards have been used, including glyphosate, propachlor, simazine and trifluralin with no reported adverse effects.

IRRIGATION

In general irrigation substantially increases tree yield (Ayerza, 1993). Large applications at monthly intervals are the most efficient once bushes are well established, and raised beds with substantial furrows are the preferred field layout. Planting on the contour will overcome the necessity for extensive levelling. A combined rainfall plus irrigation equivalent to 700 mm annually should be the target until local the optimum water requirement is established.

Jojoba has a big advantage as an integrated crop on irrigation schemes in areas where land is not the limiting factor, similarly to castor. Its major water requirement generally occurs when few other crops need irrigating, and jojoba can be used to increase the cropped area without competing with more profitable annual crops. Alternatively, it can utilise subsoil moisture from drainage which might otherwise go to waste, or rainfall collected by the small catchments developed for runoff farming (Fig. 8.20). The system of irrigation will obviously affect the plant type; if subsoil moisture is to be exploited long-rooted plants will be necessary, as opposed to fibrous-rooted types for surface irrigation.

Jojoba is moderately to highly salt tolerant, as noted in the soils section, with plants recorded as growing in the vicinity of water with a conductivity reading of 24 mmhos/cm, or within a few metres of the sea. It can thus be grown in areas where rainfall is insufficient but brackish subsoil water is available, as in Australia.

HARVESTING

Harvesting wild or small plantations of jojoba is manual, but commercial plantations require some method of mechanical harvesting. One type of harvester available straddles the row and agitates bushes to dislodge seed onto conveyers; another type is similar to a pneumatic pecan-nut harvester and collects fallen seed (Coates and Yazici, 1991). A major factor influencing mechanical harvesting is that ripe seeds are generally shed, while unripe but mature seed remains on the plant. Most pickers thus include a proportion of seed with a high moisture content, plus twigs and leaves which must be removed. This is not a major operation, but the seed must be dried which is much more expensive; thus plants may have to be pruned to suit the system of harvesting. Hedge-row pruning, as mentioned, is most popular, and this will ensure that the seed is easily available, since the fruit is borne on new growth.

The yield of seed per cultivated tree under optimum conditions is not yet known. Plants begin to yield in the third year, and 0.5 kg/bush has been recorded. Mature wild plants, and these may be from 10 to 100 years old, have yielded 0.5–2.5 kg in 1 year, the variation being due mainly seasonal conditions, although there is considerable difference between plants. Much higher yields have been reported; for instance the Office of Arid Lands Studies, University of Arizona, stated that in the Aguanga area of California, wild trees could yield 12–15 kg/tree. Fallen seeds remain sound for months in a dry environment. A 5-year-old tree in Australia shedding ripe fruit is shown in Fig. 8.22.

Seed weight is affected by season although the oil percentage remains constant; thus to estimate potential oil yield/ha on the basis of an individual plant or a specific season's seed production, can be wildly misleading. An average seed yield of 5 kg/tree annually to produce at least 2.5 tonnes of oil/ha is considered the minimum required for profitable production. This is a realistic target based on recent reports of individual

Fig. 8.22 Five year old jojoba tree shedding ripe fruit, Australia.

tree yields under optimum conditions, but should be considerably increased by selection or cloning. The productive life of a jojoba tree is unknown, but it could well be over 100 years since one productive tree in California is known to be at least 200 years of age.

A lead time of 8–10 years is common before seed production reaches sustainable high levels, and in the USA intercropping with cotton, asparagus and milo sorghums reduced establishment cost. Crops with a low water requirement are to be preferred. The cost of establishing a jojoba plantation in Australia in 1997 was A$4500/ha with a 4-year wait to first harvest. Seed yield from named cultivars in sixth year was forecast at 1000 kg/ha and average 1600 kg/ha from the 12th year. Oil was selling locally in June 1997 at A$2400/t.

STORAGE

Harvested seed will usually require cleaning and drying to around 9–10% moisture. Drying will usually be possible naturally, since plantations will normally be in warm, dry areas. Clean, dry, sound seed will keep for some years if protected from pests and vermin. Since the seed is bulky, that destined for processing should be moved with the minimum of delay.

PESTS

Jojoba in the wild appears to be basically free of major insect pests, and although a number of caterpillars, small bugs, etc. have been noted on mature plants, they are apparently of little importance. Whether this is due to environ-

Fig. 8.23 Twig wilter, *Anoplecnemis curvipes*.

mental factors or is physiological, is unclear. It will probably remain so until jojoba is sown more widely and in orchards and plantations.

The following have been mentioned as specifically damaging to jojoba: the situlapsis scale, *Situlapsis yuccae*, the false cinch bug, *Nysius ericae*, and the jojoba mealy bug *Puto simmondsia*. Larvae of a small unidentified moth bore the developing fruit and flower buds, a gall midge *Asphondylia* spp. and an unidentified *Periploca* spp. have been noted as causing minor damage on wild plants. Personal correspondence indicates that twig wilters (Fig. 8.23), stem and seed-borers, scale insects, mirid and shield bugs could become important in the Middle East and Africa.

The most damaging pests are various browsing animals who find jojoba foliage particularly attractive, and rodents which remove the seeds. Birds are not regarded as serious pests, but ants can cut off and remove leaves of seedlings.

DISEASES

As for insect pests, there are few disease records from wild jojoba, and little information on the degree of damage to cultivated plants. Various root and seedlings rots have been noted on plants grown in nurseries and under irrigation. *Fusarium* spp. and *Phytophthora nicotianae* var. *parasitica* are most commonly recorded, and can be very damaging, especially when seedlings or cuttings are over-watered. Under greenhouse conditions jojoba is susceptible to Texas rot, *Phymatotrichum omnivorum*, and wilt caused by *Verticillium dahliae* and *Pythium* species.

9
Oilseed Processing and Products

Oilseeds have long been crushed for their oil and the history of their processing and methods of extraction is fascinating. Pliny attributes the invention of oil milling to one of Apollo's sons, and describes in detail the methods employed by his Roman contemporaries to obtain olive oil. The oldest record, however, is an Assyrian clay tablet from about BC 2000 which lists components for a sesame oil press, and press owners subsequently became organised into a guild (Levey, 1959). Several extraction methods were used in Ancient Egypt and some remained almost unaltered to the present; for example a camel-powered oil press the author saw in an Egyptian village, which had been operating on the same site for generations.

In very early times the Chinese used basically the same series of operations required today, i.e. cracking and grinding seeds, heating the meal, then extracting oil under pressure. A wedge press was frequently used, wedges being driven home manually, and such presses were in use until quite recently in Manchuria. A very similar plant operated near Madras, India, using a heavy stone wheel running in a circular channel containing castor seed, and the resulting paste was boiled with water to separate the oil. The stench from this operation was most offensive, made worse by use of extracted cake for boiler fuel, and finally resulted in the operation being closed by local residents!

The most popular method of oil extraction in India is by *ghani*, basically a large pestle and mortar. The mortar is an inverted cone, the pestle a heavy baulk of timber seated at the bottom of the cone and inclined at an angle, so that seed is crushed against the mortar during rotation. The timber baulk is pulled against the side of the mortar by weight, and is rotated by a pole inserted through it at right angles. Power was frequently supplied by bullocks (Fig. 9.1), but modern *ghani* are now motor driven and constructed of lined cement or steel rather than wood.

A one-bullock *ghani* can average six charges of safflower seed/day, or 80 kg of meal (about 160 kg seed). With an oil content of 30%, 45 kg of seed would produce 23 kg hulls, 13 kg oil and 9 kg meal, the pressed cake containing some 10% oil. An improved type known as the *Nutan Ghani* was designed for use by cooperative societies, and working with one bullock for 8 hours, with 1 hour's rest in three periods, produced the results in Table 9.1. Many smallholders pay cash or a percentage of oil to have their seed pressed, and may be allowed to keep the press cake.

Very similar was the Burmese *hsi-zon* (Fig. 9.2), formerly common in the sesame producing regions but now virtually replaced by small, power driven mills. Batteries of *hzi-zon* could frequently be seen in dry zone villages and an excellent description of their manufacture and use is given by McLean (1932). One charge was approximately 12 kg of sesame seed, and three to five charges could be crushed per day with an oil yield of some 33%. The oil obtained was merely strained through cloth before use, and very popular. In Central Africa, sesame seeds were boiled to make them soft, then squeezed in a 'sausage' made from tree fibres to express the oil.

Oil is obtained commercially from oilseeds by one of three basic methods, modified or combined to suit specific conditions; batch hydraulic pressing in which the oil is expressed by hydraulic pressure from a mass of oil-bearing material; continuous mechanical pressing in which the oil-bearing material is squeezed through a tapering outlet, oil being expressed by the increasing pressure; and solvent extraction in which oil-bearing material is taken into solution with a solvent, the solution separated from insoluble residue, and the oil recovered from the solvent solution. Combinations of these processes in commercial operation are continuous mechanical screw pressing (expelling) with continuous solvent extraction, and batch hydraulic pressing and solvent extraction. With the exception of castor, which requires special treatment and equipment cleaning procedures, the other oilseeds discussed herein can be similarly processed (Fig. 9.3, courtesy of Neumanz Inc., New Jersey USA).

Fig. 9.1 Bullock-operated wooden ghani, India.

The Stamper, or Dutch press, invented in Holland in the seventeenth century, continued to be almost exclusively used in Europe up to the early years of the nineteenth century. This was a wedge-operated press, the wedges being driven home mechanically, often by a steam or water operated hammer. Hydraulic presses, based on an invention by Joseph Bramah in England in 1795, dominated the seed-crushing industry until the 1920s. The invention in the USA of the continuous high-pressure screw press, expellers, by V.D. Anderson in 1900 radically altered the structure of the industry and greatly reduced the capital investment required.

The efficiency of a mechanical screw installation with its relatively low installed cost made it possible for many smaller traders to process oilseeds. This low initial cost is of major benefit to the oilseed industry and co-operative organisations of developing countries. An example is the Anderson Lion-90 Expeller Press which processes 50–200 kg/hour and is driven by a 10 hp motor (Fig 9.4), and the larger Mega-200 (Fig. 9.5). Solvent extraction is seldom used to process small amounts of oil-seeds, and 100 t/day is probably the smallest economic unit; in the USA and the European Union 1500–2000 t/day is common (Figs 9.6. and 9.7).

New production methods are continually being introduced to reduce costs, raise oil extraction rates, or improve quality or product range, and extensively reviewed (Applewhite, 1992; Williams, 1997a & b). The efficiency of

Table 9.1 Daily working of Indian bullock ghani.

Oilseed	Quantity (kg)	Charges/day	Oil (kg)	Cake (kg)	Extraction rate (%)
Groundnut	90	10	30–40	50–58	34–43
White sesame	90	10	36–40	50–54	40–45
Coconut	136	15	75–80	50–60	55–60
Niger	90	10	25–27	63–65	28–30
Safflower (hulled)	136	15	65–68	68–71	48–50

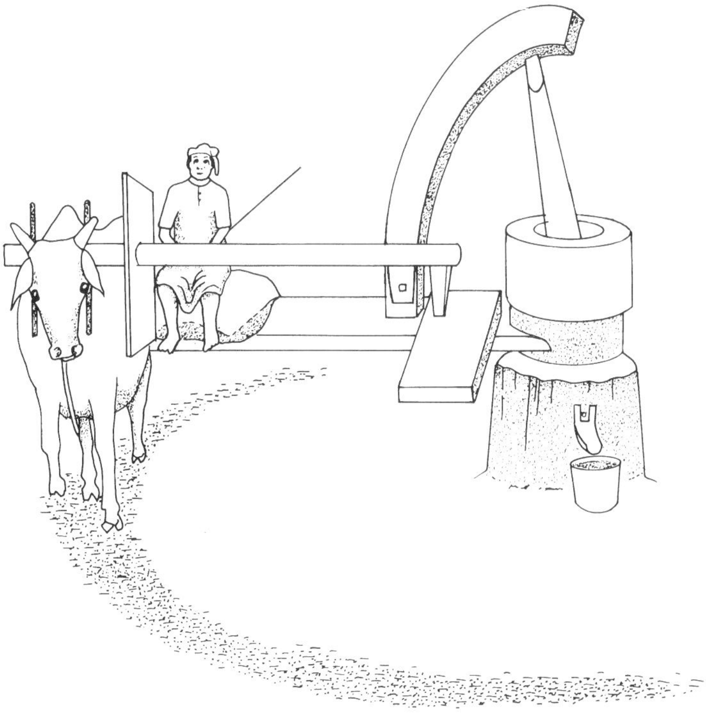

Fig. 9.2 Burmese hsi-zon mill.

commercial oilseed processing plants is epitomised by the advertising slogan of one major US crusher, 'From Seed to Salad Oil in 18 hours!'

PROCESSING

A most important aspect of oilseed processing is a strict inspection and grading unit with sufficient pre-processing storage at the receival point. Bulk seed deliveries from large growers or merchants can be accepted with the minimum inspection, since penalties can be imposed on consignments that contravene agreed standards. Deliveries are sampled on arrival, moisture content calculated electronically, and seed oil content determined by rapid nuclear magnetic resonance. The supplier is notified of the results, and his produce graded and paid for at agreed rates. New storage buildings should preferably be easily removed to alternative sites, or should have built-in pest control equipment. Either will quickly repay the higher initial investment by reducing storage loss, which can be very high in the tropics. Some common insect pests of stored products are shown in Fig. 9.8.

Large amounts of seed are grown in less-developed areas and many deliveries will be in sacks, or any handy container with the ubiquitous 18 litre paraffin can most frequent, sometimes balanced on the head of the producer's wife. Such seed may have remained for some time in the owner's possession before sufficient was collected to make worthwhile a trip to the factory or buying centre. Impurities there will almost certainly be, and in East Africa a very common castor adulterant was the very similar seed of *Jatropa* spp. Where buying is carried out at the factory a pre-cleaner to remove this extraneous material before purchasing is essential. It is useless to publish standards for buying seed, or indeed almost any other commodity produced at this level of agriculture, since the average farmer will completely ignore them in the hope that his load will not be inspected!

The greater part of this adulteration is deliberate, and a cleaning, grading and acceptance

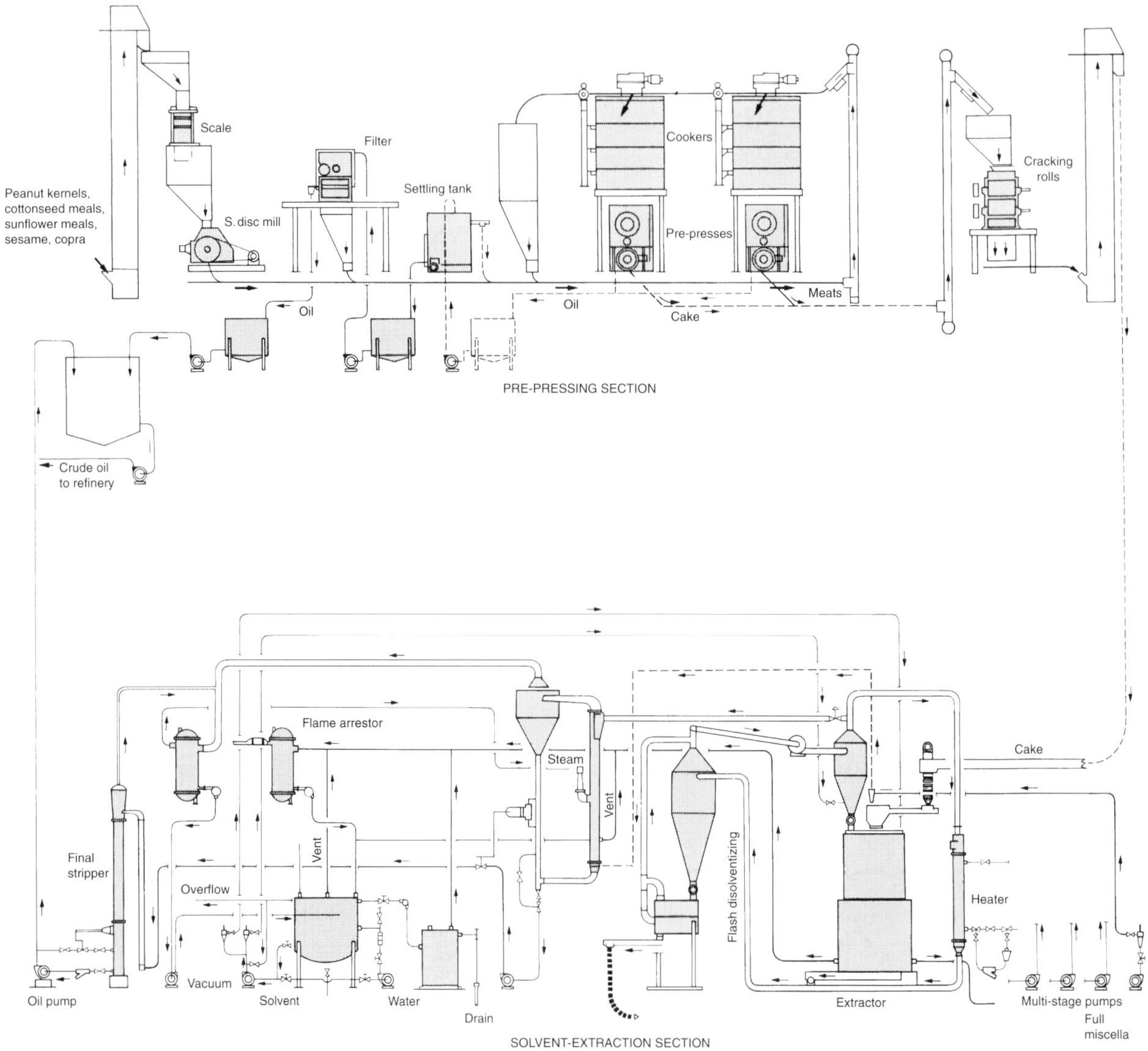

Fig. 9.3 Flow diagram of a vegetable oil mill.

routine should be established and rigourously applied, or the proportion of whole seed offered will become progressively smaller. Not only is the seed itself usually of varying degrees of maturity but it may also be composed of a number of cultivars, perhaps of different seed-coat colour. Due to the difficulty of cleaning the seed it will also invariably contain an admixture of plant debris and soil. The former can be removed by an air blast, but removal of small particles of soil, which are often identical in size and weight with sesame seed for instance, is much more difficult. Sesame produced in the southern districts of Tanzania and Fung area of the Sudan was notorious for its high soil content.

It is essential that seed delivered to the press be clean and free from all impurities. It cannot be too strongly emphasised that where oil processing plants are established in relatively remote areas, strict control of seed cleaning will do a great deal to produce good quality oil, prolong machinery life, and thus a viable mill.

Fig. 9.4 Anderson International Lion-Expeller press. Capacity 50–200 kg/h fitted with 10 hp motor.

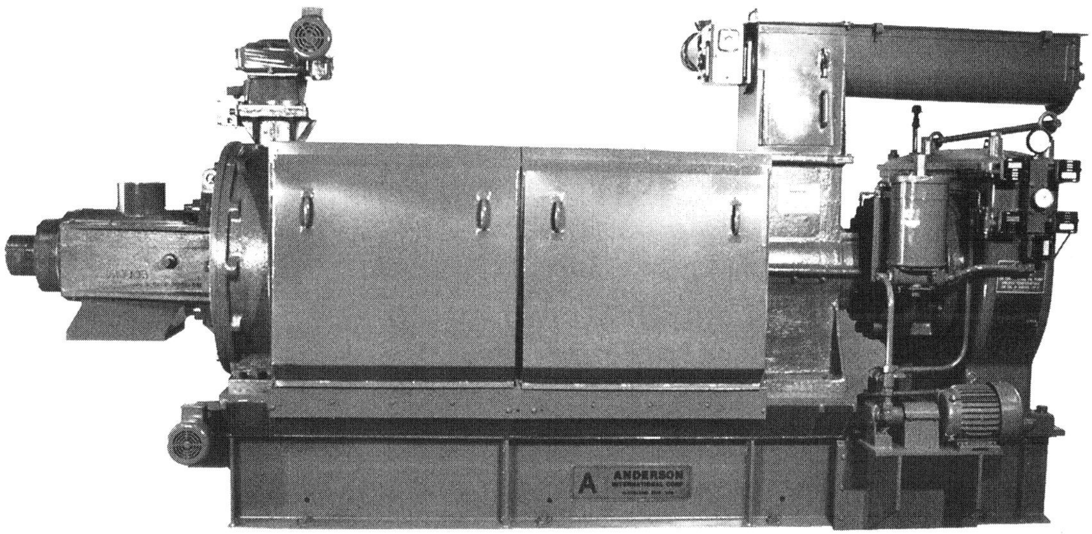

Fig. 9.5 Anderson International Mega-200 Expeller. Capacity 150–170 tonnes/h fitted with 200 hp motor.

Oilseeds deteriorate in storage, and although loss of viability is of little consequence in seed to be processed, the effect on seed oil and its constituents is important, since it will affect a factory's profitability. In India a major loss of oil quality in terms of colour, free fatty acids, iodine and other values, was due to seed kept in unsuitable or prolonged storage. The main reasons for this deterioration are high moisture, relative humidity and temperature, and fungal infection. Rapid movement from producer to processing plant is essential if full benefit is to be obtained from oilseed production, and on-farm storage should be kept to a minimum. Seed to be exported is usually governed by specifications covering quality and impurities, and if these are

Fig. 9.6 Small factory in USA operating a battery of expellers with room for expansion of operations.

Fig. 9.7 Small factory in Australia operating a solvent extraction plant. Extraction in foreground, distillation in background.

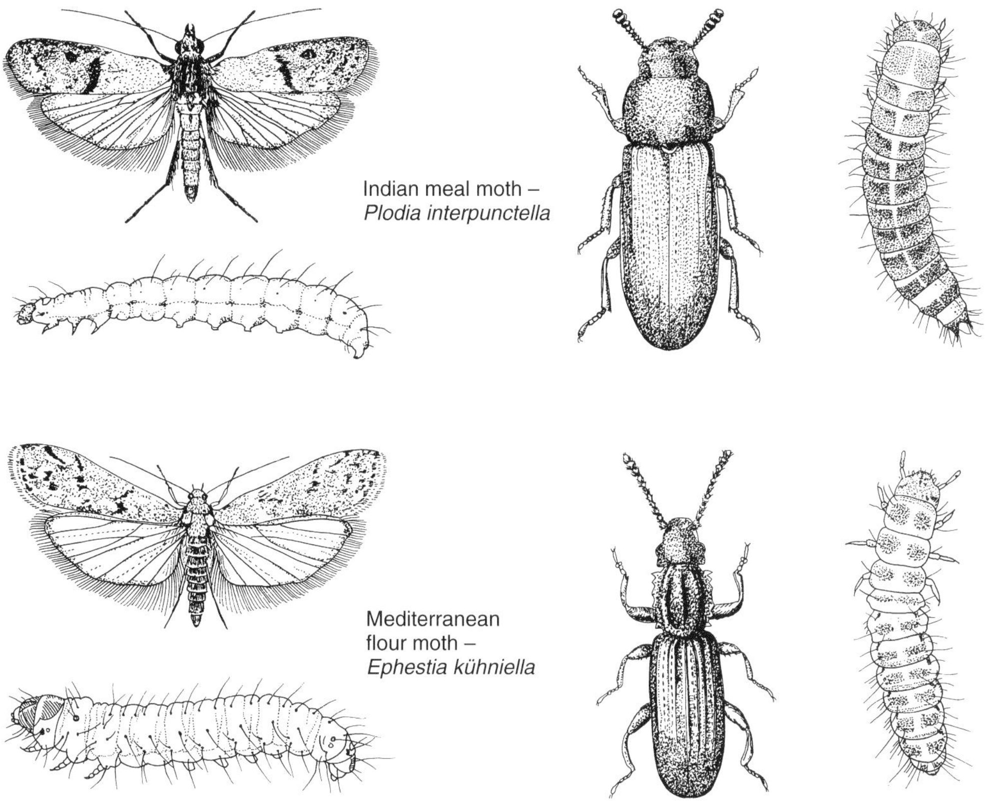

Fig. 9.8 Some common storage pests.

violated, penalties will be applied resulting in financial loss to producers.

Processing plants or marketing boards usually operate substantial storage facilities, and the main requirement is some process for cooling and drying seed, and then maintaining these conditions. Stores or silos should be constructed to store oilseeds below 25°C (preferably below 20°C) and below 7% moisture content. Under these conditions no mould develops, free fatty acid formation is minimised, and colour retained. Controlled aeration will achieve and maintain these conditions and restrict moisture migration, while painting buildings or silos white helps passive cooling and reduces solar heating. Moisture migration causes crusting and mould growth on the stored seed surface, and if unchecked the temperature of the bulk can rise to a level allowing chemical oxidation; seed may initially smoulder then ignite and burn.

Insect control is usually necessary when seed is stored for more than 3 weeks, but pesticides registered for use in stored oilseeds are currently restricted to phosphine or natural pyrethrum. For on-farm storage phosphine treatment is the simplest and most effective, although it requires the necessary installation.

HULLING

Hulling is recommended to provide high-quality edible oil, soya excepted, and also reduces the meal fibre content so increasing its marketability as stockfeed. However, hulling also increases the cost of production, and breeding thin-hulled seed or seed in which an unwanted hull ingredient has been eliminated, increases mill profitability. Hulling may be omitted if the meal is to be used for fertilizer, but the extraction rate of unhulled seed will be lower unless higher pressure is used. This will increase the extraction cost and the merits of hulling and lower pressure, or not hulling and higher pressure must be determined.

Partial hulling of some oilseeds is possible, and disc or roller hullers may be used to crack seeds which are then passed through an air blast adjusted to remove only lighter hull particles. This mixture is less liable to mush during screw pressing, and requires only a moderate increase in operating pressure. A disc huller with 60 cm discs running at 1200 rev/min, can hull up to 135 kg of safflower seed/hour, and may also be set to crack the kernels. One model of roller huller uses three sets of rollers in series, with each pair having a speed differential of 1:1.25, and a mill of this type with 22 cm diameter × 60 cm long rollers can handle up to 1800 kg of seed/hour. Safflower seed was initially difficult to decorticate owing to the hull strength and kernel softness, but newer machines overcame this problem. Jojoba seeds are also difficult to dehull or mill, for the liquid wax which forms 50% of the seed runs freely from broken seeds and clogs machinery.

COOKING AND DRYING

High protein content material usually requires cooking before pressing, since this not only coagulates the protein but frees the oil for more efficient pressing. Following cooking, which is normally carried out at a moisture content of 10–12%, the raw material must be dried to approximately 2–3% before entering the screw press. Low-protein materials with a high-oil, high-fibre content such as copra usually require drying only.

The 'art' of commercially cooking vegetable oilseeds which contain a high proportion of protein is almost as old as vegetable oil milling itself. Duration of cooking was determined by touch and visual inspection of a drawn sample, and the amount and rate of oil produced depended on the operator's skill. When hydraulic presses predominated, cookers were invariably considered necessary, and it was not until the 1950s that the cooking operation was reassessed. A modern automatic circular cooker with a diameter of 90 cm should be capable of handling 40 000–50 000 kg meal/day.

Where they are necessary, cooking and drying should be considered as two distinct functions, and cooking requires more accurate control than drying (Fig. 9.9). A cooker must be large enough to hold material up to 20 minutes under airtight conditions at 88–93°C, but drying is carried out at appreciably higher temperatures, 110–115°C. The higher temperature is critical when processing oilseeds whose oil is heat-sensitive, and in general the drying temperature must be

Fig. 9.9 Cooker, 122 cm diameter, combined with six 46 cm diameter dryers, with output of 100 t/day.

accurately controlled to avoid burning. Retention time must be as short as possible, as too long a period, even at controlled temperatures may have a deleterious effect on oil colour. Long retention is also inefficient. Removal of saturated air from the dryer is most effective when a slight negative pressure can be maintained in the unit. The development of extruders has radically altered processing procedures, reduced or eliminated the need to pre-cook, and is discussed in the next section.

EXTRACTION

A combination of press and solvent extraction methods produces a press cake with a residual oil content of 10–15%, which is delivered direct to the extractor. Meal must not be too finely ground before pressing or it tends to disintegrate in the solvent, the fine particles then remain in suspension. The most efficient method of expressing oil is by single or double screw press either hot or cold, although batch hydraulic pressing followed by solvent extraction is still fairly common. The screw press has a major advantage over batch-type hydraulic presses as it need not be stopped for recharging.

A screw press operates at greater pressure than a hydraulic press, with a consequently higher oil-extraction rate. The hot oil-bearing material is forced through a cylindrical steel cage, and then through a tapered outlet by means of a powerful rotating screw. An extremely high pressure is developed by this restricted discharge, the oil being expelled through narrow slots around the press barrel. The compressed cake is extruded through the discharge outlet. A modern type of screw press should process between 13000 and 26000kg meal/day, reducing oil content from approximately 40% to 4%–6%.

The extruders now used to pre-treat oilseeds for mechanical crushing are high-shear mechanical devices for rupturing oil cells; these extruders operate at higher moisture levels and consume less horse power/tonne than screw presses. Extruders are smaller and considerably less expensive than traditional plant used for heat treatment which require 20–40min to cook and dry an oilseed; an extruder requires about 20 seconds. Since the screw press no longer has to rupture oil cells it no longer needs to operate at a moisture content below 3%, and the same amount of oil is obtained at 5–6% moisture. The higher yield results from using a lower pressure wormshaft operating at higher speeds on the same power input; on soya a screw press with traditional preparation might press 500kg/hour at 5% residual oil; using an extruder before the press increases the throughput to 2000kg/hour.

For direct solvent extraction seeds are first rigourously cleaned, and may be hulled, although sunflower seeds for solvent extraction are not hulled. Seeds are then rough-milled, since too fine a material produces excessive fines in the oil. Prepared material is then extracted in a batch or continuous process, but the former is relatively inefficient, and a continuous counter-current flow process most common. This process is based on solids moving in one direction being washed by solvent passing in the other. Thus oil is removed more effectively as material comes into contact with increasingly purer solvent as it moves through the extractor. The main problems are separation of liquid from the solid phase, and providing efficient contact and extraction time. Extraction rates using this process are usually higher than from mechanical methods, and residues normally contain less than 2% oil. The most common solvent is food quality hexane, a petroleum derivative with a boiling-point of 65°C. Solvent extraction plants are becoming more common as the crushing industry now aims to produce meals with minimum oil content at commercially acceptable production levels, while not impairing oil or meal quality (Williams, 1997a).

An indication of the rate of working and processes involved can be obtained from the following description of a small mill combining press and solvent extraction of safflower seed. Following cleaning and conditioning, seed is pre-pressed in expellers, the resulting cake conveyed directly to extractors without flaking or granulation. Cake normally contains 10–15% oil, 6–7% moisture at 50–60°C, with a bulk density of approximately 480kg/m^3. The extractor is a horizontal basket type, with 20 baskets filled to a depth of 75–90 cm moving on an oval track, at a rate of approximately 30 cm/min. At this speed the extractor requires 1.2 kg/second of solvent, a solvent to cake ratio of 0.87 : 1.0; a more usual ratio for a vertical-column extractor would be 1.1 : 1.0. The miscella produced directly from the extractor is clear and brilliant, contains about 30% oil, which after conventional refining is nearly equivalent to high-grade non-break processed oil. The extracted meal moves through the solvent recovery unit at 70 000 kg/day, and finally contains approximately 1% oil and 8% moisture.

Extracted material passes through a solvent recovery plant and the oil is cleaned, bleached and filtered. Press oil usually contains suspended meal, 'foots', removed by settling, screening and filtering. In some small plants, especially where crude oil is utilised for soap manufacture, filtering is omitted after preliminary settling or screening. Modern plants incorporate a special settling tank and foots separator, placed beneath floor level under the oil-discharge end of the press, into which oil and foots are gravity-fed. Equipment is installed in this tank for continuously separating, draining and returning foots for reworking. Clear oil from the settling tank is pumped to a holding tank, which should ideally be fitted with warming and agitating machinery to condition the oil, which is subsequently pumped through the filter press.

The filtered crude oil will normally contain impurities; phosphatides, gums, resins, free fatty acids and colouring which must be removed. Alkali refining removes the gums, free fatty acids and some of the colouring matter, and the insoluble soap (soap stock) remaining is treated with sulphuric acid. These residues, acid oils, after washing and drying are used in the manufacture of hardboard, foundry cores, as a feedstock for certain distilled fatty acids and occasionally for soaps and surfactants.

The crude filtered oil must be refined, bleached and deodorised to produce an edible oil, an exception being olive oil. A high quality oil must be low in free fatty acids, pale in colour, and must not deposit solid material (break) on heating to about 30°C. Conversely, some oils for domestic use must be stable under refrigeration when used as a base for salad dressings etc. This operation, winterizing, consists of cooling the oil to remove components with high melting-points that may settle out at low temperatures. Sunflower oil must be so treated.

Deodorising is frequently necessary to produce a bland oil, and refined oil is commonly vacuum treated with steam at 200–250°C. Frying oils must not smoke, and the smoke-point of an oil is directly related to its free fatty acid content; thus frying oil must have the minimum of these acids. An oil with a free fatty acid content of 0.01% has a smoke-point of approximately 230°C, but this will fall as the free fatty acid content increases with use. Supercritical carbon dioxide extraction may be used to defat edible oils and thus improve their quality, and to produce various protein isolates.

CAKE AND MEAL

The limiting factor with extrusion was the substantial capital investment in the extruder and its ancillary plant including driers and specialist hammer mills, together with the high running costs, and less expensive methods were developed. One of the most important was the expander, introduced to exploit the increasing need to produce sterilised feeds, to incorporate cheaper ingredients into formulations, to extract the greatest nutritional benefit from these ingredients, all combined with sustantially lower operational costs.

The expander is similar in design to the extruder, and in oil mills increases capacity, improves oil yield and reduces solvent losses. Similarly to a single screw extruder, the expander consists of a thick-walled mixing tube divided into segments, and an over-mounted worm shaft to which mixing and kneading elements are attached. Unlike an extruder, there is no die at the outlet side but a conical discharge valve provided with an adjustable, annular gap which builds up pressure in the expander. The pressure is adjusted by a hydraulic unit at the outlet valve during operation, allowing continuous control of kneading, temperature and energy consumption.

At the outlet the material abruptly expands and the added moisture vaporises (flash evaporation), so subsequent drying is usually unnecessary. Depending on the formula, the structure of the expanded material is a dough, thick flakes or crumbles, and the size of the agglomerates can be determined by an integral cutter. High pressure conditioning is automatically controlled by predetermined pressure, temperature and energy consumption.

Cake as discharged from a press was usually at a temperature of 90–160°C, with a moisture content of 1–4% due to the frictional heat developed by pressure in the extruder barrel. It could continue to heat and begin to smoulder if accumulated in heaps, or if immediately ground into meal and sacked hot. To quickly lower the temperature of cake to a level at which it cooled gradually without danger, it was tumbled in a cooler with a cold-air blast, and sprayed with cold water to equal atmospheric humidity. This reduced the amount of moisture lost in milling and avoided large discrepancies between shipping and destination weights of cake or meal.

Cooled, humidified cake may be ground directly into meal, but it is usually stored and ground when convenient. Where 24 hour working is normal, the ratio of storage to grinding capacity should be accurately determined. This permits accumulation of all the cake produced during the night so that it can be ground together with the cake produced during the day. Cake may be ground in hammer mills or double, revolving-disc attrition mills. Power and maintenance costs for hammer mills are appreciably lower than those for attrition mills, but meal ground in the former tends to have a granular, sandy character, and cannot be reduced to the velvety fineness produced by an attrition mill. Many livestock managers, however, prefer a more granular meal to avoid injury to animals caused by inhalation of finely ground meal. Unless a definite local demand exists for finely ground meal, it is preferable to use a hammer mill with its lower power requirement and maintenance costs.

The trend to higher feed processing temperatures to improve feed hygiene, physical characteristics, and production efficiency has changed both cooling and drying methods. Some dry feed enzymes have been shown to maintain efficacy after exposure to temperatures above 90°C, but high temperature should be avoided by applying enzymes as liquids following heat treatment. Liquid addition may offer greater convenience and reduce the risk of cross-contamination between different feed types, but successful use of liquids depends on accuracy of dose and application.

Recent methods of conditioning meals include short time high temperature (STHT) conditioning as represented by the expander; and medium temperature 85°C, medium time to 2 minutes, (MTMT). All offer considerable advantages over conventional conditioning in control of bacteria and meal preparation for optimum pelleting performance. The pressure required for good conditioning has in the past been provided only by the pellet mill, and concentration on this factor led to developments in conditioning practice.

By using two pellet mills in series the conditioned meal is made denser in the first and formed in the second. This extra pressure, temperature and time exposure results in greater gelatinisation of starches in the ingredients and a molecular plasticity which gives an easier passage through the die, extending replacement time and reducing power consumption. The achievable temperature rise and pressure in the double pelleting process are however limited to that of normal roll pressure and a relatively short die thickness resistance. To increase the advantages of pre-densification new machines are being developed.

TRANSPORT

Transporting vegetable oils is not complicated by the variety of regulations that may cover movement of raw material, e.g. castor seed or pomace, and the only restrictions are those necessary for similar flammable substances. The oil can be transported in suitable drums, road tanker, railcars, or bulk ships. In less-developed countries a high proportion of petroleum products is transported by rail cars or barges; which generally return empty from up-country trips. Their use on the return journey, for castor oil in particular, can probably be arranged and such alternative loads also enable more economic use of rolling stock. Road tankers are often the most difficult for which to provide back-loading, and favourable terms can often be negotiated with the transport company. When edible oils or oil products are transported by rail or truck, security measures are usually necessary to control pilfering. Where possible, full containers which remain sealed to destination, should be shipped; part or mixed loads should be avoided.

OILS, FATS AND MEALS

Initially the derivatives from the oilseeds discused were used as human and stockfood, later also in aquaculture, but the range and variety of uses has greatly increased, including oil for fuel as later mentioned, adjuvants for pesticides, raw material for cosmetics, pharmaceuticals, engineering plastics and fibres. Derivatives such as amino acids, food proteins, flours and liquid foods are of increasing importance. The technology involved and products developed are outside the scope of this book, and the relevant literature should be consulted. Comprehensive reviews are contained in Applewhite (1992), Hudson (1994) and Lim & Sessa (1994).

An important development is generally known as genetic engineering, the artificial manipulation of genetic material of an organism. Oilseed crops have been so modified, especially soya, which was resulted in pressure from consumers to identify such material when incorporated in human food, or fed to livestock subsequently processed for food. Oilseeds could also become speciality crops producing epoxy oils if current research involving the introduction of specific genes is successful. According to Australia's CSIRO which is a leader in this field, the production of modified fatty acids used in the making of polymers and speciality chemicals is currently by chemically processing vegetable oils, or petroleum derivatives. Genes have been determined which carry out these modifications inside the plant, and enzymes have been identified that perform a range of chemical modifications to fatty acids. The aim is to design enzymes to produce fatty acids tailored to specific end uses.

The average oil content as percentage of seed weight of some common oilseeds is as follows:

Castor	50	(ND)	Olive	50	
Copra	65	(ND)	Rapeseed	50	(ND)
Cottonseed	20	(ND)	Safflower	35	(SD)
Groundnut	40		Soya	20	(SD)
Linseed	38	(D)	Sunflower	45	(SD)

Those marked ND are designated non-drying oils, SD as semi-drying, and D as drying oils. Groundnut and olive have little or no industrial use.

CASTOR

A most important factor influencing castor seed processing is the sensitivity of individuals to the seed and its derivatives. Thus it is necessary to screen prospective employees for this reaction and a simple method is to allow prospective employees to work in the mill for a short period. An individual allergic to castor seed develops congestion of the nasal passages, the eyes and nose become irritated and water; with continued exposure coughing and congestion of the nasal passages becomes severe. Those individuals who are not sensitive show very little, if any, reaction. Their eyes may water slightly, but with no congestion in the nasal passages. If no reaction occurs they may be considered non-allergic. Suitable antidotes should be easily available to treat urgent cases.

Any mill operation produces some dust, and when processing castor seed it is essential to provide effective dust-extraction equipment to control this potential allergy source. In urban areas it is also necessary to trap the dust to ensure it does not contaminate the local atmosphere, and mills that have processed castor seed for many years report no difficulty in meeting this requirement.

It is generally difficult to integrate castor oil production into edible oil mills, since special precautions are also necessary for equipment cleaning and maintenance. The toxic principle, ricin, easily contaminates other oils, or castor oil itself imparts an offensive odour or taste. The author demonstrated a simple cleaning and sterilising routine in East Africa, which also required castor to be processed last in the annual programme, after which all machinery was normally cleaned and overhauled for the next season. Accurate planning of the annual crushing programme is more important in less-developed countries, where a factory is usually unable to run year-round. In these countries castor seed is often easily obtainable, and will probably become of increasing value and importance, especially in Tanzania and Zimbabwe.

Factories that single-press castor initially clean and dry the whole seed to a moisture content of 4–5% at approximately 90°C, then condition seed by spraying with live steam to increase the moisture content to 5–6%. This seed is then fed directly to a mechanical screw press, the resulting cake containing 6–8% oil at 8% moisture. Oil from the single press operation is No. 1 quality, and after addition of a clarifier may be readily filtered at 38–48°C. Mechanical screw presses can handle up to 27000kg of castor seed per 24 hours and produce cake containing approximately 7% oil. Screw presses used for single-pressing of castor seed normally require two or three shaft and barrel changes each year, owing to the abrasive nature of the seed testa.

For double-pressing, cleaned seed is heated to 65°C and immediately passed to a mechanical screw press. Oil obtained is No. 1 quality, and if warmed to 50°C and pretreated with a small amount of clarifier, can be readily filtered. Cake from this operation contains around 20% oil, and after heating to 90°C to reduce moisture to 4–5%, is repressed to produce No. 3 quality oil and press cake with 5–7% oil. Number 1 quality oil should have a maximum acid value of 4 and a Gardner colour maximum of 3, and No. 3 quality a value of 10 and a maximum of 7 respectively.

Castor oil storage presents few problems other than selecting the most economical method; for long term storage temperature control is essential. Basically oil should be stored only long enough to accumulate the minimum quantity required for shipment, usually some 45000kg. Expensive security arrangements are usually unecessary, as there is little use for the oil and petty pilfering uncommon. A stock-proof fence is essential round the cake storage area to prevent accidental poisoning of local children and animals. Cake storage may be a problem in more humid areas, since it deteriorates rapidly with rises in temperature and humidity. Its usefulness is thus often restricted to that of fertiliser or supplementary fuel.

Crude castor oil is a pale straw colour, but after refining and bleaching is almost colourless, occasionally with a faint yellow tinge; has a distinctive odour considered offensive by many people, but is easily deodorised. The unpleasant flavour, slightly acrid with a nauseating after taste, remains in all qualities of the oil except specially treated pharmaceutical oils. Refined castor oil stores well, and does not easily become rancid. Miscible with absolute ethanol, methanol, ether, chloroform and glacial acetic acid, castor oil does not dissolve to any extent in mineral oil unless mixed with another vegetable oil. Castor oil consists almost entirely of triglycerides, principally ricinolein, the fatty acid component (in approximate %) being ricinoleic acid to 90, linoleic 4, oleic 3, stearic 1, palmitic 1, dihydroxystearic 0.7, eicosanoic 0.3 and linolenic 0.3. The presence of a hydroxyl group in ricinolein confers on castor oil the unique property of solubility in alcohol. The only other oil known to contain ricinoleic acid in any quality is ivorywood oil from *Agonandra brasiliensis*.

United Kingdom industrial castor oil is usually of two grades, the best conforming to British Standard Specification No. 654 known as English Firsts and Indian BSS is also of this quality, and English Seconds or Indian Commercial; the latter is generally darker in colour, and the only specific requirement is that it should contain a maximum of 5% free fatty acid as oleic acid. Composition and specifications for castor oil are shown in Table 9.2.

Castor oil is among the world's most versatile natural products (Weiss, 1971), and is the basic

Table 9.2 International castor oil specifications.

	British Standard First Quality*	US No 1	AOCs
Acid value	4 max.	3 max.	4 max.
Saponification value	177–187	179–185	176–187
Iodine value, Wij's	82–90	82–88	81–91
R–M value			Below 0.5
Polenske value			Below 0.5
Acetyl value	140 min.		144–150
Hydroxyl value †	156		161–169
Unsaponifiable (%)	1.0 max.	0.5 max.	Below 1
Refractive index, 20°C	1.477–1.481		1.473–1.477
Refractive index, 40°C			1.466–1.473
Specific gravity, 15.5/15.5°C	0.958–0.969	0.961–0.963	0.958–0.968
Viscosity at 25°C‡		U ± 1/2	
Colour	2.2Y–0.3R max.§	3 max.¶	
Critical solution temp. in ethanol	Below 0°C		

* Indian BSS and Brazilian BSS conform to this specification;
† Hydroxyl values corresponding with acetyl given;
‡ Gardner-Holdt; § Measured in 1 in cell, Lovibond colour scale;
¶ Gardner.
Note: Specifications of grades commonly recognised internationally, first grades normally being produced by cold pressing and other grades extracted from pressed cake and similar residues.

raw material for a common form of nylon, Nylon 11, widely used in engineering plastics and mainly produced in France, although India is developing its own production of the basic material undecenoic acid (UDA). Castor oil is a major source of raw material for the production of sebacic acid, as a lubricant, as a constituent of embalming fluids, in soap manufacture, as Turkey Red for dyeing and finishing textiles, as hydrogenated castor oil (Opalwax) in polishes, for electrical condenser impregnation, carbon paper, ointments, cosmetics and hair-dressings, as dehydrated castor oil in protective coatings, as blown oil (oxidised oil) for plasticising oilcloth, artificial leather, coated fabrics and lacquers, in brake fluids where a degree of compressibility is important, and in printing inks. This array of uses is a main reason for continuing world demand for the oil, while many other uses remain trade or military secrets.

Castor oil is traditionally associated with medical uses, but this is now a negligible proportion of total consumption, about 1% in the USA and Britain. The oil is cathartic and in the alkaline portion of the intestine is partly hydrolysed under the influence of fat splitting enzymes into glycerine and ricinoleic acid. The latter causes marked local irritation of the intestinal mucosa, thereby stimulating motor activity of the bowel, and infants require a relatively larger dose than adults. It performs a similar function in animals. Since ricinoleic acid is eventually absorbed like any other fatty acid in the small bowel, little reaches the colon following small doses of castor oil. The ricinoleic acid that is absorbed is metabolised much like other fatty acids. Medicinal castor oil production is a specialised operation, and not usually carried out in large commercial mills. Seed is decorticated, cold-pressed or pressed at 32–38°C, depending on the technique. This yields oil of a pale straw colour, which can be bleached by ultraviolet light, and after filtering provides oil of medical quality.

Castor wax is a hard, brittle, high-melting-point product with a faint characteristic fatty or waxy odour, and tasteless. It is normally supplied as uniform free-flowing white flakes. Solid wax is cream to pearly by reflected light, and bluish-white by transmitted light. When molten it is clear, transparent, very fluid, and colourless to slightly straw tinted. The solid wax has an amorphous fracture, dull to waxy lustre and white streaks. Its uses are many, including solid lubricant, and as a pressure-mould release agent in the manufacture of formed plastics and rubber goods. The characteristics of castor wax are shown in Table 9.3.

Castor meal, also known as pomace or poonac, unlike most other oilseeds contains poisonous substances; ricin, a highly toxic protein; ricinine, a relatively harmless alkaloid; and the extremely potent and incorrectly designated castor bean allergen (CBA), which should be termed castor seed allergen (CSA). These toxic principles frequently restrict its use to a fertiliser, and because of the lack of interest by stockfeed manufacturers in the USA, castor pomace is lower in price and more generally available for fertiliser than other seed meals. It is used extensively for direct application, in preparation of mixed fertilisers, and ranks high as a source of organic nitrogen (Weiss, 1971). In India where the export of seed of prohibited in order that processing should be carried out locally, pomace is a major source of nitrogen and its use widespread, particularly on paddy-fields. Trials in Sri Lanka with various organic fertilisers showed the high value of castor cake and meal, and the cake is particularly effective for sugar cane and paddy. In South Africa castor pomace is frequently used in commercial mushroom production beds. Analyses of oilseed meals used as fertiliser in the USA are shown in Table 9.4, and India in Table 9.19.

The use of castor meal as stockfeed depends on a reliable and inexpensive method of detoxification. Ricin is destroyed by heat, and castor meal can be made safe for feeding purposes by steaming, and is also detoxified during solvent recovery. However, prolonged heating or steaming to inactivate ricin does not destroy its allergenic activity, since the allergen is unaffected by heating to 110°C for 1 hour. A crushing plant at Plainview, Texas, USA, supplies detoxified meal by holding it in the solvent recovery toaster vessel after extraction, and then passing steam to heat the meal and denature the ricin. The meal is used to fatten beef cattle, and although the principle allergen, CB-1A, which is not inactivated by the steaming process does not distress cattle, sensitive persons cannot be employed in the feed lots.

Table 9.3 Characteristics of castor wax.

Specific gravity at 20°C	0.99
Melting-point	86°C
Ash content (%)	0.005
Dielectric constant (0.1 Kc) at 3°C	12
Flash-point	156°C
Fire-point	175°C
Acid value	2
Iodine value	4
Saponification value	180

Note: Penetometer needle No. 14 with 100 g weight is 0.2 mm df (0.1 for carnauba and 1.7 for beeswax). Specifications vary slightly according to manufacturer.

While the Plainview method is effective, the use of extruders as chemical reactors is relatively simple and more efficient at destroying the allergens and toxins. The extracted meal is treated with either calcium hydroxide, quicklime or calcium oxide at a moisture content of 20–24%, mixed for about half an hour, then passed through an extruder with an exit temperature of 140–160°C, dried, ground and packaged. This meal can safely be fed to poultry and livestock, but the high fibre content limits it to a maximum of 15% of the ration, which must be balanced to supply the lack of certain amino acids, threonine in particular. A comparison of castor and cottonseed meals is shown in Table 9.5.

The presence of castor oil in meal is of little importance, as the purgative effect ensures that little is absorbed by the intestines. When fed with a carrier glyceride such as olive oil, there is free absorption and no purging. In the diet of rats 48% cent of castor oil was well digested, and ricinoleic acid which composed 85% of fatty acids was readily metabolised. The protein content of decorticated meal is approximately 40%, compared with linseed 30%, rape 32%, palm kernel 17%, copra 21% and decorticated cottonseed 41%. The fibre is mainly confined to the testa, and decorticated meal is thus much less fibrous. Castor meal is slightly deficient in the essential amino acids methionine, lysine and tryptophane, which preclude its use as main source of protein in pig and poultry rations, but it is suitable for

Table 9.4 Analyses of oilseed meals used as fertiliser in USA (all as percentages).

Constituent	Castor		Rape		Soya		Cottonseed	
	Range	Mean	Range	Mean	Range	Mean	Range	Mean
N	—	—	3.7–6.4	5.4	—	—	—	—
Ammonia	0.0–0.31	0.06	—	—	—	—	—	—
Nitrate	0.12–0.35	0.18	—	—	—	—	—	—
Sol. organic	0.20–0.56	0.34	—	—	0.72–6.48	3.64	0.51–0.85	0.58
Active insol.	2.31–2.92	2.75	—	—	1.30–2.21	1.72	—	—
Total samples	1500	—	—	—	290	—	40 000	—
P_2O_5								
Available	0.6–2.7	1.83	0.85–2.55	1.75	0.96–2.47	1.61	2.07–2.90	2.51
Total samples	215	—	37	—	—	—	3000	—-
K_2O_5								
Water soluble	0.36–1.66	1.09	0.81–2.68	1.28	1.55–2.88	2.37	0.69–1.95	1.37
Total samples	212	—	34	—	—	—	3000	—
Ca	0.07–0.77	0.41	0.26–1.27	0.62	0.19–0.47	0.26	0.086–0.78	0.24
Mg	0.08–0.53	0.32	0.18–0.53	0.34	0.24–0.36	0.31	0.09–0.82	0.42
Fe and Al orders	0.19–3.10	1.21	—	—	0.005–0.012	0.01	—	—
Cu	tr.–0.04	0.005	—	—	tr.–0.003	0.002	0.002–0.015	0.004
Mn	0.0007–0.007	0.002	—	0.007	tr.–0.004	0.002	0.0013–0.0045	0.002
S	—	—	0.07–0.45	0.23	0.06–0.45	0.21	—	—
Zn	0.02–0.06	0.048	—	—	0.002–0.003	0.002	0.002–0.032	0.022
B	0.0005–0.03	0.013	—	0.001	—	—	tr.–003	0.002
Organic matter	79–87	94	76–90	82	82–87	85	—	—
Moisture	3.8–13.8	9.5	5.4–14.3	9.9	4.0–11.2	8.9	—	—
Ash	4.7–14.7	8.9	6.3–10.0	7.5	4.9–8.5	5.8	—	—

tr., trace.

Table 9.5 Comparison of castor and cottonseed meal (all figures percentages).

	Castor seed cake		Castor seed meal		Cottonseed	
	Whole	Hulled	Solvent	Expeller	Whole	Meal
Moisture	9.8	10.4	11.8	6.8	7.3	7.6
Ash	15.0	10.5	6.0	6.2	3.5	5.4
Oil	5.2	8.8	1.1	4.4	17.9	3.2
Crude protein	20.4	46.4	30.6	32.1	23.4	40.8
Crude fibre plus carbohydrates	49.4	24.0	14.0	—	—	—
N-free extract	—	—	—	27.5	23.2	29.8

cattle. The amino acid content of castor flour, pomace and other oilseed residues is shown in Table 9.6.

CRAMBE

Crambe seed is a relative newcomer to commercial crushers, and is generally not available in large enough quantities to attract industrial attention. The oil has a number of industrial uses, and where the crop will grow it provides a valuable addition to local processors. Crambe oil is normally pale yellow, but a greenish tint can occur in solvent-extracted oils. It is semi-drying and has a high erucic acid content, to 65–70%, usually 50–60%. This precludes its use as an

Table 9.6 Amino-acid content* of some oilseed residues.

	Crude protein†	Arginine		Histidine		Isoleucine		Leucine	
		A	B	A	B	A	B	A	B
Babassu meal	22.72	3.19	14.04	0.41	1.81	0.88	3.87	1.40	6.17
Castor flour	65.03	8.39	12.90	1.36	2.09	3.46	5.32	4.16	6.40
Castor pomace	39.79	3.99	10.03	0.67	1.68	1.85	4.65	2.25	5.65
Copra meal	22.22	2.54	11.43	0.38	1.71	0.89	4.00	1.40	6.30
Cottonseed meal	39.61	4.38	11.02	1.07	2.70	1.59	4.01	2.46	6.20
Linseed meal	36.73	3.41	9.28	0.66	1.80	1.68	4.57	2.19	5.96
Palm kernel meal	19.16	2.54	13.26	0.31	1.62	0.76	3.97	1.23	6.42
Groundnut meal	38.95	4.02	10.33	0.84	2.16	1.69	4.33	2.60	6.68
Safflower meal	22.10	1.72	7.78	0.44	1.99	0.85	3.85	1.22	5.52
Sesame meal	46.08	5.49	11.91	1.02	2.21	1.97	4.27	3.19	6.92
Soyabean meal	45.29	3.38	7.46	1.12	2.49	2.49	5.50	3.48	7.69
Sunflower seed meal	21.02	1.63	7.76	0.46	2.19	0.95	4.52	1.25	5.95
Tung meal	20.88	2.27	10.87	0.48	2.30	1.03	4.93	1.58	7.57

* In percentages; † Gram/100 g product: A – in sample, B – in crude protein.

edible oil, although when refined, bleached and deodorised it is similar to soyoil. The high proportion of erucic acid is industrially valuable, and the oil is used as an additive to mineral lubricants, in greases, as a mould lubricant in steel casting, and hydrogenated oil in the manufacture of specialised waxes.

The greatest potential use would appear to be as a source of erucic acid products, and crambe is an excellent alternative to high erucic acid rapeseed cultivars. Saponification of the crude oil produces unsaturated fatty acids, from which erucic acid is obtained by fractionation, and following ozonolysis also yields dibasic, brassidic and monobasic pelargonic acid, used in a wide range of plastics, resins, lubricants, surfactants, pharmaceuticals and food flavours. The fatty acid composition of average samples of crambe and high erucic acid rapeseed oil is shown in Table 9.7 and the physical and chemical properties of crambe oil in Table 9.8.

Crambe press cake is usually a medium-brown, of good aroma, readily eaten by livestock, contains 40–60% crude protein, good amino-acid balance, a high lysine content, and fat-free dehulled cake normally contains 50–55% protein. Analyses of an average cake sample are given in Table 9.9, and the amino-acid content in Table 9.10 (Van Etten *et al.*, 1965).

A major initial disadvantage of crambe meal as a stock feed was the presence of glucosinolates in the seeds. These compounds when hydrolysed by an endogenous enzyme, thioglucosidase, produce toxic compounds which limit the meal's palatability, and in fact can be toxic to non-ruminants. Defatted crambe meal usually contains 8–10% thioglucosides, primarily *epi*-progoitrin. Since the pericarp which constitutes 25–40% of the seed, contains very small amounts of thioglucosides, hulling is ineffective as a method of reducing their content, although it reduces the total fibre. Major constituents of the pericarp are lignin 30%, pentosans 23% and α-cellulose 19%. The seed-coat contains sufficient oil and protein to be of nutritional value, and is an acceptable roughage in ruminant feeds. Methods of extraction now remove thioglucosides and sinapine from the meal, and it can safely be fed to cattle and poultry. Greater utilisation of the meal as stockfeed is essential if crambe is to be commercially viable, since 1200 kg of hull-free meal is produced per 1000 kg of oil.

GROUNDNUT

Two-thirds of total world production of groundnuts is crushed for oil, and the remainder eaten as whole nuts or paste. Groundnut oil, com-

Table 9.6 *Continued*

Lysine		Methionine		Phenylalanine		Threonine		Tryplaphan		Valine	
A	B	A	B	A	B	A	B	A	B	A	B
0.98	4.31	0.53	2.33	1.35	5.94	0.71	3.13	0.24	1.06	1.19	5.24
2.20	3.38	1.16	1.78	3.38	5.19	2.50	3.84	0.93	1.43	4.38	6.74
0.12	3.02	0.58	1.46	1.86	4.68	1.29	3.24	0.44	1.11	2.15	5.40
0.68	3.06	0.35	1.59	0.94	4.23	0.84	3.78	0.21	0.94	1.22	5.49
1.67	4.20	0.59	1.49	2.09	5.25	1.38	3.47	0.63	1.59	1.98	4.98
1.33	3.62	0.61	1.66	1.65	4.49	1.39	3.78	0.64	1.74	2.04	5.55
0.66	3.44	0.41	2.14	0.82	4.28	0.60	3.13	0.20	1.04	1.03	5.38
1.38	3.53	0.41	1.04	1.94	4.97	1.16	2.98	0.47	1.22	1.88	4.82
0.60	2.71	0.34	1.54	1.16	5.25	0.65	2.94	0.26	1.18	1.09	4.93
1.27	2.76	1.22	2.65	2.18	4.73	1.68	3.64	0.88	1.91	2.33	5.06
2.79	6.17	0.63	1.39	2.20	4.86	1.82	4.03	0.76	1.69	2.45	5.40
0.80	3.81	0.46	2.19	1.76	5.12	0.72	3.43	0.29	1.38	1.03	4.90
0.87	4.17	0.47	2.24	1.52	7.28	0.96	4.60	0.35	1.68	1.74	8.33

Table 9.7 Comparison of fatty acid composition of crambe and rape oils (wt %).

Fatty acid oil	Crambe seed oil	Rapeseed
Myristic	—	trace–1.0
Palmitic	2.0–2.8	1.0–3.0
Stearic	0.5–0.8	0.2–3.0
Behenic	0.8–1.5	0.6–2.5
Lignoceric	—	0.5–2.0
Hexadecenoic	0.3–0.5	0.2–3.0
Oleic	20.1–20.9	12.0–18.0
Eicosenoic	3.5–3.8	3.0–6.0
Erucic	57.9–60.7	45.0–55.0
Linoleic	6.2–9.8	12.0–16.0
Decosadienoic	trace-0.7	trace–2.3
Linolenic	2.6–5.5	7.0–9.0

monly known as peanut oil, is a pale to medium yellow colour depending on the refining process and local consumer preference, bland or with a pleasant, slightly nutty flavour. In general, the darker the oil the poorer the quality, and a brown tinge suggests overheating of nuts during processing. For use in mayonnaise and salad oils, yellow oil is preferred, but colourless oil is used for spreads such as margarines. Groundnut oil is also used in speciality soaps, hair lotions, cosmetics and toilet preparations, since it is widely believed to be beneficial to the skin. The characteristics of groundnut oil are shown in Table 9.11, and compared with other oils in Tables 9.21 and 9.30.

The oil is of excellent quality with a high smoke-point, about 227°C, and is mainly used as a cooking and salad oil with minor amounts incorporated into margarines and similar products. As a cooking oil it has a major advantage for commercial users in that it can be easily clarified and reused for pan and deep frying without retaining any food flavours. Stability of a cooking or salad oil is an important characteristic, and is generally related to the linoleic acid content; the greater the proportion of linoleic acid the lower the stability (i.e. shorter shelf or refrigerator life). Runner types of groundnuts usually contain less linoleic acid in their oil than bunch types, and untreated oil made from runner groundnuts is thus more stable.

Groundnut oil is high in unsaturated fats, mainly oleic acid at 40–65%, and polyunsaturated linoleic acid 20–40%. Saturated fats total 14–20% and the oil thus has a polyunsaturated: saturated ratio of just over 2:1, which is considered desirable in a human diet aimed at reducing blood serum cholesterol. Bunch-type, Virginia, groundnuts normally produce oil which contains some 60% oleats, 22% linoleates and 18% glycerides. Runner-type, Spanish, nuts produce oil containing some 53% oleates, 25%

Table 9.8 Physical and chemical properties of crambe oil.

Analysis	Pre-press oil		Solvent oil	
	Crude	Refined-bleached	Crude	Refined-bleached
Moisture	0.04	0.02	0.018	0.05
Density, 25°C		0.9394		0.940
Refractive index, 20°C		1.4713		1.4719
Melting-point (°C)		6.5		6.0
Viscosity, 25°C, centipoise		85		85
Gardner colour		C		C
Ash (%)		0.005		0.020
Phosphorus (%)	0.005	0.0004	0.075	0.006
Sulphur (%)	0.031		10.026	
Iodine value		93.0		96.1
IV after hydrogenation		0.9		1.0
Free fatty acids (% as oleic)	0.62	0.29	1.85	0.14
Saponification value		173		173
Unsaponifiable matter (%)		0.47		0.59
Break (%)	0.14	0.02	0.40	0.01

Table 9.9 Average composition of air-dry crambe seed meal extracted with hexane.

Crude protein (N × 6.25) (%)	49.6
Sulphur (%)	2.2
Phosphorus (%)	1.0
Thioglucosides as progoitrin (%)	
Total by sulphate titration	10.1
Thioglucosides forming volatile isothiocyanates	1.0
Thioglucosides forming oxazolidinethiones at pH 7.0	9.1
Volatile isothocyanates (mg/g meal)	2.5
Oxazolidinethione (mg/g meal)	
By Wetter method	10.0
By conversion at pH 7.0	27.5
Nitrogen as amino acids (%)	72.5
Nitrogen as ammonia (%)	14.1
Cystine (g/16 g N)	2.6
Tryptophan (g/16 g N)	1.2

linoleates and 22% saturated glycerides, Table 9.12. The oil is easily hydrogenated, and small amounts of hydrogenated oil are often added to peanut butter and similar products to prevent oil separation. A fatty drink containing 1000 cal/225 g has been produced from homogenised oil and is used as a dietary additive in hospitals and relished by patients. A multiplicity of speciality products made from, or including groundnut oil are available from individual manufacturers.

Groundnut meal is a valuable livestock food and usually contains 2–7% oil, 45–55% protein and 4–10% fibre, depending upon extraction method. Solvent extracted cake usually has a low oil content, expeller cake the higher levels. Cake with a high oil content is normally used for livestock feeds, but if produced under hygienic conditions, cake with an oil content below 2% can be ground into flour for human consumption. Meal composition varies widely, not only due to extraction methods but also depending on cultivar and where grown. Analysis of an average sample of commercial groundnut meal is given in Table 9.13. Good-quality groundnut meal is comparable in feeding value with soyameal, with which it is generally interchangeable, but it is not so readily available.

Lysine and methionine must be added or supplied by other components of the feed for use in poultry rations, but cake may contain aflatoxins, discussed later, to which ducks and turkeys are particularly sensitive. Although an excellent feed for pigs, too high a content in the diet can produce soft pork, and in dairy cattle may influence milk fat quality, but it can be fed

Table 9.10 Amino-acid composition of crambe meal.

Amino acid	(g/16 g N)	Amino acid	(g/16 g N)
Lysine	5.3	Arginine	6.4
Methionine	1.6	Glycine	4.9
Isoleucine	3.7	Alanine	4.1
Leucine	6.1	Aspartic acid	6.3
Phenylalanine	3.7	Glutamic acid	15.5
Tyrosine	2.7	Hydroxiproline	0.4
Treonine	4.1	Proline	6.1
Valine	4.7	Serine	3.6
Histidine	2.3		

Table 9.11 Characteristics of average sample of commercial groundnut oil.

Melting-point	0–3°C
Iodine value	82–106
Thiocyanogen value	58–75.5
Saponification value	188–195
Acetyl value	8.5–9.5
R–M value	<0.5
Polenske value	<0.5
Free fatty acids	0.02–0.60%
Unsaponifiables	0.3–0.7%
Refractive index (ND^{20})	1.4697–1.4719
Density at 15°C	0.917–0.921
25°C	0.910–0.915
Mean viscosity 20°C	71.07–86.15 centipoise
Titer	26–32°C
Heat of fusion	21.7 cal/g (unhydrogenated) 24.7 cal/g (hydrogenated)
Colour (Lovibond)	16–25 yellow 1–2 red

Table 9.12 Average fatty acid composition of different types of groundnut oil.

	Erect type	Runner type
Unsaturated		
Oleic	52.9	60.0
Linoleic	24.7	21.6
Saturated		
Palmitic	8.2	6.3
Stearic	6.2	4.9
Arachidic	4.0	3.3
Lignoceric	3.1	2.6
Unsaponifiables	0.2	0.3

liberally to growing and fattening stock and horses.

Groundnuts, as peanuts, are commonly used as human food with minimum preparation; raw, boiled or roasted, shelled or unshelled. The famous Barnums Circus of the USA is credited with introducing snacks of shelled roasted peanuts to New York City, and subsequently throughout the USA and the world. A major use of processed groundnuts in human food is in peanut butter, introduced commercially in the USA around 1905. Peanut butter is prepared by removing from whole nuts the testa and germ, then roasting and grinding to a paste of varying degrees of smoothness, or it may contain small pieces of nut to give a chunky taste. Sugar, salt, maize syrup, lecithin, antioxidants and hydrogenated fat may be added during processing to improve stability, edibility or to suit local consumer preference. In addition to peanut butter, groundnuts are the base for a great variety of products, and the use of peanuts in the USA is butter 60%, salted nuts 20%, various composite bars 20%, and peanuts contribute at least 125 000 t of protein to the American diet.

In addition to this wide range of edible products, groundnut flour produced from the nuts is an excellent protein source. Flour is normally produced from sound, whole nuts with the testa and germ removed, since the latter imparts a slightly bitter taste. Typical analyses of groundnut products are shown in Table 9.14, and the essential amino acids in groundnut flour are compared with other flours in Tables 9.6. and

Table 9.13 Analysis of average solvent-extracted groundnut meal.

	(%)		(mg/kg)
Dry matter	91.5	Choline	2100
Ash	4.8	Niacine	175
Crude fibre	13.5	Pantothenic acid	55
Ether extract	1.2	Riboflavin	11.5
N-free extract	27.0	Thiamine	7.5
Crude protein	48.0	Arginine	6.1
Digestible protein		Histidine	1.25
Cattle	44.0	Isoleucine	2.1
Pigs	45.0	Leucine	3.8
Sheep	46.0	Lysine	2.4
		Methionine	0.4
Calcium (%)	0.2	Pheylalanine	2.8
Magnesium (%)	0.4	Threonine	1.6
Phosphorus (%)	0.7	Trytophan	0.5
Manganese (mg/kg)	30.0	Tyrosine	1.9
		Valine	2.9

Table 9.14 Composition of groundnut products.

	Boiled peanuts	Roasted peanuts (salted, without skins)	Peanut butter (fat, salt sugar added)	Partially defatted peanut flour (toasted)	Defatted peanut flour	Aqueous processed	
						Concentrate	Isolate
Water (%)	36.4	1.6	1.7	1.2	7.0	4.3	5.6
Calories (100 g)	376	585	589	—	—	—	—
Protein (%)	15.5	26.0	25.2	42.1	57.0	67.1	93.8
Fat (%)	31.5	49.8	50.6	34.1	0.6	2.4	0.2
Carbohydrate (%)	14.5	18.8	18.8	20.2	30.0	17.6	5.2
Fibre (%)	1.8	2.4	1.8	2.5	4.6	4.2	0.2
Ash (%)	2.1	3.8	3.7	3.4	4.6	2.4	4.1
Calcium (mg/100 g)	43	74	59	55	140	4	3.4
Phosphorus (mg/100 g)	181.0	401.0	380.0	540.0	760.0	0.8	0.6
Iron (mg/100 g)	1.3	2.1	1.9	3.6	2.1	1.9	1.8
Sodium (mg/100 g)	4	418	605	11	180	146	464
Potassium (mg/100 g)	462	674	627	1075	1290	642	248

9.26. Groundnut protein is deficient in four of the essential amino acids, and must be supplemented to provide a balanced intake. Methionine is usually the most deficient, but can easily be supplied by a synthetic product. The others can be obtained in sufficient quantity by increasing the amount of meal supplied to ensure an adequate level of trytophan. A milk suitable for feeding infants if mixed with other foods has been produced from the protein, but is presently undervalued in most countries where it could be more widely utilised.

Groundnut cake can also be used as fertiliser, and an analysis of some Indian oil cakes so used is shown in Table 9.19. However, it is usually considered too valuable for fertiliser, and is most often fed to livestock. Following decortication, groundnut hulls are usually burnt as fuel or waste, but in India they have been processed into particle board.

JOJOBA

Jojoba seeds produce a liquid wax, not an oil, although for simplicity it will be hereafter referred to as oil, and this oil has very valuable and specific characteristics. The most publicised is the direct substitution for sperm whale oil thus rendering unnecessary the hunting of these intelligent animals. Sperm whale oil is noted for its oiliness, metallic wetting properties and non-drying character that prevents tackiness and gumming, all possessed by jojoba oil. The oil lacks sperm oil's fishy odour, and the crude oil contains no stearins, thus requiring little or no treatment before use in many industrial processes. Crude jojoba oil is a clear, light yellow, with a mild, pleasant, slightly nutty taste, does not darken when sulphurised and remains liquid when highly sulphurised, unlike whale oil which requires the addition of mineral oil to remain liquid. Jojoba oil is also considered an acceptable alternative, when hydrogenated, for carnauba wax and beeswax.

Refined oil has great stability, is unsaturated, non-volatile, and can be repeatedly heated to 29°C without loss of properties. The oil is composed almost exclusively of long-chain acid-alcohol esters; predominantly C_{40-42} fatty acids and alcohols each having one double bond, and the chemical reactivity of these double bonds gives the oil great flexibility in end use. It contains no glyceride esters, which are major constituents of many other animal and vegetable oils and fats. Jojoba oil has been extensively researched and there is now a volume of literature describing its properties and applications (Wisniak, 1994).

Characteristics of the oil are shown in Table 9.15 (Wisniak, 1994), but as with the seed there may be variation due to analytical method, between cultivars or location. A brief comparison of sulphurised jojoba oil with sulphurised sperm whale oil is shown in Table 9.16 (Miwa & Rothfus, 1979).

Industrial uses for jojoba oil include cosmetics, pharmaceuticals, candles, specialised waxes and lubricants, and the range is continually increasing as more oil becomes regularly available. The oil is valuable as a carrier for medicines that must pass through the stomach to the small intestine, since it is indigestible to man and most animals. However, owing to the extremely strict health regulations governing products of the cosmetic and pharmaceutical industries and the expense of conforming, manufacturers tend to be reluctant to modify accepted formulae.

By-products from oil expression have many potential uses. Hydrogenated solid wax is one of the most important, for it has good stability and excellent hardness, with a melting-point of about 70°C. It is as hard as candelilla wax and almost as hard as carnauba and an alternative to both. X-ray studies showed the wax ester is crystallogically similar to polyethylene, and indicates additional uses. Jojoba oil reacts with sulphur chloride to form factice, of value in the preparation of rubber products including linoleum, while epoxidised oil is superior to many other epoxides in plastics manufacture. The acids, alcohols and esters have numerous industrial and chemical uses.

Table 9.15 Properties of jojoba oil.

Freezing point, °C	10.6–7.0
Melting points, °C	6.8–7.0
Boiling point at 757 mm under N_2, °C	389
Heat of fusion by DSC, cal/g	21
Refractive index at 25°C	14 650
Dielectric constant (27°C)	2.680
Specific conductivity, mho/cm (27°C)	8.86×10^{-13}
Specific gravity, 25°/25°C	0.863
Surface tension, dyne/cm (23.5°C)	34
Viscosity	
Saybolt, 1000°C, SUS	127
Saybolt, 210°C, SUS	48
Smoke point (AOCS Cc 9a-48), °C	195
Flash point (AOCS Cc 9a-48), °C	295
Fire point (COC), °C	338
Iodine value	82
Saponification value	92
Acid value	<2
Acetyl value	2
Unsaponifiable matter, %	51
Total acids, %	52
Iodine value of alcohols	77
Iodine value of acids	<76
Average molecular weight of wax esters	606

Table 9.16 Comparison of some properties of sulphurised jojoba and sperm whale oils.

Property	Jojoba		Sperm oil
	Sample 1	Sample 2	
Sulphur (wt %)	11	11	11
Viscosity at 99°C, SUS*	834	—	260
Pour-point (°C)	12	17	18
Freezing-point (°C)	9	15	15
Flash-point (°C)	223	248	240
Fire-point (°C)	293	262	280
Saponification value	81	118	167
Neutralisation value	7.0	3.5	3.0

* Saybolt universal seconds.

Extracted meal is high in protein, 25–35%, but the presence of a monoglucoside, simmondsin and a derivative benzyl cyanide, make untreated press cake dangerous as a livestock or human food. Although the seed has been traditionally eaten by local people and wildlife in jojoba's natural habitat, such small quantities would have little effect. Detoxification is simple, renders the meal a nutritious food, and readily acceptable to stock-feed manufacturers. A one-stage method of oil extraction and meal detoxification is claimed to be effective (Lazini, 1991). The pressed meal may contain 9–11% fibre and 5–15% oil depending on the extraction method. An analysis of the amino-acid composition of jojoba meal is shown in Table 9.17, but the level of a particular acid can vary with the analytical and meal extraction method (Miwa & Rothfus, 1979). The lysine content is usually adequate but not the methionine content.

NIGER

Niger oil is a clear, very pale yellow, occasionally with a bluish tint depending on the method of extraction and refining. Locally expressed oil is basically an edible oil with a slight nutty taste and little odour, and quickly becomes rancid. It is normally available in quantity only in India, and Ethiopia where it is the most commonly available oil. In India it is a substitute for the more expensive and more highly regarded sesame oil to which it is a common adulterant, and seed fried in *ghee* accompanies other foods. In Ethiopia seeds are occasionally crushed to prepare a non-alcoholic drink, but are commonly boiled with dough prepared from *ensat* (*Musa ensente* the false banana) to make a more tasty porridge. Poor quality oil is used as a general lubricant, in soap manufacture, and in lamps. It is also widely used in India, and to a lesser extent in Ethiopia, as a body oil to promote a smooth skin and supple joints.

Most of the Indian crop is expressed in local mills or by *ghani* operators, since large amounts of seed are seldom available. Oil yield and quality is therefore very variable, and reflected in analyses of this type of oil. A common extraction rate is 25–35% from seed with an oil content of 40–45%. Some characteristics of niger oil and major fatty acids are shown in Table 9.18. There is wide variation in published analyses of niger oil, since seed constituents, oil content and constituents all vary, with cultivar and locality having a major influence.

Extracted cake is very dark sometimes almost black, and in India generally contains about 14% oil, 34% crude protein, 8–10% fibre, 20% carbohydrates and 10–12% ash. Analyses of niger cake shipments from all destinations received in the United Kingdom gave the following composition: oil 2.8–10.8%, protein 6.0–30.0%, with an average of 6% oil, 32% protein and 20% fibre. In India where the greater proportion of niger is

Table 9.17 Amino-acid composition of jojoba meal.

Amino acids	% Total amino acids in jojoba meal (i.e. grams a/a per 100 g of meal)	Amino acids expressed as per cent of amino acids recovered
Lysine	1.4	5.7
Histidine	0.6	2.5
Arginine	1.9	7.8
Aspartic acid	2.6	10.6
Threonine	1.3	5.3
Serine	1.3	5.3
Glutamic acid	3.2	13.1
Proline	1.5	6.1
Glycine	2.4	9.8
Alanine	1.1	4.5
Cystine (half)	0.6	2.4
Valine	1.5	6.1
Methionine	0.1	0.4
Isoleucine	0.9	3.7
Leucine	1.8	7.3
Tryosine	1.1	4.5
Phenylalanine	1.2	4.9
Total	24.5	100.0

Source: Miwa & Rothfus, 1979.

Table 9.18 Reported characteristics of niger oil.

Acetyl value	24.1–24.9
Iodine value (Wij's)	132–139
Saponification value	280–295
R–M value	0.85–0.92
Specific gravity at 15°C	0.924–0.928
Refractive gravity at 15°C	1.471
Solidifying point (°C)	–9
Oleo-refractometer at 22°C	+26 to +30
Fatty acids	**Percentage of total**
Oleic	30–40
Linoleic	48–55
Palmitic	5.5–8.5
Stearic	2.5–5.4
Arachidic	0.3–0.5
Myristic	0.3–0.45
Linolenic	1.0–1.25

grown, the cake is fed to cattle and other livestock and is considered especially valuable for milking cows and buffalo. It is also used as a manure, and has a nitrogen content of approximately 5%, phosphoric acid 2% and potash 1.5%. Composition of niger cake compared with other common Indian oilcakes used for fertiliser is shown in Table 9.19.

Niger protein is deficient in lysine and threonine, but comparative analyses are few, and tend to differ in the proportions of amino acids. Two selected analyses with amino acids as a percentage in protein are, University of Arizona data outside and (FAO inside brackets); lysine 4.74 (5.23), methionine + cystine 3.46 (3.76), threonine 3.73 (4.56), isoleucine 4.66 (4.96), phenylalanine + tryosine 7.33 (7.06), valine 5.76 (4.26) and leucine 6.99 (7.73).

SAFFLOWER

Safflower oil is normally very pale yellow, with a bland or slightly nutty flavour depending on the method of processing. It is a drying oil, and was initially used mainly in protective coatings. Following considerable publicity regarding the possible deleterious effects of high blood cholesterol content in humans, it became a highly regarded edible oil.

Table 9.19 Average composition of Indian oilcakes used for fertiliser (all figures are percentages).

Oilcake	Nitrogen	Phos. acid	Potash
Castor	4.37	1.85	1.39
Coconut	3.02	1.90	1.77
Cottonseed (whole)	3.99	1.89	1.62
Cottonseed (hulled)	6.41	2.89	2.17
Groundnut	7.29	1.53	1.33
Linseed	5.56	1.44	1.28
Niger	4.73	1.83	1.31
Rapeseed	5.21	1.84	1.19
Safflower (whole)	4.92	1.44	1.23
Safflower (hulled)	7.88	2.20	1.92
Sesame	6.22	2.09	1.26

The high linoleic acid content, some 78% in certain cultivars, together with a small amount of saturated acids and absence of linolenic acid, enables the oil to form a fast drying non-yellowing film with low wrinkling characteristics. The oil alcoholises rapidly with polyols, and heat bleaches to very light colours during processing. Blown safflower oil has a uniform polymer structure as shown by its viscosity reduction curves. Such oils were used for high quality oil emulsion paints for external use, where their excellent drying, low wrinkling and resistance to dew damage are important qualities. Special industrial uses for safflower oil include urethane resins, caulks, putties and linoleum.

Characteristics of safflower oil from selected countries are shown in Table 9.20 and a comparison with other vegetable oils in Tables 9.21 and 9.30. The variant type rich in oleic acid was discovered in Australia (Horowitz & Winter, 1965).

Safflower oil, and occasionally the seed, is used for food in less developed countries where it is grown. In more developed countries it is a common edible oil and an ingredient in many food products, including margarine, mayonnaise, salad dressing, frozen desserts and speciality breads. Minor formulations include filled milk, simulated creams and dried eggs.

Safflower oil is particularly suitable in products where a high level of stability and flavour at low temperature is required, including margarine, emulsified cake and bread shortenings. Many food uses are a direct result of the considerable publicity given to a suspected relationship between high animal fat intake and various heart ailments. Proof that dietary fats are causative agents in heart diseases is lacking, and although certain fats do raise the level of blood cholesterol, there is no direct evidence that these fats cause atheroma or thrombosis. The mechanism by which blood cholesterol is lowered by highly unsaturated fats remains unclear, but when administered to patients with hypercholesterolemia the unsaturated fatty acids of safflower lower the serum cholesterol level, but this effect does not occur unless the total fatty acid intake is reduced.

Safflower meal from undecorticated commercial seed can be fed to cattle and sheep, but is less suitable for poultry and pigs owing to its low energy content. Partially decorticated meal is suitable for ruminants, and with lysine and methionine fortification for poultry. In a properly balanced ration, safflower meal can produce growth rates superior to those obtained from using similar amounts of soymeal. Whole ground seed can also be used as stock feed, mixed with other feed grains, can be processed into pellets which do not rapidly dry out and disintegrate thus reducing waste, and mixed with other processing by-products including apple pomace and sugarbeet pulp.

The similarity between sunflower and safflower seeds may cause confusion between

Table 9.20 Some characteristics of safflower oil.

	Usual range	Eight USA varieties	Kenya	India
Acid value	0.4–10	0.7–3.6	1.8	6.3
Saponification value	186–194	—	191.2	192
Iodine value	130–150	147.2–149.8	142.5	136–146
Thiocyanogen value	82–87	85.2–86.7	—	—
R–M value	0.1–0.5	—	—	—
Hydroxyl value	1–12	—	—	13.2*
Unsaponifiable (%)	0.3–1.3	0.5	0.8	
Specific gravity 25/25°C	0.919–0.924	—	—	0.9242†
Refractive index, nD, 25°C	1.473–1.476	1.4748–1.4752	1.4741	1.4742
Refractive index, nD, 40°C	1.467–1.470		1.4698	
Titer (°C)	—	—	—	—
Fatty acid composition				
Saturated	5–10	5.0–6.7	—	—
Palmitic	6–7	—	6.4	—
Stearic	2–3	—	3.1	—
Arachidic	—	—	0.2	—
Unsaturated				
Oleic	12–14	14.7–17.3	13.4	—
Linoleic	76–78	76.6–79.0	76.9	—
Linolenic	nil	—	nil	—

* Acetyl value; † at 27°C.

Table 9.21 Comparison of safflower with other oils.

Characteristic	Normal*	Variant†	Groundnut	Soy	Olive	Linseed	Maize
Iodine value (Wij's)	140–145	91–101	89–96	131–140	84–86	175–190	116–130
Colour (Gardner)	6–10			8–11		8–11	
Acid value	0.3–3.0		0.8–6.0	0.8–3.0	0.2–6.0	0.7–4.0	2.0–6.0
Sat. fat. acids‡	6	4–8	18	14	12	10	14
Unsaturated‡							
Oleic	10–20	74–79	61	28	82	20	30
Linoleic	70–80	11–90	21	50	6	20	56
Linolenic	0	0	0	8	0	51	0
Grammes of oil to give 100 g of linoleic acid	145		500	212	1760	524	189
Calories in oil sufficient to give 100 g of linoleic acid	130		450	164	1585	472	170

* Average analysis of USA commercial safflower seed † Australian strain ‡ Percentage of total fatty acids.

cake samples of unknown origin, but a basic structural difference in the seed pericarp is obvious when specimens of separated pericarp debris are examined under a microscope. When kept dry the cake does not become mouldy or rancid, but prolonged storage is not recommended.

Seeds of commercial cultivars contain some 60–70% kernel and 35–40% hull, and after extracting whole seed the residual cake contains about 20–22% protein and 60% hull. As the hull normally consists of 70% holocellulose, 21% lignin and 1% ash, it can be digested by ruminants only with difficulty, and for poultry is of little value. Decorticated or thin-hulled seeds have reduced fibre content, but in general hull percentage remains too high for stockfeed use.

Safflower meal from whole seed can satisfactorily supplement a grain, lucerne and silage ration for fattening cattle, and should be mixed with the grain element. It is not recommended for pig feeds except in small quantities. Hulled seedmeal is suitable for pigs when fed with a good animal protein supplement, and in a balanced ration can replace up to 25% of protein. It is also equal to good-quality cottonseed meal for feeding to cattle and sheep. Safflower seed oil is rather low in the essential amino acids lysine and methionine, and when the cake is used in poultry and pig rations these deficiencies are supplied by fishmeal.

Whole seed meal is too high in fibre content for use in chick-foods, but low-hull meal is a satisfactory soymeal substitute for layers. The amino-acid content of some oilseed residues that are used in animal and human foods is shown in Table 9.6, a comparison of safflower meal with other oilseed meals in Table 9.22, and when used for fertiliser in India, Table 9.19.

Safflower meal and flour from decorticated seed are high-protein human diet supplements. Flour from safflower kernels is light-coloured, slightly cathartic and bitter due to two phenolic glucosides, which can be extracted with little loss of nitrogen, the final product containing some 70% protein. The flavour, although distinct, is relatively bland and considered acceptable when cooked with other foods. Adding 5% to wheat flour for bread making slightly reduced loaf volume, and although the characteristic flavour remained it was considered acceptable. Edible safflower concentrates in bread and pies gave results similar to those obtained with soya concentrate. Safflower proteins, and safflower protein isolate with a protein content up to 95%, have a considerable potential as human food.

SESAME

The very large world production of sesame seed is almost wholly used for culinary purposes, the major proportion consumed in the country of origin. Because of the excellent quality of the oil it produces, sesame seed has been called 'queen of the oilseed crops' (Weiss, 1971). The whole seed may be eaten raw, either roasted or parched, or mixed with other seed before grinding to produce a blended oil. Local uses for sesame in the various countries of production are

Table 9.22 Compostion of some expeller-extracted oilseed meals.

Composition	Safflower	Sunflower	Soybean	Linseed	Cottonseed
Moisture content (approx.) (%)	7.8	8.3	10.3	10.1	8.3
Digestible energy (kcal/kg)	1590	3120	3400	3400	3130
Fibre (%)	31.0	13.0	2.8	9.0	15.7
Fat (%)	6.0	7.6	0.8	5.2	6.6
Crude protein (%)	19.7	36.8	48.9	34.0	39.7
Calcium (%)	0.23	0.43	0.26	0.44	0.19
Available prosphorus (%)	0.55	0.64	0.41	0.7	1.0
Lysine	0.47	1.30	3.03	1.34	1.61
Methionine	0.13	0.47	0.49	0.40	0.35
Cystine	0.30	0.78	0.94	0.70	0.79
Tryptophan	0.17	0.45	0.56	0.55	0.37
Threonine	0.45	1.09	1.62	1.08	1.12
Isoleucine	0.43	1.14	1.83	1.17	1.01
Estimated % of amino acids available	65	85	92	90	84

too numerous to mention, for so highly is the seed regarded that it can be found as a constituent of a wide variety of foods and sweetmeats, including the very popular halva. In India a mixture of groundnut, safflower and sesame seeds is ground together to produce the 'sweet oil' so highly regarded in the Bombay region. Sesame oil is probably the most universal constituent of Burmese dishes, and is used in the drying, roasting and stewing of meat, fish and vegetables. In developed countries it is mainly a speciality oil, but the whole seed is used in large amounts in various sweets, or to decorate breads and pastry. Because of high local demand there is little international movement of the oil.

Crude sesame oil can vary in colour from deep to pale yellow, is used directly in cooking wherever it is produced, and is usually the favoured edible oil. The refined oil is usually a clear pale yellow with a nutty flavour, and used commercially as a bland salad oil, or where an edible oil of exceptional keeping quality is required. Sesame oil consists of glycerides, the chief fatty acids being oleic and linoleic, with small proportions of stearic, palmitic and arachidic acids. It contains a larger proportion of unsaponifiable matter, approximately 1.2%, than is normally found in oilseeds, and this unsaponifiable matter includes substances not found in other fats, viz. sesamol and phytosterol. The high resistance of normal and hydrogenated sesame oil to oxidation is partially due to the sesamol content. Major fatty acids are oleic 45.3–49.4%, arachidic 0.4–1.1%, linoleic 37.7–41.2%, hexadecenoic 0.0–0.5%, palmitic 7.8–9.1%, myristic 0.1% or less, stearic 3.6–4.7%, and saturated 12–16%.

Sesame oil is also used in the pharmaceutical industry, in cosmetics, soap, and as a fixative in perfumery; in foods including shortening and margarine. The refined oil can be perfumed by placing a thin layer in a tray, spreading strongly scented flower petals such as frangipani or jasmine over its surface, and leaving undisturbed for 2 days (Weiss, 1997). In India especially, this oil is used for anointing and conditioning the skin. Minor components of the oil, sesamin and sesamolin, are effective synergists for pyrethrin, a potent insecticide.

Characteristics of commercial grades of sesame oil are shown in Table 9.23. Locally

Table 9.23 Sesame oil specifications.

Determination	Usual range	British standard	American crude
Solidifying point (°C)	–3 to –4	—	
Saponification value	188–193	188–193	185.8
Refractive index (Zeiss) at 40°C	58.5–61.0	57.9–64.0	
Iodine value	103–116	105–118	112.0
Acid value	1.6	5.0 max.	
Thiocyanogen value	74–76	—	76.3
Hydroxyl value	1–10	10	5.3
Specific gravity, 15.5/15.5°C	0.922–0.924	0.921–0.924	0.918
R–M value	0.1–1.0	—	0.51
Hehner value	95.7	—	
Free fatty acids, as oleic (%)	1–3	2.5 max.	1.0
Unsaponifiable matter, (%)	0.9–2.3	1.5 max.	2.3
Titer (°C)	22–24	—	22.0
Optical activity, ang. degrees in 200 m	+1 to +3		
Colour (Lovibond) 1 in:			
Yellow	Up to 35	30 max.	35
Red	Up to 2.5	2.5 max.	4.3
Smoke-point (°C)	—	—	165.5
Flash-point (°C)	—	—	375.0

Note: Observed Criesmer value is 60.5, or 68.1 when corrected for acidity and squalene content reported as 3–9 with 5 an average.

expressed oil varies widely in its characteristics and keeping quality, mainly due to the seed composition of the dominant local cultivar. Sesame oil is frequently adulterated with less costly oils usually groundnut, rape or cottonseed, due to its popularity and higher price.

Extracted sesame cake is a valuable stock feed, can also be treated to provide a nourishing human food and may be used as fertiliser. The cake can vary in colour from light yellow to greyish-black, depending on the predominant testa colour of the seed. Dark coloured cakes are frequently more bitter than lighter, and the latter preferred where there is a choice. Sesame cake can be identified microscopically by the presence of calcium oxalate crystals in the epidermal cells. Cake composition is very variable, and is affected not only by seed type but also by the method of oil extraction. A major loss of minerals occurs when seed is hulled, with a substantial difference in the fat content of expeller and extracted cake. The chemical composition of cake obtained from similar seed extracted by three methods in India is shown in Table 9.24 (Wankhede & Tharanathan, 1976). A comparison of sesame cake with other Indian oilcakes used as fertiliser is shown in Table 9.19 (Krishnamurthy *et al*., 1960).

In most areas of primary production sesame meal is seldom used as cattle feed, being considered too valuable, but in the main producing regions of India and Burma it is more frequently used as stockfeed, as a large proportion of the seed is crushed locally, and the cake is therefore more easily available to farmers. The free fatty acid content of Indian *ghani* cake is high, its keeping quality poor, and it must be fed to stock as soon as possible or it rapidly becomes rancid and unpalatable. In Burma the cake is usually steeped in water for several hours and the mixture then poured over chopped millet, straw or grass. In developed countries the scarcity of sesame meal in relation to the large amounts of other oilseed meal available, has restricted its use to specially compounded feeds.

The protein content of sesame meal is normally similar to that of cottonseed meal, it is high in calcium and phosphate, and correctly processed meal stores well. Sesame meal is rather low in lysine and when fed to pigs or poultry must be augmented with meat or fishmeal, soymeal or other high lysine feeds. Fortified meal is an excellent protein supplement for dairy and beef cattle, and sheep. However, if the meal forms too large a proportion of the ration of dairy cows or pigs, it tends to produce soft butter and pork. Sesame meal contains around 5% phytic acid, which reduces the biological availability of calcium, magnesium, zinc and to a lesser extent iron. It may also render proteins less soluble.

Decortication substantially reduces fibre content and increases the palatability, digestibility and protein content of the meal, but also reduces the mineral content. The bitter principle present in meal produced from whole seed is easily solvent extracted, resulting in a meal highly acceptable to consumers. Whether seed is

Table 9.24 Composition of variously extracted Indian sesame cake (all values in grams).

Constituent	Expeller-pressed		Alcohol-extracted		Hexane-extracted	
	Unhulled	Hulled	Unhulled	Hulled	Unhulled	Hulled
Moisture	8.1	8.3	8.6	8.9	8.6	8.8
Fat (ether extractives)	13.5	12.7	3.1	3.4	0.8	1.1
Protein ($N \times 6.25$)	35.1	41.3	38.2	45.8	39.6	46.7
Mineral matter	8.9	4.8	9.4	4.96	9.7	5.2
Crude fibre	5.3	3.1	5.9	3.1	6.1	3.2
Oxalic acid	3.01	0.41	3.5	0.45	3.6	0.46
Carbohydrates (by diff.)	26.09	29.39	31.3	33.39	31.6	34.54
Calcium (as Ca)	1.78	0.38	1.95	0.44	2.14	0.45
Phosphorus (as P)	0.81	0.99	0.91	1.11	0.97	1.16

decorticated before pressing or the cake is subsequently solvent extracted, depends on the amount to be used for human food. Since flour must be produced in hygienic conditions, solvent extraction is probably the most economical method.

The world population explosion combined with widespread political instability has resulted in large numbers of people living on inadequate diets. Shortage of protein is often the most physically damaging, and as sesame cake remaining after expelling the oil contains 30–50% protein, it can be a valuable supplementary food in many less developed, overpopulated areas, or countries with a large urban population. For human food, seed should be decorticated before extraction as the testa imparts a slightly bitter taste to the meal, resulting from chelation of the calcium content by oxalic acid. In Burma a piece of cake may be added to rice when it is about to boil, and in India it is heated and mixed with spices as a side dish. A wide variety of food products available in developed countries incorporate sesame oil, meal and protein derivatives.

Shortage of protein-rich and supplementary foods is particularly acute on the Indian subcontinent, and the Central Food Technological Research Institute, Mysore, has carried out considerable research into use of oilseed residues for human consumption. A number of multi-purpose or special foods containing sesame meal were readily eaten by test panels of local people. The biological value and efficiency ratio of foods containing sesame meal were higher than those without.

The comparatively simple and inexpensive production methods developed should make these supplementary foods cheap enough for large numbers of people who cannot afford to buy more expensive protein foods such as meat and milk. Introduction of low-cost protein foods would also benefit low income groups worldwide. However, where fortified diets or alternative sources of protein have been offered, those of local origin are usually preferred, and frequently no inclination to replace them with alternatives. The composition of some oilseed flours is shown in Table 9.25 and their amino-acid content in Tables 9.6 and 9.26.

Micro-atomised protein food suitable for feeding unweaned babies has been successfully produced from a mixture of sesame and other oilseed flours, fortified with vitamins and minerals. The blends were in the ratio of 4:4:2 of groundnut, soyflour and skim milk, or groundnut, soy and sesame flour. The foods kept well for 9 months in airtight tins, with a vitamin A loss over the period of 35–40%. Reconstituted in ten times their weight of boiling water and sweetened with sugar they produced a free-flowing emulsion suitable for use in feeding bottles.

Table 9.25 Composition of some Indian oilseed flours (all figures in percentages).

	Sesame low-fat	Soya low-fat	Groundnut low-fat	Chickpea	Skim-milk powder
Moisture	5.6	5.0	11.0	11.2	4.1
Protein (N × 6.25)	33.3	50.0	52.7	22.5	35.0
Fat	12.2	7.0	8.9	5.2	1.0
Carbohydrate (by diff.)	38.1	31.0	21.8	58.9	51.0
Crude fibre	4.8	2.5	1.0	—	—
Ash	6.0	5.5	4.6	2.2	6.8
Calcium	2.4	0.3	0.1	0.1	1.3
Phosphorus	0.6	0.6	0.5	0.3	1.0
Iron (mg)	19.3	20.0	2.9	8.9	0.6
Thiamine (mg)	1.0	0.7	0.9	0.5	0.4
Nicotinic acid (mg)	5.3	7.7	19.5	—	1.1
Riboflavin (mg)	0.4	0.4	0.2	—	1.4

Table 9.26 Essential amino-acid content of four common oilseed flours (all figures in mg/g N).

	Sesame	Soya	Groundnut	Cottonseed
Arginine	547	452	669	702
Histidine	121	149	152	166
Isoleucine	261	336	257	236
Leucine	461	482	380	369
Lysine	160	395	223	268
Methionine	311	195	149	188
Phenylalanine	461	508	540	498
Threonine	194	246	168	221
Tryptophan	91	86	69	74
Valine	244	328	311	308

SOYA

Soya differs from many other oilseeds in that the seeds themselves have few direct uses, are seldom used as culinary beans unless previously processed, and the major use is as meal for incorporation in stockfeeds. Soya beans are easily moved in bulk and thus huge quantities can be cheaply conveyed from producer to processor, regionally and internationally. This ease of transportation was a major factor in the expansion of soya production in the Americas. Bulk shipment to European and Asian markets was uncomplicated, and until the 1980s the availability of soya beans dominated the oilseed trade. Edible usage of soya bean derivatives is extensive, ranging from pure oil, blended oil products, fats, margarine, shortening, processed foods, meat and milk substitutes, and proteins. Flours (meals), either defatted or straight, are a basic material for a wide range of protein foods, and together with cake are a staple ingredient in livestock feeds. Soya bean husks can be used as adsorbents in laboratory solutions and for purifying industrial waste water (Marshall & Champagne, 1995). The various processes involved in extracting soya products are shown in Fig. 9.10 (Courtesy of the American Soybean Association).

Soya oil (soyoil) is highly unsaturated, classified as semi-drying, and its main uses were originally industrial, since the high linolenic acid content and unpleasant odour when used for frying limited edible acceptance. Industrial oil uses include paint, varnish, resins, plastics and similar products, but in general other materials providing the same activity can be obtained more cheaply. Soyoil is a major source of high purity stearic acid, used in food emulsifiers and flocculation agents. Currently non-food use (human and animal) of soyoil and meal is probably below 10%, but new uses continue to be developed; a water dispersable concrete sealing compound was marketed in 1998 (Anon, 1998).

The basic operations to obtain soyoil and soya meal (soymeal) were cleaning, dehulling, cracking, cooking at 50–70°C, flaking to 0.3 mm, then solvent extraction. An improvement was hot-dehulling, and the use of expanders. Instead of going directly to the solvent plant flakes are passed through the expander which produces a porous pellet, a collet, which improves the speed and rate of extraction, and cuts costs. A further improvement is a combined expeller/expander, the enhancer press. Following extraction the oil is processed, each phase resulting in different grades and uses. Soybean fatty acids are obtained when glycerine is split from triglycerides in soyoil by hydrolysis, and are usually classified according to the treatment following hydrolysis. A 1998 Unilever patent describes the low cost production of sterol esters from soya for use in low-chloresterol spreads, mayonnaise, shortenings, cooking and frying oils. The characteristics of commercial edible soyoil are shown in Table 9.27, and Tables 9.21 and 9.30.

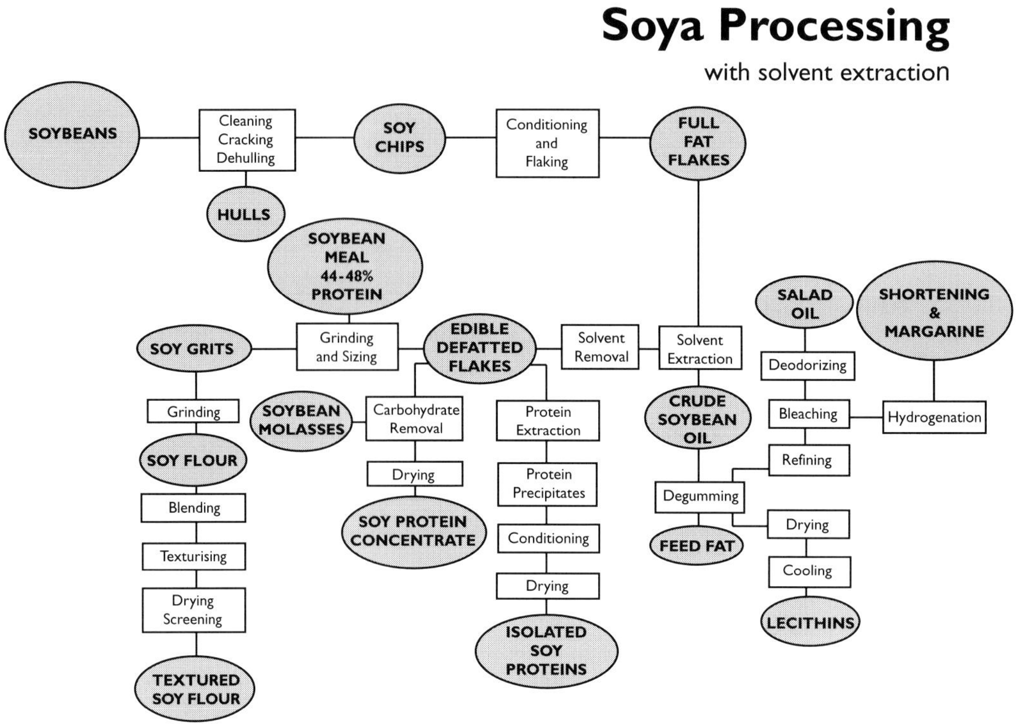

Fig. 9.10 Solvent extraction of soya beans.

Table 9.27 Characteristics of typical refined bleached and deodorised soyabean oil.

Characteristic	Value
Flavour	Bland to slightly nutty (+8.0)
Colour	2.OR/2.OY
Active oxygen method	12+ hours
Peroxide value	0.5
Free fatty acids	0.05%
Anisidine value	2.0
Oxidation value	3.0
Iodine value	132
Iron (Fe)	0.1 ppm
Copper (Cu)	0.01 ppm

Crude soyoil is a clear yellow colour with a characteristic 'green-bean' flavour which is later eliminated to produce a bland, pale-yellow oil. Following storage this odour may return, 'reversion flavour', which initially retarded wider utilisation of soyoil and its derivatives. Soyoil is not generally the culinary oil of choice, and has generally been replaced by rapeseed (canola) oil. In the USA edible crude soyoil is extracted from mature yellow soybean, but when solvent extracted the solvent must be identified. Edible refined soyoil may include treatment with alkali, bleaching, partial hydrogenation and/or winterisation.

A major cause of quality loss in soyoil occurs at harvest, and is exacerbated during storage and transport. Crude oil extracted from consignments with a significant proportion of split or damaged beans have a higher free fatty acids and iron content than oil from whole beans, and the quality falls further as storage and transit time increases. Thus strict control of harvesting procedures and efficient transport and storage facilities are very important in the production of high quality oil. These factors are of increasing importance as standards for oilseeds are raised by the growing international competition.

The meal or flakes remaining after expression are ground to produce a high protein livestock feed. The degree of grinding is related to end use, and can vary from rough grits to fine flour to be mixed directly in livestock rations or processed into pellets or cubes. Soymeal is a major protein supplement available for livestock and poultry feeds worldwide, but has suffered from competition with rapeseed. An analysis of Australian soymeal compared with other oilseed meals is shown in Table 9.22. Ground cake, ground chips or ground flakes are commonly available and sold on protein content and how manufactured. Typical composition is protein 44.0% minimum, fat 0.5% minimum, fibre 7% maximum, and moisture 12.0% maximum. High protein soymeal contains protein 47.5–49.0%, fat 0.5% minimum, fibre 3.3–3.5% maximum, and moisture 12.0% maximum.

Soymeal in the USA is described according to the method of manufacture and content; from dehulled, solvent extracted seed flour must contain a maximum of 3.5% crude fibre; from mechanical extracted seed a maximum of 7% crude fibre. Soya millfeed is composed of bean hulls and tailings resulting from the manufacture of soy grits or flour, and must contain not less than 13% crude protein and not more than 32% crude fibre. Soya millrun is composed of bean hulls and adhering residue after production of dehulled soymeal, and must contain a minimum 11% crude protein and a maximum of 35% crude fibre.

Soymeal is high in crude and digestible protein and low in fibre, has a high content of most essential amino acids (see Tables 9.6, 9.22 and 9.26) is relatively high in lysine and thus an excellent mixer for mainly grain rations. Up to 50% of some poultry and pig rations can be soymeal, but 30–40% is more common. Some soymeal with a high oil content is unsuitable for fattening stock, especially pigs, since it causes soft pork or very liquid fat. A basic factor affecting the use by compounders of soymeal is the price and availability of alternatives. For example, feed grade urea (up to 40% N) can replace a portion of soymeal in cattle rations, as can high lysine maize, while the substantial replacement of soymeal by rapeseed meal has been noted.

The oriental world had a great variety of soya flours, meals, pastes and sauces for direct human consumption, and these provided a valuable source of protein in diets often lacking this vital ingredient. Best known are probably the Japanese tofu, sufu (a cheese substitute from tofu) and shoyu (soy sauce). Indonesian tempeh is an easily produced, highly nutritious protein food. Soymilk extracted from whole beans is common in Asia and widely available in Western countries. Since these products are becoming popular outside their country of origin, brief details of the method of manufacture and composition are of interest.

Tofu is made by coagulating protein from soymilk, then pressing, and typically has a protein content of 5–15%, and there are a wide variety of local types. Tofu can be eaten raw, fried, baked, included in soups, and spray-dried and used as a dairy or meat substitute in processed foods. Tofu resembling cottage cheese with a taste more acceptable to Western palates, and packaged Japanese tofu have become available in stores worldwide.

Miso is fermented, whitish-brown, or red-brown seasoning paste made from soaked, steam-heated beans inoculated with *Aspergillus oryzae*, *Aspergillus sojae* and *Rhizopus oligosporus*. Tempeh is produced from beans soaked overnight, quickly cooked, then inoculated with *Rhizopus oryzae* and allowed to stand for 24 hours at 31°C. The finished product resembles a pressed soybean cake, and can be eaten raw, fried or roasted. Soy sauce is a popular seasoning made from fermented whole beans, meal or protein, and sometimes mixed with wheat flour. Fermentation or enzymatic action may last from 12 to 18 months, and the extract is then heated and processed to produce the sauce. Similarly to tofu, innumerable local types are available in China and Japan.

Soya flour, soyflour, is a misnomer as it does not contain gluten, and closely resembles, physically and chemically, non-fat dried milk and should be called soya powder. In the USA, soyflour is defined as finely powdered material from screened and graded meal, obtained by mechanical or solvent extraction of selected, sound, cleaned and dehulled beans. Most oil is removed, and crude fibre content must not exceed 4.0%. Full fat soyflour is obtained from

Table 9.28 Typical analyses of soyabean flours.

Product	Protein (%) (N × 6.25)	Moisture (%)	Fat (%)	Fibre (%)	Ash (%)
Soyabeans	42.6	11.0	20.0	5.3	5.0
Full-fat flour	46.6	5.0	22.1	2.1	5.2
Defatted flour	59.0	7.0	0.9	2.6	6.4
Lecithinated flour	48.6	5.5	16.4	2.2	5.3

whole beans containing 18–20% oil and may be heat processed or toasted to minimise enzyme action. Textured soyflour is made by spinning and combining soy fibres in layers to achieve the desired texture, or by a thermoplastic extrusion process, and a number of patented processes and trade marks exist. In the USA a large proportion of bread, rolls and fried foods contain 1–5% of soyflour. A range of these flours with varying fat levels is available commercially (Table 9.28). World production of full, defatted and textured flours was about 80000 tonnes in 1997–98.

High protein soyflours and soymilks are an excellent dietary additive in less-developed areas, but the author's experience is that they are not always acceptable to local people where the bean is unknown. The composition of some Indian oilseed flours is shown in Table 9.25, and the amino-acid content in Table 9.26. Soya beans are themselves a valuable source of vegetable protein, but are unpopular in many countries because of the cooking time necessary to make them tender. Germinated bean shoots are a popular vegetable. Condensed solubles are produced from washing soyflour or soyflakes with water, alkali, alcohol, and acid in various combinations; the liquid is then concentrated to a minimum of 50% solubles, and dried. Poor quality soymeal, or meal unsuitable for stock feed can be used as fertiliser, and analyses of meal so used in the USA are shown in Table 9.4.

Soymilk is a protein rich, milk-like liquid typically obtained by soaking and grinding whole beans in water, or hydrating whole, full-fat soyflour, cooking the slurry, then filtering, and there are a number of commercial production methods. The Soyfoods Association of America defines soymilk as a liquid food obtained as a result of combining aqueous-extracted whole soybean solids and water; or other edible quality soy protein solids, soyoil, and water; the product to provide a minimum 3.0% soy protein, a minimum 1.0% soybean fat and a minimum 7.0% total solids.

Soya protein, soyprotein, can replace meat and fish proteins in adult diets without reducing the utilisation of dietary nitrogen. Soyproteins also supply nutritionally valuable protein in cereal based children's diets, and with the increasing availability of soya bean on world markets should become more widely used, especially in Africa and Latin America. Soyprotein is produced from defatted soyflakes or soyflour by extraction, and commercial products can contain 95% protein, and are used in a vast range of bakery, processed meat and processed food products (Chajuss, 1998). Soyprotein can be textured to simulate meat, to which are added chicken, ham, bacon, or beef flavours. Soyprotein with crab, lobster or fish flavour and colour is widely available as seafood sticks, and is included in speciality dishes and prepared frozen meals, often without any indication that no marine constituents are involved!

In the USA, hydrolysed soyprotein is defined as produced from soyflours, concentrates or isolates, treated with acid, or an enzyme and dried. Isolated soyprotein is removed from beans and highly concentrated, and is generally produced by extracting protein from white flakes or flour with water or a mild alkali. Isolates usually have a minimum protein content of 90%. Textured soy concentrate has texture imparted by spinning and combining fibres in layers to the desired texture, or by a thermoplastic extrusion process. Chemically modified soyprotein has been processed to modify the natural protein structure by utilising acids, alkalies or other chemicals and without removing significant amounts of any nutrient. Total world production of soyprotein concentrates in 1997–98 was about 300 000 t and of protein isolates about 130 000 t.

The processes required to obtain soyprotein and isolates also provide some interesting by-products, including the phytoestrogen isoflavones genistein, genistin, and diadzain, while soya molasses are currently under investigation as another source of phytochemicals, and oligo-sugars.

SUNFLOWER

Sunflower seed (sunseed) provides an important protein source in less-developed societies, and roasted, salted, hulled or whole is a popular snack worldwide. Sunseed is also included in many petfoods, especially for birds. Since the seed characteristics for these various uses differ, specific sunflower varieties are grown to supply these markets.

In 1716 an Englishman, Arthur Bunyan, registered a patent for producing oil from sunseed, and this is the first European record of using the seed as a source of oil. The patent states that it shows:

> How from a certaine English seed might be expressed a good sweet oyle of great use to all persons concerned in the woolen manufacture, painters, leather-dressers etc. Such oyle so be made is to be expressed from the seed of the flowers commonly knowne by the name of sunflowers of all sorts both double and single.

Crude sunflower oil (sunoil) is yellow, becoming a very light or pale yellow after refining, usually bland, with little odour. It is an excellent edible oil, being relatively low in the saturated fatty acids, palmitic and stearic, and the absence or very low linolenic acid content imparts excellent storage properties. Linolenic acid readily oxidises and produces off-flavours and poor stability. The unsaturated nature of sunoil makes it most suitable for use in soft margarines and similar foods. Commercial margarine and edible oil producers generally require a minimum 62% linoleic acid content in crude oil, and this level is commonly exceeded in recent cultivars. Sunoil from seed grown in temperate regions normally contains more linoleic acid than oil from seed grown in high-temperature areas, and a difference of 50% has been recorded.

The high ratio of polyunsaturated to saturated fatty acids makes it an excellent dietary oil. As a commercial cooking oil it gives a superior flavour when used to fry potato chips and crisps, but has the disadvantage that it can darken on heating, due to an increase in the proportion of saturated fatty acids with extended use, resulting from thermal decomposition of unsaturated acids.

High linoleic acid oil is of value in the developing use of protected lipids in ruminant diets, the by-pass proteins. These protected lipids are not metabolised by organisms in the rumen and are later incorporated in milk and meat fat, and resulting products have a naturally higher level of polyunsaturated fats. Conflicting demand for sunseed with either a high linoleic or high oleic acid content has produced cultivars to satisfy this demand, for the two fatty acids normally vary inversely with each other. Characteristics of typical commercial sunoils are shown in Table 9.29 and the fatty acids compared with other common vegetable oils in Table 9.30.

Oil is the most important seed derivative, but other constituents including protein isolates, which have as many potential uses as the better known soya isolates, are gaining popularity. In general a negative correlation exists between the lipid and protein content of seed, with globulins and albumins comprising about 90% of total protein, and helianthinin the main component of globulin.

Sunflower meal (sunmeal) is a high quality protein source for stockfeed, but the high fibre content of the hull reduces its value to compounders. It is usually described as either 'sunmeal dehulled,' obtained by grinding the residue remaining after mechanical or solvent extracting dehulled seed; or 'sunmeal, whole seed,' obtained after mechanical or solvent extraction of unhulled seed. A typical analysis of solvent extracted sunmeal is shown in Table 9.31. The characteristic grey-green colour is caused by chlorogenic acid in seeds, which darkens to a green or brownish colour upon oxidation. This distinctive colour easily identifies sunmeal, but does not detract from its feeding quality or palatability. Sunmeal produced from dehulled seed commonly has a protein content around 40%, an oil content of 1–2% if solvent-extracted and 7–9% if expeller-extracted.

Sunmeal is typically low in lysine, an essential amino acid, and must be fortified for use in pig and poultry rations, but is commonly used in beef and dairy cattle feeds. When used for feeding beef-cattle, sheep and horses it must be balanced with grain and roughage, but it can be fed at high levels to milking cows. Up to 5 kg per head per day has no adverse effect on milk yield or quality in Australia. Sunmeal can be fed only in limited amounts to pigs, since it tends to produce soft pork. Sunmeal can partially replace fishmeal in poultry diets, but is a nutritionally incomplete protein for chicks. It is equal to cottonseed meal and superior to soymeal in supporting growth, and in layer rations has no influence on yolk colour or taste. A comparison of sunmeal with other oilseed meals is shown in Tables 9.6 and 9.22.

Chlorogenic acid in seeds is of no practical significance when the meal is used as stock feed, but the dark colour of flours and protein isolates limit their use for human food. Biological studies showed that sunflower protein equals soyprotein in human diets, is superior to most vegetable proteins in terms of digestibility, while the polar lipids present are of importance. The protein content of sunflower protein isolate (SPI) and promine (soyprotein) range from 88% to 91%, with a fibre content of less than 0.1%; however, the considerable potential for SPI in human diet is currently greatly underutilised.

Table 9.29 Characteristics of average sample of commercial sunflower oil.

Iodine value	125–136
Specific gravity 15/15°C	0.920–0.926
Peroxide value	3.2–4.5
Saponification value	188–194
Unsaponifiables	0.3–1.5
Titer °C	16–26
Refractive idnex 40°C	1.4663–1.684
60°C	1.4597–1.625
Solidification point	−16 to −19°C
Stability 8 h AOM	35–38
Density at 60°C	0.894–0.899
Colour (Lovibond)	1–2 Yellow
	0.1–0.3 Red
Fatty acid composition (av.)	%
Myristic	0
Palmitic	6
Stearic	4
Oleic	18
Linoleic	70
Linolenic	0
Behenic	Trace

VEGETABLE OILS AS FUELS AND LUBRICANTS

About 50 years ago in Tanzania the author used locally pressed castor oil as a gearbox and engine oil to keep his Land Rover on the road during one of the regular post-war shortages. That this was no eccentricity is supported by the subsequent multiplicity of tests on its uses as a lubricating oil, and acceptance as a jet-engine

Table 9.30 Comparison of fatty acids of sunflower oil with other vegetable oils.

Fatty acid	Sunflower	Soya	Safflower	Ground nut
Palmitic	7.2	11.5	7.2	10.1
Stearic	4.1	3.9	2.1	3.2
Oleic	16.2	24.6	9.7	53.9
Linoleic	72.5	52.0	81.0	24.2
Linolenic	0	8.0	6	0
Eicosenoic	0	0	0	3.3

Table 9.31 Typical analysis of solvent-extracted dehulled sunflower meal.

Protein	(%)		(%)
Dry matter	91.6	Aspartic acid	8.0
Protein	41.8	Threonine	3.8
NFE	26.3	Serine	3.9
Fat	1.5	Glutamic acid	21.0
Fibre	15.7	Proline	4.1
Ash	6.3	Glycine	5.5
Digest. protein	36.2	Alanine	4.2
Ruminant TDN	63.0	Valine	4.4
Ruminant ME (kcal/kg)	2505	Cystine	1.72
Ruminant DE (kcal/kg)	3055	Methionine	2.0
Calcium	0.43	Isoleucine	3.9
Phosphorus	1.04	Leucine	6.1
Potassium	1.08	Tyrosine	2.5
Manganese (mg/kg)	22.9	Phenylalanine	4.0
Riboflavin (mg/kg)	3.1	Lysine	3.3
		Histidine	2.7
		Arginine	8.1

Comparison of wholeseed meals

	Expeller (%)	Solvent (%)
Dry matter	91.6	92.5
Protein	31.1	32.1
NFE	17.9	28.2
Fat	12.6	1.3
Fibre	22.2	23.6
Ash	7.8	7.3

NFE, nitrogen free extract; TDN, total digestible nutrients; DE, digestible energy; ME, metabolisable energy.

lubricant. However, no pure vegetable or mineral oil possesses all the properties required by modern technology, and a combination is necessary. Such mixtures are often very specific in their components, but in general the lubricants most commonly used are very similar; for ordinary motor cars or farm tractors for example. It is in this field that substitution by vegetable oils becomes possible and practical, the main limiting factors being availability and price.

Price may not be the most critical factor in developing countries with locally grown oilseeds, and given the cost of liquid fossil fuels and the increasing difficulty of paying for them, import substitution that saves foreign exchange is far more important. Locally produced oil together with greater use of solar energy, could substitute for wood as the main fuel source, thus helping to preserve the world's rapidly diminishing woodlands, as described by the author in relation to essential oil production (Weiss, 1997).

Castor oil is one of the few naturally occurring oils that is very nearly a pure compound, and is the only vegetable oil composed mainly of an hydroxy acid glyceride, ricinoleic acid at 90%. Research into uses of castor oil as a lubricant probably exceeds that of any other vegetable oil, and for decades it has been an additive in special oil blends. Those marketed under the Castrol brand for use in high performance cars are well known, but the London Transport Executive also tested a castor based oil as a general and engine lubricant and found it as effective as mineral oil. The blended oil's limited solubility in petrol also resulted in a fuel saving of nearly 4%. A comparison of the viscosity of a number of fatty oils is shown in Fig. 9.11.

Sunoil and rape oil are being extensively tested as fuel for diesel engines, usually as a blend with diesel, and a 20:80 and 50:50 sunoil:diesel mixture has been successfully used under field conditions in Australia, South Africa and Brazil. This development has been extensively investigated and the results published in specialised journals. Research continues and in 1998 the Japanese company Niigata Engineering developed a method of recycling the huge amounts of waste vegetable cooking oil by mixing it with diesel to fuel power generators and similar machines (Niigata, 1997). Less well known is the use of a soyoil:diesel mixture, and a winterising technique developed at the National Centre for Agricultural Utilization Research, Peoria, USA, which allows such a mixture to be used at temperatures down to –15°C.

A most discussed substitution is that of jojoba oil for sperm whale oil as a lubricant, since the two are the only major sources of natural liquid waxes. Tests have shown that sulphurised jojoba oil is comparable to or superior to similarly treated sperm whale oil as an extreme pressure additive for use in mechanical transmission, gears and engines. Racing cars have used the oil as a general lubricant, and a normal saloon car has travelled 30000 km without requiring an oil change. Jojoba oil is soluble in benzene, petroleum ether, chlortetracycline (CTC), and hexane, mixes with alcohol and acetone, and is highly stable.

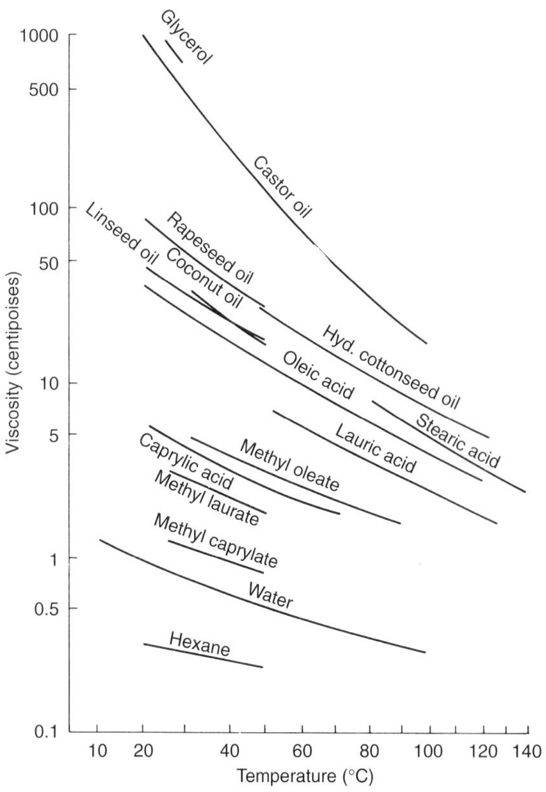

Fig. 9.11 Viscosity of some fatty oils.

TOXIC SUBSTANCES IN STOCK FEEDS

A range of chemical and other contaminants of stock feeds, other than those already mentioned, can cause injury or death of livestock and poultry, the most obvious including stones, glass and plastic. Less obvious are toxic substances which may be in the oilseeds themselves, or become contaminants following processing or storage. Probably the most common source of field contamination is misused agrochemicals; for example the accidental or ignorant incorpo-

Table 9.32 Aflatoxin content of groundnut meal received in London.

		Minimum (μg/kg)	Maximum (μg/kg)	Average (μg/kg)
Brazil	E	600	5000	1250
India	E	50	6400	920
Mozambique	S	—	—	350
	E	900	1000	950
Nigeria	S	300	1000	650
	E	250	300	270
Senegal	S	400	1530	875
	E	400	700	550
South Africa	E	—	—	45
Sudan	E	10	1200	570

S, Solvent-extracted; E, expeller-extracted.

ration of surplus or unplanted mercury dressed seed in livestock rations, which may also lead to human poisoning, while atrazine has been found in groundnuts in Australia.

A most important toxin present in oilseed cakes is aflatoxin, produced mainly by the mould *Aspergillus flavus*, which causes the condition known as aflaxtoxicosis. The mould grows on improperly dried or stored feedstuffs, or crops may be field contaminated. Water stress during maturity combined with insect damage allowing infection by *A. flavus* can lead to the formation of aflatoxin in uncut crops, and contaminated seed leads to greater infection in post-harvest storage. This damage will be exacerbated if the seed is not dried to the correct moisture level for storage. After processing, incorrect drying of cake and meal, poor storage, insect infestation and vermin can produce conditions suitable for aflatoxin production. Aflatoxins detected in groundnut meal received in London from various sources are shown in Table 9.32 (Salamon & Seaber, personal communication, 1979).

Methods of removing or neutralising aflatoxins in meals, in particular groundnut meal, was of great importance to stockfeed manufacturers. The most successful is chemical detoxification, ammoniation, using various forms of ammonia and oxidisation agents under closely controlled conditions. Meals so treated had no significant adverse effect when fed to poultry, livestock or fish. Alternatively meals can be effectively detoxified by selective chemical pre-treatment before passage through an extruder or expander at high-temperature; the machines becoming high-temperature, high-pressure chemical reactors.

Feeds produced from meals containing aflatoxin induce obvious and hidden effects in livestock varying from a general loss of condition to death, long-term degeneration of internal organs from cancer, or increased susceptibility to other diseases. Poultry, particularly ducks, are more susceptible than pigs, cattle or horses, and young stock more affected than older. The allowable levels of aflatoxin in stock feeds vary nationally, but the following may be used as a guide: ducks, turkeys, other poultry less than 1 month old and piglets less than 2 months, 10 μg/kg of feed; in all other complete stockfeeds, 20 μg/kg of feed. However, healthy adult animals can tolerate higher levels for short periods.

Groundnut is the most commonly contaminated cake, but kernels can also become infected, and if consumed by humans as nuts or peanut butter, poisoning and death usually follow. This occurred in Australia in 1997 when peanut butter was removed from national sale for some months until the source of contamination was discovered. Infected nuts in a bulk can be detected and removed using optical, electronic or similar scanners, but no method is 100% efficient. Breeding *Aspergillus* resistant strains of groundnuts has had some success, but no current commercial cultivars are resistant.

International standards on the safe level of aflatoxins in foods for human consumption remain to be agreed, with consumer groups supporting a zero level. A content of 5 parts/billion (ppb) is accepted in some countries, but most producing countries recommend a level of 15–30 ppb; 100 ppb in extracted meal is allowable for livestock feed when fed in a ratio not less than 1:4 with aflatoxin free meals. Aflatoxins have also been isolated from other field and oilseed crops, including sunflower seed, soya beans, and their meals. It has been suggested that the toxic principle of *Aspergillus* spp. could be utilised for the biological control of insect pests.

Appendix

COMPARISON OF COMMERCIAL GRADE VEGETABLE OILS

	Iodine value	Refractive index at 60°C	Sapon. value
Linseed	180	1.4655	192
Safflower	145	1.4620	191
Soya	130	1.4600	192
Sunflower	128	1.4597	191
Maize	125	1.4596	191
Sesame	111	1.4582	192
Cottonseed	108	1.4572	195
Rape	102	1.4582	175
Groundnut	90	1.4550	189
Olive	83	1.4542	192
Palm	53	1.4508	198

FATTY ACIDS CONTENT: PERCENTAGE BY WEIGHT

	Myristic	Palmitic	Stearic	Arachidic	Oleic	Linoleic	Linolenic
Linseed	—	—	9	—	23	20	48
Safflower	—	5	1	1	20	70	—
Soya	—	9	2	1	32	53	3
Sunflower	—	5	2	1	35	57	0
Maize	—	6	2	1	37	54	0
Sesame	—	8	3	1	47	41	0
Cottonseed	1	21	2	1	25	50	0
Rapeseed	—	1	1	—	22	22	3
Groundnut	—	8	4	3	55	25	0
Olive	1	9	1	1	80	8	0
Palm	2	42	4	—	42	10	0

CHARACTERISTICS OF COMMERCIAL GRADE VEGETABLE OILS

Oil	Density at 60°C	Melting-point (°C)	Titer (°C)	Unsapon. matter (%)
Babassu	0.890–0.895	22–26	22–24	0.2–0.6
Coconut	0.890–0.895	22–26	20–24	0.1–0.3
Cottonseed	0.889–0.894	5–10	32–37	0.5–2.0
Groundnut	0.885–0.890	5–10	28–32	0.4–1.0
Linseed	0.902–0.907	0	14–16	0.5–1.5
Maize	0.892–0.897	0	18–20	0.5–3.0
Olive	0.885–0.890	0–10	18–25	0.5–2.0
Palm	0.882–0.887	25–30	40–45	0.2–0.6
Rape	0.880–0.885	0	11–15	0.5–1.5
Safflower	0.898–0.903	0	15–17	0.5–1.5
Sesame	0.889–0.894	0	22–26	0.5–1.5
Soya	0.895–0.900	0	20–32	0.5–1.5
Sunflower	0.894–0.899	0	17–20	0.3–0.5

AVERAGE TOCOPHEROL CONTENT OF COMMERCIAL GRADE VEGETABLE OILS

Oil	Total tocopherol (mg/g of oil)	Individual tocopherols % of total		
		alpha-	gamma-	delta-
Castor	0.45	Trace	51.2	48.8
Cottonseed	0.92	55.4	44.6	0
Groundnut	0.93	36.4	64.1	0
Safflower	0.89	51.5	21.9	26.6
Sesame	0.66	38.9	61.5	Trace
Soya	0.96	12.0	61.8	26.2

COMPARISON OF OILSEED MEALS FROM VARIOUS COUNTRIES, RECEIVED IN LONDON

Crop/country		Oil (%)			Protein (%)			Fibre (%)		
		Min.	Max.	Av.	Min.	Max.	Av.	Min.	Max.	Av.
Groundnut										
Argentina	S	1.2	2.5	1.8	51	55	53	4.1	4.8	4.6
	E	5.2	7.6	6.4	48	52	50	3.9	4.1	4.0
Gambia	S	0.4	1.3	0.8	42	52	45	9.5	12.1	10.2
	E	4.8	7.5	5.0	45	52	49	8.9	16.6	12.8
Senegal	S	1.3	1.8	1.6	50	54	51	6.3	10.3	9.2
	E	4.7	6.2	5.7	50	53	51	7.1	8.2	7.5
S. Africa	S	0.6	3.0	1.4	46	47	46	—	—	4.3
	E	5.7	7.1	6.0	40	45	44	3.8	4.8	4.5
Rapeseed										
India	S	0.7	1.0	0.9	37	37	37	—	—	
Pakistan	S	0.5	3.2	1.4	35	42	39	8.8	10.5	10.0

COMPARISON OF OILSEED MEALS FROM VARIOUS COUNTRIES, RECEIVED IN LONDON *(Continued)*

Crop/country		Oil (%)			Protein (%)			Fibre (%)		
		Min.	Max.	Av.	Min.	Max.	Av.	Min.	Max.	Av.
Sesame										
Burma	E	7.2	9.1	8.1	38	42	40			
India	S	0.5	10.1	5.4	30	38	34			
Sudan	E	8.0	13.3	11.2	42	49	45			11.0
Sunflower										
Argentina	S	0.3	5.5	2.6	28	41	37	15.2	22.6	18.8
S. Africa	S	1.0	4.4	2.2	35	46	40	10.8	18.7	14.3
Turkey	S	1.1	2.5	1.6	39	44	41	11.7	19.0	16.1
	E	6.5	10.8	8.8	22	37	29	—	—	17.8

S = Solvent-extracted E = Expeller-extracted

FATTY ACIDS FOUND IN VEGETABLE OILS

Type	Common name	Carbon-chain length	Systematic name
Saturated	Lauric	12	Dodecanoic
	Myristic	14	Tetradecanoic
	Palmitic	16	Hexadecanoic
	Stearic	18	Octadecanoic
	Arachidic	20	Eicosanoic
	Behenic	22	Docosenoic
Mono-saturated	Oleic	18.1	*cis*-9-Octadecenoic
	Erucic	22.1	*cis*-13-Docosenoic
Di-unsaturated	Linoleic	18.2	*cis*, *cis*-9, 12-Octadecadienoic
Tri-unsaturated	Linolenic	18.3	*cis-cis-cis* 6, 9, 12-Octadecatrienoic

MAIN VEGETABLE OIL CATEGORIES

Principal fatty acid	Oil
Lauric	Coconut
	Palmkernel
	Bambassu
Palmitic	Palm
Oleic	Olive
	Groundnut
Linoleic (medium)	Soyabean
	Cottonseed
	Sesame
	Maize
Linoleic (high)	Sunflower
	Safflower
Erucic	Rapeseed
	Crambe

UNITED STATES STATISTICAL CONVERSIONS

Groundnut equivalents

1 bushel, Virginia	= 17 pounds
1 bushel, Runners	= 21 pounds
1 bushel, Spanish	= 25 pounds
1 pound of nuts, shelled	= 1.5 pounds unshelled
1 pound of nuts, shelled	= 0.95 pounds of peanut butter
Nuts, shelled	= 42.5% crude peanut butter
	= 56.6% peanut meal
	= 0.9% waste

Soya equivalents

1 bushel of soybean	= 60 pounds
	= 10.7 pounds of crude soy oil
	= 47.5 pounds of soybean meal
	= 39 pounds of soy flour
	= 20 pounds of soy protein concentrate
	= 11.8 pounds of isolated soy protein
1 metric tonne of soybeans	= 36.74 bushels
1 short ton of soybeans	= 33.33 bushels
1 short ton of soybeans	= 0.907 metric tonnes
1 long ton of soybeans	= 37.33 bushels
1 metric tonne of soybean meal	= 46.39 bushels of soybeans
1 short ton of soybean meal	= 42.08 bushels of soybeans
1 long ton of soybean meal	= 47.13 bushels of soybeans
1 metic ton of soybean oil	= 206 bushels of soybeans
Soy beans	= 79.2% soybean meal
	= 17.8% soybean oil
	= 3.0% waste

Sunflower seed equivalents

1 bushel of sunflower seed	= 28 pounds
Sunflower seed	= 42.5% sunflowerseed oil
	= 54.5% sunflowerseed meal
	= 3.0% waste

Conversion factors: US units to metric tonnes

	US unit Conversion factor
Wheat & soybeans	= bushels × 0.027216
Rice	= cwt × 0.045359
Rapeseed & sunflowerseed	= cwt × 0.045359
Corn, sorghum & rye	= bushels × 0.025401
Barley	= bushels × 0.021772
Oats	= bushels × 0.014515
Sugar	= short tons × 0.907185
Cotton	= 480 lb bales × 0.217720

Glossary

Definition of selected terms used in the text and tables. Words starred (*) are also listed.

Acetyl value. The number of milligrams of potassium hydroxide required to neutralise the acetic acid liberated by the hydrolysis of 1 g of the acetylated substance.

Acid value. The number of milligrams of potassium hydroxide required to neutralise the free acid in 1 g of the substance.

Acidulated soapstock. Residue from the complete acidulation and settling of soapstock. Contract grade should contain not less than 85% total fatty acid.

Activated earth. Bleaching earth* treated with acid to increase its pigment absorption capacity from oil.

Amino acids. A large class of organic compounds containing both the carboxyl, COOH, and the amino, NH_2 group. Their chief importance is that many proteins are composed entirely of amino-acid groupings. Eighteen commonly occur in food, and eight known as essential amino acids are not produced in the body.

Antioxidant. Any material incorporated into a fat which provides greater stability, by preventing or retarding reaction of the fat with oxygen which causes rancidity. Antioxidants for food fats must be non-toxic and edible.

Ash. Ash content of natural fats is usually due to minor contamination by metal salts. Determination is by incinerating the sample until all organic matter is consumed.

Baudouin test. A sample of the fat is boiled with concentrated hydrochloric acid. The appearance of a red colour, or addition of a few drops of an alcoholic solution of furfural, indicates the presence of sesame oil.

Biodiesel. See Methyl esters.

Biotechnology. Currently includes gene transfers and reconstruction, and commonly known as Genetic Engineering.

Bleaching earth. Special earths or clays that added to oil at about 1%, absorb unwanted pigments.

Bleaching. Treatment of a fat or oil generally with activated charcoal or diatomaceous (fuller's) earth to remove or reduce colouring materials normally present in a refined oil.

Bowman-Birk trypsin inhibitor. Soybean trypsin inhibitors* with low molecular weight, many disulfide links, and great stability to denaturation.

Break material. Precipitate (foots*) in crude vegetable oil following rapid heating to 250–300°C. Is very high in ash, rich in phosphorus, calcium and magnesium, and is apparently derived from phospholipids decomposed at these high temperatures.

Cake. Bulk resulting from expressing oilseeds.

Catalyst. Material that initiates or accelerates a chemical reaction without itself being changed. A nickel catalyst may be used to harden and stabilise an oil with the addition of hydrogen.

Cloud-point. Temperature at which the first permanent cloud, due to the crystallisation of high-melting materials, shows in the body of the oil when cooled under standardised conditions.

Cold pressed soyoil. Oil produced from undamaged, mature, cleaned yellow beans by mechanical processing and filtering, without the application of heat.

Cold test. A 125 ml bottle, filled with oil, is immersed in a water bath at 0°C. The requirement for a salad oil in the United States is that the sample remain free from cloudiness for an exposure period of $5\frac{1}{2}$ hours. The test determines how efficiently high melting point components of an oil have been removed during processing. The early appearance of cloudiness indicates that oil will separate readily into liquid and solid portions at low storage temperatures (i.e. refrigerators).

Colour. Colour evaluation of oils and fats is by comparing a well-illuminated sample with colour standards of known value, and increasingly obtained by spectrophotometric methods.

Cooking oil. A refined, bleached, and deodorised oil not further processed to remove the higher melting point components. Cooking oils tend to crystallise or become semi-solid at temperatures below 20°C. Thus heating coils must be installed in storage tanks for cooking oils in contrast to salad oils where no heating is necessary.

Cracking. Breaking whole seed into pieces to facilitate dehulling and flaking.

Crude soyoil. Unrefined oil as extracted. Is a mixture of triglycerides composed of unsaturated fatty acids (oleic, linoleic, linolenic) and saturated fatty acids, about 1.5% of free fatty acids, and 1.8% to 3.2% phospholipids, depending on bean quality, cultivar, and processing procedure.

Defatted soya flour. Produced from residue of solvent extracted beans, often by hexane; usually contains about 1% fat.

Degermed. Seed embryos wholly or partially separated from the starch endosperm.

Degummed oil. Washing crude oil with water, steam or another agent for a specified period and then separating the oil/water mixture to remove phosphatides, usually by centrifuge. In the USA commercial grade degummed soyoil must contain less than 0.02% phosphorus.

Degumming. Removal of phospholipids from vegetable oil by washing.

Dehulled. Removal of seedcoat (hull) before processing.

Deodorising. Processing fats and oils in a high vacuum with superheated steam, to remove material that produces objectionable flavours and odours in the finished product.

Diglyceride. Combination of fatty acids and glycerine in the ratio 2:1. A diglyceride may occur naturally, or result from splitting one fatty acid unit from a triglyceride during fat breakdown or hydrolysis.

Drying oil. An oil which, when applied as a thin film, readily absorbs oxygen from the air and dries to form a relatively tough, elastic substance. Such oils are usually highly unsaturated. Used as binders in paints and varnishes. Non-drying has opposite properties.

Edible crude soyoil. Oil of the following designated types produced from mature yellow soybeans: expeller expressed; expeller pressed degummed; hydraulic pressed; hydraulic pressed degummed; solvent extracted; solvent extracted degummed; mixtures of any of the above. When oil is solvent extracted the solvent used must be identified.

Edible refined soyoil. Specially refined oil for use in food products. Processes may include treatment with alkali, bleaching, partial hydrogenation and/or winterisation. The oils are classified as salad or cooking oils and shortening.

Emulsifier. Used to maintain fats as a liquid.

Esterification. Chemical reaction of an alcohol and an acid to produce an ester. A natural fat is a special type of ester made from glycerine (an alcohol) and fatty acids.

Expander. These tubular units are inserted in the processing line ahead of solvent extraction. They basically cook the proteins and starches at high temperature, rupture oil cells, and expand oil bearing material into a strong porous collet in less than 30 seconds.

Expeller. Machine for expressing oil, and basically consisting of an augur moving in a slotted barrel through which crushed seed is passed, and allowing oil to drain. Also called a continuous screw press.

Extruded. A process in which oil from crushed seed is extracted by passing the bulk through orifices under pressure, as in an extruder*.

Extruder. A jacketed augur using high pressure and temperature to produce flours, concentrates, foodstuffs or animal feeds. Also used to treat oilseed flakes before solvent extraction.

Fat. A substance composed chiefly of triglycerides of fatty acids, and solid or plastic at room temperature.

Fatty acid. A naturally occurring compound, normally consisting of strongly linked carbon and hydrogen atoms in a chain-like structure. At the end of the chain is a reactive acid group composed of carbon, hydrogen and oxygen. This acid group reacts with glycerol to construct the basic unit of a triglyceride fat molecule; a natural fat is a mixture of such molecules. See also Appendix tables.

Feed (feeds, stockfeed). Generally any processed substance intended for consumption by domestic animals, birds and fish, usually kept under intensive conditions.

Fermented. Action of yeasts, moulds, or bacteria in a controlled aerobic or anaerobic process in the manufacture of alcohol, acids, B-complex vitamins, and antibiotics. Fermentation can also occur in poorly stored oilseeds or their products.

Fire point. Temperature at which a substance, when ignited, first burns freely when the ignition agent is withdrawn. See also smoke point.

Flaking. Conversion of dehulled oilseeds into thin flakes for solvent extraction or other processing.

Flash point. Temperature, under standardised conditions, at which a liquid begins to evolve flammable vapours.

Flour (oilseed). Finely ground meal from milling oilseed products including expressed cake, but sometimes from whole hulled seed.

Flow point. Solid fat on heating becomes soft and flows downward through an orifice. Temperature at which the sample forms a hemispherical protuberance at the orifice is the flow point. The temperature at which the first drop falls from the cup is the drop-point. See also melting point.

Foots. See soapstock.

Free fatty acid (FFA). The amount of fatty acid occurring naturally in a fat, produced by processing or use. The uncombined fatty acid may result from breakdown of a fat (triglyceride) into its component fatty acid and glycerine units. A high FFA level indicates a poorly refined fat or a breakdown after manufacture or use.

Glucosinolates. Sulphur compounds present (around 4%) in some oilseeds which render the meal toxic to certain animals.

Gumming. Formation and accumulation of fat-insoluble material resulting from repeated heating of fats and oils. It is produced by oxidation and polymerisation of the fat.

Hehner number. Gives the percentage of water and fatty acid in 1 g of fat.

Hydrogen value. Measurement of unsaturation in oils containing conjugated double bonds. Although halogen addition to conjugated double bonds is incomplete, hydrogen addition proceeds quantitatively when the sample containing a catalyst, such as palladium, platinum or nickel, is heated in the presence of hydrogen gas. The reduction in gas volume serves as a measure for the hydrogen value, which is calculated in terms of its iodine equivalent.

Hydrogenated vegetable oil. During hydrogenation, vegetable oil is exposed to hydrogen gas in the presence of heat and a catalyst (nickel or copper chromite) and the hydrogen combines with certain components (unsaturated fatty acid) of the triglycerides to increase the melting point of the oil; also called hardening.

Hydrogenisation. Adding hydrogen in the presence of a catalyst to the unsaturated components of a natural fat, which increases stability and retards rancidity. Hydrogenation usually raises the melting point of a fat or changes it from a liquid to a solid.

Iodine value. A measure of the unsaturation of fatty acids and of their esters. It is based on the fact that halogen addition occurs to unsaturated bonds until these are completely saturated. There are a number of methods for determining iodine value. The most frequently used are the Wij's and Hanus procedures. The results are commonly expressed as per cent iodine absorbed, i.e. grams of iodine per 100 g of oil, whether or not the halogen used is actually iodine. Pure oleic, linoleic, linolenic acids and their esters, give the theoretical values when tested by Wij's method. With conjugated oils the Hanus method gives considerably higher values than Wij's method.

Kunitz trypsin inhibitor. Soybean trypsin inhibitors* believed to be primarily responsible for growth inhibition by raw soybeans.

Lecithin. Phospholipids obtained from vegetable oils by degumming, and included lecithin, cephalin and inositol phosphatides, glycerides, traces of tocopherols, glucosides and pigments. Dehydrated emulsion of mixed phosphatides and vegetable oil is further processed to produce the commercial grades of licithin; plastic or firm, soft, fluid, unbleached, bleached, and double bleached. High quality commercial lecithin contains 60–65% phosphatides.

Lipoxygenase. An iron containing enzyme that catalyses hydroperoxide formation in unsaturated fatty acids. Enzyme primarily responsible for creating the beany flavour in soymilk.

Margarine. A water–oil emulsion similar to butter in appearance and composition. Widely used as a butter alternative. Invented by Mege Mourie in France in 1869.

Meat analogue. Prepared from vegetable protein to resemble specific meats in texture, colour and flavour.

Meat extenders. Vegetable proteins partially substituted for meat in processed foods.

Melting point. Usually the temperature at which a natural or processed fat becomes perfectly clear and liquid, or at which a disc of the fat assumes a spherical shape under prescribed conditions. The greater the degree of unsaturation the lower the melting point. Natural fats are mixtures of various glycerides and do not have definite melting points but melt over a range of temperatures. Because of polymorphism, pure glycerides and fatty acids have more than one melting point. In the closed capillary tube melting point test, a small portion of the solidified fat is heated in a 1 mm glass tube, sealed at one end, which is suspended in a water bath near a thermometer. The temperature at which the sample becomes completely clear is the melting point. Hydrogenation raises the melting point. See also flow point*.

Methyl esters. Produced by reacting fats with methyl alcohol in the presence of an alkaline catalyst (transesterification or alcoholysis). Glycerine is released as a valuable by product. Methyl esters are also produced by reacting fatty acids with alcohol without production of glycerine. Methyl esters can be simply or fractionally distilled to higher purity. They are used as biodiesel fuel, and in the manufacture of fatty alcohols, alkanolamides, antibiotics and defoamers.

Monoglyceride. Compound of one fatty acid and one glycerine unit. May result from the combination of fatty acids and glycerine or from breakdown of di- and triglycerides. Monoglycerides contain two chemical groups, one tending toward fat solubility and the other toward water solubility. Presence of these two groups lends to the monoglyceride emulsifying properties assisting the mixture of oil and water.

Non-break oil. Edible oil that does not deposit solid material (break) when heated to at least 300°C.

Nutraceuticals. Foods containing compounds that may be beneficial for disease treatment or prevention, including isoflavones. Synonyms for nutraceuticals are functional foods and designer foods.

Oil. A substance composed chiefly of triglycerides of fatty acids, and liquid at room temperature.

Optical rotation. The angle through which the plane of polarisation of light is rotated when the polarised light passes through a layer of liquid. Unless otherwise specified, the measurement is carried out with sodium light in a layer 1 cm thick at a temperature of 20°C.

Oxidation. In fats or food products containing fat this reaction causes rancidity, objectionable flavours and odours. Hence oxidation of food fats is undesirable.

Pellets (stockfeed). Formed by compacting and forcing through die openings by a mechanical process. Also called pelleted feed, hard pellet.

Phytic acid. Is *i*-inositol hexaphosphoric acid, important in nutrition as it precipitates calcium and prevents its absorption by the body.

Polymerisation. An undesirable change in food fat composition involving agglomeration or clumping of fat and its decomposition products into larger and insoluble compounds, as in the gummy residue of frying fats. Rapid polymerisation is desirable in drying oils used as paint pigments.

Pour point. The lowest temperature at which a liquid will flow when a test container is inverted. See flow point*.

Protein. Naturally occurring combination of amino acids containing carbon, hydrogen, oxygen, nitrogen and usually sulphur. Is one of the essential constituent of all diets.

Raffinose. A soluble trisaccharide, found in soybeans, responsible for flatulence.

Rancidity. Stage in fat oxidation easily recognised by sharp, acrid and pungent off flavours and odours, but is often subjective in definition. Chemical tests may indicate progress of oxidation, but usually cannot accurately predict the onset of undesirable flavour and odour.

Refining. Treatment of a natural or processed fat to remove impurities, including washing with caustic soda, water, centrifuging, then drying by heating under vacuum.

Refractive index *(n)*. Is the ratio of the velocity of light in a vacuum to its velocity in the substance. It varies with the wavelength of the light used in its measurement. In an oil it is related to the degree of saturation, and affected by free fatty acid, oxidation and heat treatment. It increases during frying and oxidation.

Reichert–Meissel value. The R–M value of a fat is the number of millilitres of 0.1 M potassium hydroxide required to neutralise the fatty acids volatile in steam, in 5 g of fat.

Salad oil. Refined, bleached and deodorised edible oil chilled before packaging. Chilling allows removal of material which would solidify and cloud in packaged oil stored at low temperature. A high quality salad oil will store for many hours at 0°C before clouding. Some oils are natural salad oils and do not require winterisation*.

Saponification value. The number of milligrams of potassium hydroxide required to neutralise the fatty acids resulting from the complete hydrolysis of 1 g of the substance.

Shortening. Plastic or semisolid fat used in bread, cakes and other bakery products. Used for frying as it becomes liquid when heated above 38°C.

Smoke point. The smoke point is defined as the temperature at which the sample continuously gives off a thin, bluish smoke. A low smoke point indicates the presence of free fatty acids and also of mono- or diglycerides. See also fire point*.

Soapstock. By-product resulting from alkali refining vegetable oil; averages about 6% by volume of crude oil. It is often referred to as foots* since it accumulates at the bottom (foot) of the refining tank. Contract grade should contain not less than 50% total fatty acid.

Spinning. A process for texturising protein isolate by forcing a concentrated solution through a small opening into a coagulating bath.

Stachyose. A soluble tetrasaccharide found in beans (including soybeans) responsible for flatulence.

Sterols. Solid cyclic alcohols which are major constituents of the unsaponfiable portion of animal and vegetable fats and oils.

Thiocyanogen value. Is the result of determining the percentage of oleic, linoleic and linolenic acids, provided that other unsaturated acids are not present and that the iodine number and amounts of saturated acids are known.

Titer test. When molten fatty acids are cooled and begin to solidify, the latent heat of fusion is liberated and a sudden rise in temperature can be observed. The highest temperature recorded is known as the titer point.

Toasting. Cooking oilseed meal, flour or grits to increase their protein efficiency, improve their functional properties, or physical texture.

Tocopherols. Methylated derivatives of tocol, and are widely distributed in vegetable lipids. They have vitamin E activity and can protect unsaturated fats against oxidation.

Tonne. Equal to 1000 kg (2205 lbs).

Trypsin inhibitors. Proteins in soybeans believed to be responsible for growth inhibition when raw soybeans are fed to animals. (See Bowman-birk and Kunitz trypsin inhibitors*.)

Unsaponifiable matter. Includes all constituents of fats that are not saponified by alcoholic caustic potash, and that are soluble in petroleum ether or ethyl ether.

Unsaturation. Refers specifically to a shortage of hydrogen atoms in the structure of an oil. The less hydrogen, the greater the degree of unsaturation and the greater the reactivity with oxygen. Unsaturation in a fat or oil means easier formation of peroxides, easier development of rancidity and greater tendency to polymerise.

Viscosity. Bubble-tube, falling-sphere, efflux, and torsion viscosimeters are all used in this determination. In the CGS (centimetre–gram–second) system the absolute unit of viscosity is the poise. The absolute viscosity of water at 20.2°C is 1000 centipoise (one-hundredth of one poise).

Weight per millilitre. The weight expressed in grams, of 1 ml of a liquid when weighed in air at the specified temperature.

Wet-milled. Seeds may be soaked in water with or without sulphur dioxide to soften the kernel in order to facilitate separation of the various parts.

Winterised oil. Oil treated to partially remove saturated glycerides with relatively high melting points. The process involves slowly chilling the oil, maintaining it at 5°C for a specified period, then filtering to remove crystallised glycerides. Soyoil does not require winterising unless it has been partially hydrogenated to increase stability.

References

Abdel-Sabour, M.F. & Abo El-Seoud, M.A. (1996) Effects of organic-waste compost addition on sesame growth, yield and chemical composition. *Agriculture Ecosystems & Environment*, **60**(2/3), 157–64.

Abebe, M., Yermanos, D.M. & Brigham, F.T. (1978) The ecophysiology of noog. 1. Effects of photoperiod and temperature on growth. *Journal of Agricultural Science (Africa)*, **5**(1), 55–66.

Abel, G.H. (1976a) Effects of irrigation regimes, planting date nitrogen levels and row spacing on safflower. *Agronomy Journal*, **68**(3), 448–51.

Abel, G.H. (1976b) Relationship and uses of yield components in safflower. *Agronomy Journal*, **68**(3), 442–7.

Abel, G.H. & Driscoll, M.F. (1977) Sequential trait development and breeding for high yields in safflower. *Crop Science*, **16**(2), 213–16.

Abreu, C.A. de, Raij, B. van & Tanaka, R.T. (1996) Sources of manganese for soyabean and their effects on soil analysis. *Revista Brasileira de Ciencia do Solo*, **20**(1), 91–7.

Achaya, K.T. (1990) *Oilseeds and Oil Milling in India.* Oxford & IBH Publishing Co., New Delhi, India.

Adda, S., Reddy, T.P. & Kishor, P.B.K. (1994) Androclonal variation in niger (*Guizotia abyssinica Cass.*) *Euphytica*, **79**(1–2), 59–64.

Agboola, D.A. (1996) Effect of seed pretreatments on the germination of seeds of some varieties of soybeans. *Madras Agricultural Journal*, **83**(4), 212–17.

Ahmadi, M.R. & Omidi, A.H. (1997) Study and determination of natural outcrossing in winter safflower (*Carthamus tinctorius* L.). *Sesame & Safflower Newsletter*, **12**, 94–7.

Aiton, W.T. (1814) *Epitome of Hortus Kewenses* (2nd edn). R & A. Taylor, London.

Al-Aini, H., Strain, B.R. & Mooney, H.A. (1972) The physiological ecology of diverse populations of the Desert Shrub, *Simmondsia chinensis. Journal of Ecology*, **60**, 41–57.

Alemaw, G. & Wold, A.T. (1995) An agronomic and seed quality evaluation of noug (*Guizotia abyssinica. Cass.*) germplasm in Ethiopia. *Plant Breeding*, **114**(4), 375–6.

Ambuel, J., Ninomiya, S. & Takahashi, N. (1997) Fuzzy logic evaluation of soyabean plant shape. *Breeding Science*, **47**(3), 253–7.

Anac, D. *et al.* (1996) Mineral contents of broomrape (*Orobanche cernua*) and infested sunflower varieties. *Journal of Turkish Phytopathology*, **25**(3), 127–31.

Anbanandan, V. *et al.* (1993) Genetics of seed surface texture in sesame (*Sesamum indicum* L.) *Ses. Saff. Newsletter*, **8**, 61–7.

Anderson, W.K. (1978) A system approach to the adaptation of sunflower to new environments. *Field Crops Research*, **1**(2), 141–52.

Aniche, G.N. & Nwokedi, S. (1994) Biochemical changes during production of *ogiri*, a fermented food condiment from castor seeds (*Ricinus communis*). *Journal of Food Science and Technology (Mysore)*, **31**(6), 484–8.

Anon. (1998) *New soya-based concrete sealant.* Press release. Midwest Biologicals Inc., Woodburn, Indiana, USA.

Anon. (1960) *Proc. 5th Ann. Conf. Professional Officers.* Dept Research Specialist Services. Gwebi Agricultural College, Zimbabwe.

Anuar, A.R., Samsuddin, Z.H. & Yaacob, O. (1995) Contribution of legume-N by nodulated groundnut for growth of maize on an acid soil. *Soil Biology and Biochemistry*, **27**(4/5), 595–601.

Applewhite, T.H. (1992) *Proceedings World Conference on Oilseed Technology & Utilization.* American Oil Chemists Society, Champaign. Illinois, USA.

Asanome, N. & Ikeda, T. (1998) Effect of branch direction's arrangement on soyabean yield and yield components. *J. Agronomy and Crop Science*, **181**(2), 95–102.

Ashoub, A.H. (1995) Response of local and exotic safflower (*Carthamus tinctorius* L.) varieties to nitrogen application. *Annals of Agricultural Science (Cairo)*, **40**(2), 657–74.

Ashraf, M. & Fatima, H. (1995) Responses of salt-tolerant and salt-sensitive lines of safflower (*Carthamus tinctorius* L.) to salt stress. *Acta Physiologiae Plantarum*, **17**(1), 61–70.

Ashraf, M. & O'Leary, J.W. (1997) Responses of a salt-tolerant and a salt-sensitive line of sunflower to varying sodium/calcium ratios in saline sand culture. *Journal of Plant Nutrition*, **20**(2/3), 361–77.

Ashri, A. & Knowles, P.F. (1960) Cytogenetics of safflower (*Carthamus* L) species and their hybrids. *Agronomy Journal*, **52**(1), 11–17.

Ashri, A. (1995) Genetic resources of sesame: Present and future perspective. *Sesame Safflower Newsletter*, **10**, 111–12.

Australia (1994) CSIRO Division of Tropical Crops and Pastures. Application No. 92/177. *Plant Varieties Journal*, **7**(1), 13–14, 18.

Avila, J. & Montilla, D. (1997) Yield, yield components and tolerance to whitefly (*Bemisia tabacci*) incidence, as sesame (*Sesamum indicum*) selection criteria. *Sesame Safflower Newsletter*, **12**, 14–21.

AVRDC (1978) *Progress Report 1977*. AVRDC, Shanhua, Taiwan, Republic of China.

Ayerza, R. (1993) Effect of irrigation on jojoba production under arid chaco conditions: 11. Seed yields and wax quality. *Journal of the American Oil Chemists Society*, **70**(12), 1225–8.

Baagoe, J. (1974) The genus *Guizotia* (Compositae) – a taxonomic revision. *Botanisk Tidsskrift*, **69**(1), 1–39.

Bahman, Y.-S. & Zali, A.A. (1979) Comparison of winter and spring type safflower. *Crop Science*, **19**(6), 783–5.

Baldanzi, M. & Benvenuti, A. (1995) An ideotype of *Castor* (L.) for the Mediterranean environment. *Agricultura Mediterranea*, **125**(1), 16–20.

Bange, M.P., Hammer, G.L. & Rickert, K.G. (1997) Environmental control of potential yield of sunflower in the subtropics. *Australian Journal of Agricultural Research*, **48**(2), 231–40.

Banks, L.W. (1979) Harvesting and handling soybeans. *Agricultural Gazette (New South Wales)*, **90**(2), 38–9.

Bansal, K.N. & Katara, G.S. (1993) Response of safflower (*Carthamus tinctorius*) to nitrogen and irrigation. *Annals of Arid Zone* **32**(3), 193–4.

Banzato, N.V. *et al.* (1975) *Instrucoes para e cultura da mamoneira em Sao Paulo, Bol. 206* Inst. Agron. Campinas, Brazil.

Barkla, F. & Pritchard, K. (1980) The effect of time of sowing on irrigated sunflowers in Victoria. *Proceedings 4th Australian Sunflower Conference*, Shepparton, Victoria, 5, pp. 18–22.

Bascones, L. & Ritas, J.L. (1961) La nutricion mineral del ajonjonholi. *Agronomia Tropical (Maracay)*, **11**(1), 17–32; **11**(2), 93–101.

Bashir, M.E.H. *et al.* (1998) Ric c1 and Ric c3 the allergenic 2S albumin storage proteins of *Ricinus communis*; complete primary structures and phylogenetic relationships. *Inter. Archiv. Aller. Immun.*, **115**(1), 73–82.

Bataglia, O.C., Mascarenhas, H.A.A. & Miyasaka, S. (1977) Acumulo de materia seca e nutrientes em soya cultivar Santo Rosa, 57–83. In: *A Soja no Brasil Central*. Fundacao Cargill, Brazil.

Bedigian, D. (1984) Sesamum indicum L.: *Crop origin, diversity, chemistry and ethnobotany*. Ph.D.dissertation, University of Illinois, Urbana – Champaign, Illinois, USA.

Beech, D.F. (1964) The effect of leaf removal on yield attributes of safflower. *Australian Journal of Experimental Agriculture and Animal Husbandry*, **4**(14), 215–16.

Beech, D.F. & Norman, M.J.T. (1963) The effect of time of planting on yield attributes of varieties of safflower. *Australian Journal of Experimental Agriculture and Animal Husbandry*, **3**(9), 140–8.

Beech, D.F. & Norman, M.J.T. (1964) The effect of time of planting and nitrogen fertilizer on yield attributes of brassica oilseed crops in the Ord River Valley. *Australian Journal of Experimental Agriculture and Animal Husbandry*, **4**, 206–14.

Bell, M.J. & Harch, G. (1991) Effects of photoperiod on reproductive development of peanut (*Arachis hypogaea* L.) in a cool subtropical environment, I & 2. *Australian Journal of Agricultural Research*, **42**(7), 1133–49 & 1151–61.

Bell, M.J. & Wright, G.C. (1998) Groundnut Growth And Development In Contrasting Environments. Growth and Plant Density Responses. *Experimental Agriculture*, **34**, 99–112; 113–24.

Ben-Hur, M. *et al.* (1995) Surface runoff, uniformity of water distribution, and yield of peanut irrigated with a moving sprinkler system. *Agronomy Journal*, **87**(4), 609–13.

Bennett, J.M. *et al.* (1993) Single leaf carbon exchange and canopy radiation use efficiency of four peanut cultivars. *Peanut Science*, **20**(1), 1–5.

Bennett, M.R. (1996) Sesame production in Australia. *Ses. Saff. Newsletter*, **11**, 4–9.

Benzioni, A., Palzkill, D.A. & Nelson, J.M. (1992) Flower bud dormancy, ABA concentration, and survival during frost of jojoba genotypes under water stress. *Journal of the American Society for Horticultural Science*, **117**(6), 976–80.

Beratlief, C. & Iliescu, H. (1997) Highlights of proper sunflower seed storage. *Helia*, **20**(26), 121–31.

Bernstein, L. (1964) *Salt Tolerance of Plants.* Agricultural Information Bulletin No. 283. USDA, Washington D.C., USA.

Bertrand, R. *et al.* (1972) Exportation minerale par le mil et l'arachide, *Agronomie Tropicale*, **27**(12), 1287–303.

Bewley, J.D. & Downie, B. (1996) Is failure of seeds to germinate during development a dormancy-related phenomenon? In: *Plant dormancy: Physiology,*

Biochemistry and Molecular Biology. (ed. G.A.J. Land). CAB International, Wallingford, UK.

Bhende, S.M., Gangde, C.N. & Thakare, S.H. (1996) Studies on safflower harvesting by combine harvester in Maharashtra. *PKV Research Journal (India)*, **20**(1), 58–62.

Bhosale, R.S. & Patil, B.P. (1977) Response of three varieties of niger to different levels of FYM and nitrogen in Maharashtra. *Indian Journal of Agricultural Research*, **11**(4), 215–17.

Binder, R.G. *et al.* (1962) Chromatographic analyses of seed oils: fatty acid composition of castor oil. *Journal of the American Oil Chemists Society*, **39**, 513–17.

Bindir, U.B. & Kilgour, J. (1995) A groundnut harvesting machine for northern Nigeria. *Agricultural Mechanization in Asia, Africa and Latin America*, **26**(2), 35–40.

Blamey, F.P.C. (1976) Boron nutrition of sunflowers. *Agrochemophysica (South Africa)*, **8**(1), 5–9.

Blamey, F.P.C. & Chapman, J. (1979) Boron toxicity in Spanish groundnuts. *Agrochemophysica (South Africa)*, **11**(4), 57–9.

Bloem, A.A. & Barnard, R.O. (1995) Effect of residual plus applied nitrogen on grain yield of maize in the Highveld Region. *South African Journal of Plant and Soil*, **12**(3), 99–104.

Board, J.E., Wier, A.T. & Boethel, D.J. (1997) Critical light interception during seed filing for insecticide application and optimum soybean grain yield. *Agronomy Journal*, **89**(3), 369–74.

Bookerd, N., Weber, D.E. & Bezdicek, D.F. (1978) Influence of *Rhizobium japonicum* strains and inoculation methods on soybeans grown in rhizobia-populated soil. *Agronomy Journal*, **70**(4), 547–9.

Boonjan, S. & Voorasoot, N. (1992) Stomatal number, size and aperture of five sesame cultivars grown under different water regimes. *Ses. Saff. Newsletter*, **7**, 111.

Bourdette, V.R. (1979) New horizons in gas chromatography. *Agriculture Research (USA)*, **27**(11), 5.

Branch, W.D. (1995) Inheritance of peanut testa colors involved in market acceptability. *Crop Science*, **35**(1), 270–1.

Bremner, J.M. (1995) Recent research on problems in the use of urea as a nitrogen fertilizer. *Fertilizer Research*, **42**(1/3), 321–9.

Brenneman, T.B. *et al.* (1995) Suppression of foliar and soilborne peanut diseases in bahiagrass rotations. *Phytopathology*, **85**(9), 948–52.

Brooks, N.L. (1994) Characteristics and production costs of U.S. peanut farms 1991. *Agricultural Information Bulletin* No. 703. USDA, Washington DC, USA.

Burgos, P.A. *et al.* (1993) Effect of sodium chloride on growth, ion content, and hydrogen ion extrusion activity of sunflower and jojoba roots. *Journal of Plant Nutrition*, **16**(6), 1047–58.

Bukhoriev, T.A. (1997) Effectiveness of applying boron and molybdenum to soyabean crops on serozem soils in the Gissar valley. *Ivestiya Timiryazevskoi Sel'skokhozyaistvennoi Akedemii*, **2**, 192–97.

Burkhill, I.H. (1953) Habits of man and the origins of cultivated species of the Old World. *Proceedings of the Linnean Society, London*, **164**, 12–41.

Burkhill, I.H. (1966) *A Dictionary of Economic Plants of the Malay Peninsula.* Government Printer, Kuala Lumpur, Malaysia.

Burrus, E.J. (1971) Velarde's description of Primeria Alta. In: *Kino and Manje, Explorers of Sonora and Arizona*. pp. 622–56. Jesuit Hist. Inst., St Louis Univ., Miss. USA.

Burton, J.W. (1997) Soyabean (*Glycine max* (L.) Merr.). *Field Crops Research*, **53**(1/3), 171–86.

Cabrera, de Bisbal, E. & Rincon, C.A. (1997) Effect of dry periods on uptake of water by sesame cultivars during the reproductive period. *Agronomia Trop. (Maracay)*, **47**(2), 141–60.

Canvin, B.T. (1963) The formation of oil in the seed of *Ricinus communis* L. *Canadian Journal of Biochemistry*, **41**, 1879–85.

Carapetian, J. (1994) Effects of safflower sterility genes on the inflorescence and pollen grains. *Australian Journal of Botany*, **42**(3), 325–32.

Cardozo, M.S. & Li, B.W. (1994) Total dietary fiber content of selected nuts by two enzymatic-gravimetric methods. *Journal of Food Composition and Analysis*, **7**(1/2), 37–43.

Cargill Foundation (1977) *A Soya no Brasil Central.* Fundacoa Cargill. Brazil.

CCC (1980) *Research on Canola Seed, Oil, Meal and Meal Fractions*, Publication No. 57 (16th Progress Report, Canola Council of Canada), Department of Trade & Commerce, Ottawa, Canada.

Ceccardi, T.L., Heath, R.L. & Ting, I.P. (1995) Low-temperature exotherm measurement using infrared thermography. *HortScience*, **30**(1), 140–2.

Chajuss, D. (1998) Soya protein concentrate current status. *Oils and Fats International*, June Issue.

Charjan, S.K.U. & Tarar, J.L. (1993) Efficacy of systemic insecticides and fungicides on seed germination and seedling growth of sesame. *Annals Plant Physiology*, **7**(1), 126–30.

Chavan, V.M. (1961) *Niger and Safflower.* Publication of the Indian Central Oilseed Committee, Hyderabad-l. India.

Chee, Y.K. & Karim, M.Z.A. (1977) Effect of nitrogen and flower removal on the fruit characteristics of *Arachis hypogaea.* In: *Seed Technology in the Tropics,* (eds H. Chin, I. Enoch & R. Harun), pp. 97–102. Univ. Pert. Malaysia Setdang, Selangor, Malaysia.

Chitta, S.K. *et al.* (1997) Mobilisation of feed reserves in cotyledons of germinating safflower seeds. *Indian Journal of Experimental Biology*, **35**(12), 1370–72.

Choi, BuungHan & Chung, KyuYong (1997) Effect of polythene mulching on flowering and yield of groundnut in Korea. *Int. Arachis Newsl*, **17**, 49–51.

Choi, H.O. (1967) *Reports of the Crop Experiment Station (Special Crops), 1966* Suwon, Korea.

Choi, KyungJin *et al.* (1996) Dry matter production and seed yield of soyabean cultivars as affected by excessive water stress at the vegetative growth and flowering stages. *RDA Journal of Agricultural Science, Upland and Industrial Crops*, **38**(2), 117–22.

Chung, C.H. *et al.* (1995) Changes of lipid, protein, RNA and fatty acid composition in developing sesame (*Sesamum indicum* L.) seeds. *Plant Science (Limerick)*, **109**(2), 237–43.

Coates, W. & Yazici, E. (1991) Pneumatic device for harvesting jojoba seeds. *Transactions of the Australian Society of Agricultural Engineers*, **34**(6), 2304–8.

Cole, M. (1958) Castor Breeding – Tanganyika. Unpublished Report to the Dept. of Agriculture, Dar es Salaam. Tanganyika.

Coleman, C.J. (1997) Pollination techniques for seed and fruit in Zimbabwe, Central Africa. *Acta Horticulturae*, **437**, 165–8.

Covarelli, G. (1996) Weed control in sunflowers. *Informatore Fitopatologico*, **46**(6), 3–9.

Crafford, D.J. & Nott, R.W. (1978) *Effect of Soil and Rainfall on Dryland Cultivation of Groundnuts in the Highveld Region*. Pub. No. D.9 1978. Dept. Agric. Technical Serv., Pretoria. S. Africa.

Cramer, H.H. (1967) *Plant Protection and World Crop Production*. Pflanzenschutz Nachrichten, Bayer Ltd, Germany.

CSIRO (1977) *Annual Report*. Division of Plant Industry. CSIRO, Canberra, Australia.

Daffert, F.W. (1892) *Relatorio annal do Instituto Agronomico do Estado de Sao Paulo*. Campinas, Brazil.

Dagne, K. (1994) Meiosis in interspecific hybrids and genome interrelationships in *Guizotia* Cass. (Compositae). *Hereditas (Lundskrona)*, **121**(2), 119–29.

Dagne, K. & Jonsson, A. (1997) Oil content and fatty acid composition of seeds of *Guizotia* Cass (Compositae). *Journal of the Science of Food and Agriculture*, **73**(3), 274–8.

Danuso, F. (1997) Phenology of soyabeans. *Infamatore Agrario*, **53**(8), 53–8.

Dashagiriah, P. & Nagaraj, G. (1993) Seed and oil quality characteristics of some niger genotypes. *J. Oil. Tech. Ass., India.*, **25**(2), 42–44.

De Souza, P.I., Egli, D.E. & Bruening, W.P. (1997) Water stress during seed filling and leaf senescence in soybean. *Agronomy Journal*, **89**, 807–12.

Debaeke, P. & Raffaillac, D. (1996) Light interception as an indicator of leaf area index and risk of diseases in sunflower. *Helia*, **19**(24), 1–16.

Delgado, N., Layrisse, A. & Quijada, P. (1994) Inheritance of the indehiscent capsule in sesame (*Sesamum indicum* L.) *Agronomia Tropical (Maracay)*, **44**(3), 499–512.

Dembinski, F., Horodyski, A. & Teruszewka, H. (1962) A comparison of seventeen species of spring oil plants. *Pamietnik Pulawski-Prace Jung. Zesyte.*, **8**, 3–82.

Deotale, R.D. *et al.* (1996) Physiological studies in safflower (*Carthamus tinctorius* L.) as influenced by TIBA (2,3,5-triiodobenzoic acid). *PKV Research Journal*, **20**(2), 133–5.

Dharam Singh, Dagar, J.C. & Gangwar, B. (1992) Infestation by weeds and their management in oilseed crops – a review. *Agricultural Reviews (Karnal)*, **13**(3), 163–75.

Dias, D.C. & Marcos Filho, J. (1997) Electrical conductivity tests for vigour evaluation in soybean seeds. *Seed Research*, **24**(1), 1–10.

Dolya, V.S. *et al.* (1977) Oils of the seeds of nine species of the genus *Crambe. Khimiya Prirodnykh Soedinenii (USSR)*, **1**, 18–20.

Donald, C.M. (1968) The breeding of crop ideotypes. *Euphytica*, **17**, 385–403.

Dong, MuXin & Zhang, Hui (1997) Studies on the contents of major mineral elements of soyabeans, and their relationship in eight provinces of China. *Acta Agronomica Sinica*, **23**(5), 550–4.

Dubey, S.K. (1996) Response of soybean to rock phosphate applied with *Pseudomonas striata* in a Typic Chromustert. *Journal of the Indian Society of Soil Science*, **44**(2), 252–5.

Duncan, W.G. & McCloud, D.E. (1978) Assimulate partitioning and grain yield. *Agronomy Abstracts*, 9–10.

Dutta, P.C. *et al.* (1994) Variation in lipid composition of niger seed (*Guizotia abyssinica. Cass.*) samples collected in different regions of Ethiopia. *Journal of the American Oil Chemists Society*, **71**(8), 839–43.

Earle, F.R. *et al.* (1966) Compositional differences among *Crambe* samples and between seed components. *Journal of the American Oil Chemists Society*, **43**(5), 330–3.

Edrees, N. *et al.* (1976) Studies on the composition of sunflower seed heads. *Qualitas Plantarum – Plant Foods for Human Nutrition*, **25**(3/4), 247–52.

Eeden, C.F. *et al.* (1994) The effect of pod scarification by soil inhabiting insects on the quality of groundnuts (*Arachis hypogaea* L.). *South Africrican Journal of Plant and Soil*, **11**(3), 126–9.

Egli, D.B. & TeKrony, D.M. (1996) Seedbed conditions and prediction of field emergence of soybean seed. *Journal of Production Agriculture*, **9**(3), 365–70.

Egli, D.B. (1997) Cultivar maturity and response of soybean to shade stress during seed filing. *Field Crops Research*, **52**(1/2), 1–8.

Egorin, A.I. (1977) Minimum soil tillage for the cultivation of sunflower. *Byulleten Nauchotekhnicheskoi Informatsii Kul'turam*, **1**, 58–60.

El-Bially, M.E. (1997) Integrated weed management in safflower. *Annals Agricultural Science (Cairo)*, **42**(1), 135–45.

Engel, R. (1966) Untersuchungen an landwirtschaftlichen Handgeräiten in Sudan. *Zeitschrift fur Ausländische Landwirtschaft*, **5**(3), 211–30.

Estilai, A. & Hashemi, A. (1993) A four carpelled fruit mutant in jojoba. *HortScience*, **28**(7), 738–9.

Estilai, A. & Knowles, P.F. (1978) Relationships of *Carthamus leucocaulos* to other *Carthamus* species. *Canadian Journal of Genetics*, **20**, 221–33.

Eweida, M.H.T. *et al.* (1996) Effect of some intercropping treatments of soybean with sugar cane on growth, yield and quality of both components. *Annals Agricultural Science, Moshtohor*, **34**(2), 473–86.

Fei, FuHua, Tan, YuJun & Zhou, MaoFan (1997) Preliminary study on the correlation between soya bean rust infection and peroxidase isoenzymes. *Soyabean Science*, **16**(2), 107–12.

Fernandez, E.M. *et al.* (1997) Fungus incidence on peanut grains as affected by drying method and Ca nutrition. *Field Crops Research*, **52**(1/2), 9–15.

Floyd, R.O. & Cano, J.G. (1954) Effects de varias espaciementos en la producien de diferentes tipos de ajonjoli. *Turrialba*, **4**, 143–6.

Fontes, R.L.F. & Cox, F.R. (1998) Iron deficiency and zinc toxicity in soyabean grown in nutrient solution with different levels of sulfur. *J. Plant Nutrition*, **21**(8), 1715–22.

Frigerio, L. & Roberts, L.M. (1998) The enemy within; ricin and plant cells. *Journal of Experimental Botany*, **49**(326), 1473–80.

Fukushima, A. *et al.* (1997) On a color change from yellow to red in the floral tissues of a dyer's saffron cultivar. *Journal of Plant Physiology*, **150**(6), 697–706.

Futehally, S. (1982) *Inheritance of very high levels of linoleic acid in the seed oil of safflower* (Carthamus tinctorius L.). M.Sc. Thesis. University, California, Davis, USA.

Gadallah, M.A.A. & Ramadan, T. (1997) Effects of zinc and salinity on growth and anatomical structure of *Carthamus tinctorius* L. *Biol. Plantarum (Egypt)*, **39**(3), 411–18.

Gaev, V.M. & Gaponenko, L.I. (1996) Optimal density of sunflower stands. *Agrara Nauka*, 148–55.

Galbiatti, J.A. *et al.* (1995) Response of groundnuts to different dates of cessation of irrigation. *Pesquisa Agropecuaria Brasileira*, **30**(2), 195–200.

Ganesan, J. (1995) Induction of genic male sterility system in sesame. *Crop Environment*, **22**(2), 167–9.

Gangwar, S.K. *et al.* (1994) Influence of intercropping on infestation by insect pests of crops at medium high altitude of Meghalaya. *Indian Journal of Agricultural Science*, **64**(2), 137–40.

Ganguli, R.N., Shaw, S.S. & Ganguli, J. (1992) Cultural and chemical control of the niger capsule fly, *Dioxyna sorocula. Ind. J. Agric. Res.*, **26**(2), 87–90.

Garren, K.H. & Jackson, C.R. (1973) Peanut diseases. In: *Peanuts – Culture and Uses*. pp. 429–94. American Peanut Research Association, Oklahoma State University, USA.

Gautam, R.D. *et al.* (1995) Aphids infesting safflower, their predatory complex and effect on oil content. *Annals Plant Protection Science*, **3**(1), 27–30.

Gentry, H.S. (1972) Supplement to the natural history of jojoba. In: *Proceeds International Conference on Jojoba*, Tucson, Arizona, pp. 11–12.

Ghorashy, S.R., Sionit, N. & Kheradnam, M. (1972) Salt tolerance of safflower varieties during germination. *Agronomy Journal*, **64**, 256–7.

Ghorpade, G. *et al.* (1993) A decade in safflower research (1980–1990). *Sesame Safflower Newsletter*, **8**, 112–21.

Gibson, L.R. & Mullen, R.E. (1996a) Soybean seed quality reductions by high day and night temperature. *Crop Science*, **36**(6), 1615–19.

Gibson, L.R. & Mullen, R.E. (1996b) Soybean seed composition under high day and night growth temperatures. *Journal of the American Oil Chemists Society*, **73**(6), 733–7.

Gnanamurthy, P. *et al.* (1992) Selecting suitable intercrop and optimising planting ratio in sesame. *Sesame Safflower Newsletter*, **7**, 26.

Gollffer, Der & Radley, R.W. (1967) Experiment to determine the effect of defoliation and desiccation on pod maturation, seed yield and oil content of sesame. *Annual Report*, University of the West Indies, Department of Crop Production, Trinidad.

Gorbet, D.N. & Rhoads, F.M. (1975) Response of two peanut cultivars to irrigation and Kylar. *Agronomy Journal*, **67**(3), 373–6.

Goyal, S.N. & Sudhirkumar (1993) Correlation and regression studies in niger. *Gujarat Research J.*, **18**(2), 6–8, Gujarat Agric. Univ. India.

GPMB (1980) *Annual Report – 1979.* Gambia Produce Marketing Board. Government Printer, Banjul, Gambia.

Grosso, N.R. & Guzman, C.A. (1995) Chemical composition of aboriginal peanut (*Arachis hypogaea* L.) seeds from Peru. *Journal of Agricultural and Food Chemistry*, **43**(1), 102–5.

Grosso, N.R. *et al.* (1997) Proximate, fatty acid and sterol composit. *Journal of the Science of Food and Agriculture*, **73**, 186–8.

Guggari, A.K. *et al.* (1995) Response of niger to application of macro and micro nutrients in combination with farm yard manure. *Farm Systems*, **11**(3/4), 51–3.

Halder, J. *et al.* (1996) Response of niger (*Guizotia abyssinica*) to lime and sulphur application in Alfisols. *Annuls Agricultural Research (Orissa)*, **18**(3), 369–70.

Halder, J., Mohapatra, A.K. & Rao, S.V. (1997) Response of niger (*Guizotia abyssinica*) to *Azotobacter* and nitrogen. *Indian Journal Agronomy*, **42**(2), 375–6.

Halvorsen, J.D. *et al.* (1993) Density estimation for fatty acids and vegetable oils based on their fatty acids composition. *Journal of the American Oil Chemists Society*, **70**(9), 875–80.

Hammons, R.O., Soweli, G. & Smith, D.H. (1980) Registration of *Cercospora arachidicola*-resistant peanut germplasm. *Crop Science*, **20**(2), 292.

Han, S.S. *et al.* (1993) Investigation on the use of paraquat and diquat as a desiccant for sesame harvest aid. *Korean Journal of Weed Science*, **13**(1), 7–13.

Han, XiaoZeng & Xu, YanLi (1988) Study of the main factors causing yield decrease in continuously cropped soyabeans. *Soybean Science*, **17**(3), 207–12.

Hanelt, P. (1961) *Systemic study of the genus* Cartharnus *L. – a monographic review*. Ph.D. Thesis. Martin-Luther Univ., Halle-Witterburg, Germany.

Harris, H.C., McWilliam, J.R. & Mason, W.K. (1978) Influence of temperature on oil content and composition of sunflower seed, *Australian Journal of Agricultural Research*, **29**, 1203–12.

Harsong, A. (1997) Sowing time and fertilizer effects on groundnut after maize in Indonesia. *International Arachis Newsletter*, **17**, 57–9. (See also **17**, 56–7.)

Hashim, I.B., Koehler, P.E. & Kvien, C.K. (1993) Fatty acid composition, mineral content, and flavor quality of southern runner peanuts treated with herbicides and fungicides. *Peanut Science*, **20**(2), 106–11.

Hasselquist, F. (1762) *Reise nach Palastina.* Rostock, USSR.

Hayes, W.C. (1953) *The Sceptre of Egypt.* Vol 2: 191. Cambridge, Mass., USA.

Hedge, B.R., Venugopal, N. & Friesen, G.H. (1976) Chemical weed control in rainfed groundnuts. *Weed Res.*, **17**, 325–8.

He, FongFa *et al.* (1994) Karyotype of sesame (*Sesamum* L.) as related to its phylogenesis. *Journal Southwest Agricultural University (Sichuan)*, **16**(6), 573–6.

He, YongZhi (1996) Relationship between stem thickness and yield components in soyabeans. *Oil Crops of China*, **18**(2), 150–17.

He, ZhiHong (1996) The eco-graphical distribution of chemical composition of soybean seeds. *Journal Northeast Agricultural University*, **3**(2), 94–102.

Heckenberger, U., Roggatz, U. & Schurr, U. (1998) Effect of drought stress on the cytological status in *Ricinus communis*. *Journal of Experimental Botany*, **49**(319), 181–9.

Heipko, G.H., Defalla, D.A. & Nebi, Z.A.A. (1966) *Report on Castor Experiments in the Northern and Central Sudan.* Min. Agric. Khartoum, Sudan.

Heiser, C.B. (1976) *The Sunflower.* University of Oklahoma Press, USA.

Hernandez, L.F. (1997) Morphogenesis in sunflower (*Helianthus annuus* L.) as affected by exogenous application of plant growth regulators. *AgriScientia*, **13**, 3–11.

Hildebrandt, V.M. (1931) The Plant Resources of the World as Initial Material in Plant Breeding: The Sesame Plant. *Byulleten No. 41. Lenin. Academy. Agricultural. Science*. Institute of Plant Industry Moscow. Russia.

Hoque, M.S. (1993) Bradyrhizobium technology: a promising substitute for chemical nitrogen fertilizer in Bangladesh agriculture. In: *Proceedings 12th. International Plant Nutrition Colloquium* (ed. N.J. Barrow). Kluwer Academic Publishers, Dordrecht, Netherlands.

Horowitz, B. & Winter, G. (1965) A new safflower oil of low iodine value. *Nature (London)*, **179**(4459), 582–3.

Huang, S.Q. (1994) Resources and quality for sesame cultivation in China. *Scientice Agriculture Sinica*, **27**(1), 12–7.

Hudson, B.J.F. (1994) (ed.) *New And Developing Sources Of Food Proteins.* Chapman & Hall, London.

Hurt, E.F. (1948) *Sunflower for Food, Fodder and Fertility.* Faber & Faber, London.

Hussain, Y. *et al.* (1970) Charcoal rot of sesame in Iraq. *Phytopathologia Mediterranece*, **9**(1), 50–3.

Huxley, P.A. & Summerfield, R.J. (1974) Effects of night temperature and photoperiods on the reproductive ontogeny of cultivars of cowpea and soybean selected for the wet tropics. *Plant Science Letters*, **3**, 11–17.

Hwang, H.B. *et al.* (1992) Influence of waterlogging treatment at different growth stage on growth and yield in sesame. *Research Report Rural Development Administration Upland Industrial Crops (Korea)*, **34**(2), 45–53.

Hymowitz, T. (1976) Soybeans. In: *Evolution of Crop Plants*, edited by N.W. Simmonds. Longman Publishers, London.

Ibrahim, Y.M. (1994) Response of safflower to different nitrogen levels in United Arab Emirates. *Annals of Arid Zone*, **33**(1), 77–8.

Ingole, B.M. *et al.* (1998) Effect of gypsum and zinc sulphate on yield and quality of summer groundnut under broad bed and furrow systems. *J. Soils & Crops (India)*, **8**(1), 64–6.

Intodia, S.K. & Tomar, O.P. (1997) Effect of sulphur application on growth and yield of sunflower

(*Helianthus annuus*). *Indian Journal of Agricultural Science*, **67**(1), 46–7.

IPGRI (1994) Ecotypes in safflower. *IPGRI Newsletter for Asia, Pacific and Oceania*, **15**, 5–7.

Irvine, F.R. (1930) *Plants of the Gold Coast.* Oxford University Press, London.

Ishwar, P. & Ghosh, P.K. (1992) *Rodents in Indian Agriculture. Volume 1.* Scientific Publishers, Jodhpur, India.

Isoda, A. *et al.* (1996) Effects of leaf movement on radiation interception in field grown leguminous crops. IV. Relation of leaf temperature and transpiration among peanut cultivars. *Japanese Journal of Crop Science*, **65**(4), 700–6.

Jacques, S., Bacon, R.K. & Parsch, L.D. (1997) Comparison of single cropping, relay cropping and double cropping of soyabeans with wheat using cultivar blends. *Experimental Agriculture*, **33**(4), 477–86.

Jadhao, P.N. *et al.* (1994) Production potential and economics of oilseed-based double crop sequences. *Indian Journal of Agronomy*, **39**(1), 9–12.

Jadhav, R.V. & Turbatmath, P.A. (1991) Development and performance evaluation of bullock-drawn, multi-purpose hoe. *Agricultural Mechanization in Asia, Africa and Latin America*, **22**(2), 9–14.

Jagdish, P. & Bradley, I. (1997) A computer model for assessing the suitability of soils for growing soybean (*Glycine max*) in central India. *Fertilizer News*, **42**(5), 43–9.

Jana, P.K. *et al.* (1994) Effect of cobalt and *Rhizobium* on yield, oil content and nutrient concentration in irrigated summer groundnut (*Arachis hypogaea*). *Indian Journal of Agricultural Science*, **64**(9), 630–2.

Jani, S.M. & Bharodia, R.K. (1978) Varietal susceptibility of sesamum to leaf curl disease. *Gujarat Agricultural University Research Journal*, **4**(1), 162–72.

Jayaraman, S. & Subramanian, S. (1994) Lodging of soybean genotypes in understorey of coconut plantations. *Madras Agricultural Journal*, **81**(7), 369–70.

Jiang, R.F. *et al.* (1994) Effect of boric fertilizer on peanut absorption of boron and nitrogen. *Soils*, **26**(2), 83–6.

Jones, J.P. & Tucker, T.C. (1968) Effect of nitrogen fertilizer on yield nitrogen content and yield components of safflower. *Agronomy Journal*, **60**(4), 363–4.

Kackar, N.L. *et al.* (1993) In vitro regeneration of female plants of jojoba, *Simmondsia chinensis* Link., using coppice shoots. *Annals Arid Zone*, **32**(3), 175–7.

Kamal, S.S. (1976) Resistance of species of Arachis to lesser cornstalk borer. M.Sc. Thesis. Oklahoma State University, Oklahoma, USA.

Kamal-Eldin, A. & Appelqvist, L.A. (1994) Variation in fatty acid composition of the different acyl lipids in seed oils from four *Sesamum* species. *Journal of the American Oil Chemists Society*, **71**(2), 135–9.

Kandel, H.J., Schneiter, A.A. & Johnson, B.L. (1997) Intercropping legumes into sunflower at different growth stages. *Crop Science*, **37**(5), 1532–7.

Kane, M.V. *et al.* (1997) Early-maturing soybean cropping system: III. Protein and oil contents and oil composition. *Agronomy Journal*, **89**(3), 464–9.

Kane, M.V. *et al.* (1997) Early-maturing soybean cropping system. *Agronomy Journal*, **89**, Part I, pp 454–8; Part II, pp 459–64; Part III, pp 464–9.

Kang, Chul Whan *et al.* (1995) Mutation breeding for disease resistance and high yield of sesame (*Sesamum indicum* L.) in Korea. *Sesame Safflower Newsletter*, **10**, 21–36.

Kang, S.W. *et al.* (1992) Effect of continuous cropping on peanut production. *Research Report Rural Development Administration Upland Industrial Crops*, **34**(1), 109–15.

Kantolic, A. & Gimenez, P. (1996) Green soyabean: a new use for a traditional crop. *Association for the Advancement of Industrial Crops*, 424–7.

Karve, A.D. & Deshmukh, A.K. (1977) Leaf extract assay for *Alternaria* resistance screening in safflower. *Indian Journal of Genetics and Plant Breeding*, **37**(1), 154–7.

Karve, A.D., Deshmukh, A.K. & Nagekar, D.V. (1977) Hybrid safflower., In: *Proceedings International Congress on Oilseeds and Oils*. New Delhi. India.

Kasai, F.S. & Paulo, E.M. (1993) Height and date of earthing-up in a groundnut crop. *Bragantia*, **52**(1), 63–8.

Katara, G.S. & Bansal, K.N. (1995) Effect of irrigation and nitrogen on yield, uptake and water-use efficiency of safflower (*Carthamus tinctorius*). *Indian Journal of Agronomy*, **40**(2), 338–9.

Katayama, J. (1948) The progeny of female individuals in castorbeans. *Japanese Journal of Genetics*, **23**, 19.

Kaya, N. & Savran, A.F. (1996) Biochemical studies on seeds of sesame (*Sesamum indicum* L.) varieties. *Ege Univ. Ziraat Fakul. Dergisi*, **33**(2/3), 33–40.

Keimer, L. (1924) *Die Gartenpflanzen in Alten Agypten.* Hamburg, Germany.

Keisling, T.C. *et al.* (1996) Differential soybean varietal response to no-till planting in wheat straw. *Research Series – Arkansas Agricultural Experiment Station*, **449**, 26–32.

Kessavalou, A. & Walters, D.T. (1997) Winter rye as a cover crop following soybean under conservation tillage. *Agronomy Journal*, **89**, 68–74.

Khalfaoui, J.L.B. & Havard, M. (1993) Screening peanut cultivars in the field for root growth: a test by herbicide injection in the soil. *Field Crops Research*, **32**(1/2), 173–9.

Kharat, B.S., Kale, M.R. & Tatte, P.S. (1997) Effect of biogas slurry in combination with fertilizers and sulphur on growth and yield of soybean. *PKV Research Journal*, **21**(2), 211–3.

Khistaria, M.K. *et al.* (1998) Effect of intercropping of sesamum with different proportions of groundnut. *J. Maharashtra Agric. Univ.*, **22**(2), 193–5.

Kim, SokDong, Lee, SungWoo & Park, ChangHwan (1997) Seed development and germination of unripe seed in peanuts. *Korean J. Crop Sci.*, **42**(3), 333–7.

Kinman, M.L. (1970) Key to hybrid sunflower found. *Crops and Soils*, **23**, 21.

Kittock, D.L. & Williams, J.H. (1967) Date of planting tests. *Agron. J.*, **59**(5), 456–8.

Kmec, P. *et al.* (1998) Growth analysis of crambe. *Crop Science*, **38**(1), 108–12.

Knauft, D.A., Moore, K.M. & Gorbet, D.W. (1993) Further studies on the inheritance of fatty acid composition in peanut. *Peanut Science*, **20**(2), 74–6.

Knight, P. (1998) Revolution in oilseed crushing in Brazil as big boys move back in. *Oils and Fats International*, June issue.

Knowles, P.F. & Miller N.D. (1965) Safflower. *Circular 532*. Califonia Agricultural Experimental Station, University of California, Davis, California.

Kockenberger, W. *et al.* (1997) A non-invasive measurement of phloem and xylem water flow in castor bean seedlings by nuclear magnetic resonance microimaging. *Planta*, **201**(1), 53–63.

Kohyama, K. (1996) Land productivity map for soybeans, based on natural land conditions. *Japanese Journal of Soil Science and Plant Nutrition*, **67**(4), 354–61.

Kowalczuk, J. (1995) The adaptation of combine harvesters to low cutting. *Zeszyty Naukowe Akademii Rolniczej w Szczecinie, Rolnictwo*, **59**, 81–5.

Krauss, U. & Deacon, J.W. (1994) Root turnover of groundnut (*Arachis hypogaea* L.) in soil tubes. *Plant and Soil*, **166**(2), 259–70.

Krishnamurthy, K. *et al.* (1960) Nutritive value of sesame seed. *American Biochemistry*, **20**(3), 73–6.

Krishnappa, M., Srinivasan, C.N. & Sastry, J.A. (1994) Effect of macro and micronutrients on oil content in groundnut. *Current Research (Bangalore)*, **23**(1–2), 7–9.

Krug, C. (1945) Plant breeding, genetics and cytology in Latin America. *Plants and Plant Science in Latin America.* Chronica Bot. Co., New York, USA.

Kumar, H. (1996) Influence of chromosome doubling on seed filling in safflower (*Carthamus tinctorius* L.). *Ses. Saff. Newsletter*, **11**, 95–104.

Lakshmamma, P., Raj, A.S. & Rao, L.M. (1996) Effect of cobalt, phosphorus and potassium on yield and nutrient uptake in groundnut. *Annals of Agricultural Research*, **17**(3), 335–6.

Langham, D.G. (1946) Genetics of sesame. III. *Journal of Heredity*, **37**(5), 149–52.

Lanzani, A., Bondioli, P. & Brillo, A. (1991) A wet process technology applied to jojoba seed to obtain oil and detoxified protein meal. *Journal of the American Oil Chemists Society*, **68**, 772–4.

Laurence, R.C.N., Gibbons, R.W. & Young, C.T. (1976) Changes in yield, protein, oil and maturity of groundnut cultivars with the application of sulphur fertilizers and fungicides. *Journal of Agricultural Science (Camb)*, **86**, 245–50.

Laureti, D. (1995) Genotype and genotype × environment interaction effects on yield and yield components in castor (*Ricinus communis* L.). *Journal of Genetics and Breeding*, **49**(1), 27–30.

Lazzeri, L. & Malaguti, L. (1996) *Crambe abyssinica*: a new oil plant for non-food uses. *Informatore Agrario*, **52**(4), 33–6.

Lazzeri, L. *et al.* (1997) Crambe: a new possibility for set-aside. *Informatore Agrario*, **53**(4), 81–4.

Lazini, A. *et al.* (1991) A wet process technology applied to jojoba seed to obtain oil and detoxified meal. *Journal of the American Oil Chemists Society*, **68**, 772–4.

Lea, J.D. (1961) Studies on the depth and rate of root penetration of some annual tropical crops. *Tropical Agriculture (Trinidad)*, **38**(2), 93–105.

Lear, B. & Miyagawa, T. (1966) Effect of castor on soil nematodes. *Plant Disease Reporter*, **50**(2), 132–3.

Leclerc, P. (1969) Une sterilite male cytoplasmique chez le trounesol. *Annuls. Amelior. Plantes.* **19**, 99–106.

Lee, Suk Ha. *et al.* (1997) Performance and stability of brachytic and normal-stemmed isolines of soybean. *Agronomy Journal*, **89**, 275–8.

Leppik, E.E. & White, G.A. (1975) Preliminary assessment of crambe germplasm resources. *Euphytica*, **24**, 681–9.

Lessman, K.J. (1975) Variation in crambe *(C. abyssinica). Journal of the American Oil Chemists Society*, **52**(9), 386–9.

Lessman, K.J. & Meier V.D.(1972) Agronomic evaluation of crambe as a source of oil. *Crop Science*, **12**(2), 224–7.

Levey, M. (1959) *Chemistry and Chemical Technology in Ancient Mesopotamia.* Elsevier, London, England.

Li ChunQi (1995) A comparative study on the seed coat structure of soyabeans. *Acta Agricultural Boreali-Sinica*, **10** (Supplement), 64–9.

Li GuiBao, Hu Ying & Cui, Zhuan (1996) A study of the characteristics of K uptake by sesame (*Sesamum indicum* L.) plants and application techniques of K fertilizers. *Oil Crops of China*, **18**(4), 49–51.

Li LongYun *et al.* (1995) Ontogeny of *Carthamus tinctorius*. *Acta Agronomica Sinica*, **21**(6), 740–5.

Li, C & Li, Y. (1992) Study of sesame seed development. *Sesame Safflower Newsletter*, **7**, 8–17.

Li, L.R. *et al.* (1994) Analysis of seed quality of some wild genotypes of *Arachis* and hybridization experiment within the genus. *Oil Crops of China*, **16**(2), 22–6.

Liang, X.Y., Tu, L.C. & Li, A.L. (1993) Process of floral bud differentiation in sesame. *Oil Crops of China*, **15**(1), 3–6.

Lim, C. & Sessa, D.J. (1994) (Eds) *Nutrition and Utilization Technology in Aquaculture.* AOCS Press, Champaign, Illinois, USA.

Liu, Li., Zhou, JunChu & Chen, HuaKui (1998) The effect of different nitrogen compound concentrations on the nodulation and nitrogen fixing in soyabeans. *Scienta Agricultura Sinica*, **31**(4), 87–9.

Liu, WanDai (1996) The effect of covering a soyabean field with straw in preventing salinization and increasing yield. *Journal Henan Agricultural Science*, **2**, 10–11.

Liu, YuanYing *et al.* (1998) Effect of selenium on activity of glutathione peroxidase in soyabean. *Soyabean Science*, **17**(2), 157–61.

Liu, S.H. (1992) Studies on flower bud differentiation of sesame (*Sesamum indicum* L.). *Acta Agriculture Universitatis Pekinensis*, **18**(3), 257–60.

Loreto, F. *et al.* (1992) Estimation of mesophyll conductance to CO_2 flux by three different methods. *Plant Physiology*, **98**(4), 1437–43.

Lou, QingYun (1996) The main component of sunflower head pectin. *Chemistry and Industry of Forest Products (China)*, **16**(3), 45–50.

Low, D., Pistillio, P. & Mackay, J. (1975) Oilseed breeding: sunflower. *Annual Report*, CSIRO Div. Irr. Res., Griffith, NSW, Australia.

Lucia, M., de & Assennato, D. (1994) Agricultural engineering in development; post-harvest operations and management of foodgrains. *FAO Agricultural Services Bulletin*, No. 93.

Lusas, E.W. (1993) Preparation and uses of non-soy oilseed proteins. In: *Proceedings World Conference on Oilseed Technology and Utilization* (ed. T.H.J. Applewhite) Champaign, USA. Abstract in American Oil Chemists Society (1993), 268–77.

Maass, B.L. & Ocampo, C.H. (1995) Isozyme polymorphism provides fingerprints for germplasm of *Arachis glabrata* Bentham. *Genetic Resources and Crop Evolution*, **42**(1), 77–82.

Mallet, K. *et al.* (1994) Antioxidant activity of plant leaves in relation to their alpha-tocopherol content. *Food Chemistry*, **49**(1), 61–5.

Mandal, S. *et al.* (1993) Sulphur boosts the yield of summer sesamum. *Ind. Fmg.*, **43**(5), 17–18.

Manivannan, N. *et al.* (1993) Effect of spacing on seed yield in sesame (*Sesamum indicum* L.) *Sesame Safflower Newsletter*, **8**, 59–60.

Mann, H.S. & Singh, P. (1977) Oilseeds in India, with special reference to arid zone. *Annals of Arid Zone*, **16**(2), 240–56.

Manoharan, V., Dharmalingam, V. & Manjula, P.S. (1995) Studies on the extent of out-crossing in sesame. *Sesame Safflower Newsletter*, **10**, 49–51.

Mark, U. & Tevini, M. (1997) Effects of solar ultraviolet-B radiation, temperature and CO_2 on growth and physiology of sunflower and maize seedlings. *Plant Ecology*, **128**(1/2), 224–34.

Marshall, W.E. & Champagne, E.T. (1995) Agricultural byproducts as absorbents for metal ions in laboratory prepared solutions and in manufacturing wastewater. *Journal of Environmental Science and Health*, **30**(2), 241–61.

Martin, C.C. (1996) *Weed Control in Sesame.* Final report. Department of Primary Industries and Fisheries Darwin, Australia.

Marubayashi, O.M., Rosolem, C.A. & Zanotto, M.D. (1994) Root system of peanuts as affected by phosphorus fertilization. *Pesquisa Agropecuaria Brasileira*, **29**(11), 1669–73.

Mascarenhas, A.A. (1973) Acumulo de materia seca, absorcao e distribuicao de elementos durante o ciclo vegetativo da soja. *Boletim Tecnico. No. 6.* Institute Agronomy. Campinas, Brazil.

Materechera, S.A. *et al.* (1993) Field evaluation of laboratory techniques for predicting the ability of roots to penetrate strong soil and of the influence of roots on water sorptivity. *Plant and Soil*, **149**(2), 149–58.

Mathai, A.M. & Davis, T.A. (1974) Constructing the sunflower head. *Mathematical Biosciences*, **20**, 117–33.

Mathur, R.K. & Gupta, S.C. (1992) Discriminate function analysis in niger (*Guizotia abyssinica* Cass). *Crop Research (Hisar)*, **5**(1), 164–5.

Mazar, H. *et al.* (1998) Synthesis of storage reserves in developing seeds of sunflower. *Phytochemistry*, **48**(3), 429–32.

Mazzani, B. & Allievi, J.J. (1966) Effects of different harvest dates on yield and some seed characteristics in two sesame varieties. *Agronomia Tropical*, **16**(3), 223–8.

Mazzini, E. & Layrisse, A. (1998) Chemical characteristics of seeds of sesame cultivars selected from the Venezuelan germplasm collection. *Agronomia Trop. (Maracay)*, **48**(1), 5–18.

McGregor, S.E. (1976) *Insect Pollination of Cultivated Crop Plants.* USDA-ARS handbook, No. 496. Department of Agriculture, Washington. USA.

McLean, A. (1932) *Sesamum in Burma.* Agricultural Survey No. 16 of 1932. Department of Agriculture, Burma.

McPhee, J.E. & Norris C.P. (1991) Seed damage characteristics of peanut metering systems. *Agricultural Engineering Australia*, **20**(1/2), 30–3.

McPherson, R. M. *et al.* (1996) Soybean maturity group and incidence of velvetbean caterpillars (Lepidoptera: Noctuidae) and Mexican bean beetles (Coleoptera: Coccinellidae). *Journal of Economic Entomology*, **89**(6), 1601–7.

Mease, J. (1804) *Willich's Domestic Encyclopedia.* Philadelphia, USA.

Mehan, V.K. & Liao, B.S. (1994) Groundnut bacterial wilt past, present, and future. In: *Proceedings 3rd Working Group Oil Crops Research Institute,* Wuhan, China.

Mehtre, S.S., Ghatge, R.D. & Lad, S.K. (1994) Wild *Sesamum mulayanum*, a source of multiple disease resistance. *Annals of Agricultural Research*, **15**(2), 243–4.

Mehtre, S.S., Bhondve, T.S. & Dalvi, N.D. Intercropping of soyabean in different crops–a review. *Crop Research (Hissar)*, **13**(2), 519–31.

Meier, V.D. & Lessman, K.J. (1973) Breeding behaviour for crosses of *Crambe abyssinica* and a plant introduction designated *C. hispanica. Crop Science*, **13**(1), 49–5 I.

Meisner, C.A. & Karnok, K. J. (1992) Peanut root response to drought stress. *Agronomy Journal*, **84**(2), 159–65.

Mejaya, M.J. (1994) The legume collection in Malang, Indonesia. *Plant Genetic Resources Newsletter*, **97**, 49–50.

Memon, M.I. *et al.* (1995) Response of three safflower cultivars to nitrogen. *Sesame Safflower Newsletter*, **10**, 97–101.

Meyer, W.S. & Barrs, H.D. (1991) Roots in irrigated clay soils: measurement techniques and responses to rootzone conditions. *Irrigation Science*, **12**(3), 125–34.

Mir, O. (1976) Effect of nitrogen fertilizer on the performance of safflower at G.R.S. *J. Association for the Advancement of Agricultural Science Africa*, **3**(1), 52–3.

Mishra, A.K. (1994) Genetic analysis of test weight in sesame. *Ses. Saff. Newsletter*, **9**, 17–21.

Mishra, C.M. (1996) Response to sulphur of yield of groundnut under dryland conditions. *Madras Agricultural Journal*, **83**(7), 469–70.

Mishra, D. (1995) Variability and correlation studies in niger (*Guizotia abyssinica Cass.*) *Eviron. Ecolog.*, **13**(2), 439–42.

Mittal, R.K. (1996) Yield loss by foliar diseases in soybean. *Indian Journal of Mycology and Plant Pathology*, **26**(1), 107.

Miwa, T.K. & Rothfus, J.A. (1979) Extreme pressure lubricated tests on jojoba and sperm whale oils. *Journal of the American Oil Chemists Society*, **56**(8), 65–70.

Mixon, A.C. (1979) Developing groundnut lines with resistance to seed colonisation by toxin-producing strains of *Aspergillus* species. *PANS*, **25**(4), 394–400.

Moll, W. (1990) Specific nutrient needs of jojoba (*Simmondsia chinensis*) and locational limitations for cultivation in northeast Brazil. In: *Bodenfruchtbarkeit und Pflanzenernahrung* (eds Mott and Hofner). Bodenkunde und Bodenerhaltung, Justus University, Giessen, Germany.

Monika, Verma *et al.* (1997) Chemistry and biology of the oil and dye crop *Carthamus tinctorius* L: a review. *J. Medic. Aromatic Plnt Sci.*, **19**(3) 734–43.

Monotti, M. *et al.* (1997) Evaluation of sunflower varieties in diverse Italian environments in 1996. *Informatore Agrario Supplement*, **53**(9), 9–25, 28–9.

Morrison, M. J. *et al.* (1997) Yield of cool-season soybean lines differing in pubescence color and density. *Agronomy Journal*, **89**(2), 218–21.

Morse, J.W. (1950) History of soybean production. In: *Soybean and Soybean Products*, edited by K.S. Markley. Interscience Publishers. New York. USA.

Moshkin, V.A. (1986). (Ed). *Castor.* A.A. Balkema, Rotterdam, The Netherlands.

Moss, J.P. *et al.* (1993) Transfer of resistance to groundnut rosette disease from a wild *Arachis* species into cultivated groundnuts. *Int. Arachis Newsletter*, **13**, 22–3.

Mozaffari, K., Arshi, Y. & Zeinali-Khanghah, H. (1996) Research on the effects of water stress on some morphophysiological traits and yield components of sunflower (*Helianthus annuus*). *Seed and Plant*, **12**(3), 24–33.

Mundel, H.H. *et al.* (1995) Effect of soil moisture and temperature on seedling emergence and incidence of *Pythium* damping-off in safflower (*Carthamus tinctorius* L.). *Canadran Journal of Plant Science*, **75**(2), 505–9.

Murali, S., Deivanai, S.P. & Ganesan, J. (1996) Correlation and path coefficient analysis in sesame with reference to seed colour. *Sesame Safflower Newsletter*, **11**, 57–63.

Murthy, Y.L.K. and Raveesha, K.A. (1996) Seed mycoflora of soybean in Karnataka. *Plt Dis. Res.* **11**(1), 78–82.

Murthy, H.N., Hiremath, S.C. & Pyati, A.N. (1995) Genomic classification in *Guizotia* (Asteraceae). *Cytologia*, **60**(1), 67–73.

Murty, G.S.S., Salunkhe, C.K. & Joshua, D.C. (1995) Inheritance of an induced stiff stem mutant in sesame. *Sesame Safflower Newsletter*, **10**, 62–6.

Musa, G.L.C. (1993) Safflower production and research in Mexico. *Sesame Safflower Newsletter*, **8**, 109–12.

Mushtaq, A. (1996) Effect of herbicides on nodulation and *Macrophomina* blight of soybean. *Plt Dis. Research*, **11**(2), 127–31.

Musnicki, C.W., Tobola, P. & Jodlowski, M. (1996) Possibility of sunflower height limitation with aid of growth regulators. *Rosliny Oleiste*, **17**(2), 305–10.

Mustakas, G.C. *et al.* (1980) Conversion of soyabean extraction plant in Bolivia to production of flours for human consumption. *Journal of the American Oil Chemists' Society*, **47**(2), 55–8.

Muthuswamy, P. & Thangavelu, S. (1993) Capsule position and maturity stage on seed weight, oil content and yield in *Sesamum* genotypes. *Madras Agricultural Journal*, **80**(12), 706–8.

Nadagouda, V.B. *et al.* (1996) Performance of groundnut in alley cropping under irrigation in north eastern dry zone of Karnataka. *Indian Journal of Soil Conservation*, **24**(2), 132–6.

Nakamura, H., Sato, T. & Takezaki, M. (1956–58) Etude et amélioration des sésames. *Science Report Hyogo University Agricultural University* (Japan) Part 1, **2**(2), 149–56, 1956; Part 2, **3**(2), 137–45, 1958.

Nakhtore, K.L. (1952) Jangli til (wild sesame) on Adhartal Farm. *Nagpur Agricultural College Magazine*, **26**: 16–17. India.

Nalawade, B.B. & Chavan, P.D. (1990) Changes in mineral nutrients during growth of niger (*Guizotia abyssinica* Cass.). *Biovigyanam (India)*, **16**(2), 83–9.

Nalini, Pandey and Sharma, C.P. (1996) Copper effect on photosynthesis and transpiration in safflower. *Indian Journal of Experimental Biology*, **34**(8), 821–2.

Narayanasamy, P. (1993) Virus diseases of groundnut (*Arachis hypogaea* L.). *International Journal of Tropical Plant Diseases*, **11**(1), 1–16.

Natarajan, A., Jayachandran, V. & Tandavarayan, K. (1973) Schoenite – and indigenous potassic fertilizer for groundnut. *Madras Agricultural Journal*, **60**(8), 675–8.

Nautiyal, P.C., Joshi, Y.C. & Reddy, P.S. (1993) Methods to preserve seed viability in groundnut. *Indian Farming*, **43**(8), 28–30.

Neumunz, G.M. (1976) Special problems in small scale operations., *Journal of the American Oil Chemists' Society*, **53**(6), 334–7.

Nguyen, Bich Loc (1994) Comparison of the influence of inoculation with *Bradyrhizobium* sp. and nitrogen fertilizer application on yield of *Arachis hypogaea* cultivated on two different soils. *Roczniki Gleboznawcze*, **45**(3/4), 55–61.

Nielsen, J.K. (1977) Host plant relationships of *Phyllotreta nemorum* Zeit. *Journal of Economic Entomology*, **84**(4), 396–407.

Niigata (1997) *Press release.* Niigata Engineering Company 1-10-1, Kamata-Honcho, Ohta-ka. Tokyo. Japan.

Nikam, T.D. & Shitole M.G. (1997) *In vitro* plant regeneration from callus of niger (*Guizotia abyssinica Cass.*) *Plant Cell Reports* **17**(2), 155–8.

Nikitchin, D.I., Minkovskii, A.E. and Aksenov, I.V. (1995) Basic soil cultivation under sunflower. *Zemledelie (Russia)*, No. 2, 17.

Norden, A.J. (1973) Breeding of the cultivated peanut. In: *Peanuts: Culture and Uses*. pp. 175–208. Oklahoma State University Stillwater. Okla. USA.

Obeng-Ofori, D. (1995) Plant oils as grain protectants against infestations of *Cryptolestes pusillus* and *Rhyzopertha dominica* in stored grain. *Entomologica Experimentales et Applicata*, **77**(2), 133–9.

Oh, C. *et al.* (1995) Rapid gas chromatographic screening of edible seeds, nuts and beans for non-protein and protein amino acids. *Journal of Chromatography*, **708**(1), 131–41.

Oils and Fats International (1999) *Oils & Fats International.* (ed. Bill Lavers), 2 Queensway, Redhill. Surrey. England.

Ojiambo, P.S. *et al.* (1997) Comparison of several tests for detection of *Alternaria sesami* in sesame (*S. indicum* L.) seed. *Sesame Safflower Newsletter*, **12**, 1–5.

Oliveria, M.W. de, *et al.* (1997) Growth of soyabean cultivars in soils with low manganese availability. 11. Manganese content and distribution. *Revista Ceres*, **44**(251), 43–52.

Orr, E. & Adair, D. (1967) *The Production of Protein Foods and Concentrates from Oilseeds.* Report G. 31. Tropical Products Institute, London.

Osman, A.K. & Elamin, E.M. (1996) Manual weed control in rainfed groundnut in western Sudan. *International Arachis Newsletter*, **16**, 54–5.

Osman, H.E. (1985) Studies in sesame: heterosis. In: *Sesame and Safflower: Status and Potentials.* (ed. A. Ashri). pp. 157–62. FAO Plant Production and Protection Paper 566. FAO, Rome, Italy.

Osman, H.I. (1981) Genetic male sterility in sesame (*Sesamum indicum* L) reproductive characteristic and possible use in hybrid vigour. *Sesame Safflower Newsletter*, **7**, 36–41.

Pacumbaba, R.P. *et al.* (1997) Effect of fertilizers and rates of application on incidence of soyabean diseases in northern Alabama. *Plant Disease*, **81**(12), 1459–60.

Padmaja, S. & Savithri, P. (1993) Field evaluation of sesamum varieties against pest complex. *Mysore Journal of Agricultural Sciences*, **27**(3), 258–61.

Pae, SukBok *et al.* (1998) The evaluation of relationships between peanut growth and yield by mulching and non-mulching cultivation in Korea. *RDA Journal of Agricultural Science, Upland & Industrial Crops*, **40**(1), 1–6.

Paikaray, R.K. *et al.* (1997) Effect of nitrogen, phosphorus and plant density on yield and nutrient uptake in late sown niger (*Guizotia abyssinica*). *Indian Journal Agronomy*, **42**(3), 520–3.

Palmer, C.E. & Keller, W.A. (1994) In-vitro culture of oilseeds. In: *Plant Cell and Tissue Culture*. Kluwer Academic Publishers, Dordrecht, Netherlands.

Palzkill, D.A. & Feldman, W.R (1993) Optimising rooting of jojoba stem cuttings. *Journal of the American Oil Chemists Society*, **70**(12), 1221–4.

Pan, XueFeng (1996) Study on the economic threshold of *Melanagromyza sojae. Plant Protection*, **22**(1), 22–4.

Pandet, D.D. (1997) Standing crop biomass and net primary productivity of *Guizotia abyssinica* Cass. *Environment and Ecolgy*, **15**(3), 564–6.

Panchabhavi, K.S. & Krishnamoorthy, P.N. (1978) Estimation of available loss by insect pests of sunflower. *Indian Journal of Agricultural Science*, **48**(5), 264–5.

Paolini, R. *et al.* (1998) Competition between safflower and weeds as influenced by crop genotype and sowing time. *Weed Research, (Oxford)*, **38**(4), 247–55.

Park, J.M. (1967) *Annual Report*. Crop Experimental Station, Suwon, Korea.

Pascale, A.J. & Damario, E.A. (1997) Changes in agroclimatic conditions for sunflower cultivation in Argentina during the period 1961–90. *Revista Facultad Agronomia (Universidad de Buenos Aires)*, **16**(1/2), 119–25.

Pasrichuia, N.S. & Aulakh, M.S. (1991) Twenty years of sulphur research and oilseed production in Punjab, India. *Sulphur in Agriculture*, **15**, 17–23.

Patane, C. *et al.* (1996) Assessment of germination characteristics of castor bean (*Ricinus communis* L.) seed given osmotic pretreatment of autumn-winter sowing. *Rivista di Agronomia*, **30**(3), 264–9.

Pathak, A.N. & Pathak, R.K. (1972) Effect of sulphur component of some fertilizers on groundnut *(Arachis hypogaea). Indian Journal of Agricultural Research*, **6**(1), 23–6.

Pathak, R.K., Sharma, M.K. & Tripath, R.D. (1973) Quality studies in some cruficerous oilseeds. *Indian Journal of Agricultural Research*, **7**(2), 99–103.

Pathirana, R. (1995a) Comparison of selection procedures in breeding for seed yield in segregating sesame populations. *Euphytica*, **82**(1), 73–8.

Pathirana, R. (1995b) Natural cross-pollination in sesame (*Sesamum indicum* L.). *Sesame Safflower Newsletter*, **10**, 111.

Patil, B.R., Deshmukh, S.G. & Deshmukh, M.P. (1990) Studies on correlation and path analysis in safflower. *Annuls Plant Physiology*, **4**(1), 86–91.

Patil, B.B. & Chorpade, D.S. (1994) Response of advanced application of nitrogen and phosphorus on the yield of rainfed safflower. *Sesame Safflower Newsletter*, **9**, 53–8.

Patil, C.B. (1979a) Yield and yield attributes of niger as affected by different agrotechniques. *Indian Journal of Agricultural Science*, **49**(8), 633–66.

Patil, C.B. (1979b) Economics of niger oil and cake as affected by different agro-techniques. *Indian Journal of Agricultural Science*, (in press.)

Patil, C.B. & Joshi, B.P. (1978) Niger yields can be doubled. *Indian Farming*, March 1978.

Patil, N.Y. & Sabale, R.N. (1998) Response of safflower to nitrogen and irrigation under different cropping sequences. *J. Maharashtra Agricultural University*, **22**(2), 176–9.

Paul, S.R. (1995) Effect of different row spacings and seed rate on yield of niger under rainfed conditions in Assam. *Annuls Agric Research*, **16**(1), 106–7. Gossaigan, Assam.

Pazdernik, D.L., Killam, A.S. & Orf, J.H. (1997) Analysis of amino and fatty acid composition in soybean seed using near infrared reflectance spectroscopy. *Agronomy Journal*, **89**, 679–85.

Peles, J.D., Williams, C.K. & G.W. (1997) Small mammal population dynamics in strip-cropped vs monoculture agroecosystems. *Journal Sustainable Agriculture*, **9**(2/3), 51–60.

Pereira, A.S.R. (1978) Effects of artificial defoliation on field components of sunflower. *Sunflower Newletter*, **3**(2), 7–13.

Perez, M.A. & Arguello, J.A. (1995) Deterioration in peanut (*Arachis hypogaea* L. cv. Florman) seeds under natural and accelerated aging. *Seed Science and Technology*, **23**(2), 439–45.

Perucci, E. (1960) Transplanting of castor beans. *Progr. Agric. Bologna* (Italy), **6**, 1399–408.

Piggott, C.J. (1960) The effect of fertilizers on yield and quality of groundnuts in Sierra Leone. *Empire Journal Experimental Agriculture*, **28**(109), 59–64.

Piper, E.K. *et al.* (1996) The role of daily minimum temperature in modulating the development rate to flowering in soybean. *Field Crops Research*, **47**(2/3), 211–20.

PIRC (1978) *Proceedings 5th International Rape Conference*, Vol. 2. Swedish Seed Association, S-268-00. Svalov, Sweden. pp. 204–307.

Platt, B.S. (1962) *Special Report Series*, No. 302/1962. Medical Research Council. HMSO, London.

Plesnicar, M. *et al.* (1997) Boron as limiting factor in photosynthesis and growth of sunflower plants in relation to phosphate supply. *Agrochemica*, **44**(3/4), 144–54.

Poltanee, A. (1995) Effect of short-term waterlogging on growth of castorbean during early vegetative growth. *Khon Kaen Agricultural Journal* (Thailand), **23**(2), 81–6.

Ponte, da J.J. *et al.* (1997) Preliminary trial on the use of manipueira (a liquid extract of cassava root) as a foliar fertilizer. *Rev. Agricultura* (*Piracicaba*), **72**(1), 63–8.

Popelka, J.C. *et al.* (1996) Cadmium uptake and translocation during reproductive development of peanut (*Arachis hypogaea* L.). *Angewandte Botanik*, **70**(3/4), 140–3.

Popova, G.M. (1930) *Castor Oil Plant*. Lenin Acadamy Agricultural Science Indian App. (Bot. Men. Cult. Lenin) USSR, p. 63.

Porter, D.M. & Beute, M.K. (1974) *Sclerotinia* blight of peanuts. *Phytopathology*, **64**(2), 263–4.

Potter, T.I., Zanewich K.P. & Rood S.B. (1993) Gibberellin physiology of safflower: endogenous gibberellins and response to gibberellic acid. *Plant Growth Regulation*, **12**(1–2), 133–40.

Prabakaran, A.J. & Sree Rangasamy, S.R. (1995) Observations on interspecific hybrids between *Sesamum indicum* and *S. malabaricum*. L. *Sesame Safflower Newsletter*, **10**, 6–10.

Prasad, B. (1993) Effect of continuous aplication of potassium on crop yields and potassium availability under different cropping sequences in calcareous soil. *Journal of Potassium Research*, **9**(1), 48–54 & (2), 145–53.

Prasad, D. & Rangappa, T. (1994) Distribution and estimation of avoidable loss due to nematodes associated with Kalahasti malady of groundnut, *Arachis hypogaea*. *Annals of Plant Protection Science*, **2**(2), 37–42.

Pratt, P.W. & Wrather, J.A. (1998) Soyabean disease loss estimates for the southern United States, 1994–1996. *Plant Disease*, **82**(1), 114–16.

Probst, A.H. & Judd, R.W. (1973) Origin, U.S. history and development, and world distribution. In: *Soybeans: Improvement, Production and Uses*, (ed. B.E. Caldwell), pp. 1–15. Madison. Wisconsin. USA.

Punia, M.S., Sharma, G.D. & Verma, P.K. (1991) Review on jojoba (*Simmondsia chinensis*). A new yielding shrub for arid lands. *Agricultural Reviews (Karnal)*, **12**(4), 199–207.

Raghuwanshi, R.K., Sinha, M.K. & Agrawal, S.K. (1997) Effect of sulphur and zinc in soybean (*Glycine max*) – wheat (*Triticum aestivum*) cropping sequence. *Indian Journal of Agronomy*, **42**(1), 29–32.

Rahman, L. & Miah, M.N.I. (1978) Selection for number of pods and oil content on the basis of flowering pattern, Bangladesh. *Journal of Scientific and Industrial Research*, **13**(1–4), 1–8.

Rai, B.K. (1976) *Pests of Oilseed Crops in India and their Control*. Indian Council for Agricultural Research, New Delhi.

Rais, L.B. *et al.* (1993) Lipid and protein contents of jojoba leaves in relation to salt adaptation. *Plant Physiology and Biochemistry (Paris)*, **31**(40),547–57.

Rajamohan, N. (1976) Pest complex of sunflower – a bibliography. *PANS*, **22**(4), 546–63.

Rajendiran, K. (1996) Impact of ultra violet radiation on growth and chlorophyll content in *Helianthus annuus*. *Environment and Ecology*, **14**(3), 731–2.

Raju, K. & Ranganayakulu, C. (1978) Effect of saline water irrigation on salt distribution in profile, and performance of sunflower crop. *Indian Journal of Agricultural Research*, **12**(3), 183–6.

Ramamoorthy, K. (1996) Control of seed deterioration in safflower (*Carthamus tinctorius* L.) through halogenation. *Sesame Safflower Newsletter*, **11**, 91–5.

Ramamoorthy, K. & Natarajan, N. (1997) Seed invigoration by hydration–dehydration for improvement of seed quality in safflower (*Carthamus tinctorius* L.). *Journal of Agronomy and Crop Science*, **178**(2), 125–8.

Ramesh, A., Sukhwinder, K. & Dhillon, A.S. (1998) Manipulation of date of sowing for natural control of insect pests in sunflower agroecosystem. In: *Proceedings of an International Conference on Ecological Agriculture*, Vol **1**, 449–56. CRRID, Punjab Agricultural University India.

Rao, V.P. (1990) Rapid method for determination of sesame (*Sesamum indicum* L.) leaf area by linear measurements. *Journal of Research APAU*, **18**(4), 339–43.

Rasoolzadeghan, Y. (1991) Influence of seed pretreatment on salt tolerance of jojoba (*Simmondsia chinensis* (Link) Sch.) during germination. *Iranian Journal of Agricultural Science*, **22**(1–2), 33–46.

Ravetta, D.A. &Palzkill, D.A. (1992). The effect of growth regulators and apex removal on branching and flower bud production of jojoba. *Industrial Crops and Products*, **1**(1), 47–55.

Reinbott, T.M., Blevins, D.G. & Schon, M.K. (1997) Content of boron and other elements in main stem and branch leaves and seed of soybean. *Journal of Plant Nutrition*, **20**(7/8), 831–43.

Remussi, C., Saumell, H & Vid'al, G. (undated) Effects of sowing uniformity on sunflower yield. *Proceedings 6th International Sunflower Conference.* Bucharest, Romania, pp. 493–7.

Ren, Q., Pfeiffer, T.W. & Ghabrial, S.A. (1997) Soybean mosaic virus resistance improves productivity of double-cropped soybean. *Crop Science*, **37**(6), 1712–18.

Rerkasem, B. *et al.* (1997) Relationship of seed boron concentration to germination and growth of soybean (*Glycine max*). *Nutrient Cycling in Agroecosystems*, **48**(3), 217–23.

Reyes, D.M., Stolzy L.H. & Labanauskas C.K. (1977) Temperature and oxygen effects in soil on nutrient uptake in jojoba seedlings. *Agronomy Journal*, **69**(4), 647–50.

Richburg, J.S. *et al.* (1996) Weed management in southeasern peanut (*Arachis hypogaea*) with AC 263,222. *Weed Technology*, **10**(1), 145–52.

Rincon, C.A. & Salazar, N. (1997) Description of the stages of sesame development. *Agronomia Trop. (Maracay)*, **47**(4), 475–87.

Rivoira, G. (1958) *Esperienze siella cultivazione Eel ricine in Sandegna*. Nota 11, *Annuals Faculty di Agriculture University Sessare*. Vol. VI.

Robertson, J.A. & Morrison, W.H. (1977) Effect of heat and frying on sunflower oil stability. *Journal of the American Oil Chemists' Society*, **54**(1), 77A–81A.

Robertson, J.A. & Russell, R.B. (1972) Sunflower: America's neglected crop. *Journal of the American Oil Chemists' Seciety*, **49**, 239–44.

Rocha, V.S. *et al.* (1996) Effect of pod removal on yield components of soyabeans (*Glycine max* (L.) Merrill). *Revista Ceres*, **43**(246), 126–38.

Rodrigues, S.D. *et al.* (1997) Effect of calcium, boron and zinc deficiencies on yield of soyabean (*Glycine max*) cultivar. *Revista Agricultura* (Piracicaba), **72**(1), 131–45.

Rojas, P. *et al.* (1993) Variability in protein and hull content of the seed of a world collection of safflower. *Sesame and Safflower Newsletter*, **8**, 122–6.

Roseland, C.R. & Grosz, T.J. (1997) Induced responses of common annual sunflower *Helianthus annuus* L. from geographically diverse populations and deterrence to feeding by sunflower beetle. *Journal of Chemical Ecology*, **23**(20), 517–42.

Roumet, P. & Morin, F. (1997) Germination of immature soybean seeds to shorten reproductive cycle duration. *Crop Science*, **37**(2), 521–4.

Rudenko, Y.N. & Nedel'ko, Y.K. (1977) The Influence of Ramrod on Annual Weeds, and the Yield and Quality of Castor Beans. *Trudy, Vsesoyuznogo Nauchno-Issledovatelskogo Instituta Zashchity Rastenii*. Pushkinte Leningrad Obl. USSR.

Sahu, S.K. & Das, B.B. (1997) Effect of source, dose, time, and mode of sulfur application to groundnut on lateritic soils in Orissa. *Inter. Arachis Newsl*, **17**, 62–3.

Sajwan, K.S., Ornes, W.H. & Youngblood, T.V. (1996) Beryllium phytotoxicity in soybeans. *Water Air and Soil Pollution*, **86**(1/4), 117–24.

Sale, P.W.G. & Campbell, L.C. (1980) Patterns of mineral nutrient accumulation in soyabean. *Field Crops Research*, **3**(2), 157–63; **3**(2), 147–55.

Sanjukta, Parui, Mondal, A.K. & Sudhendu, Mandal (1998) Biochemical analysis of pollen of *Ricinus communis* L. *Environ. Ecology*, **16**(1), 40–3.

Santacruz-Varela, A. *et al.* (1996) Screening of sesame landraces for drought resistance in Michoacan. Mexico. *Sesame Safflower Newsletter*, **11**, 49–53.

Santos do Nascimento, R., Seidl, P.R. & Harris, R.K. (1994) Evidence for the formation of glucose (not sucrose) in the metabolism of germinating sunflower seeds. *Journal of Agricultural and Food Chemistry*, **42**(4), 882–5.

Saranga, Y. *et al.* (1998) Large sunflower seeds are characterized by low embryo vigor. *J. American Society Horticultural Science*, **123**(3), 470–4.

Saraswat, U.N. & Rabha, M.K. (1993) Major weeds of kharif crops and their control. *Indian Farming*, **43**(9), 13, 15–23.

Saraswathi, N. *et al.* (1995) Influence of pre-harvest spray of fungicides and post-harvest seed treatments on seed quality of soybean (*Glycine max* (L.) Merrill). *Farming Systems*, **11**(3–4), 21–8.

Satpute, R.A. & Kothekar, V.S. (1995) Induced alterations in safflower with reference to leaf and seed protein content. *Sesame Safflower Newsletter*, **10**, 106–10.

Schelotto, A.L.G. de (1978) Effects of artificial defoliation on the yield of sunflower. *Sunflower Newsletter*, **2**(1), 19–22.

Schweinfurth, G. (1887) Uber Pflanzenreste aus altagyptischen Grabern. *Berichte Deutschen Botanischen Gesellschaft*, **2**, 351–71.

Seguy, L. & Bouzinac, S. (1996) Soybeans in Brazil: production and cropping systems. *Documents de Travail du CIRAD-CA*, No. 2–96, 43pp.

Selim, A.K. *et al.* (1976) Breeding for disease resistance in sesame: inheritance to root rot caused by *Sclerotium bataticola. Egyptian Journal of Phytopathology*, **8**, 15–18.

Sen. N.K. & Pain, S.K. (1948) Photoperiodic effect on *Sesamum* in relation to the variations of the environmental factors of the different seasons. *Proceedings, National Institute of Science of India*, **14**, 407–20.

Shahbaz, A. & Ibrar, R. (1996) Sunflower–summer legumes intercropping systems under rainfed conditions: competition and yield advantage. *Helia*, **19**(25), 71–8.

Sharma, M. & Bichoo, S.L. (1996) Studies on foraging behaviour of honeybees on sunflower (*Helianthus annuus* L.) sown on different dates. *Crop Research (Hisar)*, **11**(1), 93–7.

Sharma, M. & Sharma, U. (1997) Effect of sowing date and sampling time on the nectar sugar content in sunflower. *Crop Research* (*Hisar*), **13**(1), 151–5.

Sharma, R.S. & Kewat, M.L. (1994) Effect of sowing date and fertilizer on yield attributes and seed yield of niger (*Guizotia abyssinica* Cass.) under rainfed conditions. *Indian Journal of Agronomy*, **39**(1), 148–50.

Sharma, S.K., Verma, R.N. & Chauhan, S. (1996) Effect of NPK doses on yield and severity of three major diseases of soybean at medium altitude of East Khasi Hills (Meghalaya). *Journal of Hill Research*, **9**(2), 279–95.

Shen, F.Y., Wang, C.T. & Duan, S.F. (1995) Aseptic culture of gynophores to obtain peanut intersectional hybrids. *Euphytica*, **81**(3), 245–9.

Shinde, S.V. *et al.* (1993) Efficacy of ammonium polyphosphate fertilizer in comparison to other P fertilizers for peanut and their residual effect on wheat. *PKV Research Journal (India)*, **17**(1), 115–16.

Shinonaga, T. & Ambe, S. (1998) Multitracer study on the absorption of radionuclides in atmosphere-plant model. *Water, Air and Soil Pollution*, **101**(1–4), 93–103.

Shlamovitz, N., Ziv, M. & Zamski, E. (1995) Light, dark and growth regulator involvement in groundnut (*Arachis hypogaea* L.) pod development. *Plant Growth Regulation*, **16**(1), 37–42.

Shockey, J.M., Rajasekharan, R. & Kemp, J.D. (1995) Photoaffinity labelling of developing jojoba seed microsomal membranes with a photoreactive analog of acyl-coenzyme A (acyl-CoA). *Plant Physiology*, **107**(1), 155–60.

Shou, HuiXia *et al.* (1998) Study on the difference of resistance to seed deterioration between cultivated and wild soyabean genotypes. *Soyabean Science*, **17**(1), 59–64.

Shukla, A.K. (1995) Effect of inter-cropping on major diseases and yield of soybean. *Legume Research*, **18**(3/4), 217–19.

Si-yin, C., Vercelloti, J.R. & Sanders, T.H. (1997) Increase of glycolytic enzymes in peanuts during peanut maturation and curing: evidence of anaerobic metabolism. *Journal of Agricultural and Food Chemistry*, **45**, 4516–21.

Sichmann, W. *et al.* (1978) *Producao de Sementes de Mamona. Boletim Technico* No. 131. CATI, Campinas, Brazil.

Simmonds, N.W. (ed.) (1976) *The Evolution of Crop Plants.* Longmans, London.

Simon, L. (1998) Cadmium accumulation and distribution in sunflower plant. *J. Plant Nutrition*, **21**(2), 331–52.

Singh, B.G. & Joythi, T. (1996) Effect of chemical soaking of safflower (*Carthamus tinctorius* L.) on dry matter production and yield. *Indian Journal of Plant Physiology*, **1**(1), 60–2.

Singh, H.V., Yadava, T.P. & Singh, R. (1978) Screening castor varieties against castor shoot and capsule borer *Dichorocis punctifera Lis. Harayana Agricultural University Journal of Research*, **7**(3), 128–32.

Singh, K.N. & Sachan, G.C. (1993) Assessment of the use of sex pheromone traps in the management of *Spodoptera litura* F. *Indian Journal of Plant Protection*, **21**(1), 7–13.

Singh, P.K. & Trivedi, H.B.P. (1993) Male sterility in niger. Birsa Agric. Univ. India, *J. Res.*, **5**(1), 81–2.

Singh, P.P. *et al.* (1991) Influence of levels and time of application of nitrogen on yield of some niger varieties. *Orissa Journal of Agricultural Research*, **4**(1–2), 34–7.

Singh, V. *et al.* (1995) An appraisal of 25 years of safflower research under irrigated conditions at Nari, India (1968–1993). *Sesame Safflower Newsletter*, **10**, 69–79.

Smartt, J.J., Gregory, W.C. & Gregory, M.P. (1978) The genome of *Arachis hypogaea. Euphytica*, **27**, 665–75.

Smith, B.W. (1950) *Arachis hypogaea:* Aerial flower and subterranean fruit. *American Journal of Botany*, **41**(8), 607–16.

Smith, D.T. (1972) Survival and emergence of volunteer castor seed in soil. *Agron. J.*, **64**(6), 799–801.

Smith, R.S. (1995) Inoculant formulations and applications to meet changing needs. In: *Proceedings 10th International Congress on Nitrogen Fixation*. St Petersburg. Russia. (See *Fertilizer Research*, **42**(1/3))

Soya Bluebook (1998) *Soyatech Inc.*, PO Box 84, Bar Harbour, Me 04609. USA.

Spears, J.F., Tekrony, D.M. & Egli, D.B. (1997) Temperature during seed filling and soybean seed germination and vigour. *Seed Science and Technology*, **25**(2), 233–44.

Spehar, C.R. & Galwey, N.W. (1997) Screening soyabeans (*Glycine max* L. Merrill) for calcium efficiency by root growth in low-Ca nutrient solution. *Euphytica*, **94**(1), 113–17.

Sprich, H. (1994) *Significance of rotation for securing the yield in predominantly cereal production systems of the Guinea savanna, with particular regard to the parasitic weed Striga hermonthica (Del.) Benth.* Ph.D. Diss. Universitat Hohenheim, Stuttgart, Germany.

Sreemannarayana, B. & Raju, A.S. (1993) Direct and residual effect of applied sulphur in sunflower based cropping systems. *Fertiliser News*, **38**(9), 39–42.

Srivastava, A.K. (1978) Effects of fertilizer on the composition and emergence of sunflower seeds. *Experimental Agriculture*, **14**, 213–16.

Staikov, G. (1965) Seed vernalisation in relation to dormancy period and degree of maturity. *Raslen nauki.*, **2**, 11–21.

Stalker, H.T. (1997) Peanut (*Arachis hypogaea* L.). *Field Crops Research*, **53**(1/3), 205–17.

Stalker, H.T. *et al.* (1994) Variation of isozyme patterns among *Arachis* species. *Theoretical and Applied Genetics*, **87**(6), 746–55.

Stanton, W.R. (1966) *Grain Legumes in Africa*. FAO, Rome, Italy.

Stecher, G. (1980) Sunflower oil as a substitute for automotive distillate. *Proceedings 4th Australian Sunflower Conference*. Section 6-3 & 6-11. Sheppar-ton, Victoria, Australia.

Stefanov, K. *et al.* (1995) Accumulation of lead, zinc and cadmium in plant seeds growing in metalliferous habitats in Bulgaria. *Food Chemistry*, **43**(3), 311–13.

Stein, H. (1963) Some evolutionary trends in *Ricinus communis. Bulletin D.7.* Israel. Tel Aiv, Israel.

Stepanova, V.M. & Korsakov, N.I. (1977) Prospects of extension of soyabean cultivation areas in European USSR. *Trudy, po Prikladnoi Botanike Genetike i Selektsii*, **59**(3), 79–83.

Stern, W.R. & Beech, N.F. (1965) The growth of safflower in a low latitude environment. *Australian Journal of Agricultural Research*, **16**(5), 801–16.

Stevenson, P.C. *et al.* (1993) The identification and characterization of resistance in wild species of *Arachis* to *Spodoptera litura* (Lepidoptera: Noctuidae). *Bulletin of Entomological Research*, **83**(3), 421–9.

Stoyanova, J.S. (1997) Growth, N fixation and transpiration in soyabean plants as affected by root temperature. *Russian Journal of Plant Physiology*, **44**(3), 357–62.

Strasil, Z. & Skala, J. (1997) The effect of site conditions and nitrogen fertilization on the yield of crambe (*Crambe abssinica* Hochst.). *Rostlinna Vyroba*, **43**(3), 143–8.

Subramanian, M. (1995) Pollen germination studies in sesame. *Annals of Agricultural Research*, **16**(2), 225–6.

Sujatha, M. (1997) In vitro adventitious shoot regeneration for effective maintenance of male sterile niger (*Guizotia abyssinica* (L.f.) Cass.). *Euphytica*, **93**(1), 89–95.

Summerfield, R.J., Minchin, F.R. & Wein, H.C. (1979) Screening soyabean and cowpea. *World Crops*, **31**(1), 21–3.

Sunarpi, A. & Anderson, J.W. (1996) Effect of sulfur nutrition on the redistribution of sulfur in vegetative soybean plants. *Plant Physiology*, **112**(2), 623–31.

Tang, GuiYing *et al.* (1996) Breeding a high-yielding and rust-resistant groundnut cultivar, Zhonghua No. 4, and its aplication. Achieving high groundnut yields. In: *Proceedings International Workshop*. Laixi, Shandong, China.

Taufiq, A. & Sudaryono, M. (1997) Role of potassium, calcium and magnesium application in increasing groundnut production on a Red Mediteran soil in Indonesia. *International Arachis Newsletter*, **17**, 59–61.

Tesu, C., Merlescu, E. and Avarvarei, I. (1975) Salinity tolerance in safflower. Agron-Hort. Lucrari Stiin. inst. Agron. Ion Ionescu. Brad, Romania, pp. 49–50.

Tewolde, H., Mulkey, J.R. & Fernandez, C.J. (1994) Recovery of sesame from defoliation and growth terminal clipping. *Agronomy Journal*, **86**(6), 1060–65.

Thomas, C.A. (1960) Relation of variety, temperature and seed immaturity to re-emergence damping-off of castorbean. *Phytopathologische Zeitschrift*, **50**, 473–4.

Thomas, C.A. (1965) Effect of photoperiod and nitrogen on reaction of sesame to *Pseudomonas sesami* and *Xanthomonas sesami. Plant Disease Reporter*, **49**(2), 119–20.

Tiwari, K.P., Namdeo, K.N. & Patel, S.B. (1996) Dry matter production and nutrient uptake by sesame (*Sesamum indicum* L.) genotypes as influenced by planting geometry and nitrogen levels. *Crop Research (Hisar)*, **12**(3), 291–9.

Tiwari, R.C., Chaplot, P.C. & Sharma, A. (1994) Effect of irrigation schedules and sulphur levels on summer groundnut (*Arachis hypogaea* L.). *Legume Research*, **17**(2), 101–4.

Tomkins, J.P. & Shipe, E.R. (1997) Environmental adaptation of long-juvenile soybean cultivars and elite strains. *Agronomy Journal*, **89**, 257–62.

Tothill, J.D. (1952) *Agriculture in the Sudan.* Oxford University Press.

Tripathy, S.K. *et al.* (1996) Effect of growth hormones on productivity and economics of summer sesamum. *Environmental Ecology*, **14**(1), 11–13.

Triplett, G.B. (1978) Weed control for double crop soybeans planted with the no-tillage method following small grain harvest. *Agronomy Journal*, **70**(4), 577–81.

Tu, LichWan (1993) Studies on sesame heterosis and its practical use in production. *Sesame & Safflower. Newsletter*, **8**, 9–17.

Turgut, I., Baydar, H. & Marquard, R. (1996) Morphogenetic and ontogenetic variability of oil and fatty acids in sesame (*Sesamum indicum* L.). *Turkish Journal of Agriculture and Forestry*, **20**(5), 459–62.

Ueda, T., Kagawa, K. & Shirai, Y. (1990) Comparison of seed quality from red- and green-stemmed castor plant strain. *Japanese Journal of Tropical Agriculture*, **34**(2), 92–3.

Ueda, T., Kagawa, K. & Shiral, Y. (1989) Morphological study on the spines of castor capsules. *Japanese Journal of Tropical Agriculture*, **33**(4), 292–4.

Ugbaja, R.A.E. (1996) Integration and utilization of rice husk as sustainable and low-input resource for castor seed production in a tropical ultisol. *Global Journal of Pure and Applied Science*, **2**(2), 123–9.

Uma, M.S., Patil, S.G. & Rao, M.R.G. (1993) Performance of safflower genotypes under varying salinity levels. *Journal of Tropical Agriculture*, **31**(2), 271–3.

Ushakova, S.A. *et al.* (1997) Photosynthesis and respiration in plants grown under red and white light. *Russian Journal of Plant Physiology*, **44**(3), 317–21.

Van Etten, C.H., Daxenbicher, M.E., Peters, J.E., Wolff, I.A. & Booth, A.N. (1965) Seed meal from *Crambe abyssinica. Journal of Agricultural and Food Chemistry*, **13**(1), 24–7.

Vance, C.P. & Graham, P.H. (1995) Nitrogen fixation in agriculture: application and perspectives. In: *Proceedings 10th International Congress on Nitrogen Fixation*. St. Peterburg, Russia. (See *Fertilizer Research*, **42**(1/3))

Vasconcelos, I.M. *et al.* (1997) Composition, toxic and antinutritional factors of newly developed cultivars of Brazilian soybean. *Journal of the Science of Food and Agriculture*, **75**, 419–26.

Vasilev, V.F. (1950) Synopsis of the species of the genus *Crambe. Zapiski Voronezhskogo Sel'skokhozyaistvennogo Instituta*, **23**(1), 41–57.

Vasishtha, A., Bajpai, M. & Singh, C. (1994) Variations in the characteristics and oil compositions in different genetic varieties of safflower. *Journal, Oil Technologists Association of India*, **26**(2), 59–63.

Vavilov, N.I. (1949) The origin, variation, immunity and breeding of cultivated plants. *Chronica Botanica*, **13**(1–6), 57.

Verdcourt, B. (1966) A proposal concerning *Glycine* L. *Taxon* **15**, 34–6.
Vidal, R., Gerbaud, A. & Vidal, D. (1996) Short-term effects of high light intensities on soybean nodule activity and photosynthesis. *Environmental and Experimental Botany*, **36**(3), 349–57.
Vyn, T.J., Opoku, G. & Swanton, C.J. (1998) Residue management and minimum tillage systems for soybean following wheat. *Agronomy Journal*, **90**(2), 131–8.
Walker, K.C. (1995) Oilseeds; The next generation. *Agro. Food Industry Hi-Tech.*, **2**(2), 37–9.
Wan, YongShan *et al.* (1997) Studies on the growth and development patterns of high-yielding groundnut sown directly in summer. *Oil Crops of China*, **19**(4), 32–6.
Wan, ChaoWen *et al.* (1998) A study on chemical composition of soyabean seeds collected from regional trials in the Yellow, Huaihe and Haihe river valleys. *Soyabean Science*, **17**(1), 10–18.
Wang, YouPing, Luo, P. & Yang, Y. (1996) Studies on the fatty acid content of crambe seed during maturation and effects of abasic acid and high osmoticum. *Acta Phytophysiologica Sinica*, **22**(4), 427–30.
Wang, Qibing *et al.* (1997) Research on the dynamics of Se uptake and accumulation in different peanut organs. *Acta Bot. Sinica*, **39**(2), 164–8.
Wang, ShiXin, Cui, JianBo & Zhuang, JiPing (1996) Effect of subsoiling on crop yield in semi-arid area of west Liaoning and its mechanism. *Chinese Journal of Applied Ecology*, **7**(3), 267–73.
Wang, ZiaoBo (1995) Geographical distribution of soyabean varieties with high protein content in Sichuan province. *Crop Gen. Res.*, **3**, 42–4.
Wang, D.Q. *et al.* (1992) The fertilizer response of crops to urea containing borate. *Journal, Shandong Agricultural University (China)*, **23**(2), 108–13.
Wang, DeXu (1995) Soybean line 84-51 with high resistance to the cyst nematode. *Crop Gen. Res.*, **3**, 46.
Wang, W.Q. *et al.* (1995) A study of the effectiveness of hybrid seed production by utilizing genic male sterility in sesame (*Sesamum indicum* L.). *Oil Crops of China*, **17**(1), 12–15.
Wang, YanFei *et al.* (1997) The dynamics of oil accumulation in sunflower (*Helianthus annuus* L.) and its relationship with temperature conditions. *Oil Crops of China*, **19**(1), 44–6.
Wang, YouPing, Luo Peng & Li XuFeng (1995) Preliminary study on *Crambe abyssinica*. *Acta Botanica Yunnanica*, **17**(2), 169–74.
Wang, YouPing, Luo, Peng & Li, XuFeng (1996) A study on plantlet regeneration from stem apices of *Crambe abyssinica*. *Oil Crops of China,* **18**(2), 4–6.
Wang, ZhiXing *et al.* (1996) Occurrence and control of *Megacopta cribraria* on soyabeans. *Plant Protection*, **22**(3), 7–9.
Wankhede, D.B. & Tharanathan, R.N. (1976) Sesame carbohydrates. *Journal of Agricultural and Food Chemistry*, **24**(3), 655–8.
Wardlaw, G.M. *et al.* (1991) Serum lipid and apolipoprotein concentrations in healthy men on diets enriched in either canola oil or safflower oil. *American Journal of Clinical Nutrition*, **54**(1), 104–10.
Watt, G. (1966) *The Commercial Products of India.* Reprinted by Today and Tomorrow Printers and Publishers (1980). New Delhi. India.
Weaver, J.E. (1926) *Root Development of Field Crops.* McGraw-Hill, New York.
Weiss, E.A. (1959) *Reports to the T.A.C. 1957–59.* Government Printer. Dar-es-Salaam. Tanzania.
Weiss, E.A. (1961) *Reports to the Rufiji Basin Survey Team.* FAO, Rome, Italy.
Weiss, E.A. (1966a) Sunflower trials in western Kenya. *East African Agricultural and Forestry Journal*, **31**(4), 405–8.
Weiss, E.A. (1966b) The Alternative. *World Crops*, **18**(2), 20–4.
Weiss, E.A. (1966c) Dwarf castor. A promising crop for East Africa. *World Crops,* **18**(12), 43–51.
Weiss, E.A. (1967a) Dwarf castor trials in western Kenya. *East African Agricultural and Forestry Journal,* **32**(3), 229–36.
Weiss, E.A. (1967b) Soya bean trials on the Uasin Gishu, Western Kenya. *East African Agricultural and Forestry Journal,* **32**(3), 223–8.
Weiss, E.A. (1969) Phosphatic fertilizer usage on West Kenya crops. *Agricultural Digest*, **18**, 29–38.
Weiss, E.A. (1971) *Castor, Sesame and Safflower.* Leonard Hill, London.
Weiss, E.A. (1979) Some indigenous plants used domestically by East African fishermen. *Economic Botany*, **33**(1), 35–51.
Weiss, E.A. (1980) An examination of some major problems affecting the establishment and administration of re-settlement projects in semi-arid areas. In: *Proceedings, International Conference on Dryland Farming*, Vol **2**, 142–6. Adelaide, South Australia.
Weiss, E.A. (1997) *Essential Oil Crops.* CAB International, Wallingford, UK.
Wheeler, T.A. *et al.* (1997) Effect of soybean cyst nematode (*Heterodera glycines*) on yield of resistant and susceptible soybean cultivars grown in Ohio. *Journal Nematology*, **24**(4 Supp), 703–9.
White, K.D. (1970) *Roman Farming.* Thames & Hudson, London.
Wightman, J.A. & Wightman, A.S. (1994) An insect, agronomic and sociological survey of groundnut fields in southern Africa. *Agriculture Ecosystems and Environment*, **51**(3), 311–31.

Williams, M.A. (1997a) Extraction of lipids from natural sources. In: *Lipid Technologies and Applications*. (eds Gunstone and Padley). Marcel Dekker, New York.

Williams, M.A. (1997b) Preparation of oil-bearing material for extraction. In: *Technology and Solvents for Extracting Oilseed and Nonpetroleum Oils.* (eds Wan and Wakelyn). AOCS Press, Champaign, Illinois, USA.

Wisniak, J. (1994) Potential uses of jojoba oil and meal – a review. *Industrial Crops and Products*, **3**(1–2), 43–68.

Wong, K.C. & Lien-Teh, W. (1936) *History of Chinese Medicine.* National Quarantine Service, Shanghai, China.

Wongyai, W. & Juttupornpong, S. (1992) Indirect selection for seed weight in sesame using capsule size as a criteria. *Sesame Safflower Newsletter*, **7**, 4–7.

Wongyai, W. (1997) Evaluation of stem growth termination in sesame. *Sesame Safflower Newsletter*, **12**, 51–8.

Worker, G.F., Jones, J.P. & Knowles, P.F. (1965) Performance of Safflower trials conducted at the Imperial Valley Station from 1957–64. *Field Crop Research Report 16.* University of California, Davis, USA.

Wu, Ying. (1998) Function of magnesium in soyabean nutrition. *Soyabean Science*, **17**(2), 162–5.

Xu, B.S. *et al.* (1992) A study of dry matter accumulation and the law of fertilizer requirement in sesame. *Acta Agriculturae Universitatis Henanensis*, **26**(4), 331–4.

Yadav, L.N., Mishra, A.K. & Verma, M.L. (1992) Sesame in Madhya Pradesh, India. *Sesame Safflower Newsletter*, **7**, 44–8.

Yadava, T. P. *et al.* (1976) CH-I. A new plant type in castor *(Ricinus communis). Current Science (India)*, **45**(19), 706.

Yahya, A. (1998) Salinity effects on growth, uptake and distribution of sodium and some essential mineral nutrients in sesame. *Journal of Plant Nutrient*, **21**(7), 1439–51.

Yermanos, D.M. (1974) Agronomic survey of jojoba in California. *Economic Botany*, **28**(4), 160–74.

Yermanos, D.M. (1975) Composition of jojoba seed during development. *Journal of the American Oil Chemists' Society*, **52**(4), 115–17.

Yermanos, D.M. (1979) Jojoba: a crop whose time has come. *California Agriculture*, **5**, 4–11.

Yermanos, D.M., Hemstreet, W., Saleeb, W. & Huszar, C.K. (1972) Oil content and composition of the seed in the world collection of sesame introduction. *Journal of the American Oil Chemists' Society*, **49**(1), 20–3.

Yermanos, D.M. *et al.* (1967) Inheritance of quantity and quality of seed oil in safflower. *Crop Science*, **7**(5), 417–22.

Yooprasert, S. & Sooksathan, I. (1992) Allelophatic effects of sesame (*Sesamum indicum* L.) on some field crops. *Sesame Safflower Newsletter*, **7**, 111.

Young, L.D. (1996) Yield loss in soybean caused by *Heterodera glycines. Journal of Nematology*, **28**(4 supplement), 604–7.

Yu, ZhangQinng *et al.* (1996) Studies on the radiation effect of ^{60}Co y-rays on the soyabean plant. *Journal Henan Agricultural Science*, **6**, 15–17.

Yuan Tong, Wang Hao & Yan WenBin (1996) A study of the development of the flower and inflorescence in sesame. *Journal Henan Agricultural Science*, **3,** 3–6.

Zambettakis, C.H. (1975) Study of the contamination of some groundnut varieties by *Aspergillus flavus. Oleagineux*, **30**(4), 161–7.

Zewdie, K. (1996) Importance of yield limiting factors on sesame under irrigation. *IAR Newsletter Agricultural Research (Ethiopia)*, **11**(2), 6.

Zhang, W.F. & Liang, Y.G. (1993) Genetic study of the pistillate characters in castor. *Hereditas (Beijing)*, **15**(2), 28–32.

Zhang, Y.X. *et al.* (1990) Cytogenetic studies in sesame (*Sesame indicum*): a new taxonomic system. *Acta Univer. Pekinensis*, **16**(1), 11–18.

Zhu Cheng *et al.* (1994) Ultra-drying treatment of oilseeds related to seed vigour and lipid peroxidation. *Oil Crops China*, **16**(4), 9–12.

Zope, R.E., Ghorpade, D.S. & Tambe, S.I. (1994a) Yield of safflower cultivars in relation to involucral bracts under dryland conditions. *Sesame Safflower Newsletter*, **9**, 45–8.

Zope, R.E., Katule, B.K. & Ghorpade, D.S. (1994b) Seed filling duration and yield in safflower. *Sesame Safflower Newsletter*, **9**, 39–42.

Zukovskyij, P.M. (1962) *Cultivated Plants and Their Wild Relatives.* Sovetskaja Nauka, Moscow 1950 (An abridged version published by the Commonwealth Agricultural Bureau, 1962).

Index